Probability Theory and Stochastic Modelling

Volume 92

The **Probability Theory and Stochastic Modelling** series is a merger and continuation of Springer's two well established series Stochastic Modelling and Applied Probability and Probability and Its Applications series. It publishes research monographs that make a significant contribution to probability theory or an applications domain in which advanced probability methods are fundamental. Books in this series are expected to follow rigorous mathematical standards, while also displaying the expository quality necessary to make them useful and accessible to advanced students as well as researchers. The series covers all aspects of modern probability theory including

- Gaussian processes
- Markov processes
- Random fields, point processes and random sets
- Random matrices
- Statistical mechanics and random media
- Stochastic analysis

as well as applications that include (but are not restricted to):

- Branching processes and other models of population growth
- Communications and processing networks
- Computational methods in probability and stochastic processes, including simulation
- Genetics and other stochastic models in biology and the life sciences
- Information theory, signal processing, and image synthesis
- Mathematical economics and finance
- Statistical methods (e.g. empirical processes, MCMC)
- Statistics for stochastic processes
- Stochastic control
- Stochastic models in operations research and stochastic optimization
- Stochastic models in the physical sciences

More information about this series at http://www.springer.com/series/13205

Hiroshi Kunita

Stochastic Flows and Jump-Diffusions

Hiroshi Kunita
Kyushu University (emeritus)
Fukuoka, Japan

ISSN 2199-3130 ISSN 2199-3149 (electronic)
Probability Theory and Stochastic Modelling
ISBN 978-981-13-3800-7 ISBN 978-981-13-3801-4 (eBook)
https://doi.org/10.1007/978-981-13-3801-4

Library of Congress Control Number: 2019930037

Mathematics Subject Classification: 60H05, 60H07, 60H30, 35K08, 35K10, 58J05

This Springer imprint is published by the registered company Springer Nature Singapore Pte Ltd.
The registered company address is: 152 Beach Road, #21-01/04 Gateway East, Singapore 189721, Singapore

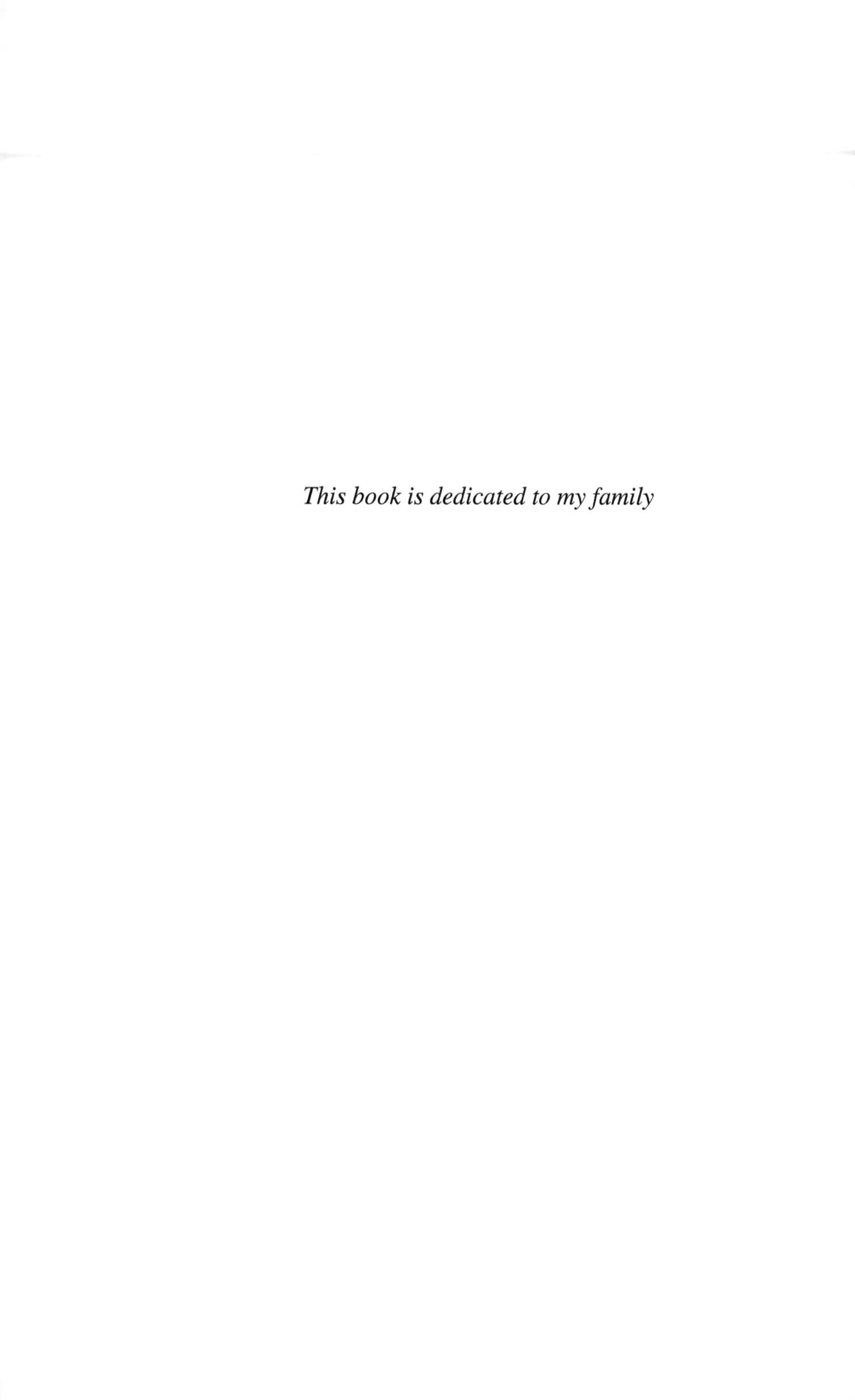

This book is dedicated to my family

Preface

The Wiener process and the Poisson random measure are fundamental to the study of stochastic processes; the former describes a continuous random evolution, and the latter describes a random phenomenon that occurs at a random time. It was shown in the 1940s that any Lévy process (process with independent increments) is represented by a Wiener process and a Poisson random measure, called the Lévy–Itô representation. Further, Itô defined a stochastic differential equation (SDE) based on a Wiener process. He defined also an equation based on a Wiener process and a Poisson random measure. In this monograph, we wish to present a modern treatment of SDE and diffusion and jump-diffusion processes. In the first part, we will show that solutions of SDE will define stochastic flows of diffeomorphisms. Then, we discuss the relation between stochastic flows and heat equations. Finally, we will investigate fundamental solutions of these heat equations (heat kernels), through the study of the Malliavin calculus.

It seems to be traditional that diffusion processes and jump processes are discussed separately. For the study of the diffusion process, theory of partial differential equations is often used, and this fact has attracted a lot of attention. On the other hand, the study of jump processes has not developed rapidly. One reason might be that for the study of jump processes, we could not find effective tools in analysis such as the partial differential equation in diffusion processes. However, recently, the Malliavin calculus for Poisson random measure has been developed, and we can apply it to some interesting problems of jump processes.

A purpose of this monograph is that we present these two theories simultaneously. In each chapter, we start from continuous processes and then proceed to processes with jumps. In the first half of the monograph, we present the theory of diffusion processes and jump-diffusion processes on Euclidean spaces based on SDEs. The basic tools are Itô's stochastic calculus. In Chap. 3, we show that solutions of these SDEs define stochastic flows of diffeomorphisms. Then in Chap. 4, relations between a diffusion (or jump-diffusion) and a heat equation (or a heat equation associated with integro-differential equation) will be studied through properties of stochastic flow.

In the latter half of the monograph, we will study the Malliavin calculus on the Wiener space and that on the space of Poisson random measure. These two types of calculus are quite different in detail, but they have some interesting things in common. These will be discussed in Chap. 5. Then in Chap. 6, we will apply the Malliavin calculus to diffusions and jump-diffusions determined by SDEs. We will obtain smooth densities for transition functions of various types of diffusions and jump-diffusions. Further, we show that these density functions are fundamental solutions for various types of heat equations and backward heat equations; thus, we construct fundamental solutions for heat equations and backward heat equations, independent of the theory of partial differential equations. Finally, SDEs on subdomains of a Euclidean space and those on manifolds will be discussed at the end of Chaps. 6 and 7.

Acknowledgements Most part of this book was written when the author was working on Malliavin calculus for jump processes jointly with Masaaki Tsuchiya and Yasushi Ishikawa in 2014–2017. Discussions with them helped me greatly to make and rectify some complicated norm estimations, which cannot be avoided for getting the smooth density. The contents of Sects. 5.5, 5.6, and 5.7 overlap with the joint work with them [46]. Further, Ishikawa read Chap. 7 carefully and gave me useful advice. I wish to express my thanks to them both. I would also like to thank the anonymous referees and a reviewer, who gave me valuable advices for improving the draft manuscript. Finally, it is my pleasure to thank Masayuki Nakamura, editor at Springer, who helped me greatly toward the publication of this book.

Fukuoka, Japan Hiroshi Kunita

Introduction

Stochastic differential equations (SDEs) based on Wiener processes have been studied extensively, after Itô's work in the 1940s. One purpose was to construct a diffusion process satisfying the Kolmogorov equation. Results may be found in monographs of Stroock–Varadhan [109], Ikeda–Watanabe [41], Oksendal [90], and Karatzas–Shreve [55]. Later, the geometric property of solutions was studied. It was shown that solutions of an SDE based on a Wiener process define a stochastic flow of diffeomorphisms [59].

In 1978, Malliavin [77] introduced an infinite-dimensional differential calculus on a Wiener space. The theory had an interesting application to solutions of SDEs based on the Wiener process. He applied the theory to the regularity of the heat kernel for hypo-elliptic differential operators. Then, the Malliavin calculus developed rapidly. Contributions were made by Bismut [9], Kusuoka–Stroock [69, 71], Ikeda–Watanabe [40, 41], Watanabe [116, 117], and many others.

At the same period, the Malliavin calculus for jump processes was studied in parallel (see Bismut [10], Bichteler–Gravereau–Jacod [7], and Leandre [74]). Later, Picard [92] proposed another approach to the Malliavin calculus for jump processes. Instead of the Wiener space, he developed the theory on the space of Poisson random measure and got a smooth density for the law of a "nondegenerate" jump Markov process. Then, Ishikawa–Kunita [45] combined these two theories and got a smooth density for the law of a nondegenerate jump-diffusion. Thus, the Malliavin calculus can be applied to a large class of SDEs.

In this monograph, we will study two types of SDEs defined on Euclidean space and manifolds. One is a continuous SDE based on a Wiener process and smooth coefficients. We will define the SDE by means of Fisk–Stratonovich symmetric integrals, since its solution has nice geometric properties. The other is an SDE with jumps based on the Wiener process and the Poisson random measure, where coefficients for the continuous part are smooth vector fields and coefficients for jump parts are diffeomorphic maps. These two SDEs are our basic objects of study. We want to show that both of these SDEs generate stochastic flows of diffeomorphisms and these stochastic flows define diffusion processes and jump-diffusion processes. In the course of the argument, we will often consider backward processes, i.e., stochastic processes describing the backward time evolution. It will be shown that inverse maps of a stochastic flow (called the inverse flow) define a backward Markov process. Further, we show that the dual process is a backward Markov process and it can be defined directly by the inverse flow through an exponential transformation. In each chapter, we will start with topics related to Wiener processes and then proceed to those related to Poisson random measures. Investigating these two subjects together, we can understand both the Wiener processes and Poisson random measures more strongly.

Chapters 1 and 2 are preliminaries. In Chap. 1, we propose a method of studying the smooth density of a given distribution, through its characteristic function (Fourier transform); it will be applied to the density problem of infinitely divisible distributions. Then, we introduce some basic stochastic processes and backward stochastic processes. These include Wiener processes, Poisson random measures, martingales, and Markov processes. In Chap. 2, we discuss stochastic integrals. Itô integrals and Fisk–Stratonovich symmetric integrals based on continuous martingales and Wiener processes are defined, and Itô's formulas are presented. Then, we define stochastic integrals based on (compensated) Poisson random measures. Further, we will give L^p-estimates of these stochastic integrals; these estimates will be used in Chaps. 3 and 6 for checking that stochastic flows have some nice properties. The backward stochastic integrals will also be discussed.

In Chap. 3, we will revisit SDEs and stochastic flows, which were discussed by the author [59, 60]. A continuous SDE on d-dimensional Euclidean space $\mathbb{R}^d$ based on a d'-dimensional Wiener process $(W_t^1, \ldots, W_t^{d'})$ is given by

$$dX_t = \sum_{k=1}^{d'} V_k(X_t, t) \circ dW_t^k + V_0(X_t, t)\, dt, \tag{1}$$

where $\circ dW_t^k$ denotes the symmetric integral based on the Wiener process W_t^k. If coefficients $V_k(x, t), k = 0, \ldots, d$ are of $C_b^{\infty,1}$-class, it is known that the family of solutions $\{X_t^{x,s}, s < t\}$ of the SDE, starting from x at time s, have a modification $\{\Phi_{s,t}(x), s < t\}$, which is continuous in s, t, x and satisfies (a) $\Phi_{s,t} : \mathbb{R}^d \to \mathbb{R}^d$ are C^∞-diffeomorphisms, (b) $\Phi_{s,u} = \Phi_{t,u} \circ \Phi_{s,t}$ for any $s < t < u$ almost surely, and (c) $\Phi_{s,t}$ and $\Phi_{t,u}$ are independent. $\{\Phi_{s,t}\}$ is called a continuous stochastic flow of diffeomorphisms defined by the SDE.

A similar problem was studied for SDE based on the Wiener process and the Poisson random measure. Let $N(dt\, dz)$ be a Poisson random measure on the space $\mathbb{U} = [0, T] \times (\mathbb{R}^{d'} \setminus \{0\})$ with the intensity measure $n(dt\, dz) = dt\nu(dz)$, where ν is a Lévy measure having a "weak drift." We consider an SDE driven by a Wiener process and Poisson random measure:

$$dX_t = \sum_{k=1}^{d'} V_k(X_t, t) \circ dW_t^k + V_0(X_t, t)\, dt + \int_{|z|>0+} (\phi_{t,z}(X_{t-}) - X_{t-}) N(dt\, dz), \tag{2}$$

where $\{\phi_{t,z}\}$ is a family of diffeomorphic maps on $\mathbb{R}^d$ with some regularity conditions (precise conditions will be stated in Sect. 3.2). It was shown that solutions define a stochastic flow of diffeomorphisms stated above (Fujiwara–Kunita [30]). In this monograph, we will prove these facts through discussions of the "master equation" and "backward SDE." By using them, some complicated arguments in previous works [30, 59] are simplified. Further, we will define a backward SDE based on a Wiener process and another based on Wiener process and Poisson random measure. These backward SDEs define backward stochastic flows of diffeomorphisms.

The solution of an SDE (or a backward SDE) based on a Wiener process defines a diffusion process (continuous strong Markov process) (or backward diffusion process). Further, the solution of an SDE (or backward SDE) based on a Wiener processes and a Poisson random measure defines a jump-diffusion (or a backward jump-diffusion). We will study these diffusion and jump-diffusion processes. Let $\{P_{s,t}\}$ be the semigroup defined by $P_{s,t}f(x) = E[f(\Phi_{s,t}(x))]$. In the case of a diffusion process on $\mathbb{R}^d$, its generator is given by a second-order differential operator

$$A(t)f(x) = \frac{1}{2}\sum_{k=1}^{d'} V_k(t)^2 f(x) + V_0(t)f(x), \tag{3}$$

where $V_k(t), k = 0, 1, \ldots, m$ are first-order partial differential operators defined by $V_k(t)f(x) = \sum_i V_k^i(x,t)\frac{\partial}{\partial x_i}f(x)$. In the case of a jump-diffusion process on $\mathbb{R}^d$, the generator is given by an integro-differential operator of the form

$$A_J(t)f = \frac{1}{2}\sum_{k=1}^{d'} V_k(t)^2 f(x) + V_0(t)f(x) + \int_{|z|>0+} \{f(\phi_{t,z}(x)) - f(x)\}\nu(dz), \tag{4}$$

where the last integral is an improper integral.

In Chap. 4, we study the relation between stochastic flows and time-dependent heat equations and backward heat equations associated with the differential operator $A(t)$ of (3) and integro-differential operator $A_J(t)$ of (4), respectively. For a given time t_1 and a bounded *smooth* function $f_1(x)$, the function $v(x,s) := P_{s,t_1}f_1(x) = E[f_1(\Phi_{s,t_1}(x))]$ is a smooth function of x. Further, in the case of diffusions, $v(x,s)$ is the unique solution of the final value problem of the time-dependent backward heat equation:

$$\frac{\partial}{\partial s}v(x,s) = -A(s)v(x,s) \quad \text{for } s < t_1, \quad v(t_1,x) = f_1(x). \tag{5}$$

This fact will be extended to a more general class of the operator $A(t)$. Consider

$$A^{\mathbf{c}}(t)f = \frac{1}{2}\sum_{k=1}^{d'}(V_k(t) + c_k(t))^2 f + (V_0 + c_0)f, \tag{6}$$

where $\mathbf{c} = (c_k(x,t); k = 0, 1, \ldots, m)$ are bounded smooth functions. We show that the semigroup with the generator $A^{\mathbf{c}}(t)$ is obtained by an exponential transformation based on $c_k(x,t)$; it is given by $P^{\mathbf{c}}_{s,t} f(x) = E[f(\Phi_{s,t}(x))G_{s,t}(x)]$, where

$$G_{s,t}(x) = \exp\Big\{\sum_{k\geq 1}\int_s^t c_k(\Phi_{s,r}(x), r) \circ dW_r^k + \int_s^t c_0(\Phi_{s,r}(x)\, dr\Big\}.$$

Further, $v(x,s) := P^{\mathbf{c}}_{s,t_1} f_1(x)$ is the unique solution of the final value problem (5) associated with the operator $A^{\mathbf{c}}(t)$. For jump-diffusion processes, we will also extend the integro-differential operator $A_J(t)$ to another one, which will be denoted by $A_J^{\mathbf{c},\mathbf{d}}(t)$. Then, we will study the final value problem of the backward heat equation associated with the operator $A_J^{\mathbf{c},\mathbf{d}}(t)$ (see Sect. 4.5).

We are also interested in the initial value problem of the time-dependent heat equation associated with the operator $A^{\mathbf{c}}(t)$ given by (6):

$$\frac{\partial}{\partial t} u(t,x) = A^{\mathbf{c}}(t)u(t,x) \ \text{ for } t > t_0, \quad u(t_0, x) = f_0(x). \tag{7}$$

For this problem, we solve SDE (1) to the backward direction and obtain a backward stochastic flow $\{\check{\Phi}_{s,t}\}$; then, we define a backward semigroup by $\check{P}^{\mathbf{c}}_{s,t} f(x) := E\big[f(\check{\Phi}_{s,t}(x))\check{G}_{s,t}(x)\big]$, where $\check{G}_{s,t}(x)$ is the exponential functional associated with the backward flow $\{\check{\Phi}_{s,t}\}$. Then, if $f_0(x)$ is a bounded *smooth* function, the solution of the forward equation (7) exists uniquely, and it is represented by $u(x,t) = \check{P}^{\mathbf{c}}_{t_0,t} f_0(x)$.

We stress that the final function f_1 (or initial function f_0) is smooth in these studies. Indeed, the smoothness of functions $v(x,s) = P^{\mathbf{c}}_{s,t} f_1(x)$, etc. with respect to x follows from the smoothness of $f_1(x)$ and the stochastic flow $\Phi_{s,t}(x)$ with respect to x, a.s. If the function f_1 is not smooth, we need additional arguments for the solution of equations (5) and (7), which will be discussed in Chap. 6 using the Malliavin calculus.

Another subject of Chap. 4 is the investigation of the dual of a given diffusion and a jump-diffusion with respect to the Lebesgue measure. It will be shown that the dual of these processes can be constructed through the change-of-variable formula concerning stochastic flows $\{\Phi_{s,t}\}$; the stochastic process defined by the inverse maps $\check{X}_s = \check{\Psi}_{s,t}(x) = \Phi_{s,t}^{-1}(x)$ should be a dual process of $X_t = \Phi_{s,t}(x)$, and it is a backward diffusion or a backward jump-diffusion with respect to s, where t is the initial time of the process. The dual semigroup of the semigroup $\{P_{s,t}\}$ is then defined by using the inverse flow $\{\check{\Psi}_{s,t}\}$ as

$$P^*_{s,t} g(x) = E[g(\check{\Psi}_{s,t}(x)) \det \nabla \check{\Psi}_{s,t}(x)],$$

where $\nabla \check{\Psi}_{s,t}$ is the Jacobian matrix of the diffeomorphism $\check{\Psi}_{s,t}; \mathbb{R}^d \to \mathbb{R}^d$.

In the latter half of this monograph, we will study the Malliavin calculus on the Wiener space and the space of Poisson random measure, called Poisson space; we will apply it for proving the existence of fundamental solutions for heat equations discussed in Chap. 4.

In Chap. 5, we will discuss the Malliavin calculus on the Wiener space and that on the Poisson space (space of Poisson random measure) separately. Then, we will combine these two. For the Malliavin calculus on the Wiener space, we will restrict our attention to the problem of finding the smooth density for laws of Wiener functionals. Our discussion is motivated by Malliavin and Watanabe (Watanabe [116, 117]), but we will take a simple and direct approach; we will not consider the Ornstein–Uhlenbeck semigroup on the Wiener space. Instead, we study the derivative operator D_t and its adjoint δ (Skorohod integral) directly. Then, we give an estimate of Skorohod integrals using L^p-Sobolev norms; a new proof is given for Theorem 5.2.1. A criterion for the smooth density of laws of Wiener functional will be given in terms of the celebrated "Malliavin covariance" in Sect. 5.3.

Another reason why we do not follow Ornstein–Uhlenbeck semigroup argument is that a similar fact is not known for Poisson space; in fact, we want to bring together the Malliavin calculus on the Wiener space and that on the Poisson space in a unified method. In Sects. 5.4, 5.5, 5.6, 5.7, and 5.8, we will discuss the Malliavin calculus on the Poisson space, which is characterized by a Lévy measure ν. A basic assumption for the Poisson random measure is that the characterizing Lévy measure is nondegenerate and satisfies the order condition at the center. Here, the origin 0 is regarded as the center of the Lévy measure defined on $\mathbb{R}^{d'} \setminus \{0\}$. We will see that the difference operator $\tilde{D}_u$ and its adjoint operator $\tilde{\delta}$ defined by Picard [92] work well as D_t and δ do on the Wiener space.

Criteria of the smooth density for Poisson functionals are more complicated. We need a family of L^p-Sobolev norms conditioned to a family of neighborhoods of the center of the Lévy measure. It will be given in Sects. 5.5, 5.6, 5.7, and 5.8. Further, in Sects. 5.9, 5.10, and 5.11, we will study the Malliavin calculus on the product of the Wiener space and the Poisson space. A criterion for the smooth density of the law of a Wiener–Poisson functional will be given after introducing the "Malliavin covariance at the center."

In the application of the Malliavin calculus to solutions of an SDE, properties of stochastic flows defined by the SDE are needed. In Chap. 6, we study the existence of smooth densities of laws of a nondegenerate diffusion and a nondegenerate jump-diffusion defined on a Euclidean space. The class of nondegenerate diffusions includes elliptic diffusions and hypo-elliptic diffusions. Further, the class of non-degenerate jump-diffusions includes pseudo-elliptic jump-diffusions. Let $P^{\mathbf{c}}_{s,t}$ (or $P^{\mathbf{c},\mathbf{d}}_{s,t}$) be the semigroup associated with the generator $A^{\mathbf{c}}(t)$ (or $A^{\mathbf{c},\mathbf{d}}_J(t)$). It will be shown that its transition functions $P^{\mathbf{c}}_{s,t}(x,\cdot)$ (or $P^{\mathbf{c},\mathbf{d}}_{s,t}(x,\cdot)$) have densities $p^{\mathbf{c}}_{s,t}(x,y)$ (or $p^{\mathbf{c},\mathbf{d}}_{s,t}(x,y)$), which are smooth with respect to both variables x and y, and further, the family of the densities is the fundamental solution of the backward heat equation associated with the operator $A^{\mathbf{c}}(t)$ (or $A^{\mathbf{c},\mathbf{d}}_J(t)$); the fundamental solution of the heat equation will be obtained as a family of density functions of a backward transition function $\check{P}^{\mathbf{c}}_{s,t}(x, E)$ associated with the semigroup $\{\check{P}^{\mathbf{c}}_{s,t}\}$, etc. Thus, initial–final value problems (5) and (7) are solved for any bounded continuous functions f_0 and f_1, respectively.

In Sects. 6.7 and 6.8, for elliptic diffusions and pseudo-elliptic jump-diffusions, we will discuss the short-time asymptotics of the transition density functions as $t \downarrow s$, making use of the Malliavin calculus. Our Malliavin calculus cannot be applied to jump processes or jump-diffusion processes which admit big jumps. Indeed, in order to apply the Malliavin calculus, solution X_t of the SDE should be at least an element of $L^{\infty-} = \bigcap_{p>1} L^p$, and the fact excludes solutions of SDEs with big jumps. In Sect. 6.9, we consider such processes: we first truncate big jumps and get the smooth density. Then, we add big jumps and show that this action should preserve the smooth density, where the short-time asymptotics of the fundamental solution will be utilized.

It is hard to apply the Malliavin calculus directly to (jump) diffusions on a bounded domain of a Euclidean space or those defined on a manifold. In Sect. 6.10, we consider a process killed outside of a bounded domain of a Euclidean space. In order to get a smooth density for the killed process, we need two facts. One is a short-time estimate of the density of a non-killed process. The other is a potential theoretic argument of a strong Markov process using hitting times. We show that the density function $q^{\mathbf{c}}_{s,t}(x, y)$ of the killed transition function is smooth with respect to x and y; further, we show that $q^{\mathbf{c}}_{s,t}(x, y)$ is the fundamental solution for the backward heat equation (5) on an arbitrary bounded domain of a Euclidean space with the Dirichlet boundary condition.

Finally, in Chap. 7, we study SDEs on a manifold. Stochastic flow generated by an SDE on a manifold will be discussed in Sect. 7.1. Diffusions, jump-diffusions, and their duals will be treated in Sects. 7.2 and 7.3. Then, the smooth density for a (jump) diffusion on a manifold will be obtained by piecing together killed densities on local charts. It will be discussed in Sects. 7.4 and 7.5.

A Guide for Readers Discussions of the Malliavin calculus for Poisson random measures contain some complicated and technical arguments. For the beginner or the reader who is mainly interested in Wiener processes and diffusion processes, we suggest to skip these arguments at the first reading. After Chap. 4, we could proceed in the following way:

Chapter 5, Sects. 5.1, 5.2, 5.3 $\longrightarrow$ Chap. 6, Sects. 6.1, 6.2, 6.3, 6.7, 6.8, 6.10 $\longrightarrow$ Chap. 7, Sects. 7.1, 7.2, 7.3, 7.4.

The author hopes that this course should be readable.

Contents

Chapter 1
Probability Distributions and Stochastic Processes

Abstract We introduce some basic facts on probability distributions and stochastic processes. Probability distributions and their characteristic functions are discussed in Sect. 1.1. Criteria for smooth densities of distributions are given by their characteristic functions. In Sect. 1.2, we consider Gaussian, Poisson and infinitely divisible distributions and give criteria for these distributions to have smooth densities. Concerning the density problem of an infinitely divisible distribution, we study the Lévy measure in detail. Regarding that the origin 0 is the center of the Lévy measure, we will give criteria for its smooth density by means of 'the order condition' at the center of the Lévy measure.

Then we consider stochastic processes. In Sect. 1.4, we consider Wiener processes, Poisson processes, Poisson random measures and Lévy processes. Among them, Poisson random measures are exposed in detail. Next, in Sects. 1.5 and 1.6, we discuss martingales, semi-martingales and their quadratic variations. These are standard tools for the Itô calculus. In Sect. 1.7, we define Markov processes. The strong Markov property will be discussed. In Sect. 1.8, we study Kolmogorov's criterion for a random field with multi-dimensional parameter to have a continuous modification.

1.1 Probability Distributions and Characteristic Functions

Let Ω be a set and let $\mathcal{F}$ be a family of subsets of Ω with the following three properties: (a) $\Omega \in \mathcal{F}$, (b) if $A \in \mathcal{F}$, then its complement set $A^c \in \mathcal{F}$, and (c) if $A_n \in \mathcal{F}, n = 1, 2, \ldots$, then $\bigcup_n A_n \in \mathcal{F}$. The set $\mathcal{F}$ is called a *σ-field* of Ω. An element of Ω is denoted by ω and is called a *sample*. Elements of $\mathcal{F}$ are called *events*. The pair $(\Omega, \mathcal{F})$ is called a *measurable space*. A map $P : \mathcal{F} \to [0, 1]$ is called a *measure* if it satisfies the following: (a) $P(A) \geq 0$ for any $A \in \mathcal{F}$, and (b) if $A_n, n = 1, 2, \ldots$ are disjoint, then $P(\bigcup_{n=1}^{\infty} A_n) = \sum_{n=1}^{\infty} P(A_n)$. Further, if (c) $P(\Omega) = 1$ is satisfied, P is called a *probability measure*. The triple $(\Omega, \mathcal{F}, P)$ is called a *probability space*.

H. Kunita, *Stochastic Flows and Jump-Diffusions*, Probability Theory and Stochastic Modelling 92, https://doi.org/10.1007/978-981-13-3801-4_1

Let $(\Omega, \mathcal{F}, P)$ be a probability space. A family of events $\{A_\lambda, \lambda \in \Lambda\}$ is called *independent*, if any finite subset $\{A_{\lambda_1}, \ldots, A_{\lambda_n}\}$ of $\{A_\lambda, \lambda \in \Lambda\}$ satisfies $P(\bigcap_{i=1}^n A_{\lambda_i}) = \prod_{i=1}^n P(A_{\lambda_i})$. Next, let $\mathcal{G}$ be a subset of $\mathcal{F}$. If $\mathcal{G}$ is a σ-field, it is called a sub σ-field of $\mathcal{F}$. Suppose that $\{\mathcal{G}_\lambda, \lambda \in \Lambda\}$ is a family of sub σ-fields of $\mathcal{F}$. It is called *independent*, if any collection of events $\{A_\lambda, \lambda \in \Lambda\}$, where each A_λ runs in $\mathcal{G}_\lambda$, is independent.

Let A, B be events. Suppose that $P(B) > 0$. The *conditional probability* of A given B is defined by $P(A|B) = \frac{P(A \cap B)}{P(B)}$. Then $P(A|B) = P(A)$ holds if and only if A and B are independent.

Let $\mathbb{R}^d$ be a d-dimensional Euclidean space. Its elements are denoted by $x = (x_1, \ldots, x_d)$. Let $\mathcal{B}(\mathbb{R}^d)$ be the Borel field of $\mathbb{R}^d$. A probability measure μ on $(\mathbb{R}^d, \mathcal{B}(\mathbb{R}^d))$ is called a *probability distribution* or simply a *distribution*. If integrals $b_i = \int_{\mathbb{R}^d} x_i d\mu(x)$ and $a_{ij} = \int_{\mathbb{R}^d} (x_i - b_i)(x_j - b_j)\mu(dx)$ exist, $b = (b_1, \ldots, b_d)$ is called the *mean vector* and $A = (a_{ij})$ is called the *covariance matrix*. For a vector $b = (b_1, \ldots, b_d)$, we set $|b| = (\sum_{i=1} |b_i|^2)^{1/2}$ and for a matrix $A = (a_{ij})$, we set $|A| = (\sum_{i,j} a_{ij}^2)^{1/2}$. When $d = 1$, $\mathbb{R}^1$ coincides with the space $\mathbb{R}$ of real numbers. b and A are called the mean and the variance, respectively.

Let S be a Hausdorff topological space with the second countability and let $\mathcal{B}(S)$ be the Borel field of S. A map $X; \Omega \to S$ is called *measurable* (or $\mathcal{G}$-measurable) if the set $\{\omega;\ X(\omega) \in E\}$ belongs to $\mathcal{F}$ (or $\mathcal{G}$) for any Borel subset E in S. A measurable map $X; \Omega \to S$ is called an S-valued *random variable*. If $S = \mathbb{R}^d$, X is called a d-dimensional random variable. Further, if $d = 1$, it is a real (real-valued) random variable or simply a random variable. Let $\{X_\lambda; \lambda \in \Lambda\}$ be a family of S-valued random variables. We denote by $\sigma(\{X_\lambda\})$ the smallest sub σ-field of $\mathcal{F}$ with respect to which $\{X_\lambda\}$ are measurable. Let $\{Y_{\lambda'}; \lambda' \in \Lambda'\}$ be another family of S'-valued random variables and let $\sigma(\{Y_{\lambda'}\})$ be the smallest σ field with respect to which $\{Y_{\lambda'}\}$ are measurable. Then two families of random variables $\{X_\lambda\}$ and $\{Y_{\lambda'}\}$ are called *mutually independent* if these sub σ-fields are independent.

Given an S-valued random variable X, we set $\mu(E) = P(\{X \in E\}),\ E \in \mathcal{B}(S)$. It is a probability measure on $(S, \mathcal{B}(S))$, called the *law* of X.

The *expectation* of a real random variable X is defined as the Lebesgue integral by its law μ; $E[X] = \int x\mu(dx)$, if the integral is well defined. It is equal to the Lebesgue integral of X by the measure P; $E[X] = \int_\Omega X(\omega)P(d\omega)$. If $E[|X|]$ is finite, X is said to be an *integrable random variable*.

Let A be an event. We denote by 1_A the indicator function of the event A, i.e., $1_A(\omega)$ is a function taking the value 1 if $\omega \in A$ and the value 0 if $\omega \notin A$. The expectation $E[X1_A]$ is often denoted by $E[X; A]$.

Let $\{X_n, n = 1, 2, \ldots\}$ be a sequence of real random variables. Let X be another real random variable. The sequence $\{X_n\}$ is said to *converge to X almost surely*, simply written as $X_n \to X$, $a.s$, if there exists $\tilde{\Omega} \in \mathcal{F}$ such that $P(\tilde{\Omega}) = 1$ and for any $\omega \in \tilde{\Omega}$, $\{X_n(\omega)\}$ converges to $X(\omega)$. If $P(|X_n - X| > \epsilon)$ converges to 0 as $n \to \infty$ for any $\epsilon > 0$, $\{X_n\}$ is said to *converge to X in probability*. Let $p \geq 1$. If $E[|X_n|^p] < \infty$ holds for any n and $E[|X_n - X|^p] \to 0$ as $n \to \infty$, $\{X_n\}$ is said to *converge to X in L^p*. A sequence of random variables $\{X_n\}$ is called *uniformly integrable*, if for any $\epsilon > 0$ there is a positive constant c such that $\sup_n \int_{|X_n|>c} |X_n| dP < \epsilon$. If $\{X_n\}$ is a sequence of a uniformly integrable random variables converging to X in probability, then $\{X_n\}$ converges to X in L^1, i.e., $\lim_{n\to\infty} E[|X_n - X|] = 0$.

Let μ_1 and μ_2 be distributions on $\mathbb{R}^d$. For a Borel subset E of $\mathbb{R}^d$, we set $\mu(E) = \int_{\mathbb{R}^d} \mu_1(E-x)\mu_2(dx)$. Then μ is a distribution on $\mathbb{R}^d$. We denote μ by $\mu_1 * \mu_2$ and call it the *convolution* of two distributions μ_1 and μ_2. It is immediate to see $\mu_1 * \mu_2 = \mu_2 * \mu_1$. Let X, Y be independent $\mathbb{R}^d$-valued random variables and let μ_X and μ_Y be their laws, respectively. Let μ_{X+Y} be the law of $X+Y$. Then it holds that $\mu_{X+Y} = \mu_X * \mu_Y$. Let X and Y be independent random variables. Then $E[f(X)g(Y)] = E[f(X)]E[g(Y)]$ holds for any bounded continuous functions f, g.

Let X be an S-valued random variable and let Z be a real integrable random variable. The *weighted law* of X with respect to Z is a bounded signed measure on $(S, \mathcal{B}(S))$ defined by $\mu_Z(E) = E[Z; \{X \in E\}]$. Its total variation measure $|\mu_Z|$ is given by $|\mu_Z|(E) = E[|Z|; \{X \in E\}]$. If a bounded signed measure μ on $\mathbb{R}^d$ satisfies $\int_{\mathbb{R}^d} |x|^n |\mu|(dx) < \infty$, where $|\mu|$ is the total variation of the measure μ, then the measure μ is said to have an *n-th moment*.

The *characteristic function* of a bounded signed measure μ on $\mathbb{R}^d$ is defined by

$$\psi(v) = \int_{\mathbb{R}^d} e^{i(v,x)} \mu(dx), \quad v = (v_1, \ldots, v_d) \in \mathbb{R}^d, \tag{1.1}$$

where $i = \sqrt{-1}$ and $(v, x) = \sum_{j=1}^d v_j x_j$. Let Z be a real integrable random variable, $X = (X_1, \ldots, X_d)$ be an $\mathbb{R}^d$-valued random variable and let μ_Z be its weighted law. Let $\psi(v)$ be its characteristic function. Then we have

$$E[e^{i(v,X)} Z] = \int_{\mathbb{R}^d} e^{i(v,x)} \mu_Z(dx) = \psi(v).$$

For a differentiable function $f(x)$ on $\mathbb{R}^d$, the partial derivatives $\frac{\partial}{\partial x_i} f(x)$ are denoted by $\partial_{x_i} f$. We set $\partial f = \partial_x f = (\partial_{x_1} f, \ldots, \partial_{x_d} f)$. Further, if f is a vector function $(f^1, \ldots, f^{d'})$, we set $\partial f = (\partial_{x_j} f^k)_{j=1,\ldots,d,k=1,\ldots,d'}$. It is called the Jacobian matrix. When $d = d'$, we will often denote it by ∇f.

The set of positive integers is denoted by $\mathbb{N}$ and the set of nonnegative integers is denoted by $\bar{\mathbb{N}}$. Let $\mathbf{j}$ be a multi-index of nonnegative integers of length d; $\mathbf{j} = (j_1, \ldots, j_d)$, where $j_1, \ldots, j_d$ are nonnegative integers. We set $|\mathbf{j}| = j_1 + \cdots + j_d$. Let $\mathbf{j}' = (j_1', \ldots, j_d')$ be another multi-index of length d. If $j_i \le j_i'$ holds for any $i = 1, \ldots, d$, then we denote the relation by $\mathbf{j} \le \mathbf{j}'$. We define the higher order differential operator $\partial^{\mathbf{j}}$ and product $x^{\mathbf{j}}$ for $x = (x_1, \ldots, x_d) \in \mathbb{R}^d$ by

$$\partial^{\mathbf{j}} = \partial_x^{\mathbf{j}} = \partial_{x_1}^{j_1} \cdots \partial_{x_d}^{j_d}, \quad x^{\mathbf{j}} = x_1^{j_1} \cdots x_d^{j_d},$$

if $|\mathbf{j}| \geq 1$. If $|\mathbf{j}| = 0$, i.e., $\mathbf{j} = (0, \ldots, 0)$, we define $\partial^0 f = f$ and $x^0 = 1$. Let $n_0 \in \mathbb{N}$. If a function $f(x)$ on $\mathbb{R}^d$ is n_0-times continuously differentiable and $|\partial^{\mathbf{j}} f(x)|$ are bounded for any $\mathbf{j}$ with $|\mathbf{j}| \leq n_0$, f is said to be a function of $C_b^{n_0}$-class, or a $C_b^{n_0}$-function.

Proposition 1.1.1 *Let $n_0 \in \bar{\mathbb{N}}$. If a bounded signed measure μ has n_0-th moment, then its characteristic function $\psi(v)$ is a $C_b^{n_0}$-function. Further, it holds that*

$$\partial_v^{\mathbf{j}} \psi(v) = i^{|\mathbf{j}|} \int_{\mathbb{R}^d} e^{i(v,x)} x^{\mathbf{j}} \mu(dx) \tag{1.2}$$

for any $\mathbf{j}$ with $|\mathbf{j}| \leq n_0$.

Proof We prove (1.2) by the induction of $n_0 \in \bar{\mathbb{N}}$. The case $n_0 = 0$ is trivial. Suppose it is valid for $\mathbf{j}$ with $|\mathbf{j}| \leq n_0 - 1$. Set $\tilde{\psi}(v) := \int e^{i(v,x)} x^{\mathbf{j}} \mu(dx)$. We have for $h \neq 0$

$$\frac{\tilde{\psi}(v + he_j) - \tilde{\psi}(v)}{h} = \int \Big\{ \frac{e^{i(v+he_j,x)} - e^{i(v,x)}}{h} \Big\} x^{\mathbf{j}} \mu(dx),$$

where e_j, $j = 1, \ldots, d$ are unit vectors in $\mathbb{R}^d$. Let h tend to 0 in the above. Since

$$\Big| \frac{e^{i(v+he_j,x)} - e^{i(v,x)}}{h} \Big| \leq |x_j|$$

holds for any h, the right-hand side converges to $\int e^{i(v,x)}(ix_j)x^{\mathbf{j}} d\mu$, by the Lebesgue convergence theorem. Therefore $\tilde{\psi}(v)$ is differentiable with respect to v_j and we get

$$\partial_{v_j} \tilde{\psi}(v) = i \int e^{i(v,x)} x_j x^{\mathbf{j}} \mu(dx).$$

It holds that $\partial_{v_j} \tilde{\psi}(v) = \partial^{\mathbf{j}+\mathbf{e_j}} \psi(v)$ and $\int e^{i(v,x)} x_j x^{\mathbf{j}} \mu(dx) = \int e^{i(v,x)} x^{\mathbf{j}+\mathbf{e_j}} \mu(dx)$. Therefore we get the formula (1.2) for $\mathbf{j} + e_j$. □

From the above proposition, the moment of the measure μ can be computed as

$$\int x^{\mathbf{j}} \mu(dx) = i^{-|\mathbf{j}|} \partial_v^{\mathbf{j}} \psi(v) \Big|_{v=0}.$$

A Borel subset E of $\mathbb{R}^d$ is called *μ-continuous*, if $\mu(\partial E) = 0$ holds. Here ∂E is the boundary set of E defined by $\bar{E} - E^o$.

Theorem 1.1.1 (Lévy's inversion formula) *The characteristic function determines the measure μ uniquely: If a rectangular set $E = \prod_{j=1}^d (a_j, b_j]$ is μ-continuous, then*

$$\mu(E) = \lim_{N\to\infty} \left(\frac{1}{2\pi}\right)^d \int_{-N}^{N} \cdots \int_{-N}^{N} \prod_{j=1}^{d} \frac{e^{-iv_j b_j} - e^{-iv_j a_j}}{-iv_j} \psi(v_1, \ldots, v_d)\, dv_1 \cdots dv_d. \tag{1.3}$$

Proof Using Fubini's theorem, we have

$$\int_{-N}^{N} \cdots \int_{-N}^{N} \prod_{j=1}^{d} \frac{e^{-iv_j b_j} - e^{-iv_j a_j}}{-iv_j} \psi(v_1, \ldots, v_d)\, dv_1 \cdots dv_d$$

$$= \int_{-N}^{N} \cdots \int_{-N}^{N} \prod_{j=1}^{d} \frac{e^{-iv_j b_j} - e^{-iv_j a_j}}{-iv_j} \left(\int \cdots \int e^{i(\sum_j v_j x_j)} \mu(dx_1 \cdots dx_d) \right) dv_1 \cdots dv_d$$

$$= \int \cdots \int \left(\int_{-N}^{N} \cdots \int_{-N}^{N} \prod_{j=1}^{d} \frac{e^{iv_j(x_j - b_j)} - e^{iv_j(x_j - a_j)}}{-iv_j} dv_1 \cdots dv_d \right) \mu(dx_1 \cdots dx_d)$$

$$= \int \cdots \int \left(\prod_{j=1}^{d} \int_{-N}^{N} \frac{e^{iv_j(x_j - b_j)} - e^{iv_j(x_j - a_j)}}{-iv_j} dv_j \right) \mu(dx_1 \cdots dx_d).$$

Set the integral $\int_{-N}^{N} \cdots dv_j$ of the above by $J(N, x_j, a_j, b_j)$. It holds that

$$J(N, x, a, b) = 2 \int_0^N \frac{\sin(x-a)v}{v} dv - 2 \int_0^N \frac{\sin(x-b)v}{v} dv.$$

Since $\lim_{N\to\infty} \int_0^N \frac{\sin t}{t} dt = \frac{\pi}{2}$ holds, we have

$$\lim_{N\to\infty} J(N, x, a, b) = \begin{cases} 0, & \text{if } x < a \text{ or } b < x, \\ \pi, & \text{if } x = a \text{ or } x = b, \\ 2\pi, & \text{if } a < x < b. \end{cases}$$

The above convergence is a bounded convergence. Therefore, by Lebesgue's bounded convergence theorem,

$$\lim_{N\to\infty} \int_{-N}^{N} \cdots \int_{-N}^{N} \prod_{j=1}^{d} J(N, x_j, a_j, b_j) \mu(dx_1 \cdots dx_d) = (2\pi)^d \mu(E),$$

if $\mu(\partial E) = 0$. We have thus proved the inversion formula (1.3). □

Corollary 1.1.1 *Let $X_1, \ldots, X_d$ be real random variables. Denote by X its vector $(X_1, \ldots, X_d)$. If*

$$E[e^{i(v,X)}] = \prod_{j=1}^{d} E[e^{iv_j X_j}] \tag{1.4}$$

holds for all v, then $X_1, \ldots, X_d$ are mutually independent.

Conversely, if $X_1, \ldots, X_d$ are mutually independent, (1.4) *is valid. Further, the characteristic function of $Y = X_1 + \cdots + X_d$ is equal to the product of characteristic functions of X_j, $j = 1, \ldots, d$, i.e., $\prod_{j=1}^d E[e^{ivX_j}]$.*

Proof Let μ be the law of X and let μ_j, $j = 1, \ldots, d$ be one-dimensional laws of X_j, $j = 1, \ldots, j$ respectively. Suppose that (1.4) holds. Then equation (1.4) is written as

$$\int e^{i(v,x)}\, d\mu(x_1, \ldots, x_d) = \int e^{i(v,x)}\, d\mu_1(x_1) \cdots d\mu(x_d).$$

Therefore we have $\mu = \mu_1 \times \cdots \times \mu_d$ by the above theorem. Hence $X_1, \ldots, X_d$ are independent.

Conversely, if $X_1, \ldots, X_d$ are independent, then $e^{iv_1X_1}, \ldots, e^{iv_dX_d}$ are also independent. Therefore (1.4) holds. The formula implies that the characteristic function of Y is equal to $\prod_{j=1}^d E[e^{ivX_j}]$. □

We study the relation for the existence of the smooth density of the measure μ and the decay property of the characteristic function $\psi(v)$ as $|v| \to \infty$.

Proposition 1.1.2 *Let $\psi(v)$ be the characteristic function of a bounded signed measure μ.*

1. *If $\int |\psi(v)|\, dv < \infty$, then μ has a bounded continuous density function $f(x)$. Further, we have the inversion formula:*

$$f(x) = \left(\frac{1}{2\pi}\right)^d \int_{\mathbb{R}^d} e^{-i(v,x)} \psi(v)\, dv. \tag{1.5}$$

2. *If $\int |\psi(v)||v|^{n_0}\, dv < \infty$ for some positive integer n_0, then the density function $f(x)$ is of $C_b^{n_0}$-class. Further, for any multi-index $\mathbf{j}$ with $|\mathbf{j}| \le n_0$, we have*

$$\partial^{\mathbf{j}} f(x) = \frac{(-i)^{|\mathbf{j}|}}{(2\pi)^d} \int_{\mathbb{R}^d} e^{-i(v,x)} \psi(v) v^{\mathbf{j}}\, dv. \tag{1.6}$$

3. *Let m_0 be a positive integer. If $\psi(v)$ is of $C_b^{m_0}$-class and functions $\partial_v^{\mathbf{i}}(\psi(v)v^{\mathbf{j}})$ are integrable and converge to 0 as $|v| \to \infty$ for any $|\mathbf{j}| \le n_0$ and $|\mathbf{i}| \le m_0$, then the density function is of $C_b^{n_0}$-class and satisfies*

$$(ix)^{\mathbf{i}} \partial^{\mathbf{j}} f(x) = \frac{(-i)^{|\mathbf{j}|}}{(2\pi)^d} \int_{\mathbb{R}^d} e^{-i(v,x)} \partial_v^{\mathbf{i}}(\psi(v)v^{\mathbf{j}})\, dv \tag{1.7}$$

for any $|\mathbf{j}| \le n_0$, $|\mathbf{i}| \le m_0$. Furthermore, there exists a positive constant c such that $|\partial^{\mathbf{j}} f(x)| \le \frac{c}{(1+|x|)^{m_0}}$ holds for all x.

Proof

1. If $|\psi(v)|$ is integrable, we can rewrite the inversion formula (1.3) as

$$\mu(E) = \Big(\frac{1}{2\pi}\Big)^d \int_{-\infty}^{\infty} \cdots \int_{-\infty}^{\infty} \prod_{j=1}^{d} \frac{e^{-iv_j b_j} - e^{-iv_j a_j}}{-iv_j} \psi(v_1, \ldots, v_d)\, dv_1 \cdots dv_d,$$

since the integrand is an integrable function on $\mathbb{R}^d$ with respect to the Lebesgue measure $dv_1 \cdots dv_d$. We consider the right-hand side. For fixed $a_1, \ldots, a_d$, it is a function of $(b_1, \ldots, b_d)$, which we denote by $F(b_1, \ldots, b_d)$. Set

$$\phi(v_j, a_j, b_j) = \frac{e^{-iv_j a_j} - e^{-iv_j b_j}}{iv_j}.$$

It holds that $|\phi(v_j, a_j, b_j)| \le |b_j - a_j|$ and $|\partial_{b_j}\phi(a_j, v_j, b_j)| \le 1$ for any v_j, a_j, b_j. Then for any $j = 1, \ldots, d$, we can change the order of the partial derivative ∂_{b_j} and the integral; we find that for any j, $F(b_1, \ldots, b_d)$ is partially differentiable with respect to b_j. Repeating this argument for $j = 1, \ldots, d$, we find that the function F is partially differentiable with respect to b_j, $j = 1, \ldots, d$ and we get

$$\partial_{b_1} \cdots \partial_{b_d} F(b_1, \ldots, b_d) = \Big(\frac{1}{2\pi}\Big)^d \int_{-\infty}^{\infty} \cdots \int_{-\infty}^{\infty} e^{-i\sum_{j=1}^{d} v_j b_j} \psi(v_1, \ldots, v_d)\, dv_1 \cdots dv_d.$$

Denote the above function by $f(b_1, \ldots, b_d)$. It is a bounded continuous function and we have $\int_{a_1}^{b_1} \cdots \int_{a_d}^{b_d} f(x)\, dx = \mu(E)$ if E is a μ-continuous rectangular set $\Pi_{j=1}^d (a_j, b_j]$. Now we can choose an increasing sequence of μ-continuous rectangular sets E_n such that $\bigcup_n E_n = \mathbb{R}^d$. Then we have $\mu(E_n) \to 1$. This implies $\int_{\mathbb{R}^d} f(x)\, dx = 1$. Therefore, all rectangular sets should be μ-continuous and the assertion is proved.
2. If $|\psi(v)||v|^{n_0}$ is integrable on $\mathbb{R}^d$, we can change the order of derivative operator $\partial^{\mathbf{j}}$, $|\mathbf{j}| \le n_0$ and the integral operator in the inversion formula (1.5). Therefore we get the assertion.
3. Suppose that $\psi(v)$ satisfies conditions of 3. The density function $f(x)$ is of $C_b^{n_0}$-class by 2. By the formula of integration by parts, we have

$$\int_{\mathbb{R}^d} e^{-i(v,x)} \partial_{v_k}(\psi(v) v^{\mathbf{j}})\, dv = ix_k \int_{\mathbb{R}^d} e^{-i(v,x)} \psi(v) v^{\mathbf{j}}\, dv.$$

Repeating this argument, we get the formula (1.7) for any $|\mathbf{j}| \le n_0$ and $|\mathbf{i}| \le m_0$. Further, since the right-hand side of (1.7) is a bounded function, $|x|^{|\mathbf{i}|}|\partial^{\mathbf{j}} f(x)|$ is a bounded function. Therefore $|\partial^{\mathbf{j}} f(x)| \le c/(1+|x|)^{m_0}$ holds. □

Let $f(x)$ be a (complex) C^{n_0}-function on $\mathbb{R}^d$. It is called *rapidly decreasing* if the function $|\partial^{\mathbf{j}} f(x)|(1+|x|^2)^i$ converges to 0 as $|x| \to \infty$ for any $|\mathbf{j}| \le n_0$ and nonnegative integer i. A C^∞-function $f(x)$ on $\mathbb{R}^d$ is said to be *rapidly decreasing*, if it is a rapidly decreasing C^{n_0}-function for any $n_0 \in \mathbb{N}$.

Theorem 1.1.2 *Let $\psi(v)$ be the characteristic function of a bounded signed measure μ. If $\psi(v)$ is a rapidly decreasing continuous function, then the measure μ has a C^∞-density function f. Further, if $\psi(v)$ is a rapidly decreasing C^∞-function, f is a rapidly decreasing C^∞-function.*

Proof If $\psi(v)$ is a rapidly decreasing continuous function, then $|\psi(v)||v|^n$ is an integrable function for any n. Therefore μ has a C_b^∞-density function. Further, if $\psi(v)$ is a rapidly decreasing C^∞-function, $\partial_v^{\mathbf{i}}(\psi(v)v^{\mathbf{j}})$ are integrable and converges to 0 as $|v| \to \infty$ for any $\mathbf{i}$ and $\mathbf{j}$. Then $|x^{\mathbf{i}}||\partial^{\mathbf{j}} f(x)|$ is a bounded function for any $\mathbf{i}$ and $\mathbf{j}$ by (1.7). Therefore $f(x)$ should be rapidly decreasing. □

1.2 Gaussian, Poisson and Infinitely Divisible Distributions

We give some distributions and their characteristic functions which will be used in this monograph.

1. *Exponential distribution.* Let μ be a distribution on $\mathbb{R}$ having the density function $f(x)$ given by $f(x) = \lambda e^{-\lambda x}$ if $x > 0$, and $f(x) = 0$ if $x \leq 0$, where $\lambda > 0$. Then μ is called the *exponential distribution* with parameter λ. Its mean is $\frac{1}{\lambda}$ and the variance is $\frac{1}{\lambda^2}$. Further, its characteristic function is $\psi(v) = \frac{\lambda}{\lambda - iv}$.

2. *Gamma distribution.* Let μ be a distribution on $\mathbb{R}$ having the density function $f(x)$ given by $f(x) = \frac{\lambda^k x^{k-1} e^{-\lambda x}}{\Gamma(k)}$ if $x > 0$, and $f(x) = 0$ if $x \leq 0$, where $\lambda > 0$, k is a positive integer and $\Gamma(k) = (k-1)!$. Then μ is called a *Gamma distribution* with parameters λ and k. Its mean is $\frac{k}{\lambda}$ and variance is $\frac{k}{\lambda^2}$. Its characteristic function is $\psi(v) = \left(\frac{\lambda}{\lambda - iv}\right)^k$. If $k = 1$, it coincides with the exponential distribution with parameter λ.

Let $\sigma_1, \ldots, \sigma_k$ be mutually independent random variables with the same exponential distribution with parameter λ. Then the law of $\tau_k := \sigma_1 + \cdots + \sigma_k$ is the Gamma distribution with parameters λ and k. Indeed, since laws of σ_j are exponential distributions with the same parameter λ, their characteristic functions are equal to $\frac{\lambda}{\lambda - iv}$. Sine σ_j are independent, the characteristic function of τ_k should be the k-fold product of $\frac{\lambda}{\lambda - iv}$, i.e., it is equal to $\left(\frac{\lambda}{\lambda - iv}\right)^k$ (Corollary 1.1.1). Hence the law of τ_k is equal to the Gamma distribution with parameters λ and k.

3. *Gaussian distribution.* Let d' be a positive integer and let $b = (b^1, \ldots, b^{d'})$ be a d'-vector. Let $A = (a^{ij})$ be a $d' \times d'$ positive definite symmetric matrix and let $\det A$ be its determinant. Assuming that $\det A \neq 0$, we define

$$f(x) = \frac{1}{(2\pi)^{\frac{d}{2}} |\det A|^{\frac{1}{2}}} \exp\Big\{ -\frac{1}{2}(x - b, A^{-1}(x - b))\Big\}. \tag{1.8}$$

Then $\mu(B) := \int_B f(x)\,dx,\ \ B \in \mathcal{B}$, is a probability distribution on $\mathbb{R}^{d'}$, called a d'-dimensional *Gaussian distribution*. Its mean vector is equal to b and covariance matrix is equal to A. It is known that its characteristic function is equal to

$$\psi(v) = \exp\left\{i(v, b) - \frac{1}{2}(v, Av)\right\}, \quad v \in \mathbb{R}^{d'}. \tag{1.9}$$

Conversely, for a given d'-vector b and a nonnegative definite symmetric $d' \times d'$-matrix A, there exists a distribution μ on $\mathbb{R}^{d'}$, whose characteristic function is equal to (1.9). Such a distribution is unique. It is also called a Gaussian distribution. Its mean and covariance are again given by b and A, respectively. If the matrix A is positive definite, μ has a density function written as in (1.8). If the rank of A is less than d', the measure μ is singular with respect to the Lebesgue measure.

4. *Poisson distribution*. Let μ be a distribution on $\mathbb{R}$ having point masses on the set $\bar{\mathbb{N}} := \{0, 1, 2, \ldots\}$ and

$$\mu(\{n\}) = \frac{\lambda^n}{n!}e^{-\lambda}, \quad n = 0, 1, 2, \ldots, \tag{1.10}$$

where λ is a positive constant. It is called a *Poisson distribution*. Its mean and the variance are equal to λ. λ is called the *parameter of the Poisson distribution*. Its characteristic function is computed as

$$\psi(v) = \exp\left\{\lambda(e^{iv} - 1)\right\}, \quad v \in \mathbb{R}. \tag{1.11}$$

5. *Compound Poisson distribution*. Let μ be a distribution on $\mathbb{R}^{d'}$, whose characteristic function is given by

$$\psi(v) = \exp\left\{\int_{\mathbb{R}_0^{d'}} (e^{i(v,z)} - 1)\nu(dz)\right\}, \quad v \in \mathbb{R}^{d'}. \tag{1.12}$$

where ν is a finite measure on $\mathbb{R}_0^{d'} = \mathbb{R}^{d'} \setminus \{0\}$. The distribution μ is called a *compound Poisson distribution (associated with measure ν)*. A Poisson distribution is a one-dimensional compound Poisson distribution associated with $\nu = \lambda\delta_1$, where δ_1 is the delta measure at the point 1.

6. *Infinitely divisible distribution*. A distribution μ on $\mathbb{R}^{d'}$ is called *infinitely divisible*, if the characteristic function of μ is represented by

$$\psi(v) = \exp\left\{i(v, b) - \frac{1}{2}(v, Av) + \int_{\mathbb{R}_0^{d'}} (e^{i(v,z)} - 1 - i(v, z)1_D(z))\nu(dz)\right\}. \tag{1.13}$$

Here, $v, b \in \mathbb{R}^{d'}$, A is a matrix as in (1.9), $D = \{z \in \mathbb{R}_0^{d'};\ |z| \leq 1\}$ and ν is a measure on $\mathbb{R}_0^{d'}$ satisfying $\int_{\mathbb{R}_0^{d'}} (|z|^2 \wedge 1)\nu(dz) < \infty$. A measure ν on $\mathbb{R}_0^{d'}$ satisfying the above property is called a *Lévy measure*. The triple (A, ν, b) is called the *characteristics of the infinitely divisible distribution*. Occasionally, we regard the Lévy measure ν is a measure on $\mathbb{R}^{d'}$ satisfying $\nu(\{0\}) = 0$. We call the point 0 the *center of the Lévy measure*.

Let μ be a distribution on $\mathbb{R}^{d'}$. Suppose that for any $n \in \mathbb{N}$, there exists a distribution μ_n such that $\mu = \mu_n^n$, where μ_n^n is the n-times convolution of the distribution μ_n. It is known that its characteristic function $\psi(v)$ is represented by (1.13). Hence μ is an infinitely divisible law. The formula (1.13) is called the *Lévy–Khintchine formula*. Conversely, if the characteristic function of a distribution μ is written as (1.13), for any $n \in \mathbb{N}$ there are distributions μ_n such that $\mu = \mu_n^n$. This is the origin of the word 'infinitely divisible'.

7. *Stable distribution*. An infinitely divisible distribution μ is called *stable* if for any $a > 0$, there are $b > 0$ and $c \in \mathbb{R}^{d'}$ such that its characteristic function $\psi(v)$ satisfies $\psi(v)^a = \psi(bv)e^{i(c,v)}$. Further, if the above holds for any $a > 0, b = a^{1/\beta}$ with some $0 < \beta \le 2$ and $c = 0$, then μ is called an *β-stable* distribution. β is called the *index* of the stable distribution. If the drift vector b and the Lévy measure ν are 0, the distribution μ satisfies the above property with $\beta = 2$. Hence a Gaussian distribution with mean 0 is 2-stable. If μ is β-stable with $0 < \beta < 2$, then $A = 0$. Further, using the polar coordinate, the Lévy measure ν is represented by

$$\nu(B) = \int_{S_{d'-1}} \lambda(d\vartheta) \int_0^\infty 1_B(r\vartheta)\frac{dr}{r^{1+\beta}}, \tag{1.14}$$

where $S_{d'-1} = \{z \in \mathbb{R}^{d'}; |z| = 1\}$ and λ is a finite measure on $S_{d'-1}$.

In the one-dimensional case, S_0 is equal to two points $\{-1, +1\}$. The Lévy measure of the one-dimensional stable distribution is given by

$$\nu(dz) = \begin{cases} c_1 z^{-1-\beta}\,dz & \text{on } (0,\infty), \\ c_2|z|^{-1-\beta}\,dz & \text{on } (-\infty, 0), \end{cases} \tag{1.15}$$

with $c_1 \ge 0, c_2 \ge 0, c_1 + c_2 > 0$. In particular if the distribution μ is symmetric, then it holds that $c_1 = c_2 > 0$. Its characteristic function is given by

$$\psi(v) = \exp\left\{ - c|v|^\beta\left(1 - i\gamma \tan\frac{\pi\beta}{2}\text{sign}v\right) + i\tau v\right\}, \tag{1.16}$$

where $c > 0, \gamma \in [-1, 1]$ and $\tau \in \mathbb{R}$, if $\beta \ne 1$.

If μ is a d'-dimensional rotation invariant β-stable distribution, its characteristic function is given by

$$\psi(v) = \exp\{-c|v|^\beta\}, \quad v \in \mathbb{R}^{d'}, \tag{1.17}$$

where $c > 0$. For further details of infinitely divisible distributions and stable distributions, we refer to Sato [99].

Let us consider whether or not an infinitely divisible distribution μ has a smooth density. If μ is rotation invariant and β-stable, its characteristic function is given by (1.17). It is a rapidly decreasing function of v and $\psi(v)|v|^n$ is an integrable function for any positive integer n. Therefore the measure μ has C_b^∞-density function by Proposition 1.1.2.

We are interested in more general infinitely divisible distributions. We start with a one-dimensional infinitely divisible distribution with the second moment; its characteristic function is represented by

$$\psi(v) = \exp\left\{ibv + \int_{\mathbb{R}} \left(\frac{e^{ivz} - 1 - ivz}{z^2}\right)\nu'(dz)\right\}, \tag{1.18}$$

where ν' is a bounded measure on $\mathbb{R}$. (At $z = 0$, we identify $\frac{e^{ivz}-1-ivz}{z^2}$ with $-\frac{1}{2}v^2$.) Setting $\nu'(\{0\}) = a$ and $\nu(dz) := \frac{1}{z^2}\nu'(dz)$ for $z \neq 0$ (Lévy measure), we get the Lévy–Khintchine formula with characteristics (a, b, ν). Now, if $a > 0$, the distribution μ has a C^∞-density, because μ is written as a convolution of two distributions μ_1, μ_2, where μ_1 is a Gaussian measure with mean b and variance a. If $a = 0$, we consider the order of the concentration of the mass of ν around the center 0. Suppose that the mass of ν' is highly concentrated near the center and satisfies the order condition

$$0 < \liminf_{\rho \to 0} \frac{\nu'(B_0(\rho))}{\rho^\alpha} \leq \limsup_{\rho \to 0} \frac{\nu'(B_0(\rho))}{\rho^\alpha} < \infty, \tag{1.19}$$

for some $0 < \alpha < 2$, where $B_0(\rho) = \{z; |z| \leq \rho\}$. Then the Lévy measure ν is said to satisfy the *order condition of exponent* α *at the center*. If the Lévy measure ν is a finite measure, it does not satisfy the order condition, since lim inf of (1.19) is 0 for any $0 < \alpha < 2$. It was shown by Orey that if the one-dimensional Lévy measure satisfies the order condition of exponent $0 < \alpha < 2$ at the center, then the law μ has a C^∞-density.

Now, let μ be a d'-dimensional infinitely divisible distribution with characteristics (A, ν, b). Suppose that the matrix A is nondegenerate. Then μ is written as the convolution of two distributions μ_1 and μ_2, where μ_1 is a Gaussian distribution with mean b and covariance A, and μ_2 is an infinitely divisible distribution with characteristics $(0, \nu, 0)$. Since μ_1 has a C^∞-density, the convolution $\mu_1 * \mu_2$ has also a C^∞-density.

Suppose next that the matrix A is degenerate. Let $\{A_0(\rho);\ 0 < \rho < 1\}$ be a family of open neighborhoods of the center in $\mathbb{R}^{d'}$, which satisfies the following two properties for any $0 < \rho < 1$:

$$(i)\ A_0(\rho) \subset B_0(\rho) = \{z \in \mathbb{R}^{d'};\ |z| < \rho\}, \quad (ii)\ \bigcup_{\rho' < \rho} A_0(\rho') = A_0(\rho).$$

It is called a family of *star-shaped neighborhoods* of the center. We define a function $\varphi_0(\rho), 0 < \rho < 1$ by

$$\varphi_0(\rho) := \int_{A_0(\rho)} |z|^2 \nu(dz) = \nu'(A_0(\rho)), \tag{1.20}$$

where ν is the Lévy measure of the infinitely divisible distribution and $\nu'(dz) = |z|^2\nu(dz)$. The function $\varphi_0(\rho), 0 < \rho < 1$ is a left continuous increasing (nondecreasing) function of ρ. Let α be a real number such that $0 < \alpha < 2$. The function $\varphi_0(\rho)$ is said to satisfy the *order condition of exponent α at the center* (with respect to the family of star-shaped neighborhoods $\{A_0(\rho)\}$), if there is a positive constant c satisfying

$$\varphi_0(\rho) \geq c\rho^\alpha, \quad \forall 0 < \rho < 1.$$

For each ρ, we define a matrix Γ_ρ of the second moment by

$$\Gamma_\rho = \left(\frac{\int_{A_0(\rho)} z^i z^j \nu(dz)}{\varphi_0(\rho)}\right), \tag{1.21}$$

where $z = (z^1, \ldots, z^{d'}) \in \mathbb{R}^{d'}$. The pair (A, ν) is called *nondegenerate at the center* (with respect to a family of star-shaped neighborhoods $\{A_0(\rho)\}$), if matrices $\{A + \Gamma_\rho; 0 < \rho < 1\}$ are uniformly positive definite. In particular, when the matrix A is equal to O, the Lévy measure ν is said to be *nondegenerate at the center.*

We define the order for symmetric $d' \times d'$ matrices. Let A_1 and A_2 be symmetric $(d' \times d')$-matrices. If $(u, A_1u) \geq (u, A_2u)$ holds for all $u \in \mathbb{R}^{d'}$, we say that A_1 is greater than or equal to A_2 and write this as $A_1 \geq A_2$. Now if $\{A + \Gamma_\rho\}$ is uniformly positive definite, there exists a symmetric positive definite matrix B such that $A + \Gamma_\rho \geq B$ holds for all ρ. B is called a lower bound of $\{A + \Gamma_\rho\}$. Taking a positive constant c such that $(u, (A + \Gamma_\rho)u) \geq c|u|^2$ holds for all $u \in \mathbb{R}^{d'}$, then $B = cI$ is what we want. Here I is the identity matrix.

We will obtain an upper estimate of the characteristic function of a nondegenerate infinitely divisible distribution. It is interesting to compare the next lemma with the equation (1.17) for rotation invariant stable distribution.

Lemma 1.2.1 *Let μ be an infinitely divisible distribution with characteristics (A, ν, b), where the matrix A is degenerate. Let ψ be its characteristic function. If, at the center, the pair (A, ν) is nondegenerate and satisfies the order condition of exponent α with respect to a suitable family of star-shaped neighborhoods $\{A_0(\rho)\}$, there exist positive constants c_1, c_2 such that*

$$|\psi(v)| \leq c_2 \exp\{-c_1|v|^{2-\alpha}\}, \quad \forall |v| \geq 1. \tag{1.22}$$

Proof From the Lévy–Khintchine formula, we have

$$|\psi(v)| = \exp\Big\{-\frac{1}{2}(v, Av) + \int_{\mathbb{R}_0^{d'}} \mathcal{R}\Big(e^{i(v,z)} - 1 - i(v,z)1_D(z)\Big)\nu(dz)\Big\},$$

where $\mathcal{R}w$ means the real part of the complex number w. We have

$$-\mathcal{R}\Big(e^{i(v,z)} - 1 - i(v,z)1_D(z)\Big) = 1 - \cos(v,z) = 2\sin^2\frac{(v,z)}{2} \geq \frac{1}{2}(v,z)^2,$$

if $|(v,z)| \leq \frac{2}{3}\pi$. Therefore,

$$-\int_{\mathbb{R}_0^{d'}} \mathcal{R}\Big(e^{i(v,z)} - 1 - i(v,z)1_D(z)\Big)\nu(dz) \geq \frac{1}{2}\int_{\{|(v,z)|\leq\frac{2\pi}{3}\}} (v,z)^2\nu(dz).$$

Let Γ_ρ the matrix defined by (1.21). Since $A_0(|v|^{-1}) \subset \{|(v,z)| \leq \frac{2\pi}{3}\}$, we have

$$\int_{\{|(v,z)|\leq\frac{2\pi}{3}\}} (v,z)^2\nu(dz) \geq \int_{A_0(|v|^{-1})} (v,z)^2\nu(dz) = \sum_{i,j} v_i v_j \Gamma^{ij}_{|v|^{-1}}\varphi_0(|v|^{-1}).$$

Since $\varphi_0(|v|^{-1}) \geq c|v|^{-\alpha}$ holds, we have

$$\frac{1}{2}(v, Av) - \int \mathcal{R}\Big(e^{i(v,z)} - 1 - i(v,z)1_D(z)\Big)\nu(dz) \geq \frac{c\wedge 1}{2}\sum_{i,j}\frac{v_i}{|v|^{\frac{\alpha}{2}}}\frac{v_j}{|v|^{\frac{\alpha}{2}}}(A+\Gamma_{|v|^{-1}})^{ij}.$$

There exists a positive constant c' such that $A + \Gamma_\rho \geq c'I$ holds for all $0 < \rho < 1$. Then the last term of the above dominates $\frac{c'(c\wedge 1)}{2}|v|^{2-\alpha}$ if $|v| > 1$. Consequently there exists $c_1 > 0$ such that $|\psi(v)| \leq \exp\{-c_1|v|^{2-\alpha}\}$ if $|v| > 1$. □

From the above lemma, $\psi(z)$ is a rapidly decreasing function. Further, if the Lévy measure ν has moments of any order, i.e., $\int |z|^n\nu(dz) < \infty$ for any $n \in \mathbb{N}$, the characteristic function given by (1.13) is infinitely differentiable. Then the density function is a rapidly decreasing C^∞-function. Therefore we have the following theorem from Theorem 1.1.2.

Theorem 1.2.1 *Let μ be an infinitely divisible distribution with characteristics (A, ν, b).*

1. *If the matrix A is nondegenerate, μ has a C_b^∞-density.*
2. *If at the center, the pair (A, ν) is nondegenerate and satisfies the order condition of exponent $0 < \alpha < 2$ with respect to a suitable family of star-shaped neighborhoods, then μ has a C_b^∞-density. Suppose further that the Lévy measure has moments of any order. Then the density is a rapidly decreasing C^∞-function.*

We can check easily that the Lévy measure of a β-stable distribution satisfies the order condition of exponent $2-\beta$ at the center, since its Lévy measure is represented by (1.14).

It should be noted that in the definition of the order condition at the center of the Lévy measure, any regularity for the Lévy measure ν is not assumed. It may or may not be absolutely continuous. It may be a measure with point mass only. However, the characteristic function of the infinitely divisible distribution has an upper estimate (1.22), which is compatible with the formula (1.17) for stable distribution.

The nondegenerate condition at the center of the Lévy measure depends on the choice of the family of neighborhoods $\{A_0(\rho)\}$. We will consider an example. Let $X^1, \ldots, X^{d'}$ be mutually independent one-dimensional random variables, whose laws are stable with indices $\beta_1, \ldots, \beta_{d'}$, respectively. Then the vector $X = (X^1, \ldots, X^{d'})$ is subject to a d'-dimensional infinitely divisible distribution. Further, the Lévy measure of the law of X is concentrated on axis $\{ze_j; z \in \mathbb{R}\}, j = 1, \ldots, d'$, where e_j are unit vectors in $\mathbb{R}^{d'}$. On each axis $\{ze_j\}$, the Lévy measure takes the form $\nu_j(dz) = \frac{c_j}{|z|^{1+\beta_j}} dz$.

If $\beta_1 = \cdots = \beta_{d'} = \beta$, the Lévy measure is nondegenerate at the center point 0 with respect to the family of open balls $A_0(\rho) = \{|z| < \rho\}$. It holds that $\Gamma_\rho = I$ and $\varphi_0(\rho) = \frac{c_j}{2-\beta}\rho^{2-\beta}$. If $\beta_j, j = 1, \ldots, d'$ are distinct, the Lévy measure is no longer nondegenerate at the center with respect to the family of open balls. We should then take $\{A_0(\rho)\}$ as a family of star-shaped neighborhood such that

$$A_0(\rho) \cap \left\{ze_j; z \in \mathbb{R}\right\} = \left\{|z|^{\frac{2-\beta_j}{\alpha}} e_j; |z| < \rho\right\}$$

holds for any j, where $\alpha = 2 - \min_j \beta_j$. Then the Lévy measure is nondegenerate at the center with respect to the above family of neighborhoods. This is a reason that we call $A_0(\rho)$ a star-shaped neighborhood. See [46].

1.3 Random Fields and Stochastic Processes

A family of S-valued random variables $\{X_\lambda; \lambda \in \Lambda\}$, where Λ is a topological space, is called an S-valued *random field* (with parameter $\lambda \in \Lambda$). A random field $\{X_\lambda, \lambda \in \Lambda\}$ is called *measurable* if $X_\lambda(\omega)$ is measurable with respect to the product σ-field $\mathcal{B}(\Lambda) \times \mathcal{F}$, where $\mathcal{B}(\Lambda)$ is the topological Borel field of Λ. When S is $\mathbb{R}^d$ or $\mathbb{R}$, it is called a d-dimensional or real-valued random field, respectively.

If a statement holds for $\omega \in \tilde{\Omega}$ with $P(\tilde{\Omega}) = 1$, we say that the statement holds for almost all ω or almost surely and it is written simply as a.s. Let X and $\tilde{X}$ be S-valued random variables on the same state space S. If $P(\{\omega; X(\omega) = \tilde{X}(\omega)\}) = 1$ holds, $\tilde{X}$ is said to be equal to X *almost surely* and both are written as $X = \tilde{X}$ a.s. Let $\{X_\lambda, \lambda \in \Lambda\}$ and $\{\tilde{X}_\lambda, \lambda \in \Lambda\}$ be two S-valued random fields with the same parameter set Λ. These two random fields are said to be *equivalent* provided that $P(X_\lambda = \tilde{X}_\lambda) = 1$ holds for all $\lambda \in \Lambda$. Further, $\{\tilde{X}_\lambda, \lambda \in \Lambda\}$ is called a *modification* of $\{X_\lambda, \lambda \in \Lambda\}$. If laws of $(X_{\lambda_1}, \ldots, X_{\lambda_n})$ and $(\tilde{X}_{\lambda_1}, \ldots, \tilde{X}_{\lambda_n})$ coincide for any finite subset $\lambda_1, \ldots, \lambda_n$ of Λ, then $\{X_\lambda\}$ and $\{\tilde{X}_\lambda\}$ are said to be *equivalent in law*.

We denote by $\mathcal{D}$ the set of all maps $\mathbf{x}; \Lambda \to S$. The value of $\mathbf{x}$ at λ is denoted by x_λ. For $\lambda_1, \ldots, \lambda_n \in \Lambda$ and Borel subsets $B_1, \ldots, B_n$ of S, we define a cylinder set of $\mathcal{D}$ by $B = \{\mathbf{x};\ x_{\lambda_1} \in B_1, \ldots, x_{\lambda_n} \in B_n\}$. Let $\mathcal{B}(\mathcal{D})$ be the smallest σ-field of $\mathcal{D}$, which contains all cylinder sets, where $\lambda_1, \ldots, \lambda_n$ run Λ and $B_1, \ldots, B_n$ run Borel sets of S.

For a given random field $\{X_\lambda, \lambda \in \Lambda\}$, there is a unique probability measure $\mathcal{P}$ on $\mathcal{B}(\mathcal{D})$ satisfying

$$\mathcal{P}(B) = P(X_{\lambda_1} \in B_1, \ldots, X_{\lambda_n} \in B_n).$$

The measure $\mathcal{P}$ is called the *law of the random field* $\{X_\lambda\}$.

If the parameter set Λ is a time interval $\mathbb{T}$ such as $[T_0, T_1]$, $[0, \infty)$ etc., then the random field $\{X_t, t \in \mathbb{T}\}$ is called an S-valued *stochastic process*. Stochastic processes are often written as $\{X_t\}$, or simply as $X_t, t \in \mathbb{T}$, X_t etc. If $X_t(\omega)$ is continuous with respect to t for almost all ω, $\{X_t\}$ is called a *continuous process*. Further, if $X_t(\omega)$ is right continuous with left-hand limits with respect to t for almost all ω, $\{X_t\}$ is called a *cadlag process*. For a cadlag process $\{X_t\}$, we denote its left limit by X_{t-}, i.e., $X_{t-} = \lim_{\epsilon \downarrow 0} X_{t-\epsilon}$. If $X_t(\omega)$ is left continuous with right-hand limits with respect to t for almost all ω, $\{X_t\}$ is called a *caglad process*. For a caglad process $\{X_t\}$, we denote its right limit by X_{t+}, i.e., $X_{t+} = \lim_{\epsilon \downarrow 0} X_{t+\epsilon}$. Further, the process $\{X_t\}$ is said to be continuous in probability if $P(d(X_{t+h}, X_t) > \epsilon) \to 0$ as $h \to 0$ for any $t \geq 0$ and $\epsilon > 0$, where d is the metric of S.

When we consider a stochastic process $\{X_t, t \in [T_0, T_1]\}$, we usually assume that the process $\{X_t\}$ describes the time evolution in the forward (positive) direction; at the initial time T_0, it starts from X_0 and moves to X_t at time $T_0 < t \leq T_1$. We could call it a *forward process*. However, in later discussions, we will consider a process $\{\check{X}_t, t \in [T_0, T_1]\}$ (with the same time parameter), which describes the time evolution in the backward (negative) direction; at the initial time T_1, it starts from $\check{X}_{T_1}$ and moves to $\check{X}_t$ at time $T_0 \leq t < T_1$. We will call it a *backward process*.

1.4 Wiener Processes, Poisson Random Measures and Lévy Processes

An $\mathbb{R}^{d'}$-valued cadlag stochastic process $\{X_t, t \in [0, \infty)\}$ is said to have *independent increments*, if $\{X_{t_{m+1}} - X_{t_m}, m = 1, 2, \ldots, n\}$ are independent for any $0 \leq t_1 < t_2 < \cdots < t_{n+1} < \infty$. It is called *time homogeneous* if the law of $X_{t+h} - X_{s+h}$ does not depend on $h > 0$. Let $\{X_t, t \in [0, \infty)\}$ be a cadlag $\mathbb{R}^{d'}$-valued process with $X_0 = 0$. It is called a *Lévy process* if it is time homogeneous, continuous in probability and has independent increments. If $\{X_t\}$ is a Lévy process, the d'-dimensional law of X_t is infinitely divisible for each t. Then the characteristic function of the law of X_t is given by the Lévy–Khintchine formula.

Let $\{X_t\}$ and $\{\hat{X}_t\}$ be two d'-dimensional Lévy processes. Let μ_t and $\hat{\mu}_t$ be laws of random variables X_t and $\hat{X}_t$, respectively. If $\mu_t = \hat{\mu}_t$ holds for all $t > 0$, these two Lévy processes are equivalent in law and hence the law $\mathcal{P}$ of the Lévy process is uniquely determined by $\{\mu_t, t > 0\}$. Indeed, for any bounded continuous function f on $\mathbb{R}^{d'n}$ and $0 \le t_1 < t_2 < \cdots < t_n$, $E[f(X_{t_1}, \ldots, X_{t_n})]$ is computed as

$$\int \cdots \int f(x_1, \ldots, x_n)\mu_{t_1}(dx_1)\mu_{t_2-t_1}(dx_2 - x_1) \cdots \mu_{t_n-t_{n-1}}(dx_n - x_{n-1}),$$

since X_t is time homogeneous and has independent increments. Hence if $\mu_t = \hat{\mu}_t$ holds for any t, we have $E[f(X_{t_1}, \ldots, X_{t_n})] = E[f(\hat{X}_{t_1}, \ldots, \hat{X}_{t_n})]$ for any bounded continuous function f. Then $\{X_t\}$ and $\{\hat{X}_t\}$ are equivalent.

Let $\{X_t\}$ be a d'-dimensional continuous Lévy process. If the law of each X_t is Gaussian, the process $\{X_t\}$ is called a *Brownian motion*. Since $\{X_t\}$ is time homogeneous, the mean vector of X_t is tb and covariance matrix is tA, where b and A are the mean and covariance of X_1. If $b = 0$ and $A = I$, then $\{X_t\}$ is called a (d'-dimensional) *Wiener process* or a *standard Brownian motion*. For Wiener processes $\{X_t\}$, laws of $\{X_t\}$ are unique, since μ_t are Gaussian distributions with means 0 and covariance matrices tI.

We will show the existence of a one-dimensional Wiener process, constructing it by the method of Wiener. We will take $\mathbb{T} = [0, 1]$. Let $\{Z_n; n = 1, 2, \ldots\}$ be an independent sequence of Gaussian random variables with means 0 and variances 1. Let $\{\phi_n(t), n = 1, 2, \ldots\}$ be a complete orthonormal system of the real $L^2([0, 1])$-space. We set $\bar{\phi}_n(t) = \int_0^t \phi_n(s)\, ds = (1_{(0,t]}, \phi_n)$. Consider the infinite sum

$$X_t = \sum_{n=1}^{\infty} \bar{\phi}_n(t) Z_n. \tag{1.23}$$

We show that it converges uniformly in $t \in [0, 1]$ a.s., if we take a good orthonormal system and $\{\phi_n\}$ be a system of functions on $[0, 1]$ defined by

$$\phi_n(t) = \begin{cases} -1, \text{ if } \dfrac{m}{2^n} \le t < \dfrac{m+1}{2^n}, & m = 0, 2, 4, \ldots, \\ 1, \text{ if } \dfrac{m}{2^n} \le t < \dfrac{m+1}{2^n}, & m = 1, 3, 5, \ldots \end{cases}$$

Then it is a complete orthonormal system and $\|\bar{\phi}_n\| \equiv \sup_t |\bar{\phi}_n(t)| \le 1/2^{n+1}$ holds for any n. The system is called the *Rademacher system*.

Proposition 1.4.1 *If $\{\phi_n\}$ is the Rademacher system, then the infinite sum* (1.23) *converges uniformly in t almost surely. Let $\{X_t, t \in [0,1]\}$ be the limit. It is a continuous process. For each t, X_t is Gaussian with mean 0 and covariance t. Further, $\{X_t, t \in [0,1]\}$ has independent increments and is time-homogeneous.*

Proof Set $X_t^n = \sum_{m=1}^n \bar{\phi}_m(t) Z_m$. Then, since $\sup_t |X_t^n - X_t^{n-1}| \le \|\bar{\phi}_n\| |Z_n|$, we have

$$\sum_{n=1}^\infty E[\sup_t |X_t^n - X_t^{n-1}|] \le \sum_{n=1}^\infty \|\bar{\phi}_n\| E[|Z_n|] \le \sum_{n=1}^\infty \|\bar{\phi}_n\| < \infty.$$

This shows $\sum_{n=1}^\infty \sup_t |X_t^n - X_t^{n-1}| < \infty$, so that X_t^n converges uniformly to X_t a.s. Therefore X_t is continuous in t a.s.

Note that the law of n-vector $(Z_1, \dots, Z_n)$ is n-dimensional Gaussian with mean 0 and covariance I. Then the linear sum $\sum_j c_j Z_j$ is one-dimensional Gaussian with mean 0 and variance $\sum_j c_j^2$. Therefore for each t, X_t^n is Gaussian with mean 0 and variance $\sum_{j=1}^n \bar{\phi}_n(t)^2$. Then the limit X_t is also Gaussian. Further, we have by Perseval's equality

$$E[X_t X_s] = \sum_{n=1}^\infty \bar{\phi}_n(t)\bar{\phi}_n(s) = \sum_{n=1}^\infty (1_{(0,t]}, \phi_n)(1_{(0,s]}, \phi_n) = (1_{(0,t]}, 1_{(0,s]}) = t \wedge s.$$

Therefore $E[X_t^2] = t$ holds for any t.

Let $0 \le 0 < t_1 < \cdots < t_n \le 1$. Then $(X_{t_1} - X_0, \dots, X_{t_n} - X_{t_{n-1}})$ is an n-dimensional Gaussian random variable. It holds that

$$E[(X_{t_{m+1}} - X_{t_m})^2] = t_{i+1} - t_m - t_m + t_m = t_{m+1} - t_m.$$

Further, if $t_l < t_m$, we have

$$\begin{aligned} &E[(X_{t_{m+1}} - X_{t_m})(X_{t_{l+1}} - X_{t_l})] \\ &= E[X_{t_{m+1}} X_{t_{m+1}}] - E[X_{t_{m+1}} X_{t_l}] - E[X_{t_m} X_{t_{l+1}}] + E[X_{t_m} X_{t_{l1}}] \\ &= t_{l+1} - t_l - t_{l+1} + t_l = 0. \end{aligned}$$

Therefore its $n \times n$-covariance matrix is diagonal. This means

$$\begin{aligned} E[e^{i \sum_m v_m (X_{t_m} - X_{t_{m-1}})}] &= \exp\Big\{ -\frac{1}{2} \sum_m (t_m - t_{m-1}) v_m^2 \Big\} \\ &= \prod_{m=1}^n E\big[e^{i v_m (X_{t_m} - X_{t_{m-1}})}\big]. \end{aligned}$$

Then $X_{t_n} - X_{t_{n-1}}, \dots, X_{t_1} - X_0$ are mutually independent (Corollary 1.1.1). Therefore $\{X_t\}$ has independent increments and is time-homogeneous. □

We can choose other orthonormal system $\{\phi_n\}$ and show that (1.23) converge uniformly in t a.s. Take $\phi_0(t) = 1, \phi_n(t) = \sqrt{2}\cos n\pi t, n = 1, 2, \dots$. Then $\bar{\phi}_0(t) = t, \bar{\phi}_n(t) = \frac{\sqrt{2}}{n\pi}\sin n\pi t, n = 1, 2 \dots$. It is known that the infinite sum (1.23) is also uniformly convergent. It is called *Wiener's construction*.

Let $\{Z'_n\}$ be another independent sequence of Gaussian random variables with mean 0 and covariance 1, which is independent of $\{Z_n\}$. We define a stochastic process by $X'_t = \sum_{n=1}^{\infty} \bar{\phi}_n(t) Z'_n$ for $t \in [0, 1]$. Then $\{X'_t\}$ and $\{X_t\}$ are equivalent in law. Further, these two processes are independent. $\{X'_t\}$ is called an *independent copy* of $\{X_t\}$. Now, take a sequence of independent copies $\{X_t^{(n)}\}, n = 1, \dots$ of the process $\{X_t\}$ of Proposition 1.4.1. Define another continuous process X_t by $\sum_{i=1}^{n} X_1^{(i)} + X_{t-n}^{(n+1)}$ if $t \in (n, n+1]$. Then $\{X_t, t \in [0, \infty)\}$ is a Wiener process.

Next, take an independent copy $\{X'_t, t \in [0, \infty)\}$ of the above one dimensional Wiener process $\{X_t, t \in [0, \infty)\}$. Then the pair $\{(X_t, X'_t), t \in [0, \infty)\}$ is a 2-dimensional Wiener process. Repeating this argument, we can also define d'-dimensional Wiener process.

Next, let $\{N_t; t \in [0, \infty)\}$ be a Lévy process with values in $\bar{\mathbb{N}}$. It is called a *Poisson process* with *intensity* λ, if the law of each N_t is a Poisson distribution with parameter $t\lambda$. The law of a Poisson process $\{N_t\}$ with intensity λ is unique, since $\{N_t\}$ is a Lévy process.

We shall construct a Poisson process with intensity λ. Let $\sigma_n; n = 1, 2, \dots$ be mutually independent positive random variables with a common exponential distribution with parameter λ. We set $\tau_0 = 0$ and $\tau_n = \sigma_1 + \dots + \sigma_n$ for $n \geq 1$. We define a stochastic process $N_t, t \in [0, \infty)$ with values in $\bar{\mathbb{N}}$ by

$$N_t(\omega) = n, \quad \text{if } \tau_n(\omega) \leq t < \tau_{n+1}(\omega). \tag{1.24}$$

Then $P(N_t = n) = P(\tau_n \leq t < \tau_n + \sigma_{n+1})$. Note that the law of τ_n is the Gamma distribution with parameters λ and n. Then $P(N_t = n)$ is computed as

$$\frac{\lambda^{n+1}}{(n-1)!} \iint_B x^{n-1} e^{-\lambda x} e^{-\lambda y}\, dx\, dy = e^{-\lambda t} \frac{(\lambda t)^n}{n!},$$

where $B = \{(x, y); 0 \leq x \leq t < x + y\}$. Therefore the law of N_t is Poisson with parameter λt. The random time τ_n defined above is called the n-th jumping time of the stochastic process N_t. The sequence of random times $\{\tau_n\}$ is called jumping times of the stochastic process $\{N_t\}$.

Proposition 1.4.2 *$\{N_t\}$ defined by (1.24) is a Poisson process with intensity λ.*

Proof We want to show that $\{N_t\}$ is time homogeneous and has independent increments. Our discussion is close to Sato [99]. We first show that the law of $(\tau_{n+1} - t, \sigma_{n+2}, \dots, \sigma_{n+m})$ under the condition $N_t = n$ is the same as the law of $(\sigma_1, \dots, \sigma_m)$. Set $a = P(N_t = n)$. Then we have

$$\begin{aligned}
&P(\tau_{n+1}-t>s_1,\sigma_{n+2}>s_2,\ldots,\sigma_{n+m}>s_m|N_t=n)\\
&=P(\tau_n\le t,\tau_{n+1}-t>s_1,\sigma_{n+2}>s_2,\ldots,\sigma_{n+m}>s_m)/a\\
&=P(\tau_n\le t,\tau_{n+1}-t>s_1)P(\sigma_{n+2}>s_2,\ldots,\sigma_{m+n}>s_m)/a\\
&=P(\tau_{n+1}-t>s_1|N_t=n)P(\sigma_2>s_2,\ldots,\sigma_m>s_m)\\
&=P(\sigma_1>s_1)P(\sigma_2>s_2,\ldots,\sigma_m>s_m)\\
&=P(\sigma_1>s_1,\sigma_2>s_2,\ldots,\sigma_m>s_m).
\end{aligned} \tag{1.25}$$

Here we used the equality $P(\tau_{n+1}-t>s|N_t=n)=e^{-\lambda s}=P(\sigma_1>s)$, which is shown as follows. Note

$$P(\tau_n<t,\tau_n+\sigma_n>t+s)=\frac{\lambda^{n+1}}{(n-1)!}\iint_{B'}x^{n-1}e^{-\lambda x}e^{-\lambda y}\,dx\,dy=e^{-\lambda(t+s)}\frac{(\lambda t)^n}{n!},$$

where $B'=\{(x,y);0<x<t+s<x+y\}$. Then we have

$$\frac{P(N_t=n,\tau_{n+1}>t+s)}{P(N_t=n)}=\frac{P(\tau_n\le t,\tau_{n+1}>t+s)}{P(N_t=n)}=e^{-\lambda s}.$$

Now we will prove that $\{N_t\}$ is time homogeneous and has independent increments. We have

$$\begin{aligned}
&P(N_{t_1}-N_{t_0}=n_1,\ldots,N_{t_m}-N_{t_{m-1}}=n_m|N_{t_0}=n_0)\\
&=P(\tau_{n_0+n_1}\le t_1<\tau_{n_0+n_1+1},\ldots,\tau_{n_0+\cdots+n_m}\le t_m<\tau_{n_0+\cdots+n_m+1}|N_{t_0}=n_0)\\
&=P(\tau_{n_1}\le t_1-t_0<\tau_{n_1+1},\ldots,\tau_{n_1+\cdots+n_m}\le t_m-t_0<\tau_{n_1+\cdots+n_m+1}).
\end{aligned}$$

In the last equality, we applied (1.25). Then we get

$$\begin{aligned}
&P(N_{t_1}-N_{t_0}=n_1,\ldots,N_{t_m}-N_{t_{m-1}}=n_m|N_{t_0}=n_0)\\
&=P(N_{t_1-t_0}=n_1,\ldots N_{t_m-t_0}-N_{t_{m-1}-t_0}=n_m).
\end{aligned}$$

Repeating this argument, we get the equality

$$\begin{aligned}
&P(N_{t_0}=n_0,N_{t_1}-N_{t_0}=n_1,\ldots,N_{t_m}-N_{t_{m-1}}=n_m)\\
&=P(N_{t_0}=n_0)\cdots P(N_{t_m-t_{m-1}}=n_m).
\end{aligned} \tag{1.26}$$

We have in particular,

$$P(N_{t_0}=n_0,N_{t_1}-N_{t_0}=n_1)=P(N_{t_0}=n_0)P(N_{t_1-t_0}=n_1).$$

Summing up the above for $n_0 = 0, 1, 2, \ldots$. Then we get

$$P(N_{t_1} - N_{t_0} = n_1) = P(N_{t_1 - t_0} = n_1).$$

This shows that N_t is time homogeneous. Further, we can rewrite (1.26) as

$$\begin{aligned} &P(N_{t_0} = n_0, N_{t_1} - N_{t_0} = n_1, \ldots, N_{t_m} - N_{t_{m-1}} = n_m) \\ &= P(N_{t_0} = n_0) \cdots P(N_{t_m} - N_{t_{m-1}} = n_m). \end{aligned} \tag{1.27}$$

This proves that $\{N_t\}$ has independent increments. □

Now, let $\{N'_t\}$ be a Poisson process with intensity λ. Then its law coincides with the law of $\{N_t\}$ constructed above. Hence $\{N'_t\}$ is a cadlag nondecreasing process with values in $\mathbb{N}$ such that $N'_t - N'_{t-}$ is 0 or 1, almost surely. Set $\tau'_0 = 0$ and define random times $\tau'_n, n = 1, 2, \ldots$ by induction as

$$\tau'_n = \begin{cases} \inf\{t > \tau'_{n-1};\ N'_t > N'_{\tau'_{n-1}}\}, & \text{if the set } \{\cdots\} \text{ is non-empty,} \\ \infty, & \text{otherwise.} \end{cases}$$

Then τ'_n is subject to the Gamma distribution with parameters λ and n, if we restrict the distribution to the interval $[0, \infty)$. The sequence $\{\tau'_n\}$ is called the *jumping times* of the Poisson process $\{N'_t\}$.

Let $\mathbb{T} = [0, \infty)$ and $\mathbb{U} = \mathbb{T} \times \mathbb{R}_0^{d'}$. Elements of $\mathbb{U}$ are denoted by $u = (r, z)$ where $r \in \mathbb{T}, z \in \mathbb{R}_0^d$. Let $(\mathbb{U}, \mathcal{B}(\mathbb{U}), n)$ be a measure space, where n is a σ-finite measure on $\mathbb{U}$. A family of $\bar{\mathbb{N}}$-valued random variables $N(B), B \in \mathcal{B}(\mathbb{U})$ is called the (abstract) *Poisson random measure on* $\mathbb{U}$ *with intensity measure* n, if the followings hold:

1. For every ω, $N(\cdot, \omega)$ is an $\bar{\mathbb{N}}$-valued measure on $\mathbb{U}$.
2. For each B, the law of $N(B)$ is a Poisson distribution with parameter $n(B)$.
3. If $B_1, \ldots, B_n$ are disjoint, $N(B_1), \ldots, N(B_n)$ are independent.

We are interested in the Poisson random measure in the case that n is the product measure $dr\nu(dz)$, where dr is the Lebesgue measure on $[0, \infty)$ and ν is a σ-finite measure on $\mathbb{R}_0^{d'}$. We shall construct it. We first consider the case where ν is a finite measure. Let $\{S_n; n = 1, 2, \ldots\}$ be a sequence of mutually independent random variables with values in $\mathbb{R}^{d'}$ with the common distribution $\mu(dz) = \nu(dz)/\lambda$, where $\lambda = \nu(\mathbb{R}_0^{d'})$. Let $\{N_t\}$ be a Poisson process with intensity λ, which is independent of $\{S_n\}$. For $0 \le s < t < \infty$ and a Borel subset E of $\mathbb{R}_0^{d'}$, we set

$$N((s, t] \times E) := \sum_{N_s < n \le N_t} 1_E(S_n) = \sum_{k; s < \tau_k \le t} 1_E(S_k), \tag{1.28}$$

where $\tau_k, k = 1, 2, \ldots$ are jumping times of the Poisson process $\{N_t\}$.

Proposition 1.4.3 *The random measure $N(dr\,dz)$ defined by* (1.28) *is a Poisson random measure on $\mathbb{U} = [0, \infty) \times \mathbb{R}_0^{d'}$ with the intensity $n(dr\,dz) = dr\nu(dz)$.*

Proof Set $N_{s,t}(E) = N((s, t] \times E)$. Let $E_k, k = 1, \ldots, n$ be disjoint subsets of $\mathbb{R}_0^{d'}$. The characteristic function of $(N_{s,t}(E_1), \ldots, N_{s,t}(E_n))$ is computed as

$$\begin{aligned} E[e^{i\sum_k v_k N_{s,t}(E_k)}] &= \sum_{m=0}^{\infty} P(N_t - N_s = m) E[e^{i\sum_k v_k (\sum_{j\le m} 1_{E_k}(S_j))}] \\ &= \sum_{m=0}^{\infty} e^{-\lambda(t-s)} \frac{(\lambda(t-s))^m}{m!} E[e^{i\sum_k v_k 1_{E_k}(S_1)}]^m, \end{aligned}$$

since $S_1, \ldots, S_m$ are independent. We have

$$E[e^{i\sum_{k=1}^n v_k 1_{E_k}(S_1)}] = \sum_k e^{iv_k}\mu(E_k) + \mu((\bigcup_k E_k)^c) = \sum_{k=1}^n (e^{iv_k} - 1)\mu(E_k) + 1.$$

Therefore,

$$\begin{aligned} E[e^{i\sum_k v_k N_{s,t}(E_k)}] &= e^{-\lambda(t-s)} \sum_{m=0}^{\infty} \frac{(\lambda(t-s))^m}{m!} \Big(\sum_{k=1}^n (e^{iv_k} - 1)\mu(E_k) + 1\Big)^m \\ &= e^{-\lambda(t-s)} \exp\Big\{\sum_{k=1}^n \lambda(t-s)\{(e^{iv_k} - 1)\mu(E_k) + 1)\}\Big\} \\ &= \prod_{k=1}^n \exp\{\lambda(t-s)(e^{iv_k} - 1)\mu(E_k)\} \\ &= \prod_{k=1}^n E[e^{iv_k N_{s,t}(E_k)}]. \end{aligned}$$

Then $N_{s,t}(E_1), \ldots, N_{s,t}(E_n)$ are mutually independent by Corollary 1.1.1. Further, each $N_{s,t}(E_k)$ is Poisson distributed with parameter $\lambda(t-s)\mu(E_k) = (t-s)\nu(E_k)$.

Next, consider $N((0, t_1] \times E_1), \ldots, N((t_{n-1}, t_n] \times E_n)$. These are independent, since $N_{t_m} - N_{t_{m-1}}, m = 1, \ldots, n$ are independent. Further, these are Poisson distributed. Consequently, $N(B_1), \ldots, N(B_n)$ are independent if $B_1, \ldots, B_n$ are disjoint rectangular sets of the form $B_m = (t_{m-1}, t_m] \times E_m$. Then the independence of $N(B_1), \ldots, N(B_n)$ holds for any disjoint Borel sets $B_1, \ldots, B_n$ of $\mathbb{U}$. □

Theorem 1.4.1 *For any given σ-finite measure ν on $\mathbb{R}_0^{d'}$ and time interval $[0, \infty)$, there exists a Poisson random measure $N(dr\,dz)$ on $\mathbb{U} = [0, \infty) \times \mathbb{R}_0^{d'}$ with the given intensity $n(dr\,dz) = dr\nu(dz)$. Further, laws of Poisson random measures on $\mathbb{U}$ with the given intensity are unique.*

Conersely, let $N(dr\,dz)$ be a Posson random measue on $\mathbb{U}$ with the intensity measure $n(dr\,dz) = dr\nu(dz)$, where ν is a finite measure. Then there exists a Poisson process $\{N_t\}$ with intensity $\lambda = \nu(\mathbb{R}_0^{d'})$ and an independent sequence of $\mathbb{R}_0^{d'}$-valued random variables $\{S_n\}$ with the common distribution $\mu(dz) = \nu(dz)/\lambda$, and the Poisson random measure is written as (1.28).

Proof If ν is a finite measure, we construct a Poisson random measure on $\mathbb{U}$ with intensity $dr\,d\nu$ as in Proposition 1.4.3. We will consider the case where ν is a σ-finite measure. Let $K_n, n = 1, 2, \ldots$ be a sequence of disjoint Borel sets of $\mathbb{R}_0^{d'}$ such that $\bigcup_n K_n = \mathbb{R}_0^{d'}$ and $\nu(K_n) < \infty$ for any n. We decompose ν as a sum of finite measures ν_n such that $\nu_n(dz) = \nu(dz)1_{K_n}$. For each ν_n there exists a Poisson random measure N^n on $\mathbb{U}$ with intensity $dt\,d\nu_n$. Then on a suitable probability space, we can construct these $N^n; n = 1, 2, \ldots$ as independent Poisson random measures. We define $N(B) = \sum_n N^n(B)$. Then $N(B)$ is also a Poisson random measure on $\mathbb{U}$. Further, its intensity is $dr\,d(\sum_n \nu_n) = dr\,d\nu$. Therefore it is the desired one.

Now, if $B_1, \ldots, B_n$ are disjoint, the law of n-vector random variable defined by $X = (N(B_1), \ldots, N(B_n))$ is the product of laws of $N(B_k), k = 1, \ldots, n$, since these are independent. Hence the law of the X is unique if $B_k, k = 1, \ldots, n$ are disjoint. Then the law of X for any Borel sets $B_1, \ldots, B_n$ is also unique. Therefore the law of the Poisson random measure is unique.

We will prove the latter assertion of the theorem. Set $N_t := N((0, t] \times \mathbb{R}_0^{d'})$. Then, N_t is a Poisson process with intensity $\lambda := \nu(\mathbb{R}_0^{d'})$. Let $\tau_1 < \cdots < \tau_n < \cdots$ be jumping times of the Poisson process. We define a sequence of $\mathbb{R}^{d'}$-valued random variables by $S_n = \int_{\{\tau_n\}\times\mathbb{R}_0^{d'}} zN(dr\,dz)$. Then $\{S_n\}$ are mutually independent with common distribution $\mu(dz) = \nu(dz)/\lambda$. Further, $N((s, t] \times E)$ is represented by (1.28). □

A d'-dimensional Lévy process $\{X_t\}$ is called a *compound Poisson process* with intensity ν, if the law of X_t is a compound Poisson distribution associated with measure $t\nu$ for each $t > 0$.

Given a finite measure ν on $\mathbb{R}_0^{d'}$, we shall construct a compound Poisson process $\{X_t\}$ with intensity ν. Let $\{N_t\}$ be a Poisson process with intensity $\lambda = \nu(\mathbb{R}_0^{d'})$ and let $\{S_n\}$ be independent $\mathbb{R}^{d'}$-valued random variables with common distribution $\mu = \nu/\lambda$. We set $S_0 = 0$ and define an $\mathbb{R}^{d'}$-valued cadlag process $\{X_t\}$ by

$$X_t = \sum_{0\le n\le N_t} S_n = \int_{(0,t]\times\mathbb{R}_0^{d'}} zN(dr\,dz). \tag{1.29}$$

Here, N is the Poisson random measure on $[0, \infty) \times \mathbb{R}_0^{d'}$, defined by (1.28) using Poisson process $\{N_t\}$ with intensity λ.

Proposition 1.4.4 *The process $\{X_t\}$ given by (1.29) is a compound Poisson process with intensity ν.*

Proof Since N is a Poisson random measure, the process $\{X_t\}$ defined by (1.29) has independent increments. We shall compute the characteristic function of X_t. We have

$$E[e^{i(v,X_t)}] = \sum_{n=0}^{\infty} E[e^{(v,\sum_{j\le n} S_j)}]P(N_t = n) = e^{-\lambda t}\sum_{n=0}^{\infty}\frac{(\lambda t)^n}{n!}\hat{\mu}(v)^n,$$

where $\hat{\mu}(v)$ is the characteristic function of the distribution μ. It is written as $\int e^{i(v,z)}\mu(dz)$. Therefore the last term of the above is equal to

$$\exp\{-\lambda t\}\exp\left\{\lambda t\int e^{i(v,z)}\mu(dz)\right\} = \exp\left\{t\int (e^{i(v,z)}-1)\nu(dz)\right\},$$

since $\lambda\mu(dz) = \nu(dz)$. Hence the law of X_t is a compound Poisson distribution associated with measure $t\nu$. Then $\{X_t\}$ is a compound Poisson process with intensity ν. □

Let $N(dr\,dz)$ be a Poisson random measure on $\mathbb{U}$ with intensity measure $n(dr\,dz) = dr\nu(dz)$. We will define a *compensated Poisson random measure* by

$$\tilde{N}(dr\,dz) = N(dr\,dz) - n(dr\,dz). \tag{1.30}$$

We will define the integral $\int_0^t\int f(z)\tilde{N}(dr\,dz)$ for a measurable function $f(z)$ such that $\int |f(z)|^2\nu(dz) < \infty$. Suppose $f(z)$ is a step function of the form $f = \sum_k c_k 1_{E_k}$, where E_k are disjoint Borel subsets of $\mathbb{R}_0^{d'}$. Then we define $\int_0^t\int f(z)\tilde{N}(dr\,dz) = \sum_k c_k\tilde{N}_t(E_k)$, where $\tilde{N}_t(E) = \tilde{N}((0,t]\times E)$. We have $E[\tilde{N}_t(E_k)\tilde{N}_t(E_l)] = 0$ for $k\neq l$, since $\tilde{N}_t(E_k)$ and $\tilde{N}_t(E_l)$ are independent. Therefore,

$$E\Big[\Big(\int_0^t\int f(z)\tilde{N}(dr\,dz)\Big)^2\Big] = t\sum_k c_k^2\nu(E_k) = t\int f^2\nu(dz). \tag{1.31}$$

For a measurable function $f(z)$ such that $\int |f|^2\nu(dz) < \infty$, choose a sequence of step functions f_n satisfying $\int |f-f_n|^2\nu(dz)\to 0$ as $n\to\infty$. Then the sequence of integrals $\int_0^t\int f_n(z)\tilde{N}(dr\,dz)$ should converge in $L^2(P)$. We denote the limit by $\int_0^t\int f(z)\tilde{N}(dr\,dz)$.

Let $\{W_t\}$ be a Brownian motion with mean 0 and covariance At. Let $N(ds\,dz)$ be a Poisson random measure on $\mathbb{U}$ with intensity $dt\nu(dz)$, which is independent of $\{W_t\}$. We will assume that ν is a Lévy measure. Then the integral $\int_0^t\int_{0<|z|\le 1} z\tilde{N}(dr\,dz)$ is well defined. We set $D = \{z\in\mathbb{R}_0^{d'};\,|z|\le 1\}$ and

$$X_t = bt + W_t + \int_0^t \int_D z\tilde{N}(dr\,dz) + \int_0^t \int_{D^c} zN(dr\,dz). \tag{1.32}$$

Then $\{X_t\}$ is a cadlag process and in fact a Lévy process.

Proposition 1.4.5 *Let $\{X_t\}$ be a Lévy process defined by* (1.32). *Then, the characteristic function of X_t is given by*

$$\exp t\Big\{i(v, b) - \frac{1}{2}(v, Av) + \int_{\mathbb{R}_0^{d'}} (e^{i(v,z)} - 1 - i(v, z)1_D(z))\nu(dz)\Big\}. \tag{1.33}$$

Proof We know that $bt + W_t$ is a Brownian motion with means bt and covariance At. Then we have $E[e^{i(v,bt+W_t)}] = \exp t\Big\{i(b, v) - \frac{1}{2}(v, Av)\Big\}$. We shall consider the characteristic function of

$$Y_t = \int_0^t \int_D z\tilde{N}(dr\,dz) + \int_0^t \int_{D^c} zN(dr\,dz).$$

Set

$$Y_t^\epsilon = \int_0^t \int_{|z|>\epsilon} zN(dr\,dz) - t\int_{1>|z|>\epsilon} z\nu(dz).$$

Then Y_t^ϵ converges to Y_t. Since $\int_0^t \int_{|z|>\epsilon} zN(ds\,dz)$ is a compound Poisson process with Lévy measure $\nu 1_{|z|>\epsilon}$, $E[e^{i(v,Y_t^\epsilon)}]$ is computed as

$$\begin{aligned}
&E\Big[\exp\Big\{i(v, \int_0^t \int_{|z|>\epsilon} zN(dr\,dz))\Big\}\Big]\exp\Big\{-it(v, \int_{1\geq|z|>\epsilon} z\nu(dz)\Big\} \\
&= \exp t\Big\{\int_{|z|>\epsilon} (e^{i(v,z)} - 1)\nu(dz)\Big\}\exp\Big\{it(v, \int_{1\geq|z|>\epsilon} z\nu(dz)\Big\} \\
&= \exp t\Big\{\int_{|z|>\epsilon} (e^{i(v,z)} - 1 - i(v, z)1_D(z))\nu(dz)\Big\}.
\end{aligned}$$

Let $\epsilon \to 0$, then we get the formula (1.33) for Y_t. □

Let $\{X_t'\}$ be a Lévy process and let μ_t' be the law of X_t'. Since $\{X_t'\}$ is time homogeneous and has independent increments, for any n, μ_t' is equal to n-times convolution of the law $\mu'_{\frac{t}{n}}$. Therefore μ_t' is infinitely divisible. Then its characteristic function is represented by (1.33) by the Lévy–Khinchin formula. Therefore, for any given Lévy process $\{X_t'\}$, the Lévy process $\{X_t\}$ constructed by (1.32) is equivalent in law with $\{X_t'\}$. Therefore we have the following theorem.

Theorem 1.4.2 *Let $\{X_t, t \in [0, \infty)\}$ be a Lévy process. Then there exist a constant b, independent Brownian motion $\{B_t, t \in [0, \infty)\}$ and Poisson random measure $N(dr\,dz)$, and X_t is represented by* (1.32).

The formula (1.32) is called the *Lévy–Itô decomposition* or *Lévy–Itô representation* of a Lévy process.

In some problems, it is convenient to use a Poisson point process instead of a Poisson random measure. We will give its definition. By a point function on $\mathbb{R}_0^{d'}$, we mean a map $q : D_q \to \mathbb{R}_0^{d'}$, where D_q is a countable set in $[0, \infty)$. A counting measure of the point function is defined by

$$N(B, q) = \sharp\{t \in D_q;\ (t, q(t)) \in B\}, \tag{1.34}$$

where B are Borel sets in $\mathbb{U} = [0, \infty) \times \mathbb{R}_0^{d'}$. A random variable $q(\omega)$ with values in the space of point functions is called a *point process*. If the counting measure of the point process $q(\omega)$ is a Poisson random measure with intensity $dt\,d\nu$, the point process is called a Poisson point process with intensity $dt\nu(dz)$ and N of (1.34) is called the associated Poisson random measure.

Given a Lévy measure ν, there exists a Poisson point process with intensity $dt\nu(dz)$. Indeed, if ν is a finite measure, consider the compound Poisson process with intensity $dt\nu(dz)$. It is represented as $X_t(\omega) = \sum_{1\le n\le N_t(\omega)} S_n(\omega)$ by (1.29). We define a point process $q = q(\omega)$ by setting

$$D_q = \{0 < \tau_1 < \cdots < \tau_n < \infty\}, \quad q(t) = (\tau_i, S_i), \quad \text{if } \tau_i = t.$$

Then it is a Poisson point process with intensity $dt\nu(dz)$. We can show the existence of the Poisson point process for any Lévy measure ν by an argument similar to the proof of Theorem 1.4.1.

1.5 Martingales and Backward Martingales

In this and the next section, we will study martingales. We will restrict our attentions to three topics of martingales, which are often used in later stochastic calculus. Three topics are the *optional sampling theorem, Doob's inequality* and *quadratic variation*.

Let $(\Omega, \mathcal{F}, P)$ be a probability space. Let $\mathcal{G}$ be a sub σ-field of $\mathcal{F}$ and let X be an integrable random variable. A $\mathcal{G}$-measurable and integrable random variable Y is called the *conditional expectation* of X given $\mathcal{G}$, if it satisfies $E[X1_G] = E[Y1_G]$ for any $G \in \mathcal{G}$. Here, 1_G is the indicator function of the set G.

The conditional expectation Y exists uniquely a.s. For the proof, consider a signed measure $\tilde{P}(A) = E[X1_A]$, $A \in \mathcal{G}$. It is absolutely continuous with respect to P, i.e., $\tilde{P}(A) = 0$ holds for any $A \in \mathcal{G}$ with $P(A) = 0$. Then, in view of the Radon–Nikodym theorem, there exists an integrable $\mathcal{G}$-measurable functional Y such that

$\tilde{P}(A) = \int_A Y\,dP$ holds for any $A \in \mathcal{G}$. The uniqueness follows from the uniqueness of the Radon–Nikodym density.

The conditional expectation of X with respect to $\mathcal{G}$ is denoted by $E[X|\mathcal{G}]$. Further, if A is an element of $\mathcal{F}$, the conditional expectation $E[1_A|\mathcal{G}]$ is denoted by $P(A|\mathcal{G})$ and is called the *conditional probability* of event A given $\mathcal{G}$.

Conditional expectations have following properties.

Proposition 1.5.1 *Let X, Y be integrable random variables.*

1. $E[c_1X + c_2Y|\mathcal{G}] = c_1E[X|\mathcal{G}] + c_2E[Y|\mathcal{G}]$ *holds a.s. for any constants* c_1, c_2.
2. $E[|X||\mathcal{G}] \geq |E[X|\mathcal{G}]|$ *holds a.s.*
3. *If* $\mathcal{G}_1 \subset \mathcal{G}_2 \subset \mathcal{F}$, *then* $E[E[X|\mathcal{G}_2]|\mathcal{G}_1] = E[X|\mathcal{G}_1]$ *holds a.s.*
4. *If* $\sigma(X)$ *and* $\mathcal{G}$ *are independent, then* $E[X|\mathcal{G}] = E[X]$ *a.s.*
5. *If* XY *is integrable and* X *is* $\mathcal{G}$*-measurable, then* $E[XY|\mathcal{G}] = XE[Y|\mathcal{G}]$.
6. *Let* $f(x)$ *be a convex function, bounded from below. If* $f(X)$ *is integrable, we have* $f(E[X|\mathcal{G}]) \leq E[f(X)|\mathcal{G}]$ *a.s.*

Proposition 1.5.2 *Let $X_1, \ldots, X_n, \ldots, X$ be integrable random variables.*

1. *(Fatou's lemma) If* $X_n \geq 0$,

$$\liminf_{n\to\infty} E[X_n|\mathcal{G}] \geq E[\liminf_{n\to\infty} X_n|\mathcal{G}], \quad a.s.$$

2. *Let* $p \geq 1$. *If* X_n *converges to* X *in* L^p, *then* $E[X_n|\mathcal{G}]$ *converges to* $E[X|\mathcal{G}]$ *in* L^p.

Proofs of these propositions are straightforward. These are omitted.

We will define a martingale with discrete time parameter. Let $\mathbb{N}$ be the set of all positive integers. An increasing sequence $\{\mathcal{F}_n, n \in \mathbb{N}\}$ of sub σ-fields of $\mathcal{F}$ is called a *filtration*. Given a filtration $\{\mathcal{F}_n\}$, we will define a martingale. Let $\{X_n; n \in \mathbb{N}\}$ be a sequence of real integrable random variables. We will assume that each X_n is $\mathcal{F}_n$-measurable. It is called a *martingale with discrete time* if $E[X_n|\mathcal{F}_m] = X_m$ holds for any $1 \leq m < n$. We denote by $\mathcal{F}_0$ the trivial σ-field $\{\emptyset, \Omega\}$, where $\emptyset$ is the empty set.

A sequence of random variables $f_n, n \in \mathbb{N}$ is called *predictable* if each f_n is $\mathcal{F}_{n-1}$-measurable. Let $\{f_n\}$ be a predictable sequence. The *martingale transform* $Y_n, n \in \mathbb{N}$ is defined by

$$Y_n = \sum_{m=1}^{n} f_m(X_m - X_{m-1}), \quad \text{where } X_0 = 0.$$

If $E[|f_m||X_m - X_{m-1}|] < \infty$ holds for any $m \geq 1$, $\{Y_n\}$ is a martingale, since Y_n is $\mathcal{F}_n$-measurable and satisfies

$$E[Y_{n+1} - Y_n|\mathcal{F}_n] = f_{n+1}E[X_{n+1} - X_n|\mathcal{F}_n] = 0.$$

Let τ be a random variable with values in $\mathbb{N} \cup \{\infty\}$. It is called a *stopping time* if $\{\tau \le n\} \in \mathcal{F}_n$ holds for any $n \in \mathbb{N}$. If τ is a stopping time, it is easily seen that $n \wedge \tau$ is also a stopping time for any positive integer n. For a stopping time τ, we set

$$\mathcal{F}_\tau = \Big\{B \in \mathcal{F};\ B \cap \{\tau \le n\} \in \mathcal{F}_n \ \text{ for any } n\Big\}.$$

It is a sub σ-field of $\mathcal{F}$.

Proposition 1.5.3 *Let $\{X_n, n \in \mathbb{N}\}$ be a martingale and let τ be a stopping time.*

1. *The stopped process $\{Y_n = X_{n\wedge\tau}, n \in \mathbb{N}\}$ is again a martingale.*
2. *Let σ be another stopping time. Then $E[X_{n\wedge\tau}|\mathcal{F}_\sigma] = X_{n\wedge\tau\wedge\sigma}$ holds for any $n \in \mathbb{N}$.*

Proof We set $f_n = 1_{\tau \ge n}$. It is a predictable sequence. Consider the martingale transform by f_n; $Y_n = \sum_{m=1}^n f_m(X_m - X_{m-1})$. It coincides with $X_{n\wedge\tau}$. Therefore $\{X_{n\wedge\tau}\}$ is a martingale.

Let σ be another stopping time. If $B \in \mathcal{F}_\sigma$, we have

$$E[Y_n; B] = \sum_{m=0}^{n} E[Y_n; B \cap \{\sigma = m\}] = \sum_{m=0}^{n} E[Y_m; B \cap \{\sigma = m\}] = E[Y_{n\wedge\sigma}; B].$$

Further, $Y_{n\wedge\sigma}$ is $\mathcal{F}_\sigma$-measurable, because

$$\{X_{n\wedge\sigma} < a\} \cap \{\sigma \le m\} = \bigcup_{l \le m}\{X_l < a, \sigma = l\} \in \mathcal{F}_m$$

holds for any m. Therefore we get the equality $E[Y_n|\mathcal{F}_\sigma] = Y_{n\wedge\sigma}$ a.s. Substituting $Y_n = X_{n\wedge\tau}$, we get $E[X_{n\wedge\tau}|\mathcal{F}_\sigma] = X_{n\wedge\tau\wedge\sigma}$ a.s. □

Proposition 1.5.4 *Let $\{X_n, n \in \mathbb{N}\}$ be a martingale and let $N \in \mathbb{N}$. Then we have for any $a > 0$*

$$aP(\sup_{n \le N} |X_n| > a) \le \int_{\sup_{n\le N}|X_n| > a} |X_N|\, dP. \tag{1.35}$$

Proof For a given $a > 0$, we set

$$\tau = \inf\{n;\ |X_n| > a\} \quad (= \infty \ \text{ if the set } \{\cdots\} \text{ is empty}).$$

Then τ is a stopping time. Indeed, we have $\{\tau \le k\} = \bigcup_{i\le k}\{|X_i| > a\} \in \mathcal{F}_k$ for any $k \in \mathbb{N}$. Then by Proposition 1.5.3, we have $E[|X_N||\mathcal{F}_\tau] \ge |X_{N\wedge\tau}|$. Therefore,

$$aP(\tau \le N) \le \int_{\tau \le N} |X_{N\wedge\tau}|\, dP \le \int_{\tau\le N} |X_N|\, dP.$$

Since $\{\tau \le N\} = \{\sup_{n\le N} |X_n| > a\}$, we get the inequality of the Proposition. □

Proposition 1.5.5 *Let $\{X_n, n \in \mathbb{N}\}$ be a martingale satisfying $E[|X_n|^p] < \infty$ for any n, where $p > 1$. Set $q = \frac{p}{p-1}$. Then we have for any $N \in \mathbb{N}$,*

$$E\Big[\sup_{n \le N} |X_n|^p\Big] \le q^p E[|X_N|^p]. \tag{1.36}$$

Proof Set $Y = \sup_{n \le N} |X_n|$ and $F(\lambda) = P(Y > \lambda)$. Then we have

$$E[Y^p] = -\int_0^\infty \lambda^p \, dF(\lambda) = \int_0^\infty F(\lambda) \, d(\lambda^p) - \lim_{\lambda \to \infty} \lambda^p F(\lambda)\big|_0^\lambda \le \int_0^\infty F(\lambda) \, d(\lambda^p).$$

Since $\lambda P(Y > \lambda) \le \int_{Y > \lambda} |X_N| \, dP$ holds by Proposition 1.5.4, we have

$$\begin{aligned} E[Y^p] &\le \int_0^\infty \frac{1}{\lambda}\Big(\int_{Y>\lambda} |X_N| dP\Big) d(\lambda^p) = \int |X_N| \Big(\int_0^Y \frac{1}{\lambda} d(\lambda^p)\Big) \, dP \\ &\le \frac{p}{p-1} \int |X_N| Y^{p-1} \, dP \le q E[|X_N|^p]^{\frac{1}{p}} E[Y^{(p-1)q}]^{\frac{1}{q}} \\ &\le q E[|X_N|^p]^{\frac{1}{p}} E[Y^p]^{\frac{1}{q}}. \end{aligned}$$

Therefore we have the inequality of the proposition. □

We give the definition of a martingale with negative parameter. Let $\mathcal{F}_{-n}, n \in \mathbb{N}$ be a sequence of sub σ-fields of $\mathcal{F}$ satisfying $\mathcal{F}_{-n} \subset \mathcal{F}_{-(n-1)}$. Let $\{X_{-n}, n \in \mathbb{N}\}$ be a sequence of integrable $\{\mathcal{F}_{-n}\}$-adapted random variables. It is called a martingale if $E[X_{-m}|\mathcal{F}_{-n}] = X_{-n}$ holds whenever $-m > -n$. A sub-martingale and a super-martingale are defined similarly.

We will next define a martingale with continuous parameter $\mathbb{T} = [0, \infty)$. Let $\{\mathcal{F}_t, t \in \mathbb{T}\}$ be a family of sub σ-fields of $\mathcal{F}$ satisfying $\mathcal{F}_s \subset \mathcal{F}_t$ for any $s < t$. We assume further that $\mathcal{F}_t$ contains all null sets of $\mathcal{F}$ and satisfies $\mathcal{F}_t = \bigwedge_{\epsilon > 0} \mathcal{F}_{t+\epsilon}$ (right continuous). Then the family $\{\mathcal{F}_t, t \in \mathbb{T}\}$ is called a *filtration*. A random variable τ with values in $\mathbb{T} \cup \{\infty\}$ is called a *stopping time* if $\{\omega; \tau(\omega) \le t\} \in \mathcal{F}_t$ holds for any $t \in \mathbb{T}$. For a stopping time τ, we set

$$\mathcal{F}_\tau = \Big\{B \in \mathcal{F};\ B \cap \{\tau \le t\} \in \mathcal{F}_t \ \text{ for any } t \in \mathbb{T}\Big\}.$$

It is a sub σ-field of $\mathcal{F}$.

Let $\{X_t\} = \{X_t, t \in \mathbb{T}\}$ be a stochastic process. $\{X_t\}$ is called *adapted* for $\{\mathcal{F}_t\}$ (or $\{\mathcal{F}_t\}$-adapted) and $\{\mathcal{F}_t\}$ is called *admissible* for $\{X_t\}$ (or $\{X_t\}$-admissible), if X_t is $\mathcal{F}_t$-measurable for each $t \in \mathbb{T}$. An $\{\mathcal{F}_t\}$-adapted cadlag process $\{X_t\}$ is called a *martingale* or $\{\mathcal{F}_t\}$-martingale, if X_t is integrable for any t and equalities $E[X_t|\mathcal{F}_s] = X_s$ hold a.s. for any $s < t$. If equality signs are replaced by $\ge$ in the above, it is called a *sub-martingale*. Further, if $\{-X_t\}$ is a sub-martingale $\{X_t\}$ is called a *super-martingale*.

Let $p > 1$. An $\{\mathcal{F}_t\}$-adapted cadlag process $\{X_t\}$ is called an *L^p-martingale* if it is a martingale and satisfies $E[|X_t|^p] < \infty$ for any t. An $\{\mathcal{F}_t\}$-adapted cadlag process $\{X_t\}$ is called a *local martingale* or *local L^p-martingale*, if there exists an increasing sequence of stopping times $\{\tau_n\}$ such that $P(\tau_n < T) \to 0$ for any positive constant T and each stopped process $X_t^{(n)} := X_{t\wedge\tau_n}, t \in \mathbb{T}$ is a martingale or L^p-martingale, respectively. A martingale is a local martingale and an L^p-martingale is a local L^p-martingale.

Theorem 1.5.1 (Doob) *Let $\{X_t\}$ be a martingale.*

1. *For any stopping time τ, the stopped process $\{Y_t = X_{t\wedge\tau}\}$ is a martingale.*
2. *For any two stopping times σ and τ, we have $E[X_{t\wedge\sigma}|\mathcal{F}_\tau] = X_{t\wedge\sigma\wedge\tau}$ for any t.*
3. *(Doob's inequality) Suppose that $\{X_t\}$ is an L^p-martingale for some $p > 1$. Set $q = \frac{p}{p-1}$. Then we have*

$$E\Big[\sup_{0\le r\le t} |X_r|^p\Big] \le q^p E\Big[|X_t|^p\Big], \quad \forall t \in \mathbb{T}. \tag{1.37}$$

Proof We prove the first assertion. For $n \in \mathbb{N}$, set $t_m^{(n)} = m/2^n, m = 1, \ldots, 2^n T$. Let τ be a stopping time. For $n \in \mathbb{N}$, we define

$$\tau_n(\omega) = t_m^{(n)}, \quad \text{if } t_{m-1}^{(n)} < \tau(\omega) \le t_m^{(n)}. \tag{1.38}$$

Then $\tau_n, n = 1, 2, \ldots$ is a sequence of stopping times decreasing to τ as $n \to \infty$. We have $E[X_{t\wedge\tau_n}; B] = E[X_{s\wedge\tau_n}; B]$ for any $B \in \mathcal{F}_s$ by Proposition 1.5.3. Therefore if $\lim_{n\to\infty} E[X_{t\wedge\tau_n}; B] = E[X_{t\wedge\tau}; B]$ holds for any t, we get $E[X_{t\wedge\tau}; B] = E[X_{s\wedge\tau}; B]$ and hence $X_{t\wedge\tau}$ is a martingale. In order to prove the above convergence, we will show that the family of random variables $\{X_{t\wedge\tau_n} - X_{t\wedge\tau}\}$ is uniformly integrable, i.e., for any $\epsilon > 0$ there exists $c > 0$ such that

$$\sup_n \int_{|X_{t\wedge\tau_n} - X_{t\wedge\tau}|>c} |X_{t\wedge\tau_n} - X_{t\wedge\tau}|\, dP < \epsilon. \tag{1.39}$$

Set $Y_{-n} = X_{t\wedge\tau_n} - X_{t\wedge\tau}$ and $\mathcal{G}_{-n} = \mathcal{F}_{t\wedge\tau_n}$. Then, in view of Proposition 1.5.3, 2, $Y_{-n}, n \in \mathbb{N}$ is a $\{\mathcal{G}_{-n}\}$-martingale with negative parameter. Therefore $|Y_{-n}|$ is a sub-martingale. Since $\{|Y_{-n}| > c\} \in \mathcal{G}_{-n}$, we have the inequality

$$\int_{|Y_{-n}|>c} |Y_{-n}|\, dP \le \int_{|Y_{-n}|>c} |Y_{-1}|\, dP$$

for any n and $c > 0$. Set $Q_1(A) := \int_A |Y_{-1}|\, dP$ on $\mathcal{G}_{-1}$. Then Q_1 is absolutely continuous with respect to P. Therefore for any $\epsilon > 0$, there exists $\delta > 0$ such that for any $A \in \mathcal{G}_{-1}$ with $P(A) < \delta$, the inequality $Q(A) < \epsilon$ holds. Further, for this δ, there exists $c > 0$ such that the inequality

$$P(|Y_{-n}| > c) \leq \frac{1}{c}E[|Y_{-n}|] \leq \frac{1}{c}E[|Y_{-1}|] \leq \delta$$

holds for any n. We have thus shown that for any $\epsilon > 0$, there exists $c > 0$ such that the inequality $\sup_n \int_{|Y_{-n}|>c} |Y_{-n}|\, dP < \epsilon$ holds. Therefore $\{Y_{-n}\}$ is uniformly integrable. Now we have

$$\int |X_{t\wedge\tau_n} - X_{t\wedge\tau}| 1_B\, dP \leq \epsilon + \int_{|Y_n|\leq c} |Y_n| 1_B\, dP.$$

Since $\{Y_n\}$ converges to 0 a.s., the last term converges to 0 as $n \to \infty$. Therefore we have $\lim_{n\to\infty} \int |X_{t\wedge\tau_n} - X_{t\wedge\tau}| 1_B\, dP = 0$. This proves the equality $\lim_{n\to\infty} E[X_{t\wedge\tau_n}; B] = E[X_{t\wedge\tau}; B]$.

The second assertion can be verified similarly, using Proposition 1.5.3, 2 and approximating both stopping times σ and τ from the above by sequences of stopping times with discrete values.

Suppose next that $\{X_t\}$ is an L^p-martingale for some $p > 1$. For $n \in \mathbb{N}$, set $t_m^{(n)} = m/2^n \wedge t, m = 1, \ldots, 2^n(T+1)$. Then $E[\sup_{m<2^n} |X_{t_m^{(n)}}|^p] \leq q^p E[|X_t|^p]$ holds by Proposition 1.5.5. Since $\sup_{r\leq t} |X_r| = \lim_{n\to\infty} \sup_m |X_{t_m^{(n)}}|$ holds a.s., we get the inequality (1.37). □

An $\{\mathcal{F}_t\}$-adapted cadlag process $A_t, t \in \mathbb{T}$ is called an *increasing process* if A_t is a nondecreasing function of t a.s. and satisfies $A_0 = 0$ a.s. If A_t is continuous in t a.s., $\{A_t\}$ is called a continuous increasing process. If A_t is written as $A_t^1 - A_t^2$ for any t, where $A_t^1, A_t^2, t \in \mathbb{T}$ are increasing processes, $\{A_t\}$ is called a *process of bounded variation*. Further, if both $\{A_t^i\}, i = 1, 2$ are continuous, $\{A_t\}$ is called a continuous process of bounded variation. If a process $\{X_t\}$ is written as the sum of a local martingale $\{M_t\}$ and a continuous process of bounded variation $\{A_t\}$, it is called a *semi-martingale*. Further, if $\{M_t\}$ is a continuous process, $\{X_t\}$ is called a *continuous semi-martingale*. If $\{M_t\}$ is a local martingale with bounded jumps, $\{X_t\}$ is called a *semi-martingale with bounded jumps*. The process $\{M_t\}$ is called the *martingale part of the semi-martingale* $\{X_t\}$.

Let $\{\mathcal{F}_t, t \in \mathbb{T}\}$ be a filtration and let τ be a stopping time with respect to the filtration. A cadlag process $\{X_t, t \in [0, \tau)\}$ is called a *local semi-martingale* if (i) τ is accessible, i.e. there exists an increasing sequence of stopping times $\{\tau_n\}$ such that $\tau_n < \tau$ and $\lim_{n\to\infty} \tau_n = \tau$ hold a.s., and (ii) stopped processes $X_t^{(n)} = X_{t\wedge\tau_n}$ are semi-martingales for all n with respect to the filtration $\{\mathcal{F}_t\}$.

A family of sub σ-fields $\{\check{\mathcal{F}}_s, s \in \mathbb{T}\}$ of $\mathcal{F}$ is called a *backward filtration* if $\check{\mathcal{F}}_t \subset \check{\mathcal{F}}_s$ for any $s < t$, $\bigcap_{\epsilon>0} \check{\mathcal{F}}_{s-\epsilon} = \check{\mathcal{F}}_s$ for any s and each $\check{\mathcal{F}}_s$ contains null sets of $\mathcal{F}$. We will consider a backward semi-martingale. We consider left continuous rather than right continuous processes, since the time evolution is backward. Let $\{\check{\mathcal{F}}_s\}$ be a backward filtration. Let $\{\check{X}_s, s \in \mathbb{T}\}$ be an integrable caglad process, adapted to a backward filtration $\{\check{\mathcal{F}}_s\}$. It is called a *backward martingale* if $E[\check{X}_s|\check{\mathcal{F}}_t] = \check{X}_t$ holds a.s. for any $s < t$. A random variable σ with values in $\mathbb{T}\cup\{-\infty\}$ is called a *backward stopping time*, if $\{\sigma \geq t\} \in \check{\mathcal{F}}_t$ holds for any $t \in \mathbb{T}$. An $\{\check{\mathcal{F}}_s\}$-adapted caglad process

$\{\check{X}_s\}$ is called a *backward local martingale* if there exists a decreasing sequence of backward stopping times σ_n such that $\lim_{n\to\infty} P(\sigma_n > 0) = 0$ and each stopped process $\{\check{X}_{s\vee\sigma_n}\}$ is a backward martingale. Further, if $\check{X}_s$ is written as the sum of a backward local martingale and a continuous process of bounded variation adapted to the backward filtration $\{\check{\mathcal{F}}_s\}$, $\{\check{X}_s\}$ is called a *backward semi-martingale*.

Let $\{\mathcal{F}_{s,t}, 0 \leq s < t < \infty\}$ be a family of sub σ-fields of $\mathcal{F}$, each of which contains null sets of $\mathcal{F}$. We assume that for any fixed s, $\mathcal{F}'_t := \mathcal{F}_{s,t}, t \geq s$ is a filtration. We assume further that for any fixed t, $\check{\mathcal{F}}'_s := \mathcal{F}_{s,t}$ is decreasing with respect to s, i.e., $\check{\mathcal{F}}'_s \subset \check{\mathcal{F}}'_{s'}$ if $s' \leq s$ and left continuous, i.e., $\bigcap_{\epsilon>0} \check{\mathcal{F}}'_{s-\epsilon} = \check{\mathcal{F}}'_s$ holds for any $s < t$. Then $\{\mathcal{F}_{s,t}\}$ is called a *two-sided filtration*. A backward local semi-martingale is defined similarly to a local semi-martingale.

Let $\{X_t, t \in [0, \infty)\}$ be a Wiener process and let $N(dsdz)$ be a Poisson random measure on $\mathbb{U} = [0, \infty) \times \mathbb{R}_0^{d'}$. Let

$$\mathcal{F}_{s,t} = \sigma(X_u - X_v, N(dr\,dz); s \leq u < v \leq t, dr \subset [s, t]).$$

Then $\{\mathcal{F}_{s,t}\}$ is a two-sided filtration generated by Wiener process and Poisson random measure. Then $\{X_t\}$ and $\{Y_t\} = \{\int_0^t \int_{\mathbb{R}_0^d} f(z)\tilde{N}(dr\,dz)\}$ are forward martingales with respect to the filtration $\mathcal{F}_t = \mathcal{F}_{0,t}$. Further, $\{\check{X}_s := X_T - X_s\}$ and $\{\check{Y}_s := Y_T - Y_s\}$ are backward martingales with respect to the backward filtration $\check{\mathcal{F}}_s = \mathcal{F}_{s,T}$.

1.6 Quadratic Variations of Semi-martingales

In this section, we will restrict time parameter to the finite interval $\mathbb{T} = [0, T]$. Martingales and stopping times with time parameter $\mathbb{T}$ are defined in the same way as those with time parameter $[0, \infty)$. Let $\Pi = \{0 = t_0 < t_1 < \cdots < t_n = T\}$ be a partition of $\mathbb{T} = [0, T]$. We set $|\Pi| = \max_m |t_m - t_{m-1}|$. We define the quadratic variation of a real cadlag process $\{X_t, t \in \mathbb{T}\}$ associated with the partition Π by

$$\langle X\rangle_t^\Pi = \sum_{m=1}^{n} (X_{t_m\wedge t} - X_{t_{m-1}\wedge t})^2, \quad t \in \mathbb{T}. \tag{1.40}$$

We are interested in the limit of quadratic variations $\{\langle X\rangle_t^\Pi\}$ as $|\Pi| \to 0$. If it converges uniformly in t in probability to a process, the limit is denoted by $\langle X\rangle_t, t \in \mathbb{T}$ and is called the *quadratic variation* of $\{X_t\}$. The quadratic variation of $\{X_t\}$ is often written as $\{\langle X_t\rangle\}$.

In this section, we show that if $\{X_t, t \in \mathbb{T}\}$ is a semi-martingale with bounded jumps, the quadratic variation exists. We will first list properties of L^2 martingales.

Proposition 1.6.1 *Let $\{X_t\}$ be an L^2-martingale.*

1. $E[(X_t - X_s)^2|\mathcal{F}_s] = E[X_t^2 - X_s^2|\mathcal{F}_s]$ *holds for any* $0 \le s < t \le T$.
2. $E[(X_v - X_u)(X_t - X_s)|\mathcal{F}_s] = 0$ *holds for any* $0 \le s < t \le u < v \le T$.
3. *For any* $0 \le s < t \le T$, *it holds that*

$$F[(X_t - X_s)^2|\mathcal{F}_s] = E[\langle X\rangle_t^{\Pi} - \langle X\rangle_s^{\Pi}|\mathcal{F}_s]. \tag{1.41}$$

Proof In view of Proposition 1.5.1, 1 and 5, we have

$$\begin{aligned} E[(X_t - X_s)^2|\mathcal{F}_s] &= E[X_t^2 - 2X_tX_s + X_s^2|\mathcal{F}_s] \\ &= E[X_t^2|\mathcal{F}_s] - 2E[X_t|\mathcal{F}_s]X_s + X_s^2 \\ &= E[X_t^2|\mathcal{F}_s] - X_s^2. \end{aligned}$$

Then we get the equality of 1. Next, we have by Proposition 1.5.1, 3,

$$E[(X_v - X_u)(X_t - X_s)|\mathcal{F}_s] = E[E[(X_v - X_u)|\mathcal{F}_u](X_t - X_s)|\mathcal{F}_s] = 0,$$

proving the orthogonal property of 2.

We have

$$X_t - X_s = \sum_{m; s < t_{m-1}} (X_{t_m \wedge t} - X_{t_{m-1} \wedge t}) + (X_{t_l} - X_s),$$

where $t_{l-1} < s < t_l$. Then using the above orthogonal property, we have

$$\begin{aligned} E[(X_t - X_s)^2|\mathcal{F}_s] &= E[\sum_{m; s < t_{m-1}} (X_{t_m \wedge t} - X_{t_{m-1} \wedge t})^2 + (X_{t_l} - X_s)^2|\mathcal{F}_s] \\ &= E[\langle X\rangle_t^{\Pi} - \langle X\rangle_s^{\Pi}|\mathcal{F}_s]. \end{aligned}$$

□

Let $\Pi_n = \{0 = t_0^{(n)} < t_1^{(n)} < \cdots < t_{t_{m_n}}^{(n)} = T\}, n = 1, 2, \ldots$ be a sequence of partitions of $\mathbb{T}$ such that $|\Pi_n| \to 0$ as $n \to \infty$. We call $\{\Pi_n\}$ regular, if $\Pi_n \subset \Pi_{n+1}$ holds for any n.

Lemma 1.6.1 *Let $\{X_t\}$ be an L^4-martingale. Let $\{\langle X\rangle_t^{\Pi_n}\}$ be the quadratic variation of $\{X_t\}$ associated with the partition Π_n. If $\{\Pi_n\}$ is regular and $|\Pi_n|$ converges to 0, the sequence $\{\langle X\rangle_t^{\Pi_n}\}$ converges to an increasing process $\{\langle X\rangle_t\}$ uniformly in $t \in \mathbb{T}$ in L^2. Further, $\{X_t^2 - \langle X\rangle_t\}$ is an L^2-martingale.*

Proof We will show first that $\{\langle X\rangle_T^{\Pi_n}, n = 1, 2, \ldots\}$ are L^2-bounded. By a direct calculation, we get the identity

$$X_t^2 - \langle X\rangle_t^{\Pi_n} = 2\Big(\sum_{t_m^{(n)}\le t} X_{t_{m-1}^{(n)}}(X_{t_m^{(n)}} - X_{t_{m-1}^{(n)}}) + X_{t_l^{(n)}}(X_t - X_{t_l^{(n)}})\Big). \tag{1.42}$$

Denote the right-hand side by $2M_t^{\Pi_n}$. We may regard that $\{M_t^{\Pi_n}, t \in \Pi_n\}$ is a martingale transform of the discrete martingale $\{X_t, t \in \Pi_n\}$. It is an L^2 martingale with time parameter Π_n. We have by Proposition 1.6.1 and Doob's inequality

$$\begin{aligned} E\big[|M_t^{\Pi_n}|^2\big] &= \sum_m E\big[(X_{t_{m-1}^{(n)}\wedge t})^2(X_{t_m^{(n)}\wedge t} - X_{t_{m-1}^{(n)}\wedge t})^2\big] \\ &\le E\big[\sup_m (X_{t_{m-1}^{(n)}\wedge t})^2\langle X\rangle_t^{\Pi_n}\big] \\ &\le 2a\,E\big[|X_t|^4\big] + \frac{1}{2a}E\Big[\Big(\langle X\rangle_t^{\Pi_n}\Big)^2\Big]. \end{aligned}$$

Here we used the inequality $|XY| \le \frac{a}{2}|X|^2 + \frac{1}{2a}Y^2$. Now, note the equality $\langle X\rangle_t^{\Pi_n} = X_t^2 - 2M_t^{\Pi_n}$. Then we get

$$E\big[(\langle X\rangle_t^{\Pi_n})^2\big] \le 2(E\big[X_t^4\big] + 4E\big[(M_t^{\Pi_n})^2\big]) \le 2(1+a)E\big[X_t^4\big] + \frac{4}{a}E\big[(\langle X\rangle_t^{\Pi_n})^2\big].$$

Taking $a > 4$, we find that $\sup_{t\in\mathbb{T}} E[(\langle X\rangle_t^{\Pi_n})^2]$ is bounded with respect to n.

Since a bounded set in the L^2-space is weakly compact, a subsequence of $\{\langle X\rangle_t^{\Pi_n}, t \in \Pi_\infty\}$ converges weakly to an increasing process $\{A_t, t \in \Pi_\infty\}$ such that $\sup_t E[A_t^2] < \infty$, where $\Pi_\infty = \bigcup_n \Pi_n$ is a countable dense subset of $\mathbb{T}$ (it is the set of dyadic rational numbers in $\mathbb{T}$). It means that there exists a subsequence $\{n'\}$ of $\mathbb{N}$ such that for any random variable F with $E[|F|^2] < \infty$, $E[\langle X\rangle_t^{\Pi_{n'}} F]$ converges to $E[A_t F]$ for all $t \in \Pi_\infty$ as $n' \to \infty$. A_t should be $\mathcal{F}_t$-measurable for any $t \in \Pi_\infty$ and nondecreasing with respect to $t \in \Pi_\infty$. Further, since $\{X_t^2 - \langle X\rangle_t^{\Pi_{n'}}, t \in \Pi_{n'}\}$ are martingales, $\{X_t^2 - A_t, t \in \Pi_\infty\}$ should also be a martingale. Then we have

$$E[(X_t - X_s)^2|\mathcal{F}_s] = E[X_t^2 - X_s^2|\mathcal{F}_s] = E[A_t - A_s|\mathcal{F}_s], \quad a.s.$$

if $s < t$ and $s, t \in \Pi_\infty$.

Using the above A_t, we will show that the sequence of martingale transforms $\{M_t^{\Pi_n}\}$ converges to a martingale uniformly in t in L^2. If $n \ge n'$, these satisfy

$$M_t^{\Pi_n} - M_t^{\Pi_{n'}} = \sum_m (X_{t_m^{(n)}\wedge t} - X_{t_{m'}^{(n')}\wedge t})(X_{t_{m+1}^{(n)}\wedge t} - X_{t_m^{(n)}\wedge t}),$$

where $t_{m'}^{(n')} \le t_m^{(n)}$. Since the equality

$$E\Big[(X_{t_{m+1}^{(n)}\wedge t} - X_{t_m^{(n)}\wedge t})^2\Big|\mathcal{F}_{t_m^{(n)}}\Big] = E\Big[A_{t_{m+1}^{(n)}\wedge t} - A_{t_m^{(n)}\wedge t}\Big|\mathcal{F}_{t_m^{(n)}}\Big], \quad a.s.$$

holds, we have

$$E\Big[(M_t^{\Pi_n}-M_t^{\Pi_{n'}})^2\Big]=\sum_m E\Big[(X_{t_m^{(n)}\wedge t}-X_{t_{m'}^{(n')}\wedge t})^2(X_{t_{m+1}^{(n)}\wedge t}-X_{t_m^{(n)}\wedge t})^2\Big]$$

$$=\sum_m E\Big[(X_{t_m^{(n)}\wedge t}-X_{t_{m'}^{(n')}\wedge t})^2(A_{t_{m+1}^{(n)}\wedge t}-A_{t_m^{(n)}\wedge t})\Big]$$

$$=E\Big[\int_0^t (X_{\phi^{(n)}(s)}-X_{\phi^{(n')}(s)})^2\,dA_s\Big],$$

where $\phi^{(n)}(s)$ are step functions of s such that $\phi^{(n)}(s)=t_m^{(n)}$ for $t_m^{(n)}<s<t_{m+1}^{(n)}$. Let $n,n'\to\infty$. Then the last term converges to 0. Therefore, because of Doob's inequality, the sequence of martingales $\{M_t^{(n)}\}$ converges uniformly in t to a right continuous L^2-martingale M_t. This means that $\{\langle X\rangle_t^{(n)}, t\in\mathbb{T}\}$ converges to an increasing process $\{\langle X\rangle_t, t\in\mathbb{T}\}$ uniformly in t with respect to L^2-norm. Further, we have $X_t^2-\langle X\rangle_t=M_t$ for all t a.s. □

Theorem 1.6.1 *Let $\{X_t\}$ be a local martingale with bounded jumps. Then its quadratic variation $\{\langle X\rangle_t\}$ exists. It is an increasing process; further, $\{X_t^2-\langle X\rangle_t\}$ is a local martingale.*

Proof If $X_t, t\in\mathbb{T}$ is a local martingale with bounded jumps, we can choose a sequence of stopping times $\{\tau_n\}$ such that each stopped process is a bounded martingale (hence L^4-martingale). Indeed, let

$$\sigma_N=\inf\{t\in\mathbb{T};\ |X_t|>N\}\ \ (=\infty\ \text{ if the set }\{\cdots\}\text{ is empty}).$$

It is called the *first leaving time* of the process X_t from the set $\{x;|x|\le N\}$ or *hitting time* of the process X_t to the set $\{x;\ |x|>N\}$. It is a stopping time, since $\{\sigma_N\le t\}=\bigcup_{0\le r\le t}\{|X_r|>N\}\in\mathcal{F}_t$ holds for any t. Here r runs rational numbers. Then $X_{t\wedge\tau_n\wedge\sigma_N}$ is a martingale by Theorem 1.5.1. Let $M>0$ be a positive constant such that $|\Delta X_t|\le M$ holds for any t a.s. Then $|X_{t\wedge\tau_n\wedge\sigma_N}|$ is bounded by $N+M$ for all n. Therefore the limit $\lim_{n\to\infty}X_{t\wedge\tau_n\wedge\sigma_N}=X_{t\wedge\sigma_N}$ is also a martingale for any N. Since $P(\sigma_N<T)\to 0$ ans $N\to\infty$, $\{X_t\}$ is a local martingale with respect to stopping times $\{\sigma_N, N=1,2,\ldots\}$.

Set $X_t^{(N)}=X_{t\wedge\sigma_N}$. It is an L^4-martingale. By the definition the quadratic variation with respect to the partition Π_n, it holds that $\langle X^{(N)}\rangle_t^{(n)}=\langle X\rangle_{t\wedge\sigma_N}^{(n)}$. Further, for any N, $\langle X^{(N)}\rangle_t^{(n)}$ converges to a cadlag increasing $\langle X^{(N)}\rangle_t$ uniformly in t in L^2-sense as $n\to\infty$. Then if $N'\ge N$ we have $\langle X^{(N')}\rangle_{t\wedge\sigma_N}=\langle X^{(N)}\rangle_t$. Therefore there exists a cadlag increasing process $\langle X\rangle_t$ such that $\langle X\rangle_{t\wedge\sigma_N}=\langle X^{(N)}\rangle_t$. Consequently, if $\{\Pi_n\}$ is regular, for any N $\langle X\rangle_t^{(n)}$ converges to $\langle X\rangle_t$ uniformly for $t\le\tau_N$ in L^2-sense. This shows that $\langle X\rangle_t^{(n)}$ converges to $\langle X\rangle_t$ uniformly in $t\in\mathbb{T}$ in probability. Further, $X_t^2-\langle X\rangle_t$ is a local martingale.

It remains to show that $\langle X\rangle_t$ does not depend on the choice of partitions $\{\Pi_n\}$. Let $\{\Pi'_n\}$ be another regular sequence of partitions. Let $\{\langle X\rangle_t^{\Pi'_n}\}$ be the sequence of the associated quadratic variations and let $\langle X\rangle'_T$ be its limit as $n\to\infty$. We can choose still another regular sequence of partitions $\{\Pi''_n\}$ such that for any $\Pi_n\in\{\Pi_n\}$ and $\Pi'_{n'}\in\{\Pi'_n\}$, there are Π''_n and $\Pi''_{n'}$ in $\{\Pi''_n\}$ such that $\Pi_n\subset\Pi''_n$ and $\Pi'_{n'}\subset\Pi''_{n'}$ holds. The sequence $\{\langle X\rangle_t^{\Pi''_n}\}$ converges to an increasing process $\langle X\rangle''_t$. It should coincide with $\langle X\rangle_t$ and $\langle X\rangle'_t$. Therefore the uniqueness of the quadratic variation follows. □

Corollary 1.6.1 *If $\{X_t\}$ be a continuous local martingale, its quadratic variation $\{\langle X\rangle_t\}$ is a continuous increasing process. Further, if $\{A_t\}$ is a continuous increasing process such that $\{X_t^2-A_t\}$ is a local martingale, the equality $A_t=\langle X\rangle_t$ holds for any $t\in\mathbb{T}$, a.s.*

Proof If $\{X_t\}$ is a continuous local martingale, its quadratic variation $\{\langle X\rangle_t\}$ is a continuous process, since $\{\langle X\rangle_t^{\Pi}\}$ are continuous processes. Suppose that $\{A_t\}$ is a continuous increasing process such that $\{X_t^2-A_t\}$ is a local martingale. Then $\{M_t:=\langle X\rangle_t-A_t\}$ is a continuous local martingale. Since M_t is a process of bounded variation, we have $\langle M\rangle_t=0$. Therefore M_t^2 is a continuous local martingale and hence $M_t=0$ for any t. Therefore $\langle X\rangle_t=A_t$ holds for ant t. □

Let $\{X_t\}$ and $\{Y_t\}$ be cadlag processes. For a partition Π, we define the *quadratic covariation* of X,Y associated with Π by

$$\langle X,Y\rangle_t^{\Pi}=\sum_{m=1}^{n}(X_{t_m\wedge t}-X_{t_{m-1}\wedge t})(Y_{t_m\wedge t}-Y_{t_{m-1}\wedge t}),\quad t\in\mathbb{T}. \tag{1.43}$$

Theorem 1.6.2 *Suppose that $\{X_t\}$ and $\{Y_t\}$ are local martingales with bounded jumps. Then quadratic covariations $\{\langle X,Y\rangle_t^{\Pi}\}$ converge to a process of bounded covariation $\{\langle X,Y\rangle_t\}$ uniformly in t in probability. Further, $\{X_tY_t-\langle X,Y\rangle_t\}$ is a local martingale.*

The process $\langle X,Y\rangle_t$ is called the quadratic covariation of X_t and Y_t.

Proof If $X_t=Y_t$ and it is a local martingale with bounded jumps, the existence of the quadratic variation $\langle X\rangle_t$ is a direct consequence of Lemma 1.6.1. We shall consider the case $X_t\neq Y_t$. It holds that

$$\langle X,Y\rangle_t^{\Pi}=\frac{1}{4}\Big(\langle X+Y\rangle_t^{\Pi}-\langle X-Y\rangle_t^{\Pi}\Big).$$

Let $|\Pi|$ tend to 0. Then the right-hand side converges to $\frac{1}{4}(\langle X+Y\rangle_t-\langle X-Y\rangle_t)$. Therefore $\langle X,Y\rangle_t^{\Pi}$ converges uniformly in t in probability. Denote the limit by $\langle X,Y\rangle_t$. It is a process of bounded variation. It satisfies

$$\langle X, Y\rangle_t = \frac{1}{4}\Big(\langle X+Y\rangle_t - \langle X-Y\rangle_t\Big).$$

Then $X_tY_t - \langle X, Y\rangle_t$ is a local martingale. □

Theorem 1.6.3 *Let $\{X_t\}$ be a continuous semimartingale and let $\{Y_t\}$ be semimartingales with bounded jumps, decomposed as $X_t = M_t + A_t$ and $Y_t = N_t + B_t$ respectively, where $\{M_t, N_t\}$ are local martingales and $\{A_t, B_t\}$ are continuous processes of bounded variation. Then $\{\langle X, Y\rangle_t^{\Pi}\}$ converges as $|\Pi| \to 0$ to a continuous process of bounded variation $\{\langle X, Y\rangle_t, t \in \mathbb{T}\}$ uniformly in t in probability. Further,*

$$\lim_{|\Pi|\to 0} \langle X, Y\rangle_t^{\Pi} = \langle X, Y\rangle_t = \langle M, N\rangle_t, \quad t \in \mathbb{T}. \tag{1.44}$$

Proof It holds that

$$\langle X, Y\rangle_t^{\Pi} = \langle M, N\rangle_t^{\Pi} + \langle M, B\rangle_t^{\Pi} + \langle A, N\rangle_t^{\Pi} + \langle A, B\rangle_t^{\Pi}.$$

Let $|\Pi| \to 0$. Then we have $|\langle M, B\rangle_t^{\Pi}| \le (\langle M\rangle_t^{\Pi})^{1/2}(\langle B\rangle_t^{\Pi})^{1/2} \to 0$ and $\langle A, B\rangle_t^{\Pi} \to 0$ since $\langle B\rangle_t^{\Pi} \to 0$. Similarly, $|\langle A, N\rangle_t^{\Pi}| \to 0$ holds. Since $\langle M, N\rangle_t^{\Pi} \to \langle M, N\rangle_t$, (1.44) holds.

Next, note

$$|\langle M, N\rangle_t - \langle M, N\rangle_s| \le (\langle M\rangle_t - \langle M\rangle_s)^{\frac{1}{2}}(\langle N\rangle_t - \langle N\rangle_s)^{\frac{1}{2}}.$$

The right-hand side converges to 0 uniformly a.s. as $t \to s$, if $\langle M\rangle_t$ is continuous in t a.s. Therefore $\langle M, N\rangle_t$ is a continuous process. □

The quadratic covariation of X_t and Y_t is often denoted by $\langle X_t, Y_t\rangle$, $\langle X_t, Y_t\rangle_t$ etc.

Let $[s, T]$ be a sub-interval of $[0, T]$. We shall consider the quadratic covariation of two semi-martingales X_t, Y_t on the interval $[s, T]$. Let $\Pi = \{t_0 < \cdots < t_m\}$ be a partition of $[s, T]$. We set $t_m' = t_{t_m \wedge t}$ and define

$$\langle X, Y\rangle_{s,t}^{\Pi} = \sum_{m=1}^{n} (X_{t_m'} - X_{t_{m-1}'})(Y_{t_m'} - Y_{t_{m-1}'}).$$

Assume the same conditions as in Theorem 1.6.2 for X_t and Y_t. Then, as $|\Pi| \to 0$, $\langle X, Y\rangle_{s,t}^{\Pi}, t \in [s, T]$ converges to a continuous process of bounded variation. We denote the limit by $\langle X, Y\rangle_{s,t}$ and call it the quadratic covariation of X_t and Y_t on the interval $[s, T]$. In particular, if both X_t, Y_t are semi-martingales for $t \in [0, T]$, then it holds that $\langle X, Y\rangle_{s,t} = \langle X, Y\rangle_t - \langle X, Y\rangle_s$ almost surely for any $0 \le s < t \le T$.

Note The quadratic covariation of two L^2 martingales X_t, Y_t was introduced by P.A. Meyer [85]. It is often denoted by $[X, Y]_t$. On the other hand, a bracket process $\langle X, Y\rangle_t$ was introduced by Kunita–Watanabe [66], which is defined as a unique natural (predictable) process of bounded variation such that $X_t Y_t - \langle X, Y\rangle_t$ is a martingale. The existence of such bracket process is due to the Doob–Meyer decomposition theorem of submartinales [84]. If one of X_t and Y_t is continuous, say X_t is continuous, then both of $[X, Y]_t$ and $\langle X, Y\rangle$ are continuous processes and coincide each other. In this monograph, we take $\langle X, Y\rangle_t$ as the definition of the quadratic covariation of X_t and Y_t.

Theorem 1.6.2 is discussed in Kunita [59], Karatzas–Shreve [55] for continuous semi-martingales. We discussed this again, without using the Doob–Meyer decomposition of a sub-martingale. In this monograph, we do not follow difficult discussions concerning natural increasing processes and predictable increasing processes, which appear in the Doob–Meyer decomposition. See [55].

1.7 Markov Processes and Backward Markov Processes

Let S be a Hausdorff topological space with the second countability. Let $\mathcal{B}(S)$ be its topological Borel field. We denote by $B_b(S)$ the set of all real bounded $\mathcal{B}(S)$-measurable functions. $C_b(S)$ is the set of all bounded continuous functions on S. A function $P(x, E)$ of two variables $x \in S, E \in \mathcal{B}(S)$ is called a *kernel* on the space $(S, \mathcal{B}(S))$ if for each $x \in S$, it is a finite measure in E and, for each $E \in \mathcal{B}(S)$, it is a $\mathcal{B}(S)$-measurable function of x. We set for $f \in B_b(S)$

$$Pf(x) = \int_S f(y)P(x, dy). \tag{1.45}$$

It is a $\mathcal{B}(S)$-measurable function. If it is a bounded function for any $f \in B_b(S)$, the kernel is called *bounded*. A bounded kernel P defines a linear transformation from $B_b(S)$ into itself.

Let $\mathbb{T} = [0, \infty)$. Let $\{P_{s,t}(x, E); 0 \leq s < t < \infty\}$ be a family of bounded kernels satisfying the following properties.

1. *Semigroup property*: $P_{s,t}P_{t,u}f = P_{s,u}f$ holds for any $s < t < u$ and $f \in B_b(S)$. The semigroup property is equivalent to
$$\int_S P_{s,t}(x, dy)P_{t,u}(y, E) = P_{s,u}(x, E), \quad \forall x \in S, \ \forall E \in \mathcal{B}(S).$$
The equation is called the *Chapman–Kolmogorov equation*.
2. *Continuity*: $\lim_{t\downarrow s} P_{s,t}f(x) = f(x)$ holds for all x for any $f \in C_b(S)$.

Then $\{P_{s,t}(x, E)\}$ is called a *transition function*. The family of linear transformations $\{P_{s,t}\}$ satisfying the above two properties is called a *semi-group (of linear transformations)*.

If $P_{s,t}(x, S) = 1$ holds for any x and $s < t$, the transition function is called *conservative* and if $P_{s,t}(x, S) \le 1$ holds for all x and $s < t$, the transition function is called *Markovian*. A conservative transition function is called a *transition probability*.

If $P_{s,t}(x, E)$ is Markovian, we adjoin a cemetery ∞ to S as a one point compactification if S is non-compact and as an isolated point if S is compact. We set $S' = S \cup \{\infty\}$ and define a transition probability $P'_{s,t}(x, E)$ on S' by $P'_{s,t}(x, E) = P_{s,t}(x, E)$ if $x \in S$, $E \subset S$ and

$$P'_{s,t}(x, \{\infty\}) = 1 - P_{s,t}(x, S) \quad \text{if } x \in S, \quad = 1 \quad \text{if } x = \infty. \tag{1.46}$$

Then $P'_{s,t}(x, E)$ is a transition probability. We denote by $B_\infty(S)$ the set of $f \in B_b(S')$ such that $f(\infty) = 0$. Further, $C_\infty(S)$ is the set of all $f \in C_b(S')$ such that $f(\infty) = 0$. Then it holds that $P'_{s,t} f(x) = P_{s,t} f(x)$ for any $x \in S$ and $s < t$ if $f \in B_\infty(S)$.

Suppose that we are given a filtration $\{\mathcal{F}_t; t \in \mathbb{T}\}$ of sub σ-fields of $\mathcal{F}$. Let $t_0 \in \mathbb{T}$. Let $X_t, t \in [t_0, \infty)$ be an $\{\mathcal{F}_t\}$-adapted S'-valued stochastic process, continuous in probability. It is called a *Markov process* of initial state (X_{t_0}, t_0) with transition function $\{P_{s,t}(x, \cdot)\}$ (or with semigroup $\{P_{s,t}\}$), if $X_{t_0} \in S$ a.s. and

$$P(X_t \in E | \mathcal{F}_s) = P'_{s,t}(X_s, E), \quad a.s.$$

holds for any Borel subset E of S' and for any $t_0 \le s < t < \infty$. The Markov property is equivalent to that $E[f(X_t)|\mathcal{F}_s] = P_{s,t} f(X_s)$ holds for any $f \in B_\infty(S)$ and $s < t$. For each $\omega \in \Omega$, $X_t(\omega), t \in [t_0, \infty)$ is called a path of the Markov process. If paths of the Markov process are continuous a.s., the process is called a continuous Markov process. If paths are right (or left) continuous a.s., it is called a right continuous (or left continuous, respectively) Markov process. If X_t is an S-valued process, i.e., $P(X_t \in S \text{ for all t}) = 1$, then X_t is called a *conservative* Markov process on S. The Markov process of the initial state $(x, s) \in S \times \mathbb{T}$ is often denoted by $X_t^{x,s}$ and the family $\{X_t^{x,s}\}$ is called the system of Markov processes with transition function $\{P_{s,t}(x, \cdot)\}$.

A transition function is *time homogeneous* if $P_{s+h,t+h}(x, E) = P_{s,t}(x, E)$ holds for any $0 \le s < t < t + h < \infty), h > 0$ and E. If it is time homogeneous, it holds that $P_{s+h,t+h} f = P_{s,t} f$ for any $f \in B_b(S)$. A time homogeneous transition function $\{P_{s,t}(x, E)\}$ is often denoted by $\{P_{t-s}(x, E)\}$ and its semigroup $\{P_{s,t}\}$ is often denoted by $\{P_{t-s}\}$. Then it holds $P_s P_t f = P_{s+t} f$. A Markov process is *time homogeneous* if the associated transition function is time homogeneous.

Let $X_t, t \in [0, \infty)$ be a d'-dimensional Lévy process. Let $\mathcal{F}_t$ be the smallest σ-field containing all null sets of $\mathcal{F}$, with respect to which $X_u, 0 \le u \le t$ are measurable. For a Borel subset E of $\mathbb{R}^{d'}$, we set

$$P_{s,t}(x, E) = P(X_t - X_s + x \in E).$$

Then, since $X_t - X_s$ is independent of $\mathcal{F}_s$, we have

$$P(X_t \in E|\mathcal{F}_s) = P(X_t - X_s + X_s \in E|\mathcal{F}_s)$$
$$= P(X_t - X_s + y \in E)|_{y=X_s} = P_{s,t}(X_s, E).$$

Therefore X_t is a time homogeneous Markov process with transition probability $\{P_{s,t}(x, E)\}$ defined above.

In particular, if $X_t = W_t, t \in [0, \infty)$ is a Wiener process, it is a time homogeneous Markov process with transition probability

$$P_{s,t}(x, E) = \frac{1}{(2\pi(t-s))^{\frac{d}{2}}} \int_E e^{-\frac{1}{2(t-s)}|x-y|^2} dy.$$

If $X_t = N_t$ is a Poisson process, it is a time homogeneous Markov process with state space $S = \{0, 1, 2, \ldots\}$. Its transition probability is given by

$$P_{s,t}(i, i+n) = \frac{(\lambda(t-s))^n}{n!} e^{-\lambda(t-s)}, \quad n = 0, 1, 2, \ldots$$

Jumping times $\tau_n, n = 1, 2, \ldots$ of the Poisson process N_t are stopping times. We show this by induction of n. Suppose that τ_{n-1} is a stopping time. Then

$$\{\tau_n \le t\} = \bigcup_{r \le t} \{N_r > N_{\tau_{n-1}}, \ \tau_{n-1} < t\} = \bigcup_{r \le t} \{N_r > n-1, \ \tau_{n-1} < t\}.$$

Since $\{\tau_{n-1} < t\} \in \mathcal{F}_t$ holds, the above set belongs to $\mathcal{F}_t$. Since this is valid for any t, τ_n is a stopping time.

Let τ be a stopping time with respect to the filtration $\{\mathcal{F}_t\}$. We set

$$\mathcal{F}_\tau = \{A \in \mathcal{F}; \ A \cap \{\tau \le t\} \in \mathcal{F}_t \text{ holds for any } t\}. \tag{1.47}$$

It is a sub σ-field of $\mathcal{F}$. A right continuous Markov process $X_t, t \in [t_0, \infty)$ is said to have the *strong Markov property*, if the equality

$$P(\{X_{\tau+t} \in E\} \cap \{\tau + t < \infty\}|\mathcal{F}_\tau) = P'_{\tau,\tau+t}(X_\tau, E)1_{\tau+t<\infty} \tag{1.48}$$

holds for all $t > 0$, $E \in \mathcal{B}(S)$ for any stopping time τ with $\tau \ge t_0$. The property is equivalent to that the equality $E[f(X_{\tau+t})1_{\tau+t<\infty}|\mathcal{F}_\tau] = P_{\tau,\tau+t}f(X_\tau)1_{\tau+t<\infty}$ holds for any $f \in C_\infty(S)$.

Proposition 1.7.1 *Assume that $P_{s,t}f(x)$ is continuous in s, t, x for any $f \in C_\infty(S)$. Then the right continuous Markov process has the strong Markov property.*

Proof Given a stopping time τ, we define for each positive integer n, a random time τ_n by

$$\tau_n = \frac{m}{2^n}, \quad \text{if } \frac{m-1}{2^n} < \tau \le \frac{m}{2^n}, \quad m = 1, 2, \ldots$$

Then $\{\tau_n, n = 1, 2, \ldots\}$ is a sequence of stopping times decreasing to the stopping time τ. Let $B \in \mathcal{F}_\tau$. Since $\tau \leq \tau_n$, we have $\mathcal{F}_\tau \subset \mathcal{F}_{\tau_n}$. Therefore, for any $B \in \mathcal{F}_\tau$, we have $B \cap \{\tau_n = \frac{m}{2^n}\} \in \mathcal{F}_{\frac{m}{2^n}}$. Then using the Markov property,

$$\begin{aligned}
&E[f(X_{\tau_n+t}); \{\tau_n + t < \infty\} \cap B] \\
&= \sum_m E\Big[f(X_{\frac{m}{2^n}+t}); \{\tau_n = \frac{m}{2^n}\} \cap B\Big] \\
&= \sum_m E\Big[P_{\frac{m}{2^n}, \frac{m}{2^n}+t} f(X_{\frac{m}{2^n}}); \{\tau_n = \frac{m}{2^n}\} \cap B\Big] \\
&= E[P_{\tau_n, \tau_n+t} f(X_{\tau_n}); \{\tau_n + t < \infty\} \cap B].
\end{aligned}$$

Let n tend to ∞. Since $P_{s,t} f(x)$ is continuous in s, t, x, we have

$$E[f(X_{\tau+t}); \{\tau + t < \infty\} \cap B] = E[P_{\tau,\tau+t} f(X_\tau); \{\tau + t < \infty\} \cap B].$$

This proves the strong Markov property. □

Let us consider again a Lévy process. Its semigroup $P_{s,t} f(x)$ is written as

$$P_{s,t} f(x) = \int_{\mathbb{R}^d} f(x + y)\mu_{s,t}(dy),$$

where $\mu_{s,t}(dy) = P(X_t - X_s \in dy)$. Therefore if f is a bounded continuous function of $\mathbb{R}^d$, $P_{s,t} f(x)$ is continuous in s, t, x. Therefore a Lévy process is a strong Markov process.

An S'-valued continuous Markov process with the strong Markov property is called a *diffusion process*.

Let $\{\check{P}_{s,t}(x, E); 0 \leq s < t < \infty\}$ be a family of bounded kernels on S satisfying the following properties.

1. *Backward semigroup property*: $\check{P}_{t,u} \check{P}_{s,t} f = \check{P}_{s,u} f$ holds for any $s < t < u$ and $f \in B_b(S)$,
2. *Continuity*: $\lim_{s \uparrow t} \check{P}_{s,t} f(x) = f(x)$ holds for all x for any $f \in C_b(S)$.

Then $\{\check{P}_{s,t}(x, E)\}$ is called a *backward transition function*. The family of linear transformations $\{\check{P}_{s,t}\}$ satisfying the above two properties is called a *backward semigroup (of linear transformations)*. Backward Markovian transition function and backward transition probability are defined similarly to the forward case. If $\check{P}_{s,t}(x, E)$ is Markovian, we define backward transition probability $\check{P}'_{s,t}(x, E)$ on $S' = S \cup \infty$ similarly to the forward case.

Suppose that we are given a backward filtration $\{\check{\mathcal{F}}_s; s \in \mathbb{T}\}$ of sub σ-fields of $\mathcal{F}$. Let $0 < t_1 < \infty$. Let $\check{X}_s, 0 \leq s \leq t_1$ be an $\{\check{\mathcal{F}}_s\}$-adapted S'-valued caglad stochastic process. It is called a *backward Markov process* on S with backward transition function $\{\check{P}_{s,t}(x, \cdot)\}$, if $P(\check{X}_s \in E|\check{\mathcal{F}}_t) = \check{P}'_{s,t}(\check{X}_t, E)$ holds a.s. for any

Borel subset E of S' and $0 \le s < t \le t_1$. The backward Markov property is equivalent to that $E[f(\check{X}_s)|\check{\mathcal{F}}_t] = \check{P}_{s,t}f(\check{X}_t)$ holds for any $f \in B_\infty(S)$ and $s < t$. A left continuous backward process $\check{X}_s$ is called a backward strong Markov process, if

$$E[f(\check{X}_{\check{\tau}-s})1_{\check{\tau}-s>0}|\check{\mathcal{F}}_{\check{\tau}}] = \check{P}_{\check{\tau}-s,\check{\tau}}f(\check{X}_{\check{\tau}})1_{\check{\tau}-s>0}$$

holds for any backward stopping time $\check{\tau}$ with $\check{\tau} \le t_1$ and any bounded continuous function f on S. If the backward semigroup $\check{P}_{s,t}$ maps $C_b(S)$ into itself and is continuous in s, t, then the left continuous backward Markov process is a backward strong Markov process.

1.8 Kolmogorov's Criterion for the Continuity of Random Field

We shall introduce Kolmogorov's criterion for a given random field to have a modification of a continuous random field. Let $\{X(x), x \in \mathbb{D}\}$ be a random field with values in a Banach space S with the norm $\| \ \|$, where $\mathbb{D}$ is a bounded domain of $\mathbb{R}^d$. Elements of $\mathbb{D}$ are denoted by $x = (x_1, \ldots, x_d)$, $y = (y_1, \ldots, y_d)$ etc. The distance of x, y is defined by $|x - y| = \max_{1 \le i \le d} |x_i - y_i|$.

The next theorem was shown first in the case $d = 1$ by Kolmogorov and Chentzov [20] and then it was extended to the case $d > 1$ by Totoki [114].

Theorem 1.8.1 (Kolmogorov–Totoki) *Let $\{X(x), x \in \mathbb{D}\}$ be a random field with values in a separable Banach space S with norm $\| \ \|$, where $\mathbb{D}$ is a bounded domain in $\mathbb{R}^d$. Assume that there exist positive constants γ, C and $\alpha > d$ satisfying*

$$E[\|X(x) - X(y)\|^\gamma] \le C|x - y|^\alpha, \quad \forall x, y \in \mathbb{D}. \tag{1.49}$$

Then there exists a continuous random field $\{\tilde{X}(x), x \in \bar{\mathbb{D}}\}$ such that $\tilde{X}(x) = X(x)$ holds a.s. for any $x \in \mathbb{D}$, where $\bar{\mathbb{D}}$ is the closure of $\mathbb{D}$.

Further, if $0 \in \mathbb{D}$ and $E[\|X(0)\|^\gamma] < \infty$, we have $E[\sup_{x\in\mathbb{D}} \|\tilde{X}(x)\|^\gamma] < \infty$.

By a linear invertible transformation, the bounded domain $\mathbb{D}$ is transformed to a bounded domain included in the cube $[0, 1]^d$. So we will discuss the case where $\mathbb{D}$ is included in the cube $[0, 1]^d$.

A real number $0 \le x \le 1$ has a dyadic expansion $x = \sum_{k=1}^\infty a_k 2^{-k}$, where a_k are 0 or 1. We set $x^n = \sum_{k=1}^n a_k 2^{-k}$ and call it the dyadic rational of length n defined from x. Let $x = (x_1, \ldots, x_d)$ be an element of $[0, 1]^d$. For x, we set $x^n = (x_1^n, \ldots, x_d^n)$. If $x = x^n$ holds for some positive integer n, x is called a dyadic rational of length n. We denote by Δ_n the set of all dyadic rationals with length n in $[0, 1]^d$ and we set $\Delta := \bigcup_{n=1}^\infty \Delta_n$. It is a dense subset of $[0, 1]^d$.

Let β be a positive number less than or equal to 1. Given a map $f : \mathbb{D} \to S$, we define, for each positive integer n, the variation of continuity and the variation of β-Hölder continuity of f by

$$\Delta_n(f) = \max_{x,y \in \Delta_n \cap \mathbb{D}, |x-y|=2^{-n}} \|f(x) - f(y)\|,$$

$$\Delta_n^\beta(f) = 2^{\beta n} \Delta_n(f).$$

Lemma 1.8.1 *Suppose that a map $f; \mathbb{D} \to S$ satisfies $\sum_{m=1}^\infty \Delta_m^\beta(f) < \infty$. Then the inequality*

$$\|f(x) - f(y)\| \leq 3\Big(\sum_{m=1}^\infty \Delta_m^\beta(f)\Big)|x - y|^\beta \tag{1.50}$$

holds for any $x, y \in \mathbb{D} \cap \Delta$. Further, the map $f; \mathbb{D} \cap \Delta \to S$ is extended to a β-Hölder continuous map $\tilde{f}; \bar{\mathbb{D}} \to S$ such that $\tilde{f}(x) = f(x)$ holds for $x \in \mathbb{D} \cap \Delta$.

Proof For a given map $f(x), x \in \mathbb{D}$ and a positive integer n, define a step function $f_n(x), x \in \bar{\mathbb{D}}$ by $f_n(x) = f(x^n)$ if $x^n \in \mathbb{D}$ and $f_n(x) = f(x)$ if $x^n \notin \mathbb{D}$, where x^n is the dyadic rational of length n defined from x. Then it holds that

$$\|f_n(x) - f_{n-1}(x)\| = \|f(x^n) - f(x^{n-1})\| \leq \Delta_n(f) \leq \Delta_n^\beta(f)$$

for any $x \in \mathbb{D}$. Therefore if $n > k$ we have

$$\|f_n(x) - f_k(x)\| \leq \sum_{m=k+1}^n \|f_m(x) - f_{m-1}(x)\| \leq \sum_{m=k+1}^n \Delta_m^\beta(f)$$

for any x. Hence the sequence of functions $\{f_n\}$ converges uniformly. Denote the limit function by $\tilde{f}$. Then it holds that $\tilde{f}(x) = f(x)$ for any $x \in \Delta = \bigcup_n \Delta_n$.

Now take any two points $x, y \in \Delta$. There exists $k \in \mathbb{N}$ such that $2^{-(k+1)} \leq |x - y| < 2^{-k}$. Further, there exists $n \in \mathbb{N}$ such that $n > k$ and $x = x^n, y = y^n \in \Delta_n$. Then we have

$$\begin{aligned}\|\tilde{f}(x) - f_k(x)\| &= \|f(x^n) - f(x^k)\| \leq \sum_{m=k+1}^n \|f(x^m) - f(x^{m-1})\| \\ &\leq \sum_{m=k+1}^n \Delta_m(f) \leq \sum_{m=k+1}^n 2^{-\beta m} \Delta_m^\beta(f) \\ &\leq \Big(\sum_{m=1}^\infty \Delta_m^\beta(f)\Big)|x - y|^\beta.\end{aligned}$$

$\|\tilde{f}(y) - f_k(y)\|$ is estimated in the same way. Since $\|f_k(x) - f_k(y)\| \le \Delta_k(f)$, we get

$$\|\tilde{f}(x) - \tilde{f}(y)\| \le \|\tilde{f}(x) - f_k(x)\| + \|f_k(x) - f_k(y)\| + \|f_k(y) - \tilde{f}(y)\|$$
$$\le 3\Big(\sum_{m=1}^{\infty} \Delta_m^{\beta}(f)\Big)|x-y|^{\beta}.$$

Therefore f satisfies the inequality (1.50) for any $x, y \in \Delta$. Further $\tilde{f}$ is β-Hölder continuous by the above inequality. Since $\tilde{f}(x) = f(x)$ holds for $x \in \mathbb{D} \cap \Delta$, $\tilde{f}(x), x \in \bar{\mathbb{D}}$ is a β-Hölder continuous extension of $f(x), x \in \mathbb{D} \cap \Delta$. Hence we get the assertion of the lemma. □

We shall apply the above lemma to the random field $X(x)$. Observe that for each ω, $X(\cdot, \omega)$ restricting x to $\mathbb{D} \cap \Delta$ can be regarded as a map form $\mathbb{D} \cap \Delta$ to S. Then we have

$$\|X(x,\omega) - X(y,\omega)\| \le 3\left(\sum_{n=1}^{\infty} \Delta_n^{\beta}(X(\omega))\right)|x-y|^{\beta}, \tag{1.51}$$

for any $x, y \in \mathbb{D} \cap \Delta$.

Lemma 1.8.2 *Let β be a positive number satisfying $\beta\gamma < \alpha - d$. Then,*

$$E\Big[\Big(\sum_{n=1}^{\infty} \Delta_n^{\beta}(X)\Big)^{\gamma}\Big]^{\frac{1}{\gamma}} \le \left(\sum_{n=1}^{\infty} (2^{-\frac{\alpha-d-\beta\gamma}{\gamma}})^n\right) \cdot (2^d C)^{\frac{1}{\gamma}} < \infty, \tag{1.52}$$

where C is the positive constant in the inequality (1.49).

Proof We will consider the case $\gamma \ge 1$ only. Observe the inequality

$$\Delta_n^{\beta}(X)^{\gamma} \le \Big(\sup_{x,y\in\Pi_n\cap\mathbb{D}, |x-y|=\frac{1}{2^n}} \|X(x) - X(y)\| 2^{n\beta}\Big)^{\gamma}$$
$$\le \sum (\|X(x) - X(y)\| 2^{n\beta})^{\gamma},$$

where the summation is taken for all $x, y \in \mathbb{D} \cap \Delta_n$ such that $|x - y| = \frac{1}{2^n}$. Then the number of summations is at most $2^{(n+1)d}$. Therefore

$$E[\Delta_n^{\beta}(X)^{\gamma}] \le 2^{(n+1)d+n\beta\gamma} E[\|X(x) - X(y)\|^{\gamma}]$$
$$\le 2^{n(d+\beta\gamma-\alpha)} 2^d C.$$

In the last inequality, we applied the inequality (1.49). Therefore we get

$$E\Big[\Big(\sum_{n=1}^{\infty}\Delta_n^\beta(X)\Big)^\gamma\Big]^{\frac{1}{\gamma}} \le \sum_{n=1}^{\infty}E[\Delta_n^\beta(X)^\gamma]^{\frac{1}{\gamma}}$$

$$\le \Big(\sum_{n=1}^{\infty}2^{-n\frac{\alpha-d-\beta\gamma}{\gamma}}\Big)\cdot(2^dC)^{\frac{1}{\gamma}} < \infty. \qquad (1.53)$$

□

Proof of Theorem 1.8.1 The random field $X(x)$ restricting x on $\mathbb{D}\cap\Delta$ satisfies the inequality (1.50), where $\sum_{n=1}^{\infty}\Delta_n^\beta(X) < \infty$ a.s. Therefore $\{X(x), x \in \mathbb{D}\cap\Delta\}$ is uniformly β-Hölder continuous a.s. Then there exists a uniformly continuous random field $\{\tilde{X}(x), x \in \bar{\mathbb{D}}\}$ such that $X(x) = \tilde{X}(x)$ holds a.s. for any $x \in \mathbb{D}\cap\Delta$. Since $X(x)$ is continuous in probability, the equality $X(x) = \tilde{X}(x)$ holds a.s for any $x \in \mathbb{D}$. Consequently, the random field $\{\tilde{X}(x), x \in \bar{\mathbb{D}}\}$ is a continuous modification of the random field $\{X(x), x \in \mathbb{D}\}$.

Now we have the inequality

$$\sup_{x\in\mathbb{D}}\|\tilde{X}\| \le \|X(0)\| + 3(\sum_n \Delta_n^\beta(X))$$

by (1.51). Therefore we have $E[\sup_{x\in\mathbb{D}}\|\tilde{X}(x)\|^\gamma] < \infty$. □

Chapter 2
Stochastic Integrals

Abstract We discuss Itô's stochastic calculus, which will be applied in later discussions. In Sects. 2.1, 2.2, 2.3, and 2.4, we discuss stochastic calculus related to integrals by Wiener processes and continuous martingales. In Sect. 2.1, we define stochastic integrals based on Wiener processes and continuous martingales. In Sect. 2.2, we establish Itô's formula. It will be applied for proving L^p-estimates of stochastic integrals, called the Burkholder–Davis–Gundy inequality, and Girsanov's theorem. The smoothness of the stochastic integral with respect to parameter will be discussed in Sect. 2.3. Fisk–Stratonovitch symmetric integrals will be discussed in Sect. 2.4.

In Sects. 2.5 and 2.6, we discuss stochastic calculus based on Poisson random measures. Stochastic integrals are defined in Sect. 2.5. The chain rules formula for jump processes and L^p-estimates of jump integrals will be discussed in Sect. 2.6. In Sect. 2.7, we discuss the backward processes and backward integrals. These topics are related to dual processes or inverse processes, which will be discussed in Chaps. 3 and 4.

2.1 Itô's Stochastic Integrals by Continuous Martingale and Wiener Process

Let $(\Omega, \mathcal{F}, P)$ be a probability space equipped with a filtration $\{\mathcal{F}_t\}$ of sub-σ-fields of $\mathcal{F}$. In this chapter, we will restrict the time parameter to a finite interval $\mathbb{T} = [0, T]$. Let $X_t, t \in \mathbb{T}$ be a continuous local martingale with respect to the filtration $\{\mathcal{F}_t\}$ and let $\phi(r), r \in \mathbb{T}$ be a real-valued process. Let $0 \leq s < t \leq T$ be a fixed time. We want to define the stochastic integral of $\phi(r)$ based on dX_r, written as $\int_s^t \phi(r)\, dX_r$. It is not a usual Lebesgue integral, since X_t is not a process of bounded variation. At the beginning, we will introduce the class of integrand $\phi(r)$, for which the integrals can be defined.

The *predictable* σ-field $\mathcal{P}$ (with respect to the filtration $\{\mathcal{F}_t\}$) is the σ-field on $\mathbb{T} \times \Omega$ generated by left continuous $\{\mathcal{F}_t\}$-adapted processes. A $\mathcal{P}$-measurable stochastic process $\varphi(t), t \in [0, T]$ is called a *predictable process* (with respect to the filtration $\{\mathcal{F}_t\}$).

H. Kunita, *Stochastic Flows and Jump-Diffusions*, Probability Theory and Stochastic Modelling 92, https://doi.org/10.1007/978-981-13-3801-4_2

Let $X_t, t \in \mathbb{T}$ be a continuous L^2-martingale adapted to the filtration $\{\mathcal{F}_t\}$ and let $\langle X\rangle_t, t \in \mathbb{T}$ be its quadratic variation. It is a unique continuous increasing process such that $X_t^2 - \langle X\rangle_t, t \in \mathbb{T}$ is a martingale (Sect. 1.6). Let $\phi(r), r \in \mathbb{T}$ be a predictable process. We define its norm by

$$\|\phi\| = E\Big[\int_0^T |\phi(r)|^2\, d\langle X\rangle_r\Big]^{\frac{1}{2}}.$$

A stochastic process $\phi(r), r \in \mathbb{T}$ is called a *simple predictable process* if it is written as $\phi(r) = \sum_{m=0}^n \phi_m 1_{(t_{m-1},t_m]}(r)$, where $0 = t_0 < t_1 < \cdots < t_n = T$ and ϕ_m are bounded and $\mathcal{F}_{t_{m-1}}$-measurable.

We shall define the stochastic integral based on a continuous L^2-martingale $X_t, t \in \mathbb{T}$. Let $\phi(r)$ be a simple predictable process. We define

$$M_t = \sum_{m=1}^n \phi_m (X_{t_m\wedge t} - X_{t_{m-1}\wedge t}), \quad t \in \mathbb{T}$$

and call it the *stochastic integral* of $\phi(r)$ by dX_r and denote it by $\int_0^t \phi(r)\, dX_r$. Then, $M_t, t \in \mathbb{T}$ may be regarded as a martingale transform with discrete time $\Pi = \{0 = t_0 < \cdots < t_n = T\}$. Therefore we have the equality $E[M_t - M_r|\mathcal{F}_r] = 0$ a.s. for $r \in \Pi$. Let $0 \le s < t \le T$. We may assume that s, t are adjoined in the partition Π. Then, using Propositions 1.5.1 and 1.6.1, we have

$$\begin{aligned}
E[(M_t - M_s)^2|\mathcal{F}_s] &= \sum_{m;t_{m-1}\ge s} E[(M_{t_m} - M_{t_{m-1}})^2|\mathcal{F}_s] \\
&= E[\sum_m \phi_m^2 (X_{t_m} - X_{t_{m-1}})^2|\mathcal{F}_s] \\
&= \sum_m E[\phi_m^2 E[(X_{t_m} - X_{t_{m-1}})^2|\mathcal{F}_{t_{m-1}}]|\mathcal{F}_s] \\
&= \sum_m E[\phi_m^2 E[\langle X\rangle_{t_m} - \langle X\rangle_{t_{m-1}}|\mathcal{F}_{t_{m-1}}]|\mathcal{F}_s] \\
&= E[\sum_m \phi_m^2 (\langle X\rangle_{t_m} - \langle X\rangle_{t_{m-1}})|\mathcal{F}_s] \\
&= E\Big[\int_s^t \phi(r)^2\, d\langle X\rangle_r \Big| \mathcal{F}_s\Big], \quad a.s.
\end{aligned}$$

Since the above holds for any $s < t$, $M_t^2 - \int_0^t \phi(r)^2\, d\langle X\rangle_r, t \in \mathbb{T}$ is a martingale. Then the quadratic variation of $M_t, t \in \mathbb{T}$ is given by $\int_0^t \phi(r)^2\, d\langle X\rangle_r, t \in \mathbb{T}$ in view of Corollary 1.6.1. We have thus shown

$$\Big\langle \int_0^t \phi(r)\,dX_r \Big\rangle = \int_0^t |\phi(r)|^2\,d\langle X\rangle_r, \quad a.s. \tag{2.1}$$

for any $0 < t \le T$. Taking expectations of both sides, we get

$$E\Big[\Big|\int_0^t \phi(r)\,dX_r\Big|^2\Big] = E\Big[\int_0^t |\phi(r)|^2\,d\langle X\rangle_r\Big] \le \|\phi\|^2$$

for any t. Now let ϕ be a predictable process such that $\|\phi\| < \infty$. Then there exists a sequence of simple predictable processes $\{\phi_n\}$ such that $\|\phi - \phi_n\| \to 0$ as $n \to \infty$. Then, using Doob's inequality for a martingale (Theorem 1.5.1), we have

$$\begin{aligned} E\Big[\sup_{t\in\mathbb{T}}\Big|\int_0^t \phi_n(r)\,dX_r - \int_0^t \phi_{n'}(r)\,dX_r\Big|^2\Big] &\le 4E\Big[\Big|\int_0^T (\phi_n(r) - \phi_{n'}(r))\,dX_r\Big|^2\Big] \\ &= 4\|\phi_n - \phi_{n'}\|^2 \to 0, \end{aligned}$$

as $n, n' \to \infty$. Hence $\lim_{n\to\infty} \int_0^t \phi_n(r)\,dX_r$ exists uniformly in t. We denote it by $\int_0^t \phi(r)\,dX_r$ and call it the *stochastic integral* or *Itô integral* of $\phi(r)$ by dX_r. The stochastic integral is a continuous L^2-martingale. It satisfies (2.1).

We will extend the stochastic integral for a continuous local martingale $X_t, t \in \mathbb{T}$. Let $\langle X\rangle_t, t \in \mathbb{T}$ be its quadratic variation. Let $L(\langle X\rangle)$ be the set of all predictable processes $\phi(t), t \in \mathbb{T}$ satisfying $\int_0^T |\phi(r)|^2\,d\langle X\rangle_r < \infty$ a.s. Let $\phi \in L(\langle X\rangle)$. For $N > 0$, define a stopping time by

$$\begin{aligned} \tau_N &= \inf\{t \in \mathbb{T};\ |X_t| + \int_0^t |\phi(r)|^2\,d\langle X\rangle_r > N\}, \\ &= \infty, \quad \text{if the set } \{\cdots\} \text{ is empty.} \end{aligned}$$

Then the stopped process $X_t^{(N)} = X_{t\wedge\tau_N}, t \in \mathbb{T}$ is a bounded continuous martingale. See the proof of Theorem 1.6.1. It holds that $\langle X^{(N)}\rangle_t = \langle X\rangle_{t\wedge\tau_N}$. Since $E[\int_0^t |\phi(r)|^2\,d\langle X^{(N)}\rangle_r] < \infty$, the family of stochastic integral $M_t^{(N)} \equiv \int_0^t \phi(r)\,dX_r^{(N)}, t \in \mathbb{T}$ is well defined as an L^2-martingale. Let $N' > N$ and consider $M_t^{(N')} \equiv \int_0^t \phi(r)\,dX_r^{(N')}$. Then it holds that $M_{t\wedge\tau_N}^{(N')} = M_t^{(N)}$. Then there exists a continuous local martingale $M_t, t \in \mathbb{T}$ such that $M_{t\wedge\tau_N} = \int_0^t \phi(r)\,dX_r^{(N)}$ holds for any $N = 1, 2, \ldots$. We set $M_t = \int_0^t \phi(r)\,dX_r$ and call it the stochastic integral or Itô integral of $\phi(r)$ by dX_r. It holds that $\langle \int_0^t \phi(r)\,dX_r\rangle = \int_0^t \phi(r)^2\,d\langle X\rangle_r$ a.s. for any $0 < t \le T$.

Let $X_t, t \in \mathbb{T}$ and $Y_t, t \in \mathbb{T}$ be continuous local martingales and let $\phi \in L(\langle X\rangle)$ and $\varphi \in L(\langle Y\rangle)$. In view of (2.1), we have the relation

$$\Big\langle \int_0^t \phi\,dX, \int_0^t \varphi\,dY \Big\rangle = \int_0^t \phi(r)\varphi(r)\,d\langle X, Y\rangle_r, \quad a.s. \tag{2.2}$$

for any $0 < t \le T$.

Let $\phi_1, \phi_2 \in L(\langle X \rangle)$ and let c_1, c_2 be constants. Then we have

$$\int_0^t (c_1\phi_1 + c_2\phi_2)(r)\, dX_r = c_1 \int_0^t \phi_1(r)\, dX_r + c_2 \int_0^t \phi_2(r)\, dX_r, \quad a.s.$$

for any $0 < t \le T$. Indeed, the equality holds if both ϕ_1, ϕ_2 are simple predictable processes. Then the equality can be extended to arbitrary predictable processes ϕ_1, ϕ_2 belonging to $L(\langle X \rangle)$.

For $0 \le s < t \le T$, the integral $\int_s^t \phi(r)\, dX_r$ is defined as

$$\int_s^t \phi(r)\, dX_r := \int_0^t \phi(r) 1_{[s,T]}(r)\, dX_r = \int_0^t \phi(r)\, dX_r - \int_0^s \phi(r)\, dX_r, \quad a.s.$$

Let $X_t, t \in \mathbb{T}$ be a continuous semi-martingale decomposed as $X_t = M_t + A_t$. Let $|A|_t$ be the process of the total variation of A_t. Let $\phi(r)$ be a predictable process satisfying $\int_0^T |\phi(r)|^2\, d\langle X \rangle_r < \infty$ and $\int_0^T |\phi(r)|\, d|A|_r < \infty$, a.s. We define the integral

$$\int_0^t \phi(r)\, dX_r := \int_0^t \phi(r)\, dM_r + \int_0^t \phi(r)\, dA_r,$$

where $\int_0^t \phi(r, \omega)\, dA(r, \omega)$ is the Lebesgue–Stieltjes integrals for each ω.

Now, we shall consider stochastic integrals by a Wiener process. Let d' be a positive integer. Let $W_t = (W_t^1, \ldots, W_t^{d'}), t \in [0, \infty)$ be a d'-dimensional Wiener process. Suppose that W_t is adapted to a given filtration $\{\mathcal{F}_t\}$ and that $W_u - W_t$ is independent of $\mathcal{F}_t$ for any $u > t$. Then $W_t, t \in [0, \infty)$ is called an $\{\mathcal{F}_t\}$-Wiener process. Components $W_t^k, k = 1, \ldots, d'$ of the $\{\mathcal{F}_t\}$-Wiener process W_t are L^2-martingales with respect to $\{\mathcal{F}_t\}$. Indeed, since $W_t^k - W_r^k$ are independent of $\mathcal{F}_r$, we have $E[W_t^k - W_r^k | \mathcal{F}_r] = E[W_t^k - W_r^k] = 0$ a.s. Further,

$$\begin{aligned} &E[(W_t^k - W_r^k)(W_t^l - W_r^l) | \mathcal{F}_r] \\ &= E[(W_t^k - W_r^k)(W_t^l - W_r^l)] = \delta_{kl}(t - r), \quad a.s. \end{aligned}$$

Therefore we have $\langle W^k, W^l \rangle_t = \delta_{kl}(t - 0)$. We define a space of d'-dimensional predictable processes by

$$\mathbf{L}_{\mathbb{T}} = \Big\{ \phi(r) = (\phi^1(r), \ldots, \phi^{d'}(r));\ \text{predictable and } \int_0^T |\phi(r)|^2\, dr < \infty \Big\},$$

where $|\phi(r)|^2 = \sum_{k=1}^{d'} |\phi^k(r)|^2$. Then for any $\phi(r) \in \mathbf{L}_{\mathbb{T}}$, integrals $\int_0^t \phi^k(r) dW_r^k$, $k = 1, \ldots, d'$ are well defined for any $0 < t \le T$. These are local martingales and satisfy

$$\left\langle \int \phi^k \, dW^k, \int \phi^l \, dW^l \right\rangle_t = \delta_{kl} \int_0^t \phi^k(r)\phi^l(r)\, dr, \quad a.s. \tag{2.3}$$

We set

$$I_t(\phi) = \int_0^t (\phi(r), dW_r) = \sum_{k=1}^{d'} \int_0^t \phi^k(r)\, dW_r^k. \tag{2.4}$$

We next consider the stochastic integral where the integrand depends on parameter. Let Λ be a parameter set. We assume that Λ is an e-dimensional Euclidean space $\mathbb{R}^e$ or its unit ball $\{\lambda \in \mathbb{R}^e; |\lambda| < 1\}$. We denote by $\mathbf{L}_{\mathbb{T}}(\Lambda)$ the set of all d'-dimensional measurable random fields $\phi_\lambda(t)$, $(t, \lambda) \in \mathbb{T} \times \Lambda$ such that for any λ, $\phi_\lambda(t)$ is predictable and satisfies $\int_\Lambda \int_{\mathbb{T}} |\phi_\lambda(r)|^2 \, dr \, d\lambda < \infty$ a.s. For $\phi_\lambda(t) \in \mathbf{L}_{\mathbb{T}}(\Lambda)$, stochastic integrals $\int_0^t (\phi_\lambda(r), dW_r)$ are well defined as continuous martingales for almost all λ. Further, the family of integrals has a modification such that it is (t, λ)-measurable. We give a Fubini theorem for changing the order of integrals by dW_r and $d\lambda$.

Proposition 2.1.1 *Let $\phi_\lambda(r) \in \mathbf{L}_{\mathbb{T}}(\Lambda)$ and $g(\lambda), \lambda \in \Lambda$ be a bounded measurable function of compact supports. Then we have for any $t \in \mathbb{T}$,*

$$\int_\Lambda \left(\int_0^t (\phi_\lambda(r), dW_r) \right) g(\lambda)\, d\lambda = \int_0^t \left(\int_\Lambda \phi_\lambda(r) g(\lambda)\, d\lambda, dW_r \right), \quad a.s. \tag{2.5}$$

Proof If $\phi_\lambda(t)$ is a simple functional written as $\sum_m \phi_\lambda(m) 1_{(t_{m-1}, t_m]}(t)$ with bounded $\mathcal{F}_{t_{m-1}} \times \mathcal{B}(\Lambda)$-measurable functional $\phi_\lambda(m)$, the equality (2.5) can be shown directly. Since $\phi_\lambda(t) \in \mathbf{L}_{\mathbb{T}}(\Lambda)$ can be approximated by a sequence of the above simple functionals, equality (2.5) is valid for any $\phi_\lambda(t) \in \mathbf{L}_{\mathbb{T}}(\Lambda)$. □

2.2 Itô's Formula and Applications

A d-dimensional continuous stochastic process $X_t = (X_t^1, \ldots, X_t^d), t \in \mathbb{T}$ is called a d-dimensional continuous semi-martingale, if all $X_t^i, t \in \mathbb{T}; i = 1, \ldots, d$ are continuous semi-martingales. Let $f(x_1, \ldots, x_d, t)$ be a smooth function. We consider the differential rule of the composite functional $f(X_t, t)$ with respect to t. The rule is called the *formula of the change of variables* or *Itô's formula*.

Theorem 2.2.1 (Itô's formula) *Let $f(x, t)$ be a $C^{2,1}$-function on $\mathbb{R}^d \times \mathbb{T}$ and let $X_t, t \in \mathbb{T}$ be a d-dimensional continuous semi-martingale. Then $f(X_t, t), t \in \mathbb{T}$ is a continuous semi-martingale. Further, we have*

$$f(X_t, t) = f(X_s, s) + \int_s^t \frac{\partial f}{\partial t}(X_r, r)\,dr \tag{2.6}$$
$$+\sum_{i=1}^{d} \int_s^t \frac{\partial f}{\partial x_i}(X_r, r)\,dX_r^i + \frac{1}{2}\sum_{i,j=1}^{d} \int_s^t \frac{\partial^2 f}{\partial x_i \partial x_j}(X_r, r)\,d\langle X^i, X^j\rangle_r,$$

a.s. for any $0 \le s < t \le T$.

Proof It is sufficient to prove the case $s = 0$. Let $\Pi_n = \{t_0 < t_1 < \cdots < t_n\}$ be a partition of $[0, t]$. Then we have by the Taylor expansion of $f(x, t)$,

$$\begin{aligned}
&f(X_t, t) - f(X_0, 0) \\
&= \sum_{m=1}^{n} \Big\{ f(X_{t_m}, t_m) - f(X_{t_{m-1}}, t_{m-1}) \Big\} \\
&= \sum_{m} \Big\{ f(X_{t_m}, t_m) - f(X_{t_m}, t_{m-1}) \Big\} \\
&\quad + \sum_{i=1}^{d} \Big\{ \sum_{m} \frac{\partial f}{\partial x_i}(X_{t_{m-1}}, t_{m-1})(X_{t_m}^i - X_{t_{m-1}}^i) \Big\} \\
&\quad + \frac{1}{2}\sum_{i,j=1}^{d} \Big\{ \sum_{m} \frac{\partial^2 f}{\partial x_i \partial x_j}(\xi_m, t_{m-1})(X_{t_m}^i - X_{t_{m-1}}^i)(X_{t_m}^j - X_{t_{m-1}}^j) \Big\} \\
&= I_1 + I_2 + I_3,
\end{aligned}$$

where ξ_m are random variables such that $|\xi_m - X_{t_{m-1}}| \le |X_{t_m} - X_{t_{m-1}}|$. Let $|\Pi_n|$ tend to 0. Then we get

$$\lim_{n\to\infty} I_1 = \int_0^t \frac{\partial f}{\partial t}(X_r, r)\,dr, \quad a.s.$$
$$\lim_{n\to\infty} I_2 = \sum_i \int_0^t \frac{\partial f}{\partial x_i}(X_r, r)\,dX_r^i, \quad a.s.$$

immediately from the definition of the stochastic integral. Further, applying Theorem 1.6.2, we get

$$\lim_{n\to\infty} I_3 = \frac{1}{2}\sum_{i,j} \int_0^t \frac{\partial^2 f}{\partial x_i \partial x_j}(X_r, r)\,d\langle X^i, X^j\rangle_r, \quad a.s.$$

Summing up these computations, we get the formula of the theorem. □

We set

$$L_{\mathbb{T}} = \Big\{ v(r);\ \text{predictable and} \int_0^T |v(r)|\, dr < \infty \Big\}.$$

Let $\phi(r) = (\phi^{ik}(r))$ be a $d \times d'$-dimensional process such that $(\phi^{i\cdot}(r)) \in \mathbf{L}_{\mathbb{T}}$ for any i and let $v(r) = (v^1(r), \ldots, v^d(r))$ be a d-dimensional process such that $v^i(r) \in L_{\mathbb{T}}$ for any i. We consider a d-dimensional semi-martingale represented by

$$X_t = X_0 + \int_0^t \phi(r)\, dW_r + \int_0^t v(r)\, dr, \tag{2.7}$$

where

$$\int_0^t \phi(r)\, dW_r = \Big(\sum_{k=1}^{d'} \int_0^t \phi^{ik}(r)\, dW_r^k, i = 1, \ldots, d \Big).$$

Corollary 2.2.1 *Let $X_t = (X_t^1, \ldots, X_t^d)$ be a continuous semi-martingale represented by* (2.7). *Then the formula* (2.6) *is rewritten as*

$$\begin{aligned} f(X_t, t) = f(X_s, s) &+ \int_s^t \frac{\partial f}{\partial t}(X_r, r)\, dr \\ &+ \sum_{i=1}^d \sum_{k=1}^{d'} \int_s^t \frac{\partial f}{\partial x_i}(X_r, r) \phi^{ik}(r)\, dW_r^k + \sum_{i=1}^d \int_s^t \frac{\partial f}{\partial x_i}(X_r, r) v^i(r)\, dr \\ &+ \frac{1}{2} \sum_{i,j=1}^d \int_s^t \frac{\partial^2 f}{\partial x_i \partial x_j}(X_r, r) (\sum_{k=1}^{d'} \phi^{ik}(r) \phi^{jk}(r))\, dr. \end{aligned} \tag{2.8}$$

Proof We have

$$\langle X^i, X^j \rangle_t = \sum_{k,l} \Big\langle \int \phi^{ik}(r)\, dW_r^k, \int \phi^{jl}(r)\, dW_r^l \Big\rangle_t = \sum_k \int_0^t \phi^{ik}(r) \phi^{jk}(r)\, dr$$

by (2.3). Therefore (2.6) implies (2.8). □

We will discuss three problems by applying the above Itô's formula. We first give Lévy's characterization of a Wiener process.

Proposition 2.2.1 *Let $X_t^1, \ldots, X_t^d$ be continuous $\{\mathcal{F}_t\}$-martingales satisfying*

$$\langle X^i, X^j \rangle_t = \delta_{ij} t, \quad i, j = 1, \ldots, d, \quad a.s.$$

for any $0 < t < T$. *Then* $X_t = (X_t^1, \ldots, X_t^d), t \in \mathbb{T}$ *is a* d*-dimensional* $\{\mathcal{F}_t\}$*-Wiener process.*

Proof We apply Itô's formula (2.6) to the function $f(x,t) = e^{\sqrt{-1}(v,x)}$. We have

$$e^{\sqrt{-1}(v,X_t-X_s)}$$

$$= 1 + \sqrt{-1}\sum_j \int_s^t e^{\sqrt{-1}(v,X_r-X_s)} v_j \, dX_r^j - \frac{1}{2}|v|^2 \int_s^t e^{\sqrt{-1}(v,X_r-X_s)} \, dr.$$

Let $A \in \mathcal{F}_s$. Multiply the functional 1_A to each term of the above equation and then take the expectation. Since $E\Big[\int_s^t e^{\sqrt{-1}(v,X_r-X_s)} v_j \, dX_r^j \Big| \mathcal{F}_s\Big] = 0$ holds, we have

$$E[e^{\sqrt{-1}(v,X_t-X_s)} 1_A] = P(A) - \frac{1}{2}|v|^2 \int_s^t E[e^{\sqrt{-1}(v,X_r-X_s)} 1_A] \, dr.$$

Differentiating the above with respect to t, $\varphi_t = E[e^{\sqrt{-1}(v,X_t-X_s)} 1_A]$ satisfies the differential equation, $\frac{d\varphi_t}{dt} = -\frac{1}{2}|v|^2 \varphi_t$. Integrating it we obtain $\varphi_t = e^{-\frac{1}{2}|v|^2(t-s)} P(A)$, or equivalently,

$$E[e^{\sqrt{-1}(v,X_t-X_s)} 1_A] = e^{-\frac{1}{2}|v|^2(t-s)} P(A).$$

Consequently, the law of $X_t - X_s$ is Gaussian with mean 0 and covariance $(t-s)I$. Further, $X_t - X_s$ is independent of $\mathcal{F}_s$. This proves that X_t is an $\{\mathcal{F}_t\}$-Wiener process. □

We next study L^p-estimations of stochastic integrals. Let $p \geq 2$. We set

$$\mathbf{L}_{\mathbb{T}}^p = \Big\{\phi \in \mathbf{L}_{\mathbb{T}};\ \|\phi\|_p < \infty\Big\}, \quad L_{\mathbb{T}}^p = \Big\{\phi \in L_{\mathbb{T}};\ \|\phi\|_p < \infty\Big\},$$

where $\|\phi\|_p = E\big[\int_0^T |\phi(r)|^p \, dr\big]^{\frac{1}{p}}$.

Proposition 2.2.2 (Burkholder–Davis–Gundy inequality) *Let* $p \geq 2$. *Suppose that* $\phi(r) \in \mathbf{L}_{\mathbb{T}}^p$. *Then the integral* $I_t(\phi) = \int_0^t (\phi(r), dW_r)$ *is* p*-th integrable. There exists a positive constant* C_p *such that for any* $0 < t \leq T$, *the inequality*

$$E[|I_t(\phi)|^p]^{\frac{1}{p}} \leq C_p t^{\frac{1}{2}-\frac{1}{p}} \|\phi\|_p \tag{2.9}$$

holds for all $\phi \in \mathbf{L}_{\mathbb{T}}^p$.

Proof We shall apply Itô's formula to the function $f(x,t) = |x|^p$ and $X_t = M_t := \int_0^t (\phi(r), dW_r)$. Then,

$$|M_t|^p = p\int_0^t |M_r|^{p-1}\text{sign}(M_r)\,dM_r + \frac{1}{2}p(p-1)\int_0^t |M_r|^{p-2}\,d\langle M\rangle_r,$$

where $\text{sign}(M_r)$ is equal to 1 if $M_r \geq 0$ and is equal to -1 if $M_r < 0$. The first term of the right-hand side is a martingale with mean 0. Therefore, taking the expectation for each term, we have

$$\begin{aligned}
E[|M_t|^p] &= \frac{1}{2}p(p-1)E\Big[\int_0^t |M_r|^{p-2}\,d\langle M\rangle_r\Big] \\
&\leq \frac{1}{2}p(p-1)E\Big[\sup_{r<t}|M_r|^p\Big]^{\frac{p-2}{p}} E\Big[\langle M\rangle_t^{\frac{p}{2}}\Big]^{\frac{2}{p}} \\
&\leq \frac{1}{2}p(p-1)q^{p-2}E[|M_t|^p]^{1-\frac{2}{p}}E[\langle M\rangle_t^{\frac{p}{2}}]^{\frac{2}{p}}.
\end{aligned}$$

Here we used Hölder's inequality and Doob's inequality for martingale. Therefore there exists a positive constant C_p such that

$$E[|M_t|^p] \leq C_p E[\langle M\rangle_t^{\frac{p}{2}}], \quad \forall 0 < t < T.$$

Since $\langle M\rangle_t = \int_0^t |\phi(r)|^2\,dr$, we have $E[\langle M\rangle_t^{\frac{p}{2}}] \leq E[\int_0^t |\phi(r)|^p\,dr]t^{\frac{p}{2}-1}$ by Hölder's inequality. Therefore we get the assertion of the proposition. □

Set $L^{\infty-} = \bigcap_{p\geq 2} L^p$ and $\mathbf{L}_{\mathbb{T}}^{\infty-} = \bigcap_{p\geq 2}\mathbf{L}_{\mathbb{T}}^p$. The proposition tells us that if $\phi \in \mathbf{L}_{\mathbb{T}}^{\infty-}$, then the stochastic integral $\int_0^t(\phi, dW)$ belongs to $L^{\infty-}$.

Next, we will consider exponential functionals of stochastic integrals given by

$$Z_t = \exp\Big\{\int_0^t(\phi(r), dW_r) - \frac{1}{2}\int_0^t |\phi(r)|^2\,dr\Big\}, \quad t \in \mathbb{T}, \tag{2.10}$$

where $\phi \in \mathbf{L}_{\mathbb{T}}$. Apply Itô's formula to the exponential function $f(x,t) = e^x$. Then we get $Z_t = 1 + \int_0^t Z_r\,dM_r$, where M_t is a local martingale written as $M_t = \int_0^t(\phi(r), dW_r)$. Therefore, $Z_t, t \in \mathbb{T}$ is a positive local martingale. We will show that $Z_t, t \in \mathbb{T}$ is a super-martingale. Set $\tau_n = \inf\{0 < t < T;\ Z_t > n\}$ ($= \infty$ if $\{\cdots\}$ is empty). Then the stopped process $Z_{t\wedge\tau_n}, t \in \mathbb{T}$ is a positive martingale. Therefore the equality $E[Z_{t\wedge\tau_n}|\mathcal{F}_r] = Z_{s\wedge\tau_n}$ holds for any $s < t$ and n, a.s. Then, using Fatou's lemma for conditional measure $P(\cdot|\mathcal{F}_r)$ (Proposition 1.5.2), we have

$$E[Z_t|\mathcal{F}_s] \leq \liminf_{n\to\infty} E[Z_{t\wedge\tau_n}|\mathcal{F}_s] = \liminf_{n\to\infty} Z_{s\wedge\tau_n} = Z_s. \tag{2.11}$$

Therefore $Z_t, t \in \mathbb{T}$ is a super-martingale.

We give a sufficient condition that $Z_t, t \in \mathbb{T}$ is an L^p-martingale.

Proposition 2.2.3 *Let* $p > 1$. *If* $\exp \int_0^T |\phi(r)|^2 dr$ *is in* L^{2p^2-p}, *then* $Z_t, t \in \mathbb{T}$ *defined by* (2.10) *is an* L^p*-martingale. Further, if* $\exp \int_0^T |\phi(r)|^2 dr$ *belongs to* $L^{\infty-}$, $Z_t, t \in \mathbb{T}$ *is an* $L^{\infty-}$*-martingale.*

Proof We shall rewrite Z_t^p as

$$Z_t^p = \exp\Big\{pM_t - p^2\langle M\rangle_t\Big\} \exp\Big\{\frac{1}{2}(2p^2 - p)\langle M\rangle_t\Big\}. \tag{2.12}$$

By the Schwartz inequality, $E[Z_t^p]$ is dominated by

$$E\Big[\Big(\exp\Big\{pM_t - p^2\langle M\rangle_t\Big\}\Big)^2\Big]^{\frac{1}{2}} E\Big[\Big(\exp\Big\{\frac{1}{2}(2p^2 - p)\langle M\rangle_t\Big\}\Big)^2\Big]^{\frac{1}{2}}.$$

Since

$$\Big(\exp\Big\{pM_t - p^2\langle M\rangle_t\Big\}\Big)^2 = \exp\Big\{2pM_t - \frac{1}{2}(2p)^2\langle M\rangle_t\Big\},$$

it is a positive local martingale and is a positive super-martingale, whose mean is less than or equal to 1. Therefore we have

$$E[Z_t^p] \le E\Big[\exp\Big\{(2p^2-p)\langle M\rangle_t\Big\}\Big]^{\frac{1}{2}} = E\Big[\exp\Big\{(2p^2-p)\int_0^t |\phi(r)|^2 dr\Big\}\Big]^{\frac{1}{2}} < \infty.$$

It remains to show that $Z_t, t \in \mathbb{T}$ is a martingale. Let $\{\tau_n\}$ be the sequence of stopping times defined above. By a similar argument, we have

$$E[Z_{t\wedge\tau_n}^p] \le E\Big[\exp\Big\{(2p^2 - p)\int_0^{t\wedge\tau_n} |\phi(r)|^2 dr\Big\}\Big]^{\frac{1}{2}}.$$

Therefore $\sup_n E[Z_{t\wedge\tau_n}^p] < \infty$. Consequently, the family of random variables $\{Z_{t\wedge\tau_n}, n = 1, 2, \ldots\}$ is uniformly integrable. Then the sequence $Z_{t\wedge\tau_n}$ converges to Z_t in L^1. Therefore, the inequality (2.11) can be replaced by the equality. This means that $Z_t, t \in \mathbb{T}$ is a martingale. The second assertion of the proposition is immediate from the first assertion. □

Now, suppose $Z_t, t \in \mathbb{T}$ of (2.10) is a positive martingale. Define

$$Q(B) = E[Z_T 1_B], \quad \forall B \in \mathcal{F}_T. \tag{2.13}$$

Then Q is a probability measure on $(\Omega, \mathcal{F}_T)$. We denote it by $Z_T \cdot P$. Since Q is equivalent (mutually absolutely continuous) to P, stochastic processes on $(\Omega, \mathcal{F}_T, P)$ can be regarded as stochastic processes on $(\Omega, \mathcal{F}_T, Q)$.

Theorem 2.2.2 (Girsanov's theorem) *$W_t^\phi := W_t - \int_0^t \phi(r)\,dr, t \in \mathbb{T}$ is an $\{\mathcal{F}_t\}$-Wiener process with respect to Q.*

Proof We have the equality

$$\begin{aligned} &Z_t \exp\{i(v, W_t^\phi - W_s^\phi)\} \\ &= Z_s \exp\Big\{\int_s^t (\phi(r), dW_r) - \frac{1}{2}\int_s^t |\phi(r)|^2\,dr + i(v, W_t - W_s) - \int_s^t \phi(r)\,dr\Big\} \\ &= Z_s \exp\Big\{\int_s^t ((\phi + iv), dW_r) - \frac{1}{2}\int_s^t |\phi + iv|^2\,dr\Big\} \exp\Big\{-\frac{1}{2}|v|^2(t-s)\Big\}. \end{aligned}$$

Take $A \in \mathcal{F}_s$ and multiply both sides of the above by 1_A and then take expectations. Since $E[\exp\{\int_s^t(\phi + iv)\,dW_r - \frac{1}{2}\int_s^t |\phi + iv|^2\,dr\}|\mathcal{F}_s] = 1$ holds by Proposition 2.2.3, we get for any $A \in \mathcal{F}_s$,

$$E[Z_t \exp\{i(v, W_t^\phi - W_s^\phi)\}1_A] = \exp\Big\{-\frac{1}{2}|v|^2(t-s)\Big\}E[Z_s 1_A].$$

Rewriting the above using the measure $dQ = Z_T\,dP$, we get

$$E_Q[\exp\{i(v, W_t^\phi - W_s^\phi)\}1_A] = \exp\Big\{-\frac{1}{2}|v|^2(t-s)\Big\}Q(A).$$

Therefore, W_t^ϕ is an $\{\mathcal{F}_t\}$-Wiener process with respect to Q. □

2.3 Regularity of Stochastic Integrals Relative to Parameters

In this section, we consider the continuity and the differentiability of stochastic integrals with respect to parameters. We assume that the parameter set Λ is equal to $\mathbb{R}^e$ or a unit ball $B_1^e = \{\lambda \in \mathbb{R}^e;\ |\lambda| < 1\}$. We denote the partial derivative $\frac{\partial}{\partial\lambda_i} f$ by $\partial_{\lambda_i} f$. We set $\partial_\lambda f = (\partial_{\lambda_1} f, \ldots, \partial_{\lambda_e} f)$. For multi-index of nonnegative integers $\mathbf{i} = (i_1, \ldots, i_e)$ we set $\partial_\lambda^{\mathbf{i}} f = \partial_{\lambda_1}^{i_1} \cdots \partial_{\lambda_e}^{i_e} f$.

Let $p > e \vee 2$. Let $\phi_\lambda(r), r \in \mathbb{T}, \lambda \in \Lambda$ be a d'-dimensional measurable random field belonging to $\mathbf{L}_\mathbb{T}(\Lambda)$. If it satisfies

$$\sup_\lambda E\Big[\int_0^T |\phi_\lambda(r)|^p\,dr\Big] < \infty,$$

$$E\Big[\int_0^T |\phi_\lambda(r) - \phi_{\lambda'}(r)|^p\,dr\Big] \le c_p|\lambda - \lambda'|^p \tag{2.14}$$

for any $\lambda, \lambda' \in \Lambda$ with a positive constant c_p, then $\phi_\lambda(r)$ is said to belong to the space $\mathbf{L}_{\mathbb{T}}^{Lip,p}(\Lambda)$. Further, $\phi_\lambda(r)$ is said to belong to the space $\mathbf{L}_{\mathbb{T}}^{n+Lip,p}(\Lambda)$, if it is n-times continuously differentiable with respect to λ for any s a.s. and for any $|\mathbf{i}| \leq n$, derivatives $\phi_\lambda^{\prime}(r) = \partial_\lambda^{\mathbf{i}}\phi_\lambda(r)$ satisfy (2.14). If $\phi_\lambda(r) \in \mathbf{L}_{\mathbb{T}}^{Lip,p}(\Lambda)$, the stochastic integral $\int_0^t(\phi_\lambda(r), dW_r)$ is well defined a.s. for each fixed λ. We are interested in the problem of finding a modification of a family of random variables $\{\int_0^t(\phi_\lambda(s), dW_s), \lambda \in \Lambda\}$ which is continuous in (t, λ) a.s. and differentiable with respect to λ a.s.

Proposition 2.3.1

1. *Suppose that* $\{\phi_\lambda(r)\}$ *is a random field belonging to the class* $\mathbf{L}_{\mathbb{T}}^{Lip,p}(\Lambda)$ *where* $p > e \vee 2$. *Then the family of Itô's stochastic integrals* $\{\int_0^t(\phi_\lambda(r), dW_r)\}$ *has a modification which is continuous in* (t, λ).
2. *Suppose further that* $\{\phi_\lambda(r)\}$ *belongs to the class* $\mathbf{L}_{\mathbb{T}}^{n+Lip,p}(\Lambda)$ *for some positive integer* n *and* $p > e \vee 2$. *Then the above family of stochastic integrals has a modification, which is* n*-times continuously differentiable with respect to* λ *a.s. and the derivatives are continuous in* (t, λ) *a.s. Further, for any* $|\mathbf{i}| \leq n$, *we have*

$$\partial_\lambda^{\mathbf{i}} \int_0^t (\phi_\lambda(r), dW_r) = \int_0^t (\partial_\lambda^{\mathbf{i}}\phi_\lambda(r), dW_r), \quad \forall (t, \lambda) \ a.s. \tag{2.15}$$

Proof

1. For a real continuous function $f(t), t \in [0, T]$, we define its norm by $\|f\| = \sup_{0 \leq t \leq T} |f(t)|$. Set $X_t^\lambda = \int_0^t(\phi_\lambda(r), dW_r), t \in \mathbb{T}$. Since it is a martingale, we have by Doob's inequality and the Burkholder–Davis–Gundy inequality

$$E\Big[\|X^\lambda - X^{\lambda'}\|^p\Big] \leq cE\Big[\int_0^T |\phi_\lambda(r) - \phi_{\lambda'}(r)|^p \, dr\Big] \leq c'|\lambda - \lambda'|.$$

Apply the Kolmogorov–Totoki theorem (Theorem 1.8.1). Then we find that X_t^λ has a modification which is continuous in λ with respect to the norm $\| \ \|$. Then the modification is continuous in (t, λ) a.s.

2. Suppose $\phi_\lambda \in \mathbf{L}_{\mathbb{T}}^{1+Lip,p}(\Lambda)$. For $\epsilon \in \Theta = (-1, 0) \cup (0, 1)$ and a positive integer $1 \leq i \leq e$, we set $Y_t^{\lambda,\epsilon} = \frac{1}{\epsilon}(X_t^{\lambda+\epsilon_i} - X_t^\lambda)$, where $\epsilon_i = \epsilon e_i$ and $e_i, i = 1, \ldots, e$ are unit vectors in $\mathbb{R}^e$. It is a stochastic process with parameter $(\lambda, \epsilon) \in \Lambda \times \Theta$. It holds by the mean value theorem that

$$\begin{aligned} Y_t^{\lambda,\epsilon} &= \int_0^t \Big(\frac{\phi_{\lambda+\epsilon_i}(r) - \phi_\lambda(r)}{\epsilon}, dW_r\Big) \\ &= \int_0^t \Big(\int_0^1 \partial_{\lambda_i}\phi_{\lambda+\theta\epsilon_i}(r) \, d\theta, dW_r\Big), \end{aligned} \tag{2.16}$$

where $|\theta| \leq 1$. Therefore

$$E[\|Y^{\lambda,\epsilon} - Y^{\lambda',\epsilon'}\|^p] \leq cE\Big[\int_0^T \Big(\int_0^1 |\partial_{\lambda_i}\phi_{\lambda+\theta\epsilon_i}(r) - \partial_{\lambda_i}\phi_{\lambda'+\theta\epsilon_i'}(r)|\,d\theta\Big)^p dr\Big]$$
$$\leq C\{|\lambda - \lambda'|^p + |\epsilon - \epsilon'|^p\}. \tag{2.17}$$

Consequently, $Y_t^{\lambda,\epsilon}$ has a modification which is continuous in t, λ, ϵ and is continuously extended at $\epsilon = 0 \in \bar{\Theta}$, by the Kolmogorov–Totoki theorem. This means that X_t^λ is continuously differentiable with respect to λ for any t and the derivative $\partial_{\lambda_i} X_t^\lambda$ coincides with $Y_t^{\lambda,0}$. It is continuous in t, λ a.s.

We have as $\epsilon \to 0$,

$$E\Big[\int_0^T \Big|\int_0^1 \partial_{\lambda_i}\phi_{\lambda+\theta\epsilon_i}(r)\,d\theta - \partial_{\lambda_i}\phi_\lambda(r)\Big|^p dr\Big] \to 0.$$

Then we get the equality $\partial_{\lambda_i} \int_0^t (\phi_\lambda(r), dW_r) = \int_0^t (\partial_{\lambda_i}\phi_\lambda(r), dW_r)$ from (2.16).

Finally, if $\phi_\lambda(r)$ is of the class $\mathbf{L}_{\mathbb{T}}^{n+Lip,p}(\Lambda)$, we can repeat the above argument inductively. Then we find that the stochastic integral is n-times differentiable with respect to the parameter and we get the equality (2.15) for any $|\mathbf{i}| \leq n$. □

For one dimensional predictable processes with parameter Λ, spaces $L_{\mathbb{T}}(\Lambda)$, $L_{\mathbb{T}}^{Lip,p}(\Lambda)$, $L_{\mathbb{T}}^{n+Lip,p}(\Lambda)$ are defined similarly. When the parameter is a spatial parameter x, we will often write $\phi_x(r)$ as $\phi(x, r)$.

Let $(f^k(x,r), k = 1, \ldots, d')$ be a d'-dimensional predictable process with spatial parameter x, belonging to the space $\mathbf{L}_{\mathbb{T}}^{2+Lip,p}(\mathbb{R}^d)$ and let $f^0(x,r)$ be a predictable process with spatial parameter x belonging to $L_{\mathbb{T}}^{2+Lip,p}(\mathbb{R}^d)$. We set

$$F(x,t) = F(x,0) + \sum_{k=0}^{d'} \int_0^t f^k(x,r)\,dW_r^k, \quad \forall (x,t) \ a.s., \tag{2.18}$$

where $F(x,0)$ is a C^2-function, $W_t^0 = t$ and $\int_0^t f^0(x,r)\,dW_r^0 = \int_0^t f^0(x,r)\,dr$. Then $F(x,t)$ has a modification of $C^{2,0}$-function of x, t by Proposition 2.3.1.

We shall obtain a differential rule for the composite of the above stochastic process $F(x,t)$ with spatial parameter x and a continuous semi-martingale X_t. The equation (2.19) below is called a *generalized Itô's formula* or the *Itô–Wentzell formula*.

Theorem 2.3.1 *Let X_t be a d-dimensional continuous semi-martingale. Then the composite $F(X_t, t)$ is a continuous semi-martingale. Further, for any $s < t$, we have*

$$F(X_t,t)=F(X_s,s)+\sum_{k=0}^{d'}\int_s^t f^k(X_r,r)\,dW_r^k$$
$$+\sum_{i=1}^{d}\int_s^t\frac{\partial F}{\partial x_i}(X_r,r)\,dX_r^i+\sum_{i=1}^{d}\sum_{k=1}^{d'}\int_s^t\frac{\partial f^k}{\partial x_i}(X_r,r)\,d\langle W^k,X^i\rangle_r$$
$$+\frac{1}{2}\sum_{i,j=1}^{d}\int_s^t\frac{\partial^2 F}{\partial x_i\partial x_j}(X_r,r)\,d\langle X^i,X^j\rangle_r. \tag{2.19}$$

Proof It suffices to prove (2.19) in the case $s=0$. Let $\Pi=\{t_0<t_1<\cdots<t_n\}$ be a partition of $[0,t]$. Then we have

$$F(X_t,t)-F(X_0,0)=\sum_{m=1}^{n}\Big\{F(X_{t_{m-1}},t_m)-F(X_{t_{m-1}},t_{m-1})\Big\}$$
$$+\sum_{m=1}^{n}\Big\{F(X_{t_m},t_m)-F(X_{t_{m-1}},t_m)\Big\}$$

almost surely. Let $|\Pi|\to 0$. Then we have

$$\sum_{m=1}^{n}\Big\{F(X_{t_{m-1}},t_m)-F(X_{t_{m-1}},t_{m-1})\Big\}=\sum_{k=0}^{d'}\Big\{\sum_{m=1}^{n}\int_{t_{m-1}}^{t_m}f^k(X_{t_{m-1}},r)\,dW_r^k\Big\}$$
$$\to\sum_{k=0}^{d'}\int_0^t f^k(X_r,r)\,dW_r^k.$$

Further, by the Taylor expansion of $F(t_m,x)$, we have

$$\sum_{m=1}^{n}\Big\{F(X_{t_m},t_m)-F(X_{t_{m-1}},t_m)\Big\}$$
$$=\sum_{i=1}^{d}\Big\{\sum_{m=1}^{n}\frac{\partial F}{\partial x_i}(X_{t_{m-1}},t_{m-1})(X_{t_m}^i-X_{t_{m-1}}^i)\Big\}$$
$$+\sum_{i=1}^{d}\Big\{\sum_{m=1}^{n}\Big(\frac{\partial F}{\partial x_i}(X_{t_{m-1}},t_m)-\frac{\partial F}{\partial x_i}(X_{t_{m-1}},t_{m-1})\Big)(X_{t_m}^i-X_{t_{m-1}}^i)\Big\}$$
$$+\frac{1}{2}\sum_{i,j=1}^{d}\Big\{\sum_{m=1}^{n}\frac{\partial^2 F}{\partial x_i\partial x_j}(\xi_m,t_m)(X_{t_m}^i-X_{t_{m-1}}^i)(X_{t_m}^j-X_{t_{m-1}}^j)\Big\}$$
$$=I_1+I_2+I_3,$$

where ξ_m are random variables such that $|\xi_m - X_{t_{m-1}}| \leq |X_{t_m} - X_{t_{m-1}}|$. Let $|\Pi|$ tend to 0. Then we get

$$\lim_{|\Pi|\to 0} I_1 = \sum_{i=1}^{d} \int_0^t \frac{\partial F}{\partial x_i}(X_r, r)\, dX_r^i$$

immediately from the definition of the stochastic integral. Further, noting Theorem 1.6.2, we get

$$\begin{aligned}\lim_{|\Pi|\to 0} I_2 &= \sum_{i=1}^{d}\sum_{k=0}^{d'} \lim_{|\Pi|\to 0} \Big\{ \sum_{m=1}^{n} \Big(\int_{t_{m-1}}^{t_m} \frac{\partial f^k}{\partial x_i}(X_{t_{m-1}}, r)\, dW_r^k \Big)(X_{t_m}^i - X_{t_{m-1}}^i) \Big\} \\ &= \sum_{i=1}^{d}\sum_{k=1}^{d'} \int_0^t \frac{\partial f^k}{\partial x_i}(X_r, r)\, d\langle W^k, X^i\rangle_r, \quad a.s.\end{aligned}$$

$$\lim_{|\Pi|\to 0} I_3 = \frac{1}{2} \sum_{i,j=1}^{d} \int_0^t \frac{\partial^2 F}{\partial x_i \partial x_j}(X_r, r)\, d\langle X^i, X^j\rangle_r, \quad a.s.$$

Summing up these computations, we get the formula of the theorem. □

2.4 Fisk–Stratonovitch Symmetric Integrals

We will define the *Fisk–Stratonovitch symmetric integral* or simply the *symmetric integral* of $\phi(r)$ by a continuous semi-martingale $X_t, t \in \mathbb{T}$ by

$$\int_0^t \phi(r) \circ dX_r = \lim_{|\Pi|\to 0} \sum_{m=1}^{n} \frac{1}{2} \{\phi(t'_m) + \phi(t'_{m-1})\}(X_{t'_m} - X_{t'_{m-1}}), \tag{2.20}$$

if the right-hand side exists. Here $\Pi = \{t_0 < \cdots < t_n\}$ are partitions of $[0, T]$ and $t'_m = t_m \wedge t$. It holds that

$$\sum_m \frac{1}{2}\{\phi(t'_m) + \phi(t'_{m-1})\}(X_{t'_m} - X_{t'_{m-1}}) = \sum_m \phi(t'_{m-1})(X_{t'_m} - X_{t'_{m-1}}) + \frac{1}{2}\langle \phi, X\rangle_t^{\Pi},$$

where $\langle \phi, X\rangle_t^{\Pi}$ is the quadratic covariation of processes ϕ and X associated with the partition Π. If $\phi(t)$ is a semi-martingale with bounded jumps, then $\langle \phi, X\rangle_t^{\Pi}, t \in [0, T]$ converges to a continuous process of bounded variation, which we denote by $\langle \phi, X\rangle_t$ (see Sect. 1.6). Then we have:

Proposition 2.4.1 *Let $\{\phi(r)\}$ be a semi-martingale with bounded jumps. Then, the symmetric integral $\int_0^t \phi(r) \circ dX_r$ is well defined for any $t \in \mathbb{T}$. Further, we have*

$$\int_0^t \phi(r) \circ dX_r = \int_0^t \phi(r)\, dX_r + \frac{1}{2}\langle \phi, X\rangle_t, \quad a.s., \tag{2.21}$$

where $\int_0^t \phi(r)\, dX_r$ is the Itô integral.

For $0 \le s < t \le T$, we define the symmetric integral by

$$\int_s^t \phi(r) \circ dX_r = \int_0^t \phi(r) \circ dX_r - \int_0^s \phi(r) \circ dX_r.$$

Then it holds that

$$\int_s^t \phi(r) \circ dX_r = \int_s^t \phi(r) \circ dX_t + \frac{1}{2}(\langle \phi, X\rangle_t - \langle \phi, X\rangle_s).$$

Remark It is known that the symmetric integral is well defined if the martingale part of $\phi(r)$ is a local L^2-martingale. In such a case, the quadratic covariation $\langle \phi, X\rangle_t$ exists and the equality (2.21) holds. For the proof, we need additional arguments for the quadratic covariation. See Protter [96].

A feature of the symmetric integral is that it works similarly as the usual differential calculus. Let $X_t = (X_t^1, \ldots, X_t^d), t \in \mathbb{T}$ be a d-dimensional continuous semi-martingale represented by

$$X_t^i = \sum_{k=1}^{d'} \int_0^t \phi^{ik}(r) \circ dW_r^k + \int_0^t v^i(r)\, dr. \tag{2.22}$$

Here, $(\phi^{ik}(r))$ is a d'-dimensional continuous semi-martingale written as

$$\phi^{ik}(r) = \phi^{ik}(0) + \sum_{l=1}^{d'} \int_0^r \phi_1^{ikl}(s)\, dW_s^l + \int_0^r \phi_2^{ik}(s)\, ds,$$

where $\phi_1^{ikl}(r) \in \mathbf{L}_{\mathbb{T}}$, and $\phi_2^{ik}(r) \in L_{\mathbb{T}}$. Then, using Itô integral, X_t is written as

$$X_t^i = \sum_{k=1}^{d'} \int_0^t \phi^{ik}(r)\, dW_r^k + \int_0^t v^i(r)\, dr + \frac{1}{2}\int_0^t \bar{\phi}_1^i(r)\, dr,$$

where $\bar{\phi}_1^i(r) = \sum_{k=1}^{d'} \phi_1^{ikk}(r), i = 1, \ldots, d$. Set $\phi(r) = (\phi^{ij}(r))$ and $v(r) = (v^i(r))$. Using vector and matrix notations, we rewrite (2.22) as

$$X_t = \int_0^t \phi(r) \circ dW_r + \int_0^t v(r)\, dr.$$

We shall reconsider the differential rule of the composite functional $f(X_t, t)$, when $X_t, t \in \mathbb{T}$ is a semi-martingale represented by the symmetric integral. It will be seen that the differential rule for the symmetric integral is similar to the usual differential rule.

Theorem 2.4.1 *Let $\{X_t\}$ be a d-dimensional continuous semi-martingale represented by (2.22). Let $f(x, t)$ be a $C^{3,1}$-function on $\mathbb{R}^d \times \mathbb{T}$. Then we have for any $0 \le s < t \le T$,*

$$f(X_t, t) = f(X_s, s) + \int_s^t \frac{\partial f}{\partial t}(X_r, r)\, dr \tag{2.23}$$
$$+ \sum_{i=1}^{d}\sum_{k=1}^{d'} \int_s^t \frac{\partial f}{\partial x_i}(X_r, r)\phi^{ik}(r) \circ dW_r^k + \sum_{i=1}^{d} \int_s^t \frac{\partial f}{\partial x_i}(X_r, r) v^i(r)\, dr.$$

Proof It is sufficient to prove the case $s = 0$. We shall rewrite the term expressed by symmetric integral, using Itô integral. If f is a $C^{1,3}$-function, Theorem 2.2.1 tells us that $\frac{\partial f}{\partial x_i}(X_t, t)$ is a continuous semi-martingale and it satisfies

$$\frac{\partial f}{\partial x_i}(X_t, t) = \frac{\partial f}{\partial x_i}(X_0, 0) + \sum_{j=1}^{d} \int_0^t \frac{\partial^2 f}{\partial x_i \partial x_j}(X_r, r)\, dX_r^j + C_t,$$

where C_t is a continuous process of bounded variation. Then the martingale part of the semi-martingale $\frac{\partial f}{\partial x_i}(X_t, t)\phi^{ik}(t)$ is

$$\sum_{j=1}^{d}\sum_{l=1}^{d'} \int_0^t \frac{\partial^2 f}{\partial x_i \partial x_j}(X_r, r)\phi^{ik}(r)\phi^{jl}(r)\, dW_r^l + \sum_{l=1}^{d'} \int_0^t \frac{\partial f}{\partial x_i}(X_r, r)\phi_1^{ikl}(r)\, dW_r^l.$$

Therefore

$$\left\langle \frac{\partial f}{\partial x_i}(X_t, t)\phi^{ik}(t), W_t^k \right\rangle_t$$
$$= \sum_{j=1}^{d} \int_0^t \frac{\partial^2 f}{\partial x_i \partial x_j}(X_r, r)\phi^{ik}(r)\phi^{jk}(r)\, dr + \int_0^t \frac{\partial f}{\partial x_i}(X_r, r)\phi_1^{ikk}(r)\, dr.$$

Then $\sum_{k=1}^{d'} \int_0^t \frac{\partial f}{\partial x_i}(X_r, r)\phi^{ik}(r) \circ dW_r^k$ is equal to

$$\sum_{k=1}^{d'} \int_0^t \frac{\partial f}{\partial x_i}(X_r, r)\phi^{ik}(r)\, dW_r^k + \frac{1}{2}\sum_{j=1}^{d} \int_0^t \frac{\partial^2 f}{\partial x_i \partial x_j}(X_r, r)\Big(\sum_{k=1}^{d'} \phi^{ik}(r)\phi^{jk}(r)\Big)\, dr$$
$$+ \frac{1}{2}\int_0^t \frac{\partial f}{\partial x_i}(X_r, r)\bar{\phi}_1^i(r)\, dr.$$

Therefore the right-hand side of (2.23) is equal to the right-hand side of (2.8) if we replace $\upsilon(r)$ in (2.8) by $\upsilon(r) + \frac{1}{2}\bar{\phi}_1(r)$. Therefore the formula (2.23) holds. □

Let Λ be the e-dimensional parameter set. Let $\{\phi_\lambda(r), \lambda \in \Lambda, t \in \mathbb{T}\}$ be a measurable random field such that for any λ, $\phi_\lambda(t)$ is a continuous semi-martingale. We assume

$$\int_\Lambda \int_\mathbb{T} \Big(|\phi_\lambda(r)|^2 + |\langle \phi_\lambda(r), X_r\rangle| \Big)\, d\lambda\, dr < \infty. \tag{2.24}$$

We give a Fubini theorem for the change of the order of the symmetric integrals $\circ dW_r$ and $d\lambda$.

Proposition 2.4.2 *Let $\{\phi_\lambda(r), \lambda \in \Lambda, r \in \mathbb{T}\}$ be a measurable random field such that for any λ, $\phi_\lambda(r), r \in \mathbb{T}$ is a continuous semi-martingales satisfying* (2.24). *Then we have for any $t \in \mathbb{T}$*

$$\int_\Lambda \Big(\int_0^t \phi_\lambda(r) \circ dW_r \Big) g(\lambda)\, d\lambda = \int_0^t \Big(\int_\Lambda \phi_\lambda(r) g(\lambda)\, d\lambda \Big) \circ dW_r, \quad a.s. \tag{2.25}$$

for any bounded measurable function $g(\lambda)$ with compact supports.

The proof is straightforward from Proposition 2.1.1.

We shall study the regularity of the symmetric integrals with respect to parameters.

Proposition 2.4.3 *Let $\phi_\lambda(r) = (\phi_\lambda^1(r), \ldots, \phi_\lambda^{d'}(r)), \lambda \in \Lambda, r \in \mathbb{T}$ be a d'-dimensional measurable random field such that for any λ, $\phi_\lambda(r)$ is a continuous semi-martingale written as*

$$\phi_\lambda(r) = \int_0^r \phi_{1,\lambda}(s)\, dW_s + \int_0^r \phi_{2,\lambda}(s)\, ds, \tag{2.26}$$

where $\phi_\lambda^i(r), \phi_{1,\lambda}^i(r) \in \mathbf{L}_\mathbb{T}^{Lip,p}(\Lambda)$ for $i = 1, \ldots, d$ for some $p > e \vee 2$.

1. *Then the symmetric integral $\int_0^t \phi_\lambda(r) \circ dW_r$ has a modification which is continuous in (t, λ).*
2. *Assume further that $\phi_\lambda^i(r), \phi_{1,\lambda}^i(r), i = 1, \ldots, d$ belong to $\mathbf{L}_\mathbb{T}^{n+Lip,p}(\Lambda)$ for some positive integer n and $p > e \vee 2$. Then the symmetric integral is n-times continuously differentiable with respect to λ a.s. and the derivative is continuous in (t, λ) a.s. Further, we have for any $|\mathbf{i}| \le n$,*

$$\partial_\lambda^\mathbf{i} \int_0^t \phi_\lambda(r) \circ dW_r = \int_0^t \partial_\lambda^\mathbf{i} \phi_\lambda(r) \circ dW_r, \quad \forall (t, \lambda) \;\; a.s. \tag{2.27}$$

Proof

1. The symmetric stochastic integral $\int_0^t \phi_\lambda(r) \circ dW_r$ is equal to the Itô integral $\int_0^t \phi_\lambda(r)\, dW_r + \frac{1}{2}\int_0^t \bar{\phi}_{1,\lambda}(r)\, dr$. If $\phi_\lambda(r)$ satisfies (2.14), the first term of the right-hand side has a modification which is continuous in (t, λ) by Propositions 2.3.1. The second term of the right-hand side is continuous in (t, λ) obviously. Therefore the symmetric integral has a modification which is continuous in (t, λ).
2. Under the assumption, integrals $\int_0^t \phi_\lambda(r)\, dW_r$ and $\frac{1}{2}\int_0^t \bar{\phi}_{1,\lambda}(r)\, dr$ are continuously differentiable with respect to λ and we can change the order of ∂_λ and $\int$, in view of Proposition 2.3.1. Then we get the formula (2.27). □

We shall rewrite the generalized Itô's formula (Theorem 2.3.1) in the previous section using symmetric integrals. We will assume slightly stronger conditions for $f^k(x,t)$. Let $f^k(x,t), k = 1, \ldots, d'$ be predictable processes with spatial parameter x represented by $f^k(x,t) = \int_0^t (f_1^k(x,r), dW_r) + \int_0^t f_2^k(x,r)\, dr$, where $f_1^k(x,r) \in \mathbf{L}_{\mathbb{T}}^{3+Lip,p}(\mathbb{R}^d)$ and $f_2^k(x,r) \in L_{\mathbb{T}}^{3+Lip,p}(\mathbb{R}^d)$ for $p > 2 \vee d$. Let $f^0(x,r) \in L_{\mathbb{T}}^{3+Lip,p}(\mathbb{R}^d)$. Let $F(x,t), t \in \mathbb{T}$ be a stochastic process with a spatial parameter $x \in \mathbb{R}^d$ defined by

$$F(x,t) = F(x,0) + \sum_{k=0}^{d'} \int_0^t f^k(x,r) \circ dW_r^k, \tag{2.28}$$

where $F(x,0)$ is a C^3-function of x. Then $F(x,t)$ is a $C^{3,0}$-function of x, t a.s. and it is a continuous semi-martingale for any x by Proposition 2.4.3.

Theorem 2.4.2 *For any d-dimensional continuous semi-martingale $X_t, t \in \mathbb{T}$, we have*

$$\begin{aligned} F(X_t, t) &= F(X_s, s) \\ &+ \sum_{k=0}^{d'} \int_s^t f^k(X_r, r) \circ dW_r^k + \sum_{i=1}^{d} \int_s^t \frac{\partial F}{\partial x_i}(X_r, r) \circ dX_r^i, \end{aligned} \tag{2.29}$$

for any $0 \le s < t \le T$.

Proof It is sufficient to prove it in the case $s = 0$. Since $f^k(X_t, t)$ is a continuous semi-martingale, we have, similarly to the proof of Theorem 2.4.1,

$$\int_0^t f^k(X_r, r) \circ dW_r^k = \int_0^t f^k(X_r, r)\, dW_r^k + \frac{1}{2}\sum_{i=1}^{d} \int_0^t \frac{\partial f^k}{\partial x_i}(X_r, r)\, d\langle W^k, X^i\rangle_r,$$

if $k = 1, \ldots, d'$. Next, $\frac{\partial F}{\partial x_i}(X_t, t)$ is a continuous semi-martingale represented by

$$\frac{\partial F}{\partial x_i}(X_t, t) - \frac{\partial F}{\partial x_i}(X_0, 0)$$
$$= \sum_{k=0}^{d'} \int_0^t \frac{\partial f^k}{\partial x_i}(X_r, r)\, dW_r^k + \sum_{j=1}^{d} \int_0^t \frac{\partial^2 F}{\partial x_i \partial x_j}(X_r, r)\, dX_r^j + A_t^c,$$

where A_t^c is a continuous process of bounded variation, in view of Theorem 2.3.1. Therefore,

$$\int_0^t \frac{\partial F}{\partial x_i}(X_r, r) \circ dX_r^i = \int_0^t \frac{\partial F}{\partial x_i}(X_r, r)\, dX_r^i + \frac{1}{2}\sum_{k=1}^{d'} \int_0^t \frac{\partial f^k}{\partial x_i}(X_r, r)\, d\langle W^k, X^i\rangle_r$$
$$+ \frac{1}{2}\sum_{j=1}^{d} \int_0^t \frac{\partial^2 F}{\partial x_i \partial x_j}(X_r, r)\, d\langle X^j, X^i\rangle_r.$$

Then (2.19) implies (2.29). □

2.5 Stochastic Integrals with Respect to Poisson Random Measure

We will define stochastic integrals based on Poisson random measures and compensated Poisson random measures. Let $N(du) \equiv N(dr\,dz)$ be a Poisson random measure on $\mathbb{U} = \mathbb{T} \times \mathbb{R}_0^{d'}$ with intensity $n(du) = n(dr\,dz) = dr\,\nu(dz)$, where ν is a Lévy measure. Here, elements of $\mathbb{T}$ are denoted by r and elements of $\mathbb{R}_0^{d'}$ are denoted by $z = (z^1, \ldots, z^{d'})$. We define the *compensated Poisson random measure* $\tilde{N}$ by

$$\tilde{N}(du) = N(du) - n(du). \tag{2.30}$$

Let $\{\mathcal{F}_t, t \in \mathbb{T}\}$ be a filtration such that for any $0 \le s < t < t' \le T$, $N((s,t] \times E)$ is $\{\mathcal{F}_t\}$-adapted and $N((t,t'] \times E)$ are independent of $\mathcal{F}_t$. Then $N(du)$ is called an $\{\mathcal{F}_t\}$-Poisson random measure. Let $\psi(u) = \psi(r,z)$ be a measurable random field with parameter $u \in \mathbb{U}$. It is called *predictable*, if for any z, $\psi(r,z)$ is a predictable process with respect to the filtration $\{\mathcal{F}_t\}$. We set

$$\mathbf{L}_{\mathbb{U}}^2 = \Big\{\psi(u);\ \text{predictable and } E\Big[\int_{\mathbb{U}} |\psi(u)|^2 n(du)\Big] < \infty\Big\}. \tag{2.31}$$

For $0 < t \le T$, we set $\mathbb{U}_t = \{u = (r,z) \in \mathbb{U};\ 0 \le r \le t,\ z \in \mathbb{R}_0^{d'}\}$. It is a subdomain of $\mathbb{U}$. We will define the stochastic integral $\int_{\mathbb{U}_t} \psi(u)\tilde{N}(du)$ for $\psi \in \mathbf{L}_{\mathbb{U}}^2$.

We first consider the integral for a simple predictable random field $\psi(r, z)$ written as

$$\psi(r, z) = \sum_{m=1}^{n} \psi_m(z) 1_{(t_{m-1}, t_m]}(r),$$

where $0 = t_0 < \cdots < t_n = T$ and $\psi_m(z)$ are $\mathcal{F}_{t_{m-1}} \times \mathcal{B}(\mathbb{R}_0^{d'})$-measurable functionals such that $E[\int (|\psi_m(z)| + |\psi_m(z)|^2) \nu(dz)] < \infty$. We define

$$M_t := \int_{\mathbb{U}_t} \psi(u) \tilde{N}(du) = \sum_{m=1}^{n} \int_{\mathbb{R}_0^{d'}} \psi_m(z) \tilde{N}((t'_{m-1}, t'_m], dz), \tag{2.32}$$

where $t'_m = t_m \wedge t$ and $\tilde{N}((s, t], A) = N((s, t] \times A) - (t - s)\nu(A)$ are signed measures on $\mathbb{R}_0^{d'}$ for almost all ω.

Lemma 2.5.1 *$M_t, t \in \mathbb{T}$ defined by (2.32) is an L^2-martingale. Further,*

$$M_t^2 - \int_{\mathbb{U}_t} |\psi(u)|^2 n(du), \quad t \in \mathbb{T} \tag{2.33}$$

is also a martingale.

Proof We denote $\int \psi_m(z) \tilde{N}((t_{m-1}, t_m], dz)$ by Z_m. We want to show that for any m

$$E[Z_m | \mathcal{F}_{t_{m-1}}] = 0, \tag{2.34}$$
$$E[Z_m^2 | \mathcal{F}_{t_{m-1}}] = (t_m - t_{m-1}) E\Big[\int \psi_m(z)^2 \nu(dz) \Big| \mathcal{F}_{t_{m-1}} \Big].$$

Suppose that $\psi_m(z)$ is a deterministic step function $\psi_m(z) = \sum_i c_i 1_{E_i}(z)$, where c_i are constants and E_i are disjoint subsets of $\mathbb{R}_0^{d'}$ with $\nu(E_i) < \infty$. Since Z_m and $\mathcal{F}_{t_{m-1}}$ are independent, we have

$$E[Z_m | \mathcal{F}_{t_{m-1}}] = E[Z_m] = \sum_i c_i E[\tilde{N}((t_{m-1}, t_m] \times E_i)] = 0,$$

$$E[Z_m^2 | \mathcal{F}_{t_{m-1}}] = E[Z_m^2] = \sum_{i,j} c_i c_j E[N((t_{m-1}, t_m] \times E_i) N((t_{m-1}, t_m] \times E_j)].$$

Since the law of $N((t_{m-1}, t_m] \times E_i)$ is Poisson with parameter $(t_m - t_{m-1})\nu(E_i)$, its variance $E[\tilde{N}((t_{m-1}, t_m], E_i)^2]$ is equal to $(t_m - t_{m-1})\nu(E_i)$. Further, if $i \neq j$, $\tilde{N}((t_{m-1}, t_m], E_i)$ and $\tilde{N}((t_{m-1}, t_m], E_j)$ are independent, so that the expectation of the product of these two random variables is equal to 0. Consequently, we get

$$E[Z_m^2|\mathcal{F}_{t_{m-1}}] = (t_m - t_{m-1})\sum_i c_i^2 \nu(E_i) = (t_m - t_{m-1})\int_{\mathbb{R}_0^{d'}} \psi_m(z)^2 \nu(dz).$$

Then, equalities (2.34) hold for any deterministic function $\psi_m(z)$ satisfying $\int(|\psi_m(z)| + |\psi_m(z)|^2)\nu(dz) < \infty$. Next, if $\psi_m(z)$ is $\mathcal{F}_{t_{m-1}}$-measurable, we may regard that it is a deterministic function under the conditional measure $P(\cdot|\mathcal{F}_{t_{m-1}})$ a.s. Therefore equalities (2.34) hold a.s.

Further, if $m > l$, $E[Z_m Z_l|\mathcal{F}_{t_{l-1}}] = E[E[Z_m|\mathcal{F}_{t_{m-1}}]Z_l|\mathcal{F}_{t_{l-1}}] = 0$. Then, since $M_t = \sum_{m;t_m \le t} Z_m$, $M_t - M_s$ satisfies

$$E\Big[M_t - M_s\Big|\mathcal{F}_s\Big] = E\Big[\sum_{m;s<t_m\le t} Z_m\Big|\mathcal{F}_s\Big] = 0, \quad a.s., \tag{2.35}$$

$$\begin{aligned} E\Big[(M_t - M_s)^2|\mathcal{F}_s\Big] &= E\Big[\Big(\sum_{m;s<t_m\le t} Z_m\Big)^2\Big|\mathcal{F}_s\Big] \\ &= E\Big[\int_s^t\int_{\mathbb{R}_0^{d'}} \Big|\psi(u)|^2 n(du)\Big|\mathcal{F}_s\Big], \quad a.s. \end{aligned} \tag{2.36}$$

Equation (2.35) shows that M_t is a martingale and (2.36) shows that M_t^2 is integrable and $M_t^2 - \int_{\mathbb{U}_t} |\psi(u)|^2 n(du)$ is a martingale. □

For $\psi(u) \in \mathbf{L}^2_{\mathbb{U}}$, there exists a sequence $\psi_n(u)$ of simple predictable random fields such that $E\Big[\int |\psi(u) - \psi_n(u)|^2 n(du)\Big] \to 0$. Set $M_t^n = \int_{\mathbb{U}_t} \psi_n(u)\tilde{N}(du)$. Then we have by Doob's inequality and (2.36)

$$\begin{aligned} E[\sup_{0\le t\le T} |M_t^n - M_t^{n'}|^2] &\le 4E[|M_T^n - M_T^{n'}|^2] \\ &\le 4E\Big[\int_{\mathbb{U}} |\psi_n(u) - \psi_{n'}(u)|^2 n(du)\Big] \to 0, \end{aligned}$$

as $n, n' \to \infty$. Therefore M_t^n converges to an L^2-martingale M_t uniformly in $0 < t < T$ in L^2-norm. We denote it by $\int_{\mathbb{U}_t} \psi(u)\tilde{N}(du)$. It is a right continuous L^2-martingale. It satisfies (2.35) and (2.36). The equality (2.35) shows that $M_t = \int_{\mathbb{U}_t} \psi(u)\tilde{N}(du)$ is an L^2- martingale and (2.36) shows that $M_t^2 - \int_{\mathbb{U}_t} |\psi(u)|^2 n(du)$ is an L^1-martingale. For $0 < s < t \le T$ and a Borel subset B of $\mathbb{R}_0^{d'}$, we define the stochastic integral based on the compensated Poisson random measure by

$$\int_s^t\int_B \psi(r,z)\tilde{N}(dr\,dz) := \int_{\mathbb{U}_t} \psi 1_{\mathbb{T}\times B}\tilde{N}(du) - \int_{\mathbb{U}_s} \psi 1_{\mathbb{T}\times B}\tilde{N}(du).$$

Set

$$\mathbf{L}_{\mathbb{U}} = \left\{ \psi(u);\ \text{predictable and} \int_{\mathbb{U}} |\psi(u)|^2 n(du) < \infty,\ a.s. \right\}. \tag{2.37}$$

We can extend the stochastic integral for ψ belonging to $\mathbf{L}_{\mathbb{U}}$. In fact, if $\psi \in \mathbf{L}_{\mathbb{U}}$, the stochastic integral $\int_s^t \int_{\mathbb{R}_0^{d'}} \psi(r,z)\tilde{N}(dr\,dz)$ can be defined as a local L^2-martingale. Since the discussion is similar to that for continuous martingales, it is omitted. The integral is called the *stochastic integral based on compensated Poisson random measure* $\tilde{N}$. We have the following.

Theorem 2.5.1 *For any $\psi \in \mathbf{L}_{\mathbb{U}}$, $M_t := \int_0^t \int_{\mathbb{R}_0^{d'}} \psi(r,z)\tilde{N}(dr\,dz), t \in \mathbb{T}$ is defined as a local L^2-martingale. Further, $M_t^2 - \int_0^t \int_{\mathbb{R}_0^{d'}} \psi(r,z)^2 n(dr\,dz), t \in \mathbb{T}$ is a local martingale.*

2.6 Jump Processes and Related Calculus

Suppose that the Wiener process $W_t = (W_t^1, \ldots, W_t^{d'}), t \in \mathbb{T}$ and Poisson random measure $N(dr\,dz), (r,z) \in \mathbb{T} \times \mathbb{R}_0^{d'}$ are defined on the same probability space $(\Omega, \mathcal{F}, P)$ with the filtration $\{\mathcal{F}_t\}$, are independent of each other, and that W_t is an $\{\mathcal{F}_t\}$-Wiener process and $N(dr\,dz)$ is an $\{\mathcal{F}_t\}$-Poisson random measure. Then functional spaces of predictable processes $\mathbf{L}_{\mathbb{T}}$, $\mathbf{L}_{\mathbb{U}}$ are defined on the probability space.

Let D be a Borel subset of $\mathbb{R}_0^{d'}$ including the set $\{z \in \mathbb{R}_0^{d'};\ |z| \le 1\}$. Let $\psi(u)$ be a predictable random field such that $|\psi(u)| < \infty$ a.e. $n(du)P(d\omega)$ and $\psi(u)1_D(z) \in \mathbf{L}_{\mathbb{U}}$. We define the integral of ψ by $\tilde{N}_D(du)$ by

$$\int_s^t \int_B \psi(r,z)\tilde{N}_D(dr\,dz) = \int_s^t \int_{D\cap B} \psi(r,z)\tilde{N}(dr\,dz) + \int_s^t \int_{D^c \cap B} \psi(r,z)N(dr\,dz),$$

where the last integral by dN is the Lebesgue–Stieltjes integral. Let $X_t = (X_t^1, \ldots, X_t^d), t \in \mathbb{T}$ be a d-dimensional cadlag process represented by

$$X_t^i = X_0^i + \int_0^t (\phi^i(r), dW_r) + \int_0^t v^i(r)\,dr + \int_0^t \int_{\mathbb{R}_0^{d'}} \psi^i(r,z)\tilde{N}_D(dr\,dz). \tag{2.38}$$

Here $\phi^i \in \mathbf{L}_{\mathbb{T}}$, $v^i \in L_{\mathbb{T}}$ and $\psi^i(\cdot, z)1_D(z) \in \mathbf{L}_{\mathbb{U}}$. We call it an *Itô process*. The second, the third and the fourth term of the right-hand side of (2.38) are called, the diffusion part, the drift part and the jump part of the Itô process X_t. Any Lévy process is an Itô process, since a Lévy process has a Lévy–Itô decomposition (Theorem 1.4.2).

Let X_t be an Itô process represented above. Let $\varphi(r) = (\varphi^{ij}(r))$ be a predictable process such that components of $\varphi(r)\phi(r)$, $\varphi(r)v(r)$, $\varphi(r)\psi(r,z)1_D(z)$ belong to $\mathbf{L}_{\mathbb{T}}$, $L_{\mathbb{T}}$, $\mathbf{L}_{\mathbb{U}}$, respectively. Then the stochastic integral

$$\int_0^t \varphi(r)\,dX_r := \int_0^t \varphi(r)\phi(r)\,dW_r + \int_0^t \varphi(r)\upsilon(r)\,dr + \int_0^t\int_{\mathbb{R}_0^{d'}} \varphi(r)\psi(r,z)\tilde{N}_D(dr\,dz)$$

is well defined. It is again an Itô process.

We give the differential rule for the above Itô process (Itô's formula).

Theorem 2.6.1 ([66]) *Let $f(x^1, \ldots, x^d, t)$ be a function of $C^{2,1}$-class. Let X_t be an Itô process given by (2.38). Then, $f(X_t, t), t \in [0, T]$ is an Itô process and satisfies*

$$\begin{aligned} f(X_t,t) &= f(X_s,s) + \int_s^t \frac{\partial f}{\partial t}(X_r,r)\,dr \qquad (2.39)\\ &+\sum_{i=1}^d\sum_{k=1}^{d'}\int_s^t \frac{\partial f}{\partial x_i}(X_{r-},r)\phi^{ik}(r)\,dW_r^k + \sum_{i=1}^d \int_s^t \frac{\partial f}{\partial x_i}(X_r,r)\upsilon^i(r)\,dr\\ &+\frac{1}{2}\sum_{i,j=1}^d \int_s^t \frac{\partial^2 f}{\partial x_i \partial x_j}(X_r,r)\Big(\sum_{k=1}^{d'}\phi^{ik}(r)\phi^{jk}(r)\Big)\,dr\\ &+\int_s^t\int_D \Big\{f(X_r+\psi(r,z),r) - f(X_r,r) - \sum_i \frac{\partial f}{\partial x_i}(X_r,r)\psi^i(r,z)\Big\}n(dr\,dz)\\ &+\int_s^t\int_{\mathbb{R}_0^{d'}} \{f(X_{r-}+\psi(r,z),r) - f(X_{r-},r)\}\tilde{N}_D(dr\,dz), \quad a.s. \end{aligned}$$

for any $0 \le s < t \le T$, Here $X_{r-} := \lim_{\epsilon\downarrow 0} X_{r-\epsilon}$ is a predictable process.

Proof We will prove the formula (2.39) in the case $d = 1, d' = 1$ only. We first assume that $\psi(r, z)$ satisfies $\psi(r, z) = \psi(r, z)1_{|z|\ge\epsilon}$ a.s., where $\epsilon > 0$. Let $J_t = \int_0^t \int_{|z|\ge\epsilon} z\tilde{N}_D(dr\,dz)$. Its jumping times are defined by $\tau_0 = 0$ and for $n \ge 1$ by induction as

$$\begin{aligned} \tau_n &= \inf\{t;\ T > t > \tau_{n-1};\ J_t - J_{\tau_{n-1}} > 0\},\\ &= \infty, \quad \text{otherwise}. \end{aligned}$$

Then X_t has possible jumps at these stopping times. We have

$$\begin{aligned} f(X_t,t) - f(X_0,0) &= \sum_{n;\tau_n\le t}\{f(X_{\tau_n-},\tau_n) - f(X_{\tau_{n-1}},\tau_{n-1})\}\\ &\quad+\sum_{n;\tau_n\le t}\{f(X_{\tau_n},\tau_n) - f(X_{\tau_n-},\tau_n)\}\\ &= I_1 + I_2. \end{aligned}$$

We shall calculate I_1. If $\tau_{n-1} \le t$, we set

$$X'_t = X_{\tau_{n-1}} + \int_{\tau_{n-1}}^t \phi(r)\, dW_r + \int_{\tau_{n-1}}^t \upsilon'(r)\, dr,$$

where $\upsilon' = \upsilon - \int_D \psi \nu(dz)$. Then it holds that $X_t = X'_t$ for $\tau_{n-1} \le t < \tau_n$ and it has no jumps for $t \ge \tau_n$. We apply Itô's formula for the continuous Itô process X'_t. We have

$$\begin{aligned} f(X_{\tau_n-}, \tau_n) - f(X_{\tau_{n-1}}, \tau_{n-1}) &= \int_{\tau_{n-1}}^{\tau_n} \frac{\partial f}{\partial t}(X_r, r)\, dr \\ + \int_{\tau_{n-1}}^{\tau_n} \frac{\partial f}{\partial x}(X_{r-}, r)\phi(r)\, dW_r &+ \frac{1}{2}\int_{\tau_{n-1}}^{\tau_n} \frac{\partial^2 f}{(\partial x)^2}(X_r, r)\phi(r)^2\, dr \\ + \int_{\tau_{n-1}}^{\tau_n} \frac{\partial f}{\partial x}(X_r, r)\upsilon(r)\, dr &- \int_{\tau_{n-1}}^{\tau_n}\int_D \frac{\partial f}{\partial x}(X_r, r)\psi(r, z) n(dr\, dz). \end{aligned}$$

Summing up these for n such that $\tau_n \le t$, we get

$$\begin{aligned} I_1 = &\int_0^t \frac{\partial f}{\partial t}(X_r, r)\, dr + \int_0^t \frac{\partial f}{\partial x}(X_{r-}, r)\phi(r)\, dW_r + \frac{1}{2}\int_0^t \frac{\partial^2 f}{(\partial x)^2}(X_r, r)\phi(r)^2\, dr \\ &+ \int_0^t \frac{\partial f}{\partial x}(X_r, r)\upsilon(r)\, dr - \int_0^t\int_D \frac{\partial f}{\partial x}(X_r, r)\psi(r, z) n(dr\, dz). \end{aligned}$$

Next, we have

$$\begin{aligned} I_2 &= \sum_{0<r\le t, \Delta X_r \ne 0} \{f(X_r, r) - f(X_{r-}, r)\} \\ &= \int_0^t \int_{\mathbb{R}_0^{d'}} \big\{f(X_{r-} + \psi(r, z), r) - f(X_{r-}, r)\big\} N(dr\, dz) \\ &= \int_0^t \int_{\mathbb{R}_0^{d'}} \big\{f(X_{r-} + \psi(r, z), r) - f(X_{r-}, r)\big\} \tilde{N}_D(dr\, dz) \\ &\quad + \int_0^t \int_D \Big\{f(X_r + \psi(r, z), r) - f(X_r, r)\Big\} n(dr\, dz), \end{aligned}$$

almost surely, since $X_r = X_{r-}$ holds a.e. $dn\, dP$. From these two computations of I_1 and I_2, we get (2.39).

We next consider the general case. In the definition of Itô process, we replace $\psi(r, z)$ by $\psi_\epsilon(r, z) = \psi(r, z)1_{|z|>\epsilon}$. Let X_t^ϵ be the Itô process associated with α, υ and ψ_ϵ. Then X_t^ϵ has finite jumps, so that Itô's formula is valid for X_t^ϵ. Further, each term of (2.39) converges as $\epsilon \to 0$. Hence we get (2.39) for X_t. □

We will give L^p-estimations of the integral $\int_{\mathbb{U}_t} \psi(u)\tilde{N}(du)$. We transform the intensity measure n on $\mathbb{U}$ to a bounded measure m on $\mathbb{U}$ by setting

$$m(du) = \gamma(u)^2 n(du), \quad \text{where } \gamma(u) = |z| \wedge 1 \text{ for } u = (r, z).$$

For $p \geq 2$, we define

$$\mathbf{L}_{\mathbb{U}}^p = \Big\{\psi \in \mathbf{L}_{\mathbb{U}};\ \|\psi\|_p < \infty\Big\}, \tag{2.40}$$

where $\|\psi\|_p = E\big[\int_{\mathbb{U}} \big|\frac{\psi(u)}{\gamma(u)}\big|^p m(du)\big]^{\frac{1}{p}}$. In the case $p = 2$, the above $\mathbf{L}_{\mathbb{U}}^2$ coincides with $\mathbf{L}_{\mathbb{U}}^2$ defined by (2.31). We set $\mathbf{L}_{\mathbb{U}}^{\infty-} = \bigcap_{p>2} \mathbf{L}_{\mathbb{U}}^p$.

Proposition 2.6.1 *Let $p \geq 2$. Let $\psi(u)$ be an element of $\mathbf{L}_{\mathbb{U}}^p$. Then the stochastic integral $\int_{\mathbb{U}_t} \psi(u)\tilde{N}(du) \equiv \int_0^t \int_{\mathbb{R}_0^{d'}} \psi(r, z)\tilde{N}(dr\,dz)$ is p-th integrable. Further, there exists a positive constant C_p such that*

$$E\Big[\Big|\int_{\mathbb{U}_t} \psi(u)\tilde{N}(du)\Big|^p\Big]^{\frac{1}{p}} \leq C_p \|\psi\|_p \tag{2.41}$$

holds for any $0 < t \leq T$ and $\psi \in \mathbf{L}_{\mathbb{U}}^p$.

The equation (2.41) is rewritten as

$$E\Big[\Big|\int_0^t\int_{\mathbb{R}^{d'}} \psi(r, z)\tilde{N}(dr\,dz)\Big|^p\Big] \leq C_p E\Big[\int_0^t\int_{\mathbb{R}^{d'}} \Big|\frac{\psi(r, z)}{\gamma(z)}\Big|^p dr\mu(dz)\Big],$$

where $\gamma(z) = |z| \wedge 1$ and $\mu(dz) = \gamma(z)^2\nu(dz)$.

Proof We consider the process $Y_t = \int_0^t \int \psi\tilde{N}(dr\,dz)$. We apply the formula (2.39) for $f(x, t) = |x|^p$ and $D = \mathbb{R}_0^{d'}$. Then,

$$\begin{aligned} |Y_t|^p = &\int_0^t\int \big\{|Y_{r-} + \psi|^p - |Y_{r-}|^p\big\}\tilde{N}(dr\,dz) \\ &+ \int_0^t\int \big\{|Y_r + \psi|^p - |Y_r|^p - p|Y_r|^{p-2}Y_r\psi\big\}n(dr\,dz). \end{aligned} \tag{2.42}$$

Denote the first term of the right-hand side by Z_t. Then Z_t is a local martingale. We may choose an increasing sequence of stopping times τ_n such that $P(\tau_n < T) \to 0$ as $n \to \infty$, $Z_{t\wedge\tau_n}$ is a martingale with mean 0 and the last term in (2.42) stopped at τ_n is integrable. To make the notation simple, we denote $t \wedge \tau_n$ by t'. Since

$$\begin{aligned} |Y_r + \psi|^p - |Y_r|^p - p|Y_r|^{p-2}Y_r\psi &= \frac{1}{2}p(p-1)|Y_r + \theta\psi|^{p-2}\psi^2 \\ &\leq c_1|Y_r|^{p-2}\psi^2 + c_2|\psi|^p \end{aligned}$$

holds for some $|\theta| < 1$, we get from (2.42),

$$E[|Y_{t'}|^p] \le c_1 E\Big[\int_0^{t'}\int |Y_r|^{p-2}\psi^2 n(dr\,dz)\Big] + c_2 E\Big[\int_0^{t'}\int |\psi|^p n(dr\,dz)\Big]. \tag{2.43}$$

We shall compute the first term of the right-hand side. It holds that

$$\begin{aligned}\int_0^{t'}\int |Y_r|^{p-2}\psi^2 n(dr\,dz) &\le \sup_r |Y_r|^{p-2}\int_0^{t'}\int \psi^2 n(dr\,dz)\\ &\le c_3 \sup_r |Y_r|^p + c_4\Big(\int_0^{t'}\int \psi^2 n(dr\,dz)\Big)^{\frac{p}{2}},\end{aligned}$$

where $c_3 = \frac{p-2}{p}\lambda^{\frac{p}{p-2}}$, $c_4 = \frac{2}{p}\lambda^{-\frac{p}{2}}$ and λ is a positive constant. In the last inequality, we used the inequality $ab \le \frac{(\lambda a)^{p'}}{p'} + \frac{(b/\lambda)^{q'}}{q'}$, where $a, b > 0$, $\lambda > 0$ $p', q' > 1$ and $\frac{1}{p'} + \frac{1}{q'} = 1$. Therefore,

$$E\Big[\int_0^{t'}\int |Y_r|^{p-2}\psi^2 n(dr\,dz)\Big] \le c_3 E\Big[\sup_{0<r\le t'} |Y_r|^p\Big] + c_4 E\Big[\Big(\int_0^{t'}\int \psi^2 n(dr\,dz)\Big)^{\frac{p}{2}}\Big].$$

Substitute the above to right-hand side of (2.43). Then, using Doob's inequality, we get

$$\begin{aligned}E\Big[\sup_{0<r\le t'} |Y_r|^p\Big] &\le q^p c_1 c_3 E\Big[\sup_{0<r\le t'} |Y_r|^p\Big]\\ &+ q^p c_1 c_4 E\Big[\Big(\int_0^{t'}\int |\psi|^2 n(dr\,dz)\Big)^{\frac{p}{2}}\Big] + q^p c_2 E\Big[\int_0^{t'}\int |\psi|^p n(dr\,dz)\Big].\end{aligned}$$

We can choose λ such that $q^p c_1 c_3 < 1$. Then move $q^p c_1 c_3 E[\sup_{0<r\le t'} |Y_r|^p]$ to the left-hand side. Then we get the inequality

$$\begin{aligned}&E\Big[\sup_{0<r\le t'} |Y_r|^p\Big] \\ &\le c_5 E\Big[\Big(\int_0^{t'}\int |\psi|^2 n(dr\,dz)\Big)^{\frac{p}{2}}\Big] + c_6 E\Big[\int_0^{t'}\int |\psi|^p n(dr\,dz)\Big].\end{aligned} \tag{2.44}$$

So far we have chosen constants c_5, c_6 not depending on the form of martingales Y_t and stopping times τ_n. Therefore the inequality (2.44) holds for any $t' = t \wedge \tau_n$ with the common constants c_5, c_6. Now let n tend to infinity. Then we find that (2.44) holds for any t.

Further, we have for $p \geq 2$,

$$E\Big[\Big(\int_0^t\int |\psi(r,z)|^2 n(dr\,dz)\Big)^{\frac{p}{2}}\Big] \leq cE\Big[\int_{\mathbb{U}_t} \Big|\frac{\psi(u)}{\gamma(u)}\Big|^p m(du)\Big],$$

$$E\Big[\int_0^t\int |\psi(r,z)|^p n(dr\,dz)\Big] \leq E\Big[\int_{\mathbb{U}_t} \Big|\frac{\psi(u)}{\gamma(u)}\Big|^p m(du)\Big].$$

Therefore the assertion of the proposition follows. □

The above proposition is an analogue of the Burkholder–Davis–Gundy inequality (Proposition 2.2.2). The proposition tells us that if $\psi \in \mathbf{L}_{\mathbb{U}}^{\infty-}$, then the stochastic integral $\int_0^t \psi \tilde{N}(dr\,dz)$ belongs to $L^{\infty-}$.

We will study the regularity of stochastic integrals based on Poisson random measures with respect to parameter $\lambda \in \Lambda$, by applying the above proposition. Let Λ be $\mathbb{R}^e$ or its unit ball. Let $\psi_\lambda(r,z), r \in \mathbb{T}, z \in \mathbb{R}^{d'}, \lambda \in \Lambda$ be a measurable random field such that for any λ, z, $\psi_\lambda(r,z)$ is a predictable process. We assume that it is continuously differentiable in z and satisfies $\psi_\lambda(r,0) = 0$. Suppose that for some $p > e \vee 2$, there exists a positive constant c such that

$$\sup_{\lambda,r} E\Big[\int \Big|\frac{\psi_\lambda(r,z)}{\gamma(z)}\Big|^p \mu(dz)\Big] < \infty, \tag{2.45}$$

$$\sup_r E\Big[\int \Big|\frac{\psi_\lambda(r,z)}{\gamma(z)} - \frac{\psi_{\lambda'}(r,z)}{\gamma(z)}\Big|^p \mu(dz)\Big] \leq c_p|\lambda - \lambda'|^p \tag{2.46}$$

for any $\lambda, \lambda' \in \Lambda$ with a positive constant c_p. Then $\psi_\lambda(r,z)$ is said to belong to the space $\mathbf{L}_{\mathbb{U}}^{Lip,p}(\Lambda)$. Assume further that $\psi_\lambda(r,z)$ is n-times differentiable with respect to λ for any r, z a.e. and the derivatives $\tilde{\psi}_\lambda(r,z) = \partial_\lambda^{\mathbf{i}}\psi_\lambda(r,z)$ satisfy (2.45) and (2.46) for any multi-index $\mathbf{i}$ with $|\mathbf{i}| \leq n$, then ψ_λ is said to belong to the class $\mathbf{L}_{\mathbb{U}}^{n+Lip,p}(\Lambda)$.

Proposition 2.6.2 *If $\psi_\lambda \in \mathbf{L}_{\mathbb{U}}^{Lip,p}(\Lambda)$ for some $p > e \vee 2$, then the stochastic integral $\int_0^t\int_{\mathbb{R}_0^{d'}} \psi_\lambda(r,z)\tilde{N}(dr\,dz)$ has a modification which is continuous in λ and cadlag in t. If $\psi_\lambda \in \mathbf{L}_{\mathbb{U}}^{n+Lip,p}(\Lambda)$, then the integral has a modification which is n-times continuously differentiable in λ a.s. Further, for any $\mathbf{i}$ with $|\mathbf{i}| \leq n$, we have*

$$\partial_\lambda^{\mathbf{i}} \int_0^t\int_{\mathbb{R}_0^{d'}} \psi_\lambda(r,z)\tilde{N}(dr\,dz) = \int_0^t\int_{\mathbb{R}_0^{d'}} \partial_\lambda^{\mathbf{i}}\psi_\lambda(r,z)\tilde{N}(dr\,dz), \quad \forall t, \ a.s. \tag{2.47}$$

The proof of the above proposition can be carried out similarly to the proof of Proposition 2.3.1, making use of Proposition 2.6.1 instead of the Burkholder–Davis–Gundy inequality.

Note Stochastic integrals by Wiener processes and continuous martingales are discussed in many books. See Ikeda–Watanabe [41], Karatzas–Shreve [55], Revuz–

Yor [97], Rogers–Williams [98] and Oksendall [90]. Generalized Itô's formula or the Itô–Wentzell formula (Theorem 2.2.1) is taken from Kunita [59].

Stochastic integrals by Poisson random measures are discussed in Ikeda–Watanabe [40], Applebaum [1] and Kunita [60]. L^p-estimates of the integrals given in Proposition 2.6.2 is an improvement of L^p-estimates given in [1] and [60]. These L^p-estimates are needed for the study of the stochastic differential equation and the application of the Malliavin calculus to the solution. These problems will be discussed in Chaps. 3, 4 and 6.

2.7 Backward Integrals and Backward Calculus

We defined the stochastic integral by a Wiener process in Sect. 2.1. We could regard it as the forward Itô integral. We will define the backward Itô integral. We first show that a Wiener process is a backward martingale with respect to a two-sided filtration. Let $W_t, t \in \mathbb{T} = [0, T]$ be a Wiener process. It is called an $\{\mathcal{F}_{s,t}\}$-Wiener process if for any $0 \leq s < t \leq T$, $W_t - W_s$ is $\mathcal{F}_{s,t}$-measurable and the following three σ-fields are mutually independent:

$$\mathcal{F}_{0,s}, \quad \sigma(W_u - W_v; s \leq u < v \leq t), \quad \mathcal{F}_{t,T}.$$

An example of two-sided filtration is given by $\mathcal{F}_{s,t} = \sigma(W_u - W_v; s \leq u < v \leq t)$.

Let $W_t = (W_t^1, \ldots, W_t^{d'}), t \in \mathbb{T}$ be an $\{\mathcal{F}_{s,t}\}$-Wiener process. We consider a Backward process $\check{W}_t = (W_t^1, \ldots, W_t^{d'}), t \in \mathbb{T}$, where

$$\check{W}_t^k := W_t^k - W_T^k, \quad t \in \mathbb{T}.$$

Then $\check{W}_t^k, t \in \mathbb{T}, k = 1, \ldots, d'$ are *backward martingale* with respect to the backward filtration $\{\mathcal{F}_{t,T}\}$. i.e., $\check{W}_t^k$ satisfy $E[\check{W}_s^k|\mathcal{F}_{t,T}] = \check{W}_t^k$, for $s < t$, since $\check{W}_t^k - \check{W}_s^k = (W_t^k - W_s^k)$ is independent of $\mathcal{F}_{t,T}$.

A *backward predicable σ-field* is the σ-field generated by right continuous $\{\mathcal{F}_{t,T}\}$-adapted processes. Let $\check{\mathbf{L}}_{\mathbb{T}}$ be the set of all *backward predictable process* $\check{\phi}(r) = (\check{\phi}^1(r), \ldots, \check{\phi}^{d'}(r)), r \in \mathbb{T}$ satisfying $\int_0^T |\check{\phi}(r)|^2 dr < \infty$, a.s. We can define *backward Itô integrals* of $\check{\phi}$ based on the backward martingales $\check{W}_t$. In fact, if $\check{\phi}^k(r), r \in \mathbb{T}$ is a continuous process, we define

$$\begin{aligned}\int_s^T \check{\phi}^k(r)\, d\check{W}_r^k &:= \lim_{|\Pi|\to 0} \sum_{m=1}^n \check{\phi}^k(t_{m-1})(\check{W}_{t_{m-1}}^k - \check{W}_{t_m}^k) \\ &= \lim_{|\Pi|\to 0} \sum_{m=1}^n \check{\phi}^k(t_{m-1})(W_{t_{m-1}}^k - W_{t_m}^k),\end{aligned}$$

where $\Pi = \{s = t_n < \cdots < t_0 = T\}$ are partitions of the interval $[s, T]$. It is a backward local L^2-martingale with respect to $s \in [0, T]$. The backward Itô integral coincides with the forward Itô integral $\int_s^T \check{\phi}^k(r)\,dW_r^k$, provided that $\check{\phi}^k(r)$ is a forward–backward predictable process of bounded variation. The backward Itô integral can be extended for any $\check{\phi}(r)$ of $\check{\mathbf{L}}_{\mathbb{T}}$. It is a continuous backward local martingale. We define $\int_s^t \check{\phi}^k(r)\,dW_r^k = \int_s^T \check{\phi}^k(r)\,dW_r^k - \int_t^T \check{\phi}^k(r)\,dW_r^k$. Further, we set $\int_s^t (\check{\phi}(r), dW_r) = \sum_{k=1}^{d'} \int_s^t \check{\phi}^k(r)\,dW_r^k$.

A *backward symmetric integral* is well defined for a backward semi-martingale $\check{\phi}(r)$ with bounded jumps. In fact, the following limit exists:

$$\lim_{|\Pi|\to 0} \sum_m \frac{1}{2}\{\check{\phi}^k(t_m) + \check{\phi}^k(t_{m-1})\}(W_{t_{m-1}}^k - W_{t_m}^k)$$

$$= \lim_{|\Pi|\to 0} \Big\{\sum_m \check{\phi}(t_{m-1})(W_{t_{m-1}}^k - W_{t_m}^k) - \frac{1}{2}(\langle \check{\phi}, W^k\rangle_T^\Pi - \langle \check{\phi}, W^k\rangle_s^\Pi)\Big\}.$$

We denote it by $\int_s^T \check{\phi}^k(r) \circ dW_r^k$ and call it the backward symmetric integral of $\check{\phi}^k(r)$ by dW_t^k. For $0 \le s < t \le T$, the backward symmetric integral $\int_s^t \check{\phi}^k(r) \circ dW_r^k$ is defined by $\int_s^T \check{\phi}^k(r) \circ dW_r^k - \int_t^T \check{\phi}^k(r) \circ dW_r^k$. Then it holds that

$$\int_s^t \check{\phi}^k(r) \circ dW_r^k = \int_s^t \check{\phi}(r)\,dW_r^k - \frac{1}{2}\Big(\langle \check{\phi}, W^k\rangle_t - \langle \check{\phi}, W^k\rangle_s\Big). \tag{2.48}$$

Further, we set $\int_s^t (\check{\phi}(r), \circ dW_r) = \sum_{k=1}^{d'} \int_s^t \phi^k(r) \circ dW_r^k$.

We next consider the backward integral by Poisson random measure. Let $\{\mathcal{F}_{s,t}\}$ be a two sided filtration. Let $N(dr\,dz)$ be a Poisson random measure with the intensity measure $n(dr\,dz) = dr\nu(dz)$. We call it an $\{\mathcal{F}_{s,t}\}$-Poisson random measure if (1) for any $s < t$,

$$\mathcal{F}_{s,t} \supset \sigma(N((u, v] \times E); s \le u \le v \le t),$$

and (2) for any $s < t$, the following three σ-fields are independent;

$$\mathcal{F}_{0,s}, \quad \sigma(N((u, v] \times E); s \le u \le v \le t), \quad \mathcal{F}_{t,T}.$$

Let N be an $\{\mathcal{F}_{s,t}\}$-Poisson random measure with the intensity measure $n(dr\,dz) = dr\nu(dz)$. Let $\check{\mathbf{L}}_{\mathbb{U}}$ be the set of *backward predictable* processes $\check{\psi}(r, z)$ satisfying $\int_0^T \int |\check{\psi}(r, z)|^2 n(dr\,dz) < \infty$ a.s. For $\check{\psi} \in \check{\mathbf{L}}_{\mathbb{U}}$, we will define the *backward integral* of $\check{\psi}$ based on the compensated Poisson random measure. In Sect. 1.5, we defined a backward martingale as a caglad (left continuous with right-hand limit) process. We can define the backward integral of $\check{\psi}$ by the compensated Poisson random measure $\tilde{N}$ as a caglad backward martingale, in the same way as the forward integral discussed in Sect. 2.5. Denote it by $\check{X}_s$. Set

$\check{X}_{s+} = \lim_{\epsilon \downarrow 0} \check{X}_{s+\epsilon}$. Then it holds that $\check{X}_s = \check{X}_{s+}$ a.s. for any s, since the Poisson random measure satisfies $N(\{s\} \times \mathbb{R}_0^{d'}) = 0$ almost surely for any s. Then $\check{X}_{s+}$ is a cadlag backward martingale. We denote it as $\int_s^T \int_{\mathbb{R}_0^{d'}} \check{\psi}(r,z)\tilde{N}(dr\,dz)$. We can also define the backward integrals $\int_s^T \int_B \check{\psi}(r,z)\tilde{N}_D(dr\,dz)$ as cadlag processes, where D and B are Borel subsets of $\mathbb{R}_0^{d'}$ as in the previous section.

Let $\check{X}_s = (\check{X}_s^1, \ldots, \check{X}_s^d), s \in [0,T]$ be a backward cadlag process such that

$$\check{X}_s^i = \check{X}_T + \int_s^T (\check{\phi}^i(r), dW_r) + \int_s^T \check{v}^i(r)\,dr + \int_s^T \int_{\mathbb{R}_0^{d'}} \check{\psi}^i(r,z)\tilde{N}_D(dr\,dz) \tag{2.49}$$

holds for any $0 \le s < T$, where $\check{X}_T$ is a constant, $\check{\phi}^i(r) \in \check{\mathbf{L}}_{\mathbb{T}}$, $\check{v}^i(r) \in \check{L}_{\mathbb{T}}$ and $\check{\psi}^i(r,z)1_D(z) \in \check{\mathbf{L}}_{\mathbb{U}}$. Here $\check{L}_{\mathbb{T}}$ is the set of backward predictable processes $\check{v}(r)$ satisfying $\int_0^T |\check{v}(r)|\,dr < \infty$ a.s. We call it a *backward Itô process*.

Below, we give a backward differential rule for the backward Itô Process. Since the proof is parallel to the proof of Itô's formula for a forward Itô process (Theorem 2.6.1), it will be omitted.

Theorem 2.7.1 *Let $\check{X}_s = (\check{X}_s^1, \ldots, \check{X}_s^d), s \in [0,T]$ be a backward Itô process given by (2.49). Let $f(x^1, \ldots, x^d, t)$ be a function of $C^{2,1}$-class. Then, $f(\check{X}_s, s), s \in [0,T]$ is a one-dimensional backward Itô process. Further, for any $0 \le s < t \le T$, it is written as*

$$\begin{aligned} f(\check{X}_s, s) &= f(\check{X}_t, t) - \int_s^t \frac{\partial f}{\partial t}(\check{X}_r, r)\,dr \\ &+ \sum_{i=1}^d \sum_{k=1}^{d'} \int_s^t \frac{\partial f}{\partial x_i}(\check{X}_r, r)\check{\phi}^{ik}(r)\,dW_r^k + \sum_{i=1}^d \int_s^t \frac{\partial f}{\partial x_i}(\check{X}_r, r)\check{v}^i(r)\,dr \\ &+ \frac{1}{2}\sum_{i,j=1}^d \int_s^t \frac{\partial^2 f}{\partial x_i \partial x_j}(\check{X}_r, r)\Big(\sum_{k-1}^{d'} \check{\phi}^{ik}(r)\check{\phi}^{jk}(r)\Big)\,dr \\ &+ \int_s^t \int_D \Big\{ f(\check{X}_r + \check{\psi}(r,z), r) - f(\check{X}_r, r) - \sum_{i=1}^d \frac{\partial f}{\partial x_i}(\check{X}_r, r)\check{\psi}^i(r,z) \Big\} n(dr\,dz) \\ &+ \int_s^t \int_{\mathbb{R}_0^{d'}} \{ f(\check{X}_r + \check{\psi}(r,z), r) - f(\check{X}_r, r)\}\tilde{N}_D(dr\,dz), \quad a.s. \end{aligned} \tag{2.50}$$

Chapter 3
Stochastic Differential Equations and Stochastic Flows

Abstract In this chapter, we show that solutions of a continuous symmetric stochastic differential equation (SDE) on a Euclidean space define a continuous stochastic flow of diffeomorphisms and that solutions of an SDE with diffeomorphic jumps define a right continuous stochastic flow of diffeomorphisms. Sections 3.1 and 3.2 are introductions. Definitions of these SDEs and stochastic flows will be given and the geometric property of solutions will be explained as well as basic facts. Rigorous proof of these facts will be given in Sects. 3.3, 3.4, 3.5, 3.6, 3.7, 3.8 and 3.9. In Sect. 3.3, we study another Itô SDE with parameter, called the master equation. Applying results of Sect. 3.3, we show in Sect. 3.4 that solutions of the original SDE define a stochastic flow of C^∞-maps. For the proof of the diffeomorphic property, we need further arguments. In Sect. 3.5 we consider backward SDE and backward stochastic flow of C^∞-maps. Further, the forward–backward calculus for stochastic flow will be discussed in Sects. 3.5, 3.6 and 3.8. These facts will be applied in Sects. 3.7 and 3.9 for proving the diffeomorphic property of solutions.

3.1 Geometric Property of Solutions I; Case of Continuous SDE

Let $\mathbb{T} = [0, \infty)$. Let d be a positive integer and l, m be nonnegative integers. Let us introduce some classes of functions on $\mathbb{R}^d \times \mathbb{T}$. A function $f(x, t)$ on $\mathbb{R}^d \times \mathbb{T}$ is said to be of $C^{l,m}$-class (or $C^{l,m}$-function), if it is l-times continuously differentiable with respect to x, and m-times continuously differentiable with respect to t. Further, if f and its derivatives are all bounded continuous functions of x, t, the function f is said to be of $C_b^{l,m}$-class (or $C_b^{l,m}$-function). Further, if f is of class $C_b^{l,m}$ for any positive integer l, f is said to be of $C_b^{\infty,m}$-class. Let d' be another positive integer. A function $f(x, t, z)$ on $\mathbb{R}^d \times \mathbb{T} \times \mathbb{R}^{d'}$ is said to be of $C_b^{l,m,n}$-class if it is l-times continuously differentiable with respect to x, m-times continuously differentiable with respect to t, and n-times continuously differentiable with respect to z, and f

H. Kunita, *Stochastic Flows and Jump-Diffusions*, Probability Theory and Stochastic Modelling 92, https://doi.org/10.1007/978-981-13-3801-4_3

and its derivatives are bounded continuous functions of x, t, z. Further, if f is of class $C_b^{l,m,n}$ for any positive integer l, f is said to be of $C_b^{\infty,m,n}$-class.

Let ϕ be a map from $\mathbb{R}^d$ into itself. Suppose that the map ϕ is 'one to one' and 'onto'. Then the inverse map ϕ^{-1}; $\mathbb{R}^d \to \mathbb{R}^d$ is well defined. If both maps ϕ and ϕ^{-1} are C^∞-maps, ϕ is called a *diffeomorphism* of $\mathbb{R}^d$ or a *diffeomorphic map* on $\mathbb{R}^d$.

Let $W_t = (W_t^1, \ldots, W_t^{d'})$, $t \in \mathbb{T}$ be a d'-dimensional Wiener process. For $0 \leq s < t < \infty$, we denote by $\mathcal{F}_{s,t}$ the smallest σ-field containing all null sets of $\mathcal{F}$ with respect to which random variables $W_u - W_v, s \leq u, v \leq t$ are measurable. Then $\{\mathcal{F}_{s,t}, 0 \leq s < t < \infty\}$ is called a *two-sided filtration* generated by the Wiener process. We set $\mathcal{F}_t = \mathcal{F}_{0,t}$. Then $\{\mathcal{F}_t\}$ is a filtration, with respect to which W_t^i are $\{\mathcal{F}_t\}$-Wiener processes.

Let $V_k(x,t) = (V_k^1(x,t), \ldots, V_k^d(x,t))$, $k = 0, 1, \ldots, d'$ be a family of vector fields on $\mathbb{R}^d$ with parameter $t \in \mathbb{T}$ of $C_b^{\infty,1}$-class. We consider a continuous symmetric SDE on $\mathbb{R}^d$ with coefficients $V_k(x,t), k = 0, 1, \ldots, d'$. Suppose we are given $t_0 \in \mathbb{T}$ and a $\mathcal{F}_{t_0}$-measurable random variable X_0 with values in $\mathbb{R}^d$. Suppose that there exists a continuous $\mathbb{R}^d$-valued $\{\mathcal{F}_t\}$-semi-martingale $X_t = (X_t^1, \ldots, X_t^d)$, $t \geq t_0$ satisfying

$$X_t^i = X_0^i + \sum_{k=1}^{d'} \int_{t_0}^t V_k^i(X_r, r) \circ dW_r^k + \int_{t_0}^t V_0^i(X_r, r)\, dr, \quad i = 1, \ldots, d$$

or in vector notation,

$$X_t = X_0 + \sum_{k=1}^{d'} \int_{t_0}^t V_k(X_r, r) \circ dW_r^k + \int_{t_0}^t V_0(X_r, r)\, dr \tag{3.1}$$

for all $t \geq t_0$ almost surely. Then the semi-martingale $X_t, t \geq t_0$ is called a solution of the *symmetric SDE* with coefficients $V_k(x,t), k = 0, \ldots, d'$, starting from X_0 at time t_0.

The equation is said to have a (path-wise) unique solution if two solutions $X_t, t \geq t_0$ and $X_t', t \geq t_0$ of equation (3.1) having the same initial condition should satisfy $X_t = X_t'$ for all $t \geq t_0$ a.s.

We will discuss the existence and the uniqueness of the solution of the equation. However, instead of this equation, it is often convenient to rewrite it as Itô's stochastic differential equation. Suppose that X_t is a solution of equation (3.1). By Proposition 2.4.1, the symmetric integral in equation (3.1) can be rewritten as

$$\int_{t_0}^t V_k(X_r, r) \circ dW_r^k = \int_{t_0}^t V_k(X_r, r)\, dW_r^k + \frac{1}{2}\langle V_k(X_t, t), W_t^k \rangle_{t_0,t}. \tag{3.2}$$

Here $V_k(X_t, t)$ is a continuous semi-martingale and it is represented by Itô's formula (Corollary 2.2.1) as

$$V_k(X_t,t)-V_k(X_0,t_0)$$
$$=\sum_{i,l}\int_{t_0}^t \frac{\partial V_k}{\partial x_i}(X_r,r)V_l^i(X_r,r)\,dW_r^l+\int_{t_0}^t \frac{\partial V_k}{\partial t}(X_r,r)\,dr$$
$$+\frac{1}{2}\sum_{i,j}\int_{t_0}^t \frac{\partial^2 V_k}{\partial x_i \partial x_j}(X_r,r)(\sum_l V_l^i(X_r,r)V_l^j(X_r,r))\,dr.$$

Then quadratic covariation of $V_k(X_t,t)$ and W_t^k on the interval $[t_0,t]$ is computed as

$$\langle V_k(X_t,t),W_t^k\rangle_{t_0,t}=\sum_l\Big\langle\int_{t_0}^t V_l(r)V_l(X_r,r)\,dW_r^l,W_t^k\Big\rangle$$
$$=\int_{t_0}^t V_k(r)V_k(X_r,r)\,dr,$$

where $V_k(r)V_k(x,r)=\sum_i V_k^i(x,r)\frac{\partial}{\partial x_i}V_k(x,r)$. Consequently, setting

$$V_0'(x,r)=V_0(x,r)+\frac{1}{2}\sum_{k=1}^{d'}V_k(r)V_k(x,r), \tag{3.3}$$

a solution $X_t, t\geq t_0$ of equation (3.1) satisfies the following continuous *Itô SDE*:

$$X_t=X_0+\sum_{k=1}^{d'}\int_{t_0}^t V_k(X_r,r)\,dW_r^k+\int_{t_0}^t V_0'(X_r,r)dr. \tag{3.4}$$

Conversely, if $X_t, t\geq t_0$ is a solution of Itô's equation (3.4), $X_t, t\geq t_0$ is a continuous semi-martingale and satisfies the symmetric equation (3.1). It is known that the equation (3.4) has a path-wise unique solution, if coefficients $V_0', V_i, i=1,\ldots,d'$ are Lipschitz continuous with respect to x. Therefore, equation (3.1) has also a unique solution.

For a given $x\in\mathbb{R}^d$ and $s\in\mathbb{T}$, let $X_t^{x,s}, t\geq s$ be the solution of equation (3.2) starting from x at time s. We are interested in the geometric property of the family of solutions $\{X_t^{x,s}, t\geq s\}$. For this, we need the definition of the stochastic flow. Let C be the set of continuous maps from $\mathbb{R}^d$ into itself. Let $\{\Phi_{s,t}; 0\leq s<t<\infty\}$ (or simply denoted by $\{\Phi_{s,t}\}$) be a family of C-valued random variables. It is called a *stochastic flow of* C^∞*-maps* if it satisfies:

(1) Maps $\Phi_{s,t};\mathbb{R}^d\to\mathbb{R}^d$ are C^∞ a.s. for any $s<t$.
(2) $\Phi_{t,u}\circ\Phi_{s,t}=\Phi_{s,u}$ a.s. for any $s<t<u$. Here $\phi\circ\psi$ is the composite of two maps ϕ,ψ of $\mathbb{R}^d$ into itself.

Further, if it satisfies the following (1′) and the above (2), then it is called a *stochastic flow of diffeomorphisms*.

(1′) Maps $\Phi_{s,t}; \mathbb{R}^d \to \mathbb{R}^d$ are diffeomorpshims a.s. for any $s < t$.

A stochastic flow $\{\Phi_{s,t}\}$ is called *continuous* if $\Phi_{s,t}(x), x \in \mathbb{R}^d$ and its derivatives $\partial^{\mathbf{i}}\Phi_{s,t}(x)$ are continuous with respect to (x, s, t) $(s < t)$ a.s. for any multi-index $\mathbf{i}$.

Let $\{\Phi_{s,t}\}$ be a continuous stochastic flow of diffeomorphisms. If for any x, s, $X_t^{x,s} := \Phi_{s,t}(x)$ is a solution of equation (3.1) starting from x at time s, it is said that the stochastic flow is generated by the SDE, or the SDE defines the stochastic flow.

We want to show the following two facts:

1. *Solutions of equation* (3.1) *generate a continuous stochastic flow* $\{\Phi_{s,t}\}$ *of diffeomorphisms.*
2. *Let* $\check{\Psi}_{s,t}$ *be the inverse map of* $\Phi_{s,t}$. *Then for any* x, t, $\check{X}_s = \check{\Psi}_{s,t}(x), s < t$ *is a backward semi-martingale with respect to the two-sided filtration* $\{\mathcal{F}_{s,t}\}$ *and satisfies the backward continuous symmetric stochastic differential equation:*

$$\check{\Psi}_{s,t}(x) = x - \sum_{k=1}^{d'} \int_s^t V_k(\check{\Psi}_{r,t}(x), r) \circ dW_r^k - \int_s^t V_0(\check{\Psi}_{r,t}(x), r)\, dr. \tag{3.5}$$

$\{\check{\Psi}_{s,t}\}$ is called the *inverse flow* of the stochastic flow $\{\Phi_{s,t}\}$. The fact that equation (3.1) defines a continuous stochastic flow of C^∞-maps will be proved in Sect. 3.4. The proof of the diffeomorphic property is difficult. It will be given in Sect. 3.7, after the discussion of the backward symmetric stochastic differential equation in Sect. 3.5.

It should be noted that equation (3.1) for the flow $\{\Phi_{s,t}\}$ and equation (3.5) for the inverse flow $\{\check{\Psi}_{s,t}\}$ are symmetric, because these two equations are written by using symmetric integrals. We write $W_t^0 = t$ conventionally and we will write the symmetric equation (3.1) for the solution $\Phi_{s,t}(x)$ by

$$\Phi_{s,t}(x) = x + \sum_{k=0}^{d'} \int_s^t V_k(\Phi_{s,r}(x), r) \circ dW_r^k. \tag{3.6}$$

Then the inverse flow satisfies the *backward symmetric SDE*.

$$\check{\Psi}_{s,t}(x) = x - \sum_{k=0}^{d'} \int_s^t V_k(\check{\Psi}_{r,t}(x), r) \circ dW_r^k. \tag{3.7}$$

Using Itô integrals, the symmetry will not be valid. See the discussion after Theorem 3.7.1. Another reason that we consider the symmetric equation is that the extension of the definition of (3.1) to manifolds is direct forward. It will be discussed in Chap. 7.

We are also interested in the following two properties. One is the L^p-estimate of the flow $\{\Phi_{s,t}\}$. The other is the property of independent increments.

3. For any $p \geq 2$, $s < t$ and $|\mathbf{i}| > 0$, we have

$$\sup_x \frac{E[|\Phi_{s,t}(x)|^p]}{(1+|x|)^p} < \infty, \quad \sup_x E[|\partial^{\mathbf{i}}\Phi_{s,t}(x)|^p] < \infty. \tag{3.8}$$

$$\sup_x \frac{E[|\check{\Psi}_{s,t}(x)|^p]}{(1+|x|)^p} < \infty, \quad \sup_x E[|\partial^{\mathbf{i}}\check{\Psi}_{s,t}(x)|^p] < \infty. \tag{3.9}$$

4. *The flow $\{\Phi_{s,t}\}$ has independent increments; for any $0 \leq t_0 < \cdots t_n \leq T$, $\Phi_{t_{i-1},t_i}, i = 1, \ldots, n$ are independent.*

Let $\mathbb{G}$ be the set of all diffeomorphic maps on $\mathbb{R}^d$. It is a group by the multiplication $\phi\psi := \phi \circ \psi$ and the inverse $\phi^{-1} :=$ inverse map of ϕ. We define a metric d on $\mathbb{G}$ by $d(\phi, \psi) = \sum_{N=1}^{\infty} \frac{1}{2^N} \frac{d_N(\phi,\psi)}{1+d_N(\phi,\psi)}$, where

$$d_N(\phi, \psi) = \sum_{|\mathbf{i}|\leq N} \Big(\sup_{|x|\leq N} |\partial^{\mathbf{i}}(\phi(x) - \psi(x))| + \sup_{|x|\leq N} |\partial^{\mathbf{i}}(\phi^{-1}(x) - \psi^{-1}(x))| \Big).$$

Then $\mathbb{G}$ is a complete metric space. The flow $\{\Phi_{s,t}\}$ may be regarded as a $\mathbb{G}$-valued random field, satisfying $\Phi_{s,t} = \Phi_{t_n,t}\Phi_{t_{n-1},t_n} \cdots \Phi_{t_1,s}$ for any $t_0 < t_1 < \cdots < t_n$. Hence it has independent increments with respect to the multiplication. Then the continuous stochastic flow $\{\Phi_{s,t}\}$ may be considered as a (time inhomogeneous) Brownian motion (Wiener process) on the group $\mathbb{G}$.

3.2 Geometric Property of Solutions II; Case of SDE with Jumps

Discussions for the flow property are often parallel to those for a more general class of SDE having jumps. We shall introduce SDEs on $\mathbb{R}^d$ governed by a Wiener process and a Poisson random measure. These SDEs may have jumps and can be regarded as an extension of a continuous SDE.

Let $\mathbb{R}_0^{d'} = \mathbb{R}^{d'} \setminus \{0\}$. Its elements are denoted by $z = (z^1, \ldots, z^{d'})$. Let $N(dr\,dz)$ be a Poisson random measure on $\mathbb{U} = \mathbb{T} \times \mathbb{R}_0^{d'}$ with the intensity measure $n(dr\,dz) := dr\nu(dz)$, where ν is a Lévy measure on $\mathbb{R}_0^{d'}$. We assume that it is independent of the Wiener process W_t. For $0 \leq s < t < \infty$, we denote by $\mathcal{F}_{s,t}$ the smallest σ-field containing all null sets of $\mathcal{F}$ with respect to which random variables $W_u - W_v, s \leq u, v \leq t$ and $N((u, v] \times E), s \leq u < v \leq t, E \in \mathcal{B}(\mathbb{R}_0^{d'})$ are measurable. Then $\{\mathcal{F}_{s,t}\}$ is a two-sided filtration. It is called a *two-sided filtration* generated by the Wiener process W_t and the Poisson random measure $N(dr\,dz)$. We set $\mathcal{F}_t = \mathcal{F}_{0,t}$. Then $\{\mathcal{F}_t\}$ is a filtration of sub σ-fields of $\mathcal{F}$.

Let $V_k(x,t), k = 1, \dots, d'$ be a family of vector fields of $C_b^{\infty,1}$-class defined on $\mathbb{R}^d$. Let $g(x,t,z)$ be a map from $\mathbb{R}^d \times \mathbb{T} \times \mathbb{R}^{d'}$ to $\mathbb{R}^d$ satisfying the following two conditions.

Condition (J.1). Components $g^i(x,t,z), i = 1, \dots, d$ of $g(x,t,z)$ are functions of $C_b^{\infty,1,2}$-class on $\mathbb{R}^d \times \mathbb{T} \times \mathbb{R}^{d'}$. Further, $g(x,t,0) = 0$ for any x,t.

Condition (J.2). Set $\phi_{t,z}(x) = g(x,t,z)+x$. For any t, z, the map $\phi_{t,z}; \mathbb{R}^d \to \mathbb{R}^d$ is a diffeomorphism of $\mathbb{R}^d$. Further, $h(x,t,z) := x - \phi_{t,z}^{-1}(x)$ satisfy Condition (J.1).

Let $z = (z^1, \dots, z^{d'}) \in \mathbb{R}^{d'}$. We set

$$b_\epsilon^k = \int_{\epsilon \le |z| \le 1} z^k \nu(dz), \quad \tilde{V}_k(x,t) = \partial_{z^k} \phi_{t,z}(x)\Big|_{z=0}, \quad k = 1, \dots, d'.$$

$\tilde{V}_k(x,t), k = 1, \dots, d'$ are called *tangent vectors of maps* $\phi_{t,z}(x)$ at $z = 0$.

Suppose we are given time $0 \le t_0 < \infty$ and $\mathcal{F}_{t_0}$-adapted $\mathbb{R}^d$-valued random variable X_0. Suppose that there exists an $\mathbb{R}^d$-valued semi-martingale $X_t, t \ge t_0$ (with bounded jumps) satisfying

$$X_t = X_0 + \sum_{k=0}^{d'} \int_{t_0}^t V_k(X_r, r) \circ dW_r^k \tag{3.10}$$

$$+ \lim_{\epsilon \to 0} \Big\{ \int_{t_0}^t \int_{|z| \ge \epsilon} g(X_{r-}, r, z) N(dr\,dz) - \sum_{k=1}^{d'} b_\epsilon^k \int_{t_0}^t \tilde{V}_k(X_{r-}, r)\, dr \Big\}.$$

Then $X_t, t \ge t_0$ is called a solution of a *symmetric SDE with characteristics* $(V_k(x,t), k = 0, \dots, d', g(x,t,z), \nu)$, starting from X_0 at time t_0. We should check that the last term is well defined for any semi-martingale $X_t, t \ge t_0$. It holds that

$$\begin{aligned} &\int_{t_0}^t \int_{|z| \ge \epsilon} g(X_{r-}, r, z) N(dr\,dz) - \sum_{k=1}^{d'} b_\epsilon^k \int_{t_0}^t \tilde{V}_k(X_{r-}, r)\, dr \\ &= \int_{t_0}^t \int_{|z| \ge \epsilon} g(X_{r-}, r, z) \tilde{N}(dr\,dz) \\ &\quad + \int_{t_0}^t \int_{|z| \ge \epsilon} \Big\{ g(X_{r-}, r, z) - 1_D(z) \sum_{k=1} z^k \tilde{V}_k(X_r, r) \Big\} n(dr\,dz), \end{aligned} \tag{3.11}$$

where $D = \{z \in \mathbb{R}_0^{d'};\ |z| \le 1\}$. Let $\epsilon \to 0$. The first term of the right-hand side converges to an L^2-martingale, since $E[\iint |g(X_r, r, z)|^2 n(dr\,dz)] < \infty$. Since $g(x,t,z) - \sum_k z^k \tilde{V}_k(x,t) = O(|z|^2)$ holds by the Taylor expansion, the second term converges as $\epsilon \to 0$. Consequently, for any cadlag process $X_t, t \ge t_0$, the left-hand side of the above converges in L^2 as $\epsilon \to 0$.

In order to solve the equation, we will consider the *Itô SDE with jumps*

$$X_t = X_0 + \int_{t_0}^t \alpha(X_{r-}, r)\, dW_r + \int_{t_0}^t \beta(X_r, r)\, dr + \int_{t_0}^t \int \chi(X_{r-}, r, z)\tilde{N}(dr\, dz), \tag{3.12}$$

where $\alpha(x, r) = (V_k^i(x, r))$,

$$\beta(x, r) = V_0(x, r) + \frac{1}{2}\sum_{k=1}^{d'} V_k(r)V_k(x, r) + \lim_{\epsilon \to 0} \int_{|z| \geq \epsilon} \Big\{ g(x, r, z) - 1_D(z)(\sum_{k=1}^{d'} z^k \tilde{V}_k(x, r)) \Big\}\, d\nu, \tag{3.13}$$

and $\chi(x, r, z) = g(x, r, z)$.

It will be shown in Sect. 3.4 that the solution of the Itô equation (3.12) exists uniquely. Further, for the solution X_t, the symmetric integral of equation (3.10) is well defined and in fact is the unique solution of equation (3.10).

Let $\{\Phi_{s,t}\}$ be a stochastic flow of C^∞-maps (or of diffeomorphisms). If $X_t^{x,s} := \Phi_{s,t}(x), t \geq s$ is a solution of equation (3.10) starting from x at time s for any x, s, the stochastic flow is said to be generated by SDE (3.10) and SDE (3.10) defines the stochastic flow $\{\Phi_{s,t}\}$. One of main subjects of this chapter is to prove that under Conditions (J.1) and (J.2), SDE (3.10) generates a stochastic flow $\{\Phi_{s,t}\}$ of diffeomorphisms.

We begin with some preliminary observations for solutions of some special equations. If $g(x, t, z) \equiv 0$, equation has no jump part. It is a continuous SDE written as (3.1). We next consider an SDE which does not contain a diffusion part. We assume that the Lévy measure is a finite measure. Then the jump part of equation (3.10) is written simply as

$$\int_{t_0}^t \int_{\mathbb{R}^{d'}} g(X_{r-}, r, z)N(dr\, dz) - \sum_k b_0^k \int_{t_0}^t \tilde{V}_k(X_{r-}, r)\, dr,$$

where $b_0^k = \lim_{\epsilon \to 0} b_\epsilon^k$. Hence setting $\tilde{V}_0(x, t) = V_0(x, t) - \sum_k b_0^k \tilde{V}_k(x, t)$, the equation is written as

$$X_t = X_0 + \int_{t_0}^t \tilde{V}_0(X_r, r)\, dr + \int_{t_0}^t \int_{\mathbb{R}_0^{d'}} g(X_{r-}, r, z)N(dr\, dz). \tag{3.14}$$

Suppose first that $\tilde{V}_0(x, t) = 0$. The above Poisson random measure $N(dr\, dz)$ is constructed from a Poisson process N_t with the intensity $\lambda = \nu(\mathbb{R}_0^{d'})$ and a sequence of i.i.d. random variables $\{S_n\}$ with the common law $\mu(E) = \nu(E)/\lambda$. In fact, $N((t_0, t], E) := \sum_{0<n\le N_t - N_{t_0}} 1_E(S_n)$ is a Poisson random measure on $\mathbb{U} = [t_0, \infty) \times \mathbb{R}_0^d$ with intensity $dr\nu(dz)$ (Proposition 1.4.3). Now, let $t_0 \equiv \tau_0 < \tau_1 < \cdots \tau_n < \cdots$ be a sequence of jumping times of the process $N_t - N_{t_0}, t \in [t_0, \infty)$. We set

$$\Phi_{t_0,t}(x) = \begin{cases} x, & \text{if } t < \tau_1, \\ \phi_{\tau_n,S_n} \circ \cdots \circ \phi_{\tau_1,S_1}(x), & \text{if } \tau_n \le t < \tau_{n+1},\ n = 1, 2, \ldots \end{cases} \tag{3.15}$$

Here, $\phi_{\tau_n,S_n} \circ \cdots \circ \phi_{\tau_1,S_1}$ are composites of maps $\phi_{\tau_m,S_m}; \mathbb{R}^d \to \mathbb{R}^d, m = 1, \ldots, n$. Then the map $\Phi_{t_0,t}; \mathbb{R}^d \to \mathbb{R}^d$ is a diffeomorphism a.s., since $\phi_{\tau_m,S_m}, m = i, \ldots, n$ are diffeomorphic. Further, the stochastic process $X_t := \Phi_{t_0,t}(X_0), t \ge t_0$ satisfies

$$\begin{aligned} X_t &= X_0 + \sum_{m; t_0 < \tau_m \le t} \left\{\phi_{\tau_m,S_m}(X_{\tau_{m-1}}) - X_{\tau_{m-1}}\right\} \\ &= X_0 + \sum_{m; t_0 < \tau_m \le t} g(X_{\tau_{m-1}}, \tau_m, S_m) \\ &= X_0 + \int_{t_0}^t \int_{\mathbb{R}_0^{d'}} g(X_{r-}, r, z) N(dr\, dz). \end{aligned} \tag{3.16}$$

Therefore, $\{\Phi_{t_0,t}\}$ defined by (3.15) is a stochastic flow of diffeomorphisms generated by the jump SDE (3.16). It may be considered as a compound Poisson process with values in the group $\mathbb{G}$ of diffeomorphisms.

In the case $\tilde{V}_0(x, t) \ne 0$, let $\varphi_{s,t}(x)$ be the *deterministic flow* generated by the vector field $\tilde{V}_0(x, t)$; i.e., $\varphi_{s,t}(x) = x + \int_s^t \tilde{V}_0(\varphi_{s,r}(x), r)\, dr$. We set

$$\Phi_{t_0,t} = \begin{cases} \varphi_{t_0,t}, & \text{if } t < \tau_1, \\ \varphi_{\tau_n,t} \circ \phi_{\tau_n,S_n} \circ \varphi_{\tau_{n-1},\tau_n} \circ \cdots \circ \phi_{\tau_1,S_1} \circ \varphi_{t_0,\tau_1}, & \text{if } \tau_n \le t < \tau_{n+1}. \end{cases}$$

Then $\Phi_{t_0,t}; \mathbb{R}^d \to \mathbb{R}^d$ are diffeomorphisms a.s. for any $t_0 < t$. Further, $X_t := \Phi_{t_0,t}(X_0), t \ge t_0$ is a solution of equation (3.14). Hence the above $\{\Phi_{t_0,t}\}$ is also a stochastic flow of diffeomorphisms generated by SDE (3.14).

Now let us return to the SDE (3.10). We assume that the Lévy measure is finite. Then the equation is rewritten as

$$\begin{aligned} X_t = X_0 &+ \sum_{k=1}^{d'} \int_{t_0}^t V_k(X_r, r) \circ dW_r^k + \int_{t_0}^t \tilde{V}_0(X_r, r)\, dr \\ &+ \int_{t_0}^t \int_{|z|>0} g(X_{r-}, r, z) N(dr\, dz). \end{aligned} \tag{3.17}$$

Let $\{\Phi_{t_0,t}^0\}$ be the stochastic flow generated by the continuous SDE (3.1), replacing the drift vector field V_0 by $\tilde{V}_0$. Set

$$\Phi_{t_0,t}(x) \tag{3.18}$$
$$= \begin{cases} \Phi^0_{t_0,t}(x), & \text{if } t < \tau_1, \\ \Phi^0_{\tau_n,t} \circ \phi_{\tau_n,S_n} \circ \Phi^0_{\tau_{n-1},\tau_n} \circ \cdots \circ \phi_{\tau_1,S_1} \circ \Phi^0_{t_0,\tau_1}(x), & \text{if } \tau_n \le t < \tau_{n+1}. \end{cases}$$

Then $X_t = \Phi_{t_0,t}(X_0), t \ge t_0$ satisfies

$$X_t - X_{\tau_{m-1}} = \sum_{k=1}^{d'} \int_{\tau_{m-1}}^t V_k(X_r, r) \circ dW_r^k + \int_{\tau_{m-1}}^t \tilde{V}_0(X_r, r)\, dr,$$

if $\tau_{m-1} \le t < \tau_m$. At $t = \tau_m$, we have $X_{\tau_m} - X_{\tau_m-} = g(X_{\tau_m-}, \tau_m, S_m)$. Therefore, if $\tau_n \le t < \tau_{n+1}$,

$$\begin{aligned} X_t - X_0 &= X_t - X_{\tau_n} + \sum_{m=1}^n (X_{\tau_m-} - X_{\tau_{m-1}}) + \sum_{m=1}^n (X_{\tau_m} - X_{\tau_m-}) \\ &= \sum_{k=1}^{d'} \int_{t_0}^t V_k(X_r, r) \circ dW_r^k + \int_{t_0}^t \tilde{V}_0(X_r, r)\, dr \\ &\quad + \sum_{m=1}^n g(X_{\tau_m-}, \tau_m, S_m). \end{aligned} \tag{3.19}$$

The last term of the right-hand side is equal to $\int_{t_0}^t g(X_{r-}, r, z) N(dr\, dz)$. Therefore the above $X_t, t \ge t_0$ is a solution of equation (3.17). Consequently, $\{\Phi_{t_0,t}\}$ defined by (3.18) is the stochastic flow generated by the SDE (3.17).

Remark In Proposition 1.4.4, we constructed a compound Poisson process on $\mathbb{R}^d$, regarding it as an Abelian group. The expression (3.18) tells us that the stochastic flow $\{\Phi_{s,t}\}$ is a compound Poisson process with values in the non-Abelian group $\mathbb{G}$ of diffeomorphisms on $\mathbb{R}^d$. Since it has independent increments in the group $\mathbb{G}$, the stochastic flow may be considered as a Lévy process with values in the group $\mathbb{G}$.

Finally, consider the SDE (3.10) where the associated Lévy measure is an infinite measure. The solution cannot be written as the above, since jumps may occur infinitely many times in a finite interval. It should be the limit of a sequence of above solutions. For a given $\epsilon > 0$, consider an SDE

$$\begin{aligned} X_t = X_0 &+ \sum_{k=1}^{d'} \int_{t_0}^t V_k(X_r, r) \circ dW_r^k + \int_{t_0}^t \tilde{V}_0^\epsilon(X_r, r)\, dr \\ &+ \int_{t_0}^t \int_{|z| \ge \epsilon} g(X_{r-}, r, z) N(dr\, dz), \end{aligned} \tag{3.20}$$

where $\tilde{V}_0^\epsilon(r) = V_0(r) - \sum_k b_\epsilon^k V_k(r)$ and $b_\epsilon^k = \int_{\epsilon \le |z| \le 1} z^k \nu(dz)$. Let $\{\Phi_{s,t}^\epsilon\}$ be the stochastic flow of diffeomorphisms generated by SDE (3.20). We will show in Sect. 3.9 that $X_t^{\epsilon,x,t_0} := \Phi_{t_0,t}^\epsilon(x), t \ge t_0$ converges to a stochastic process X_t^{x,t_0} as $\epsilon \to 0$ and that the limit is a solution of the SDE (3.10). Further, it has a modification $\Phi_{t_0,t}(x)$ which are diffeomorphic maps on $\mathbb{R}^d$ for any $t \ge t_0$.

A stochastic flow $\{\Phi_{s,t}\}$ is called *right continuous*, if $\Phi_{s,t}(x)$ is right continuous with respect to t and s almost surely. Here a C^∞-function with parameter t denoted by $\phi_t(x)$ is called right continuous at t_0 if $\sup_{|x| \le M} |\partial^{\mathbf{i}}\phi_t(x) - \partial^{\mathbf{i}}\phi_{t_0}(x)|$ converges to 0 as $t \downarrow t_0$ for any $M > 0$ and multi-index $\mathbf{i}$. It is expected that the SDE (3.10) satisfying (J.1) and (J.2) should define a right continuous stochastic flow of diffeomorphisms.

For the rigorous proof of these facts, we need long discussions. In Sect. 3.3 we consider another SDE, which we will call the *master equation*. Coefficients of the equation are parameterized by λ. We will discuss the continuity and the differentiability of the solution with respect to parameter λ. Then results of the master equation will be applied to solutions of the SDE (3.10). Let X_t^{x,t_0} be the solution of the SDE starting from x at time t_0. By applying results to the master equation, we show in Sect. 3.4 that the family of solutions $\{X_t^{x,t_0}; x \in \mathbb{R}^d\}$ has a modification such that it is infinitely differentiable with respect to x for any $t_0 \le t < \infty$ a.s. Then, we may regard that for each $t_0 < t < \infty$, X_t^{x,t_0} is a C^∞-map from $\mathbb{R}^d$ to itself a.s.

3.3 Master Equation

Let us consider another Itô SDE on $\mathbb{R}^d$ with parameter $\lambda \in \Lambda$ given by

$$X_t = c_\lambda + \int_{t_0}^t \alpha_\lambda(X_{r-}, r)\,dW_r + \int_{t_0}^t \beta_\lambda(X_{r-}, r)\,dr + \int_{t_0}^t \int_{\mathbb{R}_0^{d'}} \chi_\lambda(X_{r-}, r, z)\tilde{N}(dr\,dz). \tag{3.21}$$

Here, for the parameter set Λ we take a Euclidean space $\mathbb{R}^e$ or its unit ball $B_1^e = \{\lambda \in \mathbb{R}^e; |\lambda| < 1\}$. Let X_t be a cadlag $\{\mathcal{F}_t\}$-adapted process with values in $\mathbb{R}^d$. If each component X_t^i of $X_t = (X_t^1, \ldots, X_t^d)$ satisfies

$$X_t^i = c_\lambda^i + \sum_{k=1}^{d'} \int_{t_0}^t \alpha_\lambda^{ik}(X_{r-}, r)\,dW_r^k + \int_{t_0}^t \beta_\lambda^i(X_{r-}, r)\,dr + \int_{t_0}^t \int_{\mathbb{R}_0^{d'}} \chi_\lambda^i(X_{r-}, r, z)\tilde{N}(dr\,dz),$$

$X_t, t \ge t_0$ is called a solution of equation (3.21) starting from c_λ at time t_0.

For the initial condition $c_\lambda = (c_\lambda^1, \ldots, c_\lambda^d)$, we assume that it is non-random, bounded and Lipschitz continuous with respect to λ. Coefficients

$$\alpha_\lambda(x,r) = (\alpha_\lambda^{ik}(x,r)), \quad \beta_\lambda(x,r) = (\beta_\lambda^1(x,r), \ldots, \beta_\lambda^d(x,r)),$$

$$\chi_\lambda(x,r,z) = (\chi_\lambda^1(x,r,z), \ldots, \chi_\lambda^d(x,r,z)),$$

where $x \in \mathbb{R}^d, r \in \mathbb{T}, z \in \mathbb{R}^{d'}$, are possibly random, jointly measurable and predictable with respect to the filtration $\{\mathcal{F}_t\}$. Further, $\chi_\lambda(x,r,0) = 0$ and $\chi_\lambda(x,r,z)$ is continuously differentiable with respect to $z \in S(\nu)$, where $S(\nu)$ is the support of the Lévy measure ν.

Further, $\alpha_\lambda, \beta_\lambda, \chi_\lambda$ and $\partial_z \chi_\lambda$ satisfy the following three conditions:

1. *At $x = 0$, coefficients are uniformly L^p-bounded with respect to λ and r;*

$$\sup_{\lambda,r} \Big\{ E[|\alpha_\lambda(0,r)|^p] + E[|\beta_\lambda(0,r)|^p] + E\Big[\int_{\mathbb{R}_0^{d'}} \Big|\frac{\chi_\lambda(0,r,z)}{\gamma(z)}\Big|^p \mu(dz)\Big]\Big\} < \infty \tag{3.22}$$

holds for any $p \geq 2$. Here $\gamma(z) = |z| \wedge 1$ and $\mu(dz) = \gamma(z)^2 \nu(dz)$.

2. *Coefficients are uniformly Lipschitz continuous with respect to x; there exist positive constants c and c_p such that*

$$|\alpha_\lambda(x,r) - \alpha_\lambda(x',r)| \leq c|x - x'|, \tag{3.23}$$

$$|\beta_\lambda(x,r) - \beta_\lambda(x',r)| \leq c|x - x'|,$$

$$\Big(\int_{\mathbb{R}_0^{d'}} \Big|\frac{\chi_\lambda(x,r,z)}{\gamma(z)} - \frac{\chi_\lambda(x',r,z)}{\gamma(z)}\Big|^p \mu(dz)\Big)^{\frac{1}{p}} \leq c_p|x - x'|$$

hold for any x, x', r, λ and $p \geq 2$ a.s.

3. *With respect to parameter λ, coefficients are weakly Lipschitz continuous in the following sense: For any $T > 0$ and $p \geq 2$, there exists a positive random field $K_{\lambda,\lambda'}(r)$ such that*

$$E\Big[\Big(\int_0^T K_{\lambda,\lambda'}(r)\,dr\Big)^p\Big]^{\frac{1}{p}} < c_p|\lambda - \lambda'| \tag{3.24}$$

holds with a positive constant c_p, and coefficients satisfy

$$|\alpha_\lambda(x,r) - \alpha_{\lambda'}(x,r)| \leq K_{\lambda,\lambda'}(r)(1 + |x|), \tag{3.25}$$

$$|\beta_\lambda(x,r) - \beta_{\lambda'}(x,r)| \leq K_{\lambda,\lambda'}(r)(1 + |x|),$$

$$\Big(\int_{\mathbb{R}_0^{d'}} \Big|\frac{\chi_\lambda(x,r,z)}{\gamma(z)} - \frac{\chi_{\lambda'}(x,r,z)}{\gamma(z)}\Big|^p \mu(dz)\Big)^{\frac{1}{p}} \leq K_{\lambda,\lambda'}(r)(1 + |x|)$$

a.s. for any x, r, λ, λ'.

Equation (3.21) is called a *master equation* (with parameter λ). If $\chi_\lambda(x, r) \equiv 0$, the equation is called a *continuous master equation.*

For the analysis of the master equation, we will often use an inequality called *Gronwall's inequality*:

Let $f(t)$ and $g(t)$ be nonnegative continuous functions on $[t_0, \infty)$ satisfying the inequality

$$f(t) \le g(t) + c \int_{t_0}^{t} f(r)\, dr, \quad \forall t \ge t_0,$$

where c is a positive constant. Then the inequality

$$f(t) \le g(t) + c \int_{t_0}^{t} g(r) e^{c(t-r)}\, dr \tag{3.26}$$

holds for any $t \ge t_0$. In particular, if $f(t) \le c \int_{t_0}^{t} f(r)\, dr$ holds for any $t \ge 0$, then $f(t) \equiv 0$ holds.

The inequality (3.26) is called *Gronwall's inequality*. The proof is straightforward and it is omitted.

For an $\mathbb{R}^d$-valued cadlag stochastic process $X_t, t \in [t_0, \infty)$, we define norms $\| \ \|_t$ depending on parameter $t_0 \le t \le T$ by $\|X\|_t = \sup_{t_0 \le s \le t} |X_{r-}|$.

Lemma 3.3.1 *Let $p \ge 2$. For any λ, equation* (3.21) *has a solution belonging to L^p.*

Proof The existence of the solution can be verified by a standard argument using the successive approximation. Set $X_t^{\lambda,0} = c_\lambda$ and define $X_t^{\lambda,n}$ for $n \ge 1$ by induction as follows:

$$\begin{aligned} X_t^{\lambda,n+1} &= c_\lambda + \int_{t_0}^{t} \alpha_\lambda(X_{r-}^{\lambda,n}, r)\, dW_r + \int_{t_0}^{t} \beta_\lambda(X_{r-}^{\lambda,n}, r)\, dr \\ &\quad + \int_{t_0}^{t} \int \chi_\lambda(X_{r-}^{\lambda,n}, r, z) \tilde{N}(dr\, dz). \end{aligned}$$

Then, using Doob's inequality for martingales (Theorem 1.5.1), we have

$$\begin{aligned} & E\Big[\|X^{\lambda,n+1} - X^{\lambda,n}\|_t^p\Big] \\ & \le C_0 \Big\{ E\Big[\Big| \int_{t_0}^{t} (\alpha_\lambda(X_{r-}^{\lambda,n}, r) - \alpha_\lambda(X_{r-}^{\lambda,n-1}, r))\, dW_r \Big|^p\Big] \\ & \quad + E\Big[\Big| \int_{t_0}^{t} |\beta_\lambda(X_{r-}^{\lambda,n}, r) - \beta_\lambda(X_{r-}^{\lambda,n-1}, r)|\, dr \Big|^p\Big] \\ & \quad + E\Big[\Big| \int_{t_0}^{t} \int \{\chi_\lambda(X_{r-}^{\lambda,n}, r, z) - \chi_\lambda(X_{r-}^{\lambda,n-1}, r, z)\} \tilde{N}(dr\, dz) \Big|^p\Big]\Big\}. \end{aligned}$$

Using the Burkholder–Davis–Gundy inequality (Propositions 2.2.2 and 2.6.1) and the uniform Lipschitz continuity for coefficients, we have for any $t_0 \leq t \leq T$

$$E\Big[\Big(\int_{t_0}^{t} (\alpha_\lambda(X_{r-}^{\lambda,n}, r) - \alpha_\lambda(X_{r-}^{\lambda,n-1}, r))\, dW_r\Big)^p\Big]$$

$$\leq C_p E\Big[\int_{t_0}^{t} |\alpha_\lambda(X_{r-}^{\lambda,n}, r) - \alpha_\lambda(X_{r-}^{\lambda,n-1}, r)|^p \, dr\Big]$$

$$\leq C_p' \int_{t_0}^{t} E[\|X^{\lambda,n} - X^{\lambda,n-1}\|_r^p]\, dr,$$

$$E\Big[\Big|\int_{t_0}^{t} |\beta_\lambda(X_{r-}^{\lambda,n}, r) - \beta_\lambda(X_{r-}^{\lambda,n-1}, r)|\, dr\Big|^p\Big]$$

$$\leq C_1 \int_{t_0}^{t} E[\|X^{\lambda,n} - X^{\lambda,n-1}\|_r^p]\, dr,$$

$$E\Big[\Big(\int_{t_0}^{t}\int \{\chi_\lambda(X_{r-}^{\lambda,n}, r, z) - \chi(X_{r-}^{\lambda,n-1}, r, z)\}\tilde{N}(dr\, dz)\Big)^p\Big]$$

$$\leq C' E\Big[\int_{t_0}^{t} \Big(\int \Big|\frac{\chi_\lambda(X_{r-}^{\lambda,n}, r, z)}{\gamma(z)} - \frac{\chi_\lambda(X_{r-}^{\lambda,n-1}, r, z)}{\gamma(z)}\Big|^p \mu(dz)\Big)\, dr\Big]$$

$$\leq C'' \int_{t_0}^{t} E[\|X^{\lambda,n} - X^{\lambda,n-1}\|_r^p]\, dr.$$

Therefore there exists a positive constant C such that

$$E[\|X^{\lambda,n+1} - X^{\lambda,n}\|_t^p] \leq C \int_{t_0}^{t} E[\|X^{\lambda,n} - X^{\lambda,n-1}\|_r^p]\, dr.$$

Repeating this argument, we arrive at

$$E[\|X^{\lambda,n+1} - X^{\lambda,n}\|_t^p] \leq \frac{(C(t - t_0))^n}{n!} E[\|X^{\lambda,1} - X^{\lambda,0}\|_t^p].$$

By the definition of $X_t^{\lambda,1}$, it holds that $\sup_\lambda E[\|X^{\lambda,1} - X^{\lambda,0}\|_t^p] < \infty$. Then there exists a cadlag process X_t^λ such that $\sup_\lambda E[\|X^\lambda - X^{\lambda,0}\|_t^p] < \infty$ and $\sup_\lambda E[\|X^{\lambda,n} - X^\lambda\|_t^p]$ converges to 0 for any $t_0 \leq t \leq T$. The process X_t^λ is a solution of the above stochastic differential equation. □

Lemma 3.3.2 *Any solution X_t^λ of the master equation (3.21) belongs $L^{\infty-} = \bigcap_{p>1} L^p$. Furthermore, for any $T > t_0$ and $p \geq 2$, there is a positive constant c such that*

$$E[\|X^\lambda\|_T^p] < c(1 + |c_\lambda|)^p \tag{3.27}$$

holds for all $\lambda \in \Lambda$.

Proof For $n > 0$, define a stopping time by

$$\tau^n = \begin{cases} \inf\{t \geq t_0;\ |X_t^\lambda| \geq n\}, \text{ if } \{\cdots\} \neq \emptyset, \\ \infty, \qquad\qquad\qquad\qquad\ \text{if } \{\cdots\} = \emptyset. \end{cases}$$

Then setting $t' = t \wedge \tau^n$ and using the Burkholder–Davis–Gundy's inequality, we have the following L^p-estimate:

$$\begin{aligned} &E[\|X^\lambda\|_{t'}^p] \\ &\leq c_p \Big\{ |c_\lambda|^p + E\Big[\int_{t_0}^{t'} |\alpha_\lambda(X_{r-}^\lambda, r)|^p\, dr\Big] + E\Big[\int_{t_0}^{t'} |\beta_\lambda(X_{r-}^\lambda, r)|^p\, dr\Big] \\ &\quad + E\Big[\int_{t_0}^{t'} \Big(\int \Big|\frac{\chi_\lambda(X_{r-}^\lambda, r, z)}{\gamma(z)}\Big|^p \mu(dz)\Big)\, dr\Big]. \end{aligned}$$

We have

$$|\alpha_\lambda(X_{r-}^\lambda, r)| \leq |\alpha_\lambda(0, r)| + c|X_{r-}^\lambda|, \quad |\beta_\lambda(X_{r-}^\lambda, r)| \leq |\beta_\lambda(0, r)| + c|X_{r-}^\lambda|,$$

$$\Big(\int \Big|\frac{\chi_\lambda(X_{r-}^\lambda, r, z)}{\gamma(z)}\Big|^p \mu(dz)\Big)^{\frac{1}{p}} \leq \Big(\int \Big|\frac{\chi_\lambda(0, r, z)}{\gamma(z)}\Big|^p \mu(dz)\Big)^{\frac{1}{p}} + c|X_{r-}^\lambda|.$$

Therefore, there exist positive constants C_1, C_2 such that

$$E\big[\|X^\lambda\|_{t'}^p\big] \leq C_1(1 + |c_\lambda|)^p + C_2 E\Big[\int_{t_0}^{t'} |X_{r-}^\lambda|^p\, dr\Big].$$

We have $|X_{r'}^\lambda| \leq n$ a.s. for $r' < t'$. Then the last integral is finite. Therefore,

$$E[\|X^\lambda\|_{t'}^p] \leq C_1(1 + |c_\lambda|)^p + C_2 \int_{t_0}^{t} E[\|X^\lambda\|_{r'}^p]\, dr.$$

By Gronwall's inequality, there exists a positive constant c_p such that

$$E[\|X^\lambda\|_{t\wedge\tau^n}^p] \leq c_p(1 + |c_\lambda|)^p$$

holds for all n and λ. Let n tend to ∞. The above inequality indicates that the set $\{\tau^n \leq t\}$ shrinks to the empty set a.s., so that we get $E[\|X^\lambda\|_t^p] < c_p(1+|c_\lambda|)^p$. The inequality is valid for all λ, t. Hence we get the inequality of the lemma. □

Lemma 3.3.3 *Let* $\{X_t^\lambda, \lambda \in \Lambda\}$ *be any family of solutions of master equations* (3.21). *Then for any* $T > t_0$ *and* $p \geq 2$, *there exists a positive constant* C *such that*

$$E[\|X^\lambda - X^{\lambda'}\|_T^p] \leq C|\lambda - \lambda'|^p \tag{3.28}$$

holds for all $\lambda, \lambda' \in \Lambda$.

Proof Set $Y_t = X_t^\lambda - X_t^{\lambda'}$. Then we have

$$\begin{aligned} Y_t = c_\lambda - c_{\lambda'} &+ \int_{t_0}^t \big(\alpha_\lambda(X_{r-}^\lambda, r) - \alpha_{\lambda'}(X_{r-}^{\lambda'}, r)\big)\, dW_r \\ &+ \int_{t_0}^t \big(\beta_\lambda(X_{r-}^\lambda, r) - \beta_{\lambda'}(X_{r-}^{\lambda'}, r)\big)\, dr \\ &+ \int_{t_0}^t \int \big\{\chi_\lambda(X_{r-}^\lambda, r, z) - \chi_{\lambda'}(X_{r-}^{\lambda'}, r, z)\big\} \tilde{N}(dr\, dz). \end{aligned}$$

We shall consider terms of the right-hand side. We have by (3.25)

$$\begin{aligned} &E\Big[\Big|\int_{t_0}^t (\alpha_\lambda(X_{r-}^\lambda, r) - \alpha_{\lambda'}(X_{r-}^{\lambda'}, r))\, dW_r\Big|^p\Big] \\ &\leq c_0 E\Big[\int_{t_0}^t |X_{r-}^\lambda - X_{r-}^{\lambda'}|^p\, dr\Big] + c_0 E\Big[\int_{t_0}^t K^{\lambda,\lambda'}(r)^p (1 + |X_{r-}^\lambda|)^p\, dr\Big]. \end{aligned}$$

Since $K_r^{\lambda,\lambda'}$ satisfies (3.24) and since $|X_t^\lambda|$ are L^p-bounded with respect to x, λ by Lemma 3.3.2, the above is dominated by

$$c_1 \int_{t_0}^t E[\|X^\lambda - X^{\lambda'}\|_r^p]\, dr + c_1'|\lambda - \lambda'|^p.$$

Similarly, we can show that both

$$E\Big[\Big|\int_{t_0}^t (\beta_\lambda(X_{r-}^\lambda, r) - \beta_{\lambda'}(X_{r-}^{\lambda'}, r))\, dr\Big|^p\Big],$$

$$E\Big[\Big|\int_{t_0}^t \int \big\{\chi_\lambda(X_{r-}^\lambda, r, z) - \chi_{\lambda'}(X_{r-}^{\lambda'}, r, z)\big\} \tilde{N}(dr\, dz)\Big|^p\Big]$$

satisfy the similar inequality. Further, note that c_λ is Lipschitz continuous. Then there exist positive constants C, C' such that

$$E[\|Y\|_t^p] \le C|\lambda - \lambda'|^p + C' \int_{t_0}^t E[\|Y\|_r^p]\,dr$$

holds for all λ, λ' and t. Then, Gronwall's inequality implies the inequality of the lemma. □

Theorem 3.3.1 *Assume Conditions 1–3 for the master equation. For any c_λ, the master equation* (3.21) *has a unique solution X_t^λ, which belongs to $L^{\infty-}$. The family of solutions $\{X_t^\lambda, \lambda \in \Lambda\}$ has a modification which is continuous with respect to λ and cadlag with respect to t a.s. Further, for any $T > t_0$ and $p \ge 2$ there is a positive constant c such that $\{X_t^\lambda, \lambda \in \Lambda\}$ satisfies*

$$E[\|X^\lambda\|_T^p] \le c(1 + |c_\lambda|)^p \tag{3.29}$$

for all $\lambda \in \Lambda$.

Proof The existence of a solution belonging to $L^{\infty-}$ is proved in Lemma 3.3.1. Suppose that X_t^λ and $\tilde{X}_t^\lambda$ are solutions of the master equation with the same initial value c_λ. Then both belong to $L^{\infty-}$ by Lemma 3.3.2. Then Lemma 3.3.3 tells us $X_t^\lambda = \tilde{X}_t^\lambda$ a.s., since $\lambda = \lambda'$.

Now the family of solutions $\{X_t^\lambda; \lambda \in \Lambda\}$ satisfies (3.28). Then X_t^λ has a modification which is continuous in λ by the Kolmogorov-Totoki theorem (Theorem 1.8.1). Further, it satisfies (3.29) for any $p \ge 2$ by Lemma 3.3.1. □

We shall next consider the differentiability of the solution with respect to parameter λ. We assume that coefficients $\alpha_\lambda, \beta_\lambda, \chi_\lambda$ and the initial value c_λ are differentiable with respect to λ. Let $\lambda = (\lambda_1, \ldots, \lambda_e)$ and let $\partial_{\lambda_i}\alpha_\lambda$ be the partial derivative of α_λ with respect to λ_i. We set $\partial_\lambda\alpha_\lambda = (\partial_{\lambda_1}\alpha_\lambda, \ldots, \partial_{\lambda_e}\alpha_\lambda)$.

Theorem 3.3.2 *Assume that coefficients of the master equation are continuously differentiable with respect to x and λ, and derivatives $\partial_x\alpha_\lambda(x, r)$, $\partial_x\beta_\lambda(x, r)$, $\partial_x\chi_\lambda(x, r, z)$, $\partial_\lambda\alpha_\lambda(x, r)$, $\partial_\lambda\beta_\lambda(x, r)$ and $\partial_\lambda\chi_\lambda(x, r, z)$ satisfy Conditions 1–3 of the master equation. Assume further that $\partial_\lambda c_\lambda$ exists and is Hölder continuous.*

1. *The family of solutions $\{X_t^\lambda, \lambda \in \Lambda\}$ has a modification which is continuously differentiable with respect to λ for any t and the derivative is continuous in λ and cadlag in t a.s. Further, for any $T > t_0$ and $p \ge 2$, it holds that*

$$E\left[\left\|\frac{1}{\epsilon}(X^{\lambda+\epsilon e_i} - X^\lambda) - \partial_{\lambda_i}X^\lambda\right\|_T^p\right] \to 0 \tag{3.30}$$

as $\epsilon \to 0$ uniformly in λ. Here e_i are unit vectors in $\mathbb{R}^e$.

2. *The derivative* $\partial_\lambda X_t^\lambda$ *satisfies the following equation:*

$$\partial_\lambda X_t^\lambda = \partial_\lambda c_\lambda + \int_{t_0}^t \{\partial_\lambda \alpha_\lambda(X_{r-}^\lambda, r) + \partial_x \alpha_\lambda(X_{r-}^\lambda, r)\partial_\lambda X_{r-}^\lambda\}\, dW_r$$
$$+\int_{t_0}^t \{\partial_\lambda \beta_\lambda(X_{r-}^\lambda, r) + \partial_x \beta_\lambda(X_{r-}^\lambda, r)\partial_\lambda X_{r-}^\lambda\}\, dr \tag{3.31}$$
$$+\int_{t_0}^t \int_{\mathbb{R}_0^d} \{\partial_\lambda \chi_\lambda(X_{r-}^\lambda, r, z) + \partial_x \chi_\lambda(X_{r-}^\lambda, r, z)\partial_\lambda X_{r-}^\lambda\}\tilde{N}(dr\, dz).$$

3. *For any* $T > t_0$ *and* $p \geq 2$ *there is a positive constant* c *such that*

$$E[\|\partial_\lambda X^\lambda\|_T^p] \leq c(1 + |\partial_\lambda c_\lambda|)^p \tag{3.32}$$

holds for all λ.

Proof Let $\lambda \in \Lambda$ and $\epsilon \in \Theta = \mathbb{R} \setminus \{0\}$. For a positive integer i with $1 \leq i \leq e$, we set $\epsilon_i = \epsilon e_i$, where e_i is an unit vector in $\mathbb{R}^e$. We set

$$Y_t^{\lambda,\epsilon} = \frac{1}{\epsilon}\Big\{X_t^{\lambda+\epsilon_i} - X_t^\lambda\Big\}.$$

We want to prove that $\lim_{\epsilon \to 0} Y_t^{\lambda,\epsilon}$ exists uniformly in t for all λ almost surely and it is continuous with respect to λ almost surely. For this purpose, we will find an SDE which governs $Y_t^{\lambda,\epsilon}$. Since $X_t^{\lambda+\epsilon_i}$ and X_t^λ are solutions of (3.21), we find that $x = Y_t^{\lambda,\epsilon}$ is a solution of a master equation on $\mathbb{R}^d$ with parameter λ, ϵ starting from $c_{\lambda,\epsilon} := (c_{\lambda+\epsilon_i} - c_\lambda)/\epsilon$ at time t_0, whose coefficients are given by

$$\alpha_{\lambda,\epsilon}(x, r) = \frac{1}{\epsilon}\Big\{\alpha_{\lambda+\epsilon_i}(X_{r-}^\lambda + \epsilon x, r) - \alpha_\lambda(X_{r-}^\lambda, r)\Big\},$$
$$\beta_{\lambda,\epsilon}(x, r) = \frac{1}{\epsilon}\Big\{\beta_{\lambda+\epsilon_i}(X_{r-}^\lambda + \epsilon x, r) - \beta_\lambda(X_{r-}^\lambda, r)\Big\},$$
$$\chi_{\lambda,\epsilon}(x, r, z) = \frac{1}{\epsilon}\Big\{\chi_{\lambda+\epsilon_i}(X_{r-}^\lambda + \epsilon x, r, z) - \chi_\lambda(X_{r-}^\lambda, r, z)\Big\},$$

because $X_{r-}^{\lambda+\epsilon_i} = X_{r-}^\lambda + \epsilon Y_{r-}^{\lambda,\epsilon}$. We will show that these coefficients fulfill three conditions of the master equation. We will consider $\alpha_{\lambda,\epsilon}$ only, since other coefficients can be handled similarly. It holds that

$$|a_{\lambda,\epsilon}(x, r) - a_{\lambda,\epsilon}(x', r)| \leq \frac{1}{\epsilon}\Big|a_{\lambda+\epsilon_i}(X_{r-}^\lambda + \epsilon x, r) - a_{\lambda+\epsilon_i}(X_{r-}^\lambda + \epsilon x', r)\Big| \leq c|x - x'|.$$

We will next consider $|\alpha_{\lambda,\epsilon}(x, r) - \alpha_{\lambda',\epsilon'}(x, r)|$. Rewrite first the right-hand side of $a_{\lambda,\epsilon}(x, r)$ using the mean value theorem. Then it is written as

$$\alpha_{\lambda,\epsilon}(x,r) = \partial_x \alpha_{\lambda+\theta\epsilon_i}(X_{r-}^{\lambda} + \epsilon\theta x, r)x + \partial_{\lambda_i}\alpha_{\lambda+\theta\epsilon_i}(X_{r-}^{\lambda} + \epsilon\theta x, r),$$

where $0 \le |\theta| \le 1$. Therefore, $|\alpha_{\lambda,\epsilon}(x,r) - \alpha_{\lambda',\epsilon'}(x,r)|$ is dominated by the sum of the following two;

$$|\partial_x \alpha_{\lambda+\theta\epsilon_i}(X_{r-}^{\lambda} + \epsilon\theta x, r) - \partial_x \alpha_{\lambda'+\theta\epsilon_i'}(X_{r-}^{\lambda'} + \epsilon'\theta x, r)|, \tag{3.33}$$

$$|\partial_{\lambda_i} \alpha_{\lambda+\theta\epsilon_i}(X_{r-}^{\lambda} + \epsilon\theta x, r) - \partial_{\lambda_i} \alpha_{\lambda'+\theta\epsilon_i'}(X_{r-}^{\lambda'} + \epsilon'\theta x, r)|.$$

The first term of (3.33) is dominated by

$$\begin{aligned}
&|\partial_x \alpha_{\lambda+\theta\epsilon_i}(X_{r-}^{\lambda} + \epsilon\theta x, r) - \partial_x \alpha_{\lambda'+\theta\epsilon_i'}(X_{r-}^{\lambda} + \epsilon\theta x, r)| \\
&+ |\partial_x \alpha_{\lambda'+\theta\epsilon_i'}(X_{r-}^{\lambda} + \epsilon\theta x, r) - \partial_x \alpha_{\lambda'+\theta\epsilon_i'}(X_{r-}^{\lambda'} + \epsilon'\theta x, r)| \\
&\le K_{\lambda+\theta\epsilon_i,\lambda'+\theta\epsilon_i'}(1 + |x| + |X_{r-}^{\lambda}|) + c(|X_{r-}^{\lambda} - X_{r-}^{\lambda'}| + |\epsilon - \epsilon'|(1 + |x|)),
\end{aligned}$$

where $K_{\lambda,\lambda'}$ is a positive functional satisfying $E[(\int_0^T K_{\lambda,\lambda'}(r)\,dr)^p] \le c_p|\lambda - \lambda'|^p$ for any $p \ge 2$. A similar estimate is valid for the second term of (3.33). Consequently, we obtain the inequality

$$|\alpha_{\lambda,\epsilon}(x,r) - \alpha_{\lambda',\epsilon'}(x,r)| \le 2K_{(\lambda,\epsilon),(\lambda',\epsilon')}(r)(1 + |x|),$$

where

$$K_{(\lambda,\epsilon),(\lambda',\epsilon')}(r) = K_{\lambda+\theta\epsilon_i,\lambda'+\theta\epsilon_i'}(r)(1 + |X_r^{\lambda'}|) + c(|X_{r-}^{\lambda} - X_{r-}^{\lambda'}| + |\epsilon - \epsilon'|).$$

It satisfies

$$E\Big[\Big(\int_0^T K_{(\lambda,\epsilon),(\lambda',\epsilon')}(r)\,dr\Big)^p\Big] \le c_p'(|\lambda - \lambda'|^p + |\epsilon - \epsilon'|^p)$$

for any $p \ge 2$. Therefore, $\alpha_{\lambda,\epsilon}(x,r)$ fulfills Conditions 1–3 of the master equation. In the same way, we can verify that $\beta_\lambda(x,r)$ and $\chi_\lambda(x,r,z)$ fulfill Conditions 1–3 of the master equation.

Now Lemmas 3.3.2 and 3.3.3 tell us that for any $K > 0$, $T > t_0$ and $p \ge 2$, there exists a positive constant C such that inequalities

$$E[\|Y^{\lambda,\epsilon}\|_T^p] \le C(1 + |c_{\lambda,\epsilon}|)^p, \tag{3.34}$$

$$E[\|Y^{\lambda,\epsilon} - Y^{\lambda',\epsilon'}\|_T^p] \le C\{|\lambda - \lambda'|^p + |\epsilon - \epsilon'|^p\} \tag{3.35}$$

hold for any $|\lambda|, |\lambda'| \le K$ and $0 < |\epsilon|, |\epsilon'| < 1$. Then, we apply the Kolmogorov–Totoki theorem to $Y^{\lambda,\epsilon}$, regarding it as a random field with values in

$$B = \{X_t;\ \mathbb{R}^d\text{-valued cadlag function of } t \in \mathbb{T}\}.$$

Then the random field $Y^{\lambda,\epsilon}$ has a modification such that it is uniformly continuous on the set $\{|\lambda| < K, 0 < |\epsilon| < 1\}$. It means that $Y_t^{\lambda,\epsilon}$ can be extended continuously at $\epsilon = 0$, and for any t, $Y_t^{\lambda,0} = \exists \lim_{\epsilon\to 0} Y_t^{\lambda,\epsilon}$ (uniformly in t) is continuous in λ a.s. This means that X_t^λ is continuously differentiable with respect to λ for any t and $\partial_{\lambda_i} X_t^\lambda = Y_t^{\lambda,0}$ holds for all λ, t a.s. Further, set $\lambda' = \lambda$ and let ϵ' tend to 0 for the inequality (3.35). Then we get

$$E[\|Y^{\lambda,\epsilon} - \partial_{\lambda_i} X^\lambda\|_T^p] \le C_p \epsilon^p.$$

Therefore (3.30) holds uniformly in λ.

Next we will prove (3.31). The solution X_t^λ satisfies

$$\begin{aligned} X_t^\lambda = c_\lambda &+ \int_{t_0}^t \alpha_\lambda(X_{r-}^\lambda, r)\, dW_r + \int_{t_0}^t \beta_\lambda(X_{r-}^\lambda, r)\, dr \\ &+ \int_{t_0}^t \int \chi_\lambda(X_{r-}^\lambda, r, z)\tilde{N}(dr\, dz). \end{aligned}$$

We want to show that each term of the right-hand side is continuously differentiable with respect to λ and we can change the order of the derivative operator ∂_λ and the integral operator $\int_{t_0}^t$. We will check this for the Itô integral $\int_{t_0}^t \alpha_\lambda(X_{r-}^\lambda, r)\, dW_r$. Since α_λ and X_r^λ are continuously differentiable with respect to λ, the composite $\alpha_\lambda(X_{r-}^\lambda, r)$ is also continuously differentiable with respect to λ. Further, the derivative $\partial_\lambda X_{r-}^\lambda$ is Lipschitz continuous in L^p, i.e., it satisfies

$$\sup_r E[|\partial_\lambda X_{r-}^\lambda - \partial_\lambda X_{r-}^{\lambda'}|^p] \le c_p |\lambda - \lambda'|^p,$$

in view of (3.35). Therefore $\alpha_\lambda(X_{r-}^\lambda, r)$ belongs to the class $\mathbf{L}_{\mathbb{T}}^{1+Lip,p}(\Lambda)$ (Sect. 2.3). Consequently, the Itô integral is continuously differentiable in λ and we get

$$\begin{aligned} \partial_\lambda \int_{t_0}^t \alpha_\lambda(X_{r-}^\lambda, r)\, dW_r &= \int_{t_0}^t \partial_\lambda(\alpha_\lambda(X_{r-}^\lambda, r))\, dW_r \\ &= \int_{t_0}^t \{\partial_\lambda \alpha_\lambda(X_{r-}^\lambda, r) + \partial_x \alpha_\lambda(X_{r-}^\lambda, r)\partial_\lambda X_{r-}^\lambda\}\, dW_r, \end{aligned}$$

in view of Proposition 2.3.1. We can show similar facts for $\int_{t_0}^t \beta_\lambda(X_{r-}^\lambda, r)\, dr$ and $\int_{t_0}^t \int \chi_\lambda(X_{r-}^\lambda, r)\tilde{N}(dr\, dz)$. Then we get the equality (3.31).

Finally, apply Lemma 3.3.2 to the master equation (3.31). Then we get (3.32).

□

Remark Later, we will be concerned with a more general master equation. Let $\delta(x,r,z), (x,r,z) \in \mathbb{R}^d \times \mathbb{T} \times \mathbb{R}^{d'}$ be a $C_b^{\infty,0,0}$ function. Let $(r_i, z_i), i = 1, \ldots, n$ be points in $\mathbb{T} \times \mathbb{R}_0^{d'}$. We consider an SDE:

$$X_t = c_\lambda + \int_{t_0}^t \alpha_\lambda(X_{r-}, r)\, dW_r + \int_{t_0}^t \beta_\lambda(X_{r-}, r)\, dr \tag{3.36}$$

$$+ \int_{t_0}^t \int_{\mathbb{R}_0^{d'}} \chi_\lambda(X_{r-}, r, z)\tilde{N}(dr\, dz) + \sum_{i; t_0 \le r_i \le t} \delta(X_{r_i-}, r_i, z_i).$$

If coefficients $\alpha_\lambda, \beta_\lambda, \chi_\lambda$ satisfy Conditions 1–3, the equation has a unique solution X_t^λ. Further, assertions of Theorems 3.3.1 and 3.3.2 are valid.

3.4 L^p-Estimates and Regularity of Solutions; C^∞-Flows

Let us return to SDE (3.10). For its coefficients, we assume that $V_0(x,t), \ldots, V_{d'}(x,t)$ are $C_b^{\infty,1}$-functions and $g(x,t,z)$ satisfies Condition (J.1). Instead of the above equation, consider the Itô SDE;

$$X_t = \lambda + \int_{t_0}^t \alpha(X_{r-}, r)\, dW_r + \int_{t_0}^t \beta(X_{r-}, r)\, dr$$

$$+ \int_{t_0}^t \int_{\mathbb{R}_0^{d'}} \chi(X_{r-}, r, z)\tilde{N}(dr\, dz), \tag{3.37}$$

where $\lambda \in \mathbb{R}^d$, $\alpha^{ik}(x,r) = V_k^i(x,r)$, $\chi(x,r,z) = g(x,r,z)$ and $\beta(x,r)$ is defined by (3.13). It is in fact a $C_b^{\infty,1}$-function.

The functions α and β are uniformly Lipschitz continuous with respect to x. Further, since $\partial_{x_i}\chi(x,r,z)$ and $\partial_{z^k}\partial_{x_i}\chi(x,r,z)$ are bounded, for any $p \ge 2$, there exists a positive constant c_p such that

$$\Big(\int \Big|\frac{\chi(x,r,z)}{\gamma(z)} - \frac{\chi(x',r,z)}{\gamma(z)}\Big|^p \mu(dz)\Big)^{\frac{1}{p}} \le c_p|x - x'|$$

holds for any x, x' and r. Therefore $\chi(x,r,z)$ satisfies the inequality (3.23). Then (3.37) is a master equation with parameter $\lambda \in \mathbb{R}^d$ and initial condition $c_\lambda = \lambda$. It has a unique solution $X_t^\lambda, t \ge t_0$ and the solution belongs to $L^{\infty-}$ by Theorem 3.3.1.

Theorem 3.4.1 *The family of solutions* $\{X_t^\lambda; \lambda \in \mathbb{R}^d\}$ *of equation* (3.37) *has a modification, which is continuous with respect to* λ *and cadlag with respect to* t *a.s.*

The modification X_t^λ is continuously differentiable with respect to λ for all t and the derivative is cadlag with respect to t a.s. Its derivative $\partial_\lambda X_t^\lambda$ satisfies

$$\partial_\lambda X_t^\lambda = I + \int_{t_0}^t \partial_x \alpha(X_{r-}^\lambda, r)\partial_\lambda X_{r-}^\lambda \, dW_r + \int_{t_0}^t \partial_x \beta(X_{r-}^\lambda, r)\partial_\lambda X_{r-}^\lambda \, dr$$
$$+ \int_{t_0}^t \int_{\mathbb{R}_0^{d'}} \partial_x \chi(X_{r-}^\lambda, r, z)\partial_\lambda X_{r-}^\lambda \tilde{N}(dr\, dz). \tag{3.38}$$

Further, for any $T > t_0$ and $p \geq 2$ there is a positive constant c_p such that

$$E[\|X^\lambda\|_T^p] \leq c_p(1 + |\lambda|)^p, \quad E[\|\partial_\lambda X^\lambda\|_T^p] < c_p. \tag{3.39}$$

Proof Theorem 3.3.1 tells us that the solution X_t^λ has a modification such that it is continuous in λ and cadlag with respect to t. For the differentiability, we will apply Theorem 3.3.2. We will again regard λ as a parameter. It satisfies properties of the master equation, by setting $\alpha_\lambda(x, r) = \alpha(x, r)$ etc. These coefficients satisfy the condition of Theorem 3.3.2. Therefore X_t^λ has a modification which is continuously differentiable with respect to λ a.s. for any t. Further, the derivative $\partial_\lambda X_t^\lambda$ satisfies (3.38), since $\partial_\lambda \alpha = \partial_\lambda \beta = \partial_\lambda \chi = 0$ holds in the formula (3.31). Equation (3.39) follows from (3.29) and (3.32). □

We will next show the existence of higher derivatives of X_t^λ.

Theorem 3.4.2 *For SDE* (3.10), *assume that diffusion and drift coefficients $V_k, k = 0, \ldots, d'$ are $C_b^{\infty,1}$-functions of x, r and jump coefficient g satisfies Condition (J.1). Then the solution starting from λ at time t_0 has a modification $X_t^\lambda, t \in \mathbb{T}, \lambda \in \Lambda$ which is infinitely differentiable with respect to λ and derivatives are cadlag with respect to t a.s. Further, for any $|\mathbf{i}| \geq 2$, $T > t_0$ and $p \geq 2$ there is a positive constant c such that*

$$\sup_\lambda E[\|\partial_\lambda^{\mathbf{i}} X^\lambda\|_T^p] < c. \tag{3.40}$$

Proof We saw in the previous lemma that $X_t^\lambda, t \in \mathbb{T}, \lambda \in \Lambda$ is continuously differentiable with respect to λ and the derivative $\partial X_t^\lambda := \partial_\lambda X_t^\lambda$ satisfies the SDE (3.38). Further, it satisfies

$$E[\|\partial X^\lambda - \partial X^{\lambda'}\|_T^p] \leq C_p|\lambda - \lambda'|^p, \tag{3.41}$$

because (3.35) holds with $\lambda = \lambda, \lambda' = \lambda'$ and $\epsilon = \epsilon' = 0$. We may regard (3.38) as an SDE for $\bar{x} = \bar{X}_t^\lambda := \partial X_t^\lambda$ with initial data and random coefficients with parameter λ given by

$$c_\lambda = 1, \quad \alpha_\lambda(\bar{x}, r) = \partial_x \alpha(X_{r-}^\lambda, r)\bar{x}, \tag{3.42}$$
$$\beta_\lambda(\bar{x}, r) = \partial_x \beta(X_{r-}^\lambda, r)\bar{x}, \quad \chi_\lambda(\bar{x}, r, z) = \partial_x \chi(X_{r-}^\lambda, r, z)\bar{x}.$$

We want to show the differentiability of $\bar{X}_t^\lambda$ with respect to λ. Note that $\partial_x\alpha(X_{r-}^\lambda, r)$ etc. are bounded functional. We will consider the stopped process $\bar{X}_{t\wedge\tau}^\lambda$, where $\tau = \tau_N$ is the stopping time defined by

$$\tau_N = \inf\{t \geq t_0;\ \sup_{|\lambda|\leq K} |\bar{X}_t^\lambda| > N\}\ (= \infty\ \text{ if } \{\cdots\} \text{ is empty}),$$

where K is a positive constant. It holds that $P(\tau_N \leq t) \to 0$ as $N \to \infty$. Further, $\bar{x} = \bar{X}_t^\lambda$ satisfies a master equation with coefficients $\alpha'_\lambda(\bar{x}, r)$, $\beta'_\lambda(\bar{x}, r)$ and $\chi'_x(\bar{x}, r, z)$, where

$$\alpha'_\lambda(\bar{x}, r) = \begin{cases} \alpha_\lambda(\bar{x}, r), & \text{if } r < \tau, \\ 0, & \text{if } r \geq \tau, \end{cases}$$

etc. These coefficients are differentiable with respect to $\bar{x}$ and parameter λ and further derivatives $\partial_{\bar{x}}\alpha'_\lambda(\bar{x}, r)$ and $\partial_\lambda\alpha'_\lambda(\bar{x}, r)$ satisfy Conditions 1–3 of the master equation. The first term is easily verified. We consider the second term. It holds that

$$\partial_\lambda\alpha'_\lambda(\bar{x}, r) = \partial_x\partial_x\alpha(X_{r\wedge\tau}^\lambda, r)\partial_\lambda X_{r\wedge\tau}^\lambda\bar{x}.$$

It is uniformly Lipschitz continuous with respect to $\bar{x}$, since $\partial_x\partial_x\alpha(X_{r\wedge\tau}^\lambda)\partial_x X_{r\wedge\tau}^\lambda$ is bounded a.s. Further, it is weakly Lipschitz continuous with respect to the parameter λ. Hence they satisfy conditions of Theorem 3.3.2. Then $\bar{X}_{t\wedge\tau}^\lambda$ is continuously differentiable with respect to λ by Theorem 3.3.2. Let N tend to ∞. Then we find that $\bar{X}_t^\lambda$ is continuously differentiable with respect to λ a.s.

Now, the second derivative $\bar{x} = \bar{X}_t = \partial_\lambda^{\mathbf{i}} X_t^\lambda$ ($|\mathbf{i}| = 2$) satisfies an another linear SDE with random coefficients. Its initial data is 0 and diffusion coefficient is given by

$$\alpha_\lambda(\bar{x}, r) := \partial_x^{\mathbf{i}}\alpha(X_{r-}^\lambda, r)(\partial_\lambda X_{r-}^\lambda)^2 + \partial_x\alpha(X_{r-}^\lambda, r)\bar{x}. \tag{3.43}$$

Drift coefficient and jump coefficient are given similarly. These coefficients satisfy Conditions 1–3 of the master equation. Then we get (3.40) in the case $|\mathbf{i}| \leq 2$.

By a similar device, we can show that $\partial_\lambda^{\mathbf{i}} X_t^\lambda$ is again continuously differentiable with respect to λ. Further, its derivatives satisfy (3.40) for $|\mathbf{i}| \leq 3$. Repeating this argument inductively, we find that X_t^λ is infinitely continuously differentiable with respect to λ for any t a.s. and (3.40) holds for any $\mathbf{i}$. □

For a continuous SDE, we get stronger assertions in Theorems 3.4.1 and 3.4.2.

Theorem 3.4.3 *Let $X_t^{\lambda,s}, t \in \mathbb{T}, \lambda \in \Lambda$ be a solution of a continuous symmetric SDE* (3.1) *starting from λ at time s. Then for any $T > 0$ and $p \geq 2$, there is a positive constant c such that the inequality*

$$E[|X_t^{\lambda,s} - X_{t'}^{\lambda',s'}|^p] \leq c\Big\{|\lambda - \lambda'|^p + |s - s'|^{\frac{p}{2}} + |t - t'|^{\frac{p}{2}}\Big\} \tag{3.44}$$

holds for any $s, s', t, t' \in [0, \infty)$ such that $s \vee s' < t \wedge t'$ and $\lambda, \lambda' \in \mathbb{R}^d$. Further, $X_t^{\lambda,s}$ has a modification such that it is continuous in (λ, s, t), infinitely differentiable with respect to λ and derivatives $\partial_\lambda^{\mathbf{i}} X_t^{\lambda,s}$ are continuous in (λ, s, t) a.s. for any **i**.

Proof We will show the inequality (3.44). Suppose $s' < s < t < t'$. Then $X_{t'}^{\lambda',s'} = X_{t'}^{X_s^{\lambda',s'},s}$ holds a.s. Therefore,

$$
\begin{aligned}
& E[|X_t^{\lambda,s} - X_{t'}^{\lambda',s'}|^p] \\
& \leq c_1\{E[E[|X_t^{\lambda,s} - X_t^{z,s}|^p]|_{z=X_s^{\lambda',s'}}] + E[|X_t^{\lambda',s'} - X_{t'}^{\lambda',s'}|^p]\} \\
& \leq c_2\{E[|\lambda - X_s^{\lambda',s'}|^p] + E[|X_t^{\lambda',s'} - X_{t'}^{\lambda',s'}|^p]\} \\
& \leq c_3\{|\lambda - \lambda'|^p + E[|\lambda' - X_s^{\lambda',s'}|^p] + E[|X_t^{\lambda',s'} - X_{t'}^{\lambda',s'}|^p]\}.
\end{aligned}
$$

Here we used the inequality $E[|X_t^{\lambda,s} - X_t^{z,s}|^p] \leq c_2|\lambda - z|^p$. Since

$$
\begin{aligned}
X_s^{\lambda',s'} - \lambda' &= \int_{s'}^{s} \alpha(r)\, dW_r + \int_{s'}^{s} \beta(r)\, dr, \\
X_t^{\lambda',s'} - X_{t'}^{\lambda',s'} &= \int_t^{t'} \alpha'(r)\, dW_r + \int_t^{t'} \beta'(r)\, dr
\end{aligned}
$$

hold with bounded functionals $\alpha(r), \beta(r)$ etc., we get by the Burkholder–Davis–Gundy inequality,

$$
E[|X_s^{\lambda',s'} - \lambda'|^p] \leq c_4|s - s'|^{\frac{p}{2}}, \quad E[|X_t^{\lambda',s'} - X_{t'}^{\lambda',s'}|^p] \leq c_5|t - t'|^{\frac{p}{2}}.
$$

Summing up these computations, we get (3.44). Then by the Kolmogorov–Totoki theorem, $X_t^{s,\lambda}$ has a modification which is continuous in (λ, s, t).

We can apply the similar argument for the processes $\partial_\lambda^{\mathbf{i}} X_t^{\lambda,s}$ and we find that $\partial_\lambda^{\mathbf{i}} X_t^{\lambda,s}$ is continuous in λ, s, t. □

We saw in Theorem 3.4.2 that the solution $X_t^{\lambda,s}$ of equation (3.37) starting from λ at time s has a modification of a C^∞-function of λ. We denote the modification by $\Phi_{s,t}(\lambda)$. Then the maps $\lambda \in \mathbb{R}^d \to \Phi_{s,t}(\lambda) \in \mathbb{R}^d$ are C^∞-maps a.s. for any $s < t$. Further, the equality $X_u^{\lambda,s} = X_u^{X_t^{\lambda,s},t}$ holds a.s. for $s < t < u$ by the path-wise uniqueness of the solution. Therefore the family of maps $\{\Phi_{s,t}\}$ satisfies the equality $\Phi_{s,u}(\lambda) = \Phi_{t,u}(\Phi_{s,t}(\lambda))$ for any λ a.s. for any $s < t < u$.

In the following discussion, we write the spatial parameter $\lambda \in \mathbb{R}^d$ as $x \in \mathbb{R}^d$. Then the family of maps $\{\Phi_{s,t}; 0 \leq s < t < \infty\}$ is a *stochastic flow of C^∞-maps* or simply C^∞-flow.

Theorem 3.4.4 *Consider a symmetric SDE with jumps given by* (3.10)*. Solutions of the SDE define a stochastic flow of C^∞-maps $\{\Phi_{s,t}\}$. Further, for any $T > 0$ and $p \geq 2$ and multi-index* $\mathbf{i}$,

$$E[|\Phi_{s,t}(x) - x|^p], \quad E[|\partial^{\mathbf{i}}\Phi_{s,t}(x)|^p] \tag{3.45}$$

are bounded with respect to $0 < s < t < T$ and x.

If the SDE is continuous, it defines a continuous stochastic flow of C^∞-maps.

We want to prove that maps $\Phi_{s,t}; \mathbb{R}^d \to \mathbb{R}^d$ are diffeomorphic a.s. for any $s < t$. However, discussions are not simple. The proof will be given in Sect. 3.6 for continuous SDEs and in Sect. 3.8 for SDEs with jumps. Before we proceed to the proof, we will develop the backward calculus. In the next section, we discuss stochastic flows of C^∞-maps generated by backward SDEs.

3.5 Backward SDE, Backward Stochastic Flow

We have so far discussed a continuous SDE (3.1) and an SDE with jumps (3.10); these equations describe the time evolution to the forward direction. However, we are also interested in solving these equations to the backward direction: For this purpose, we need to define the backward SDEs. Solutions of backward SDEs will be written as $\check{X}_s$, while solutions of forward ones have been written as X_t.

We will first consider a continuous backward symmetric SDE, which has the same coefficients $V_k(x,t), k = 0, \ldots, d'$ of $C^{\infty,1}$-class as the forward SDE (3.1). Given $t_1 > 0$ and $x \in \mathbb{R}^d$, if a continuous backward $\mathcal{F}_{s,t_1}$-semi-martingale $\check{X}_s, s \leq t_1$ satisfies

$$\check{X}_s = x + \sum_{k=0}^{d'} \int_s^{t_1} V_k(\check{X}_r, r) \circ d W_r^k \tag{3.46}$$

for $0 \leq s < t_1$, $\check{X}_s, 0 \leq s \leq t_1$ is called a solution of the symmetric backward equation with coefficients $V_k(x,t)$ starting from x at time t_1. Here $\circ d W_r^k, k = 1, \ldots, d'$ denote the backward symmetric integrals with respect to $\check{W}_r^k = W_t^k - W_T^k, k = 1, \ldots, d'$ and $\circ d W_r^0 = dr$. See Sect. 2.7.

The existence and the uniqueness of the solution of the backward equation can be verified in the same way as those of the forward equation. Let $\check{X}_s^{x,t_1}, 0 \leq s \leq t_1$ be the solution of the equation (3.46). Then it has a modification such that it is a C^∞-function of x. We denote it by $\check{\Phi}_{s,t_1}(x)$. Then $x \to \check{\Phi}_{s,t_1}(x); \mathbb{R}^d \to \mathbb{R}^d$ are C^∞-maps a.s. It satisfies the backward flow property $\check{\Phi}_{s,u} = \check{\Phi}_{s,t} \circ \check{\Phi}_{t,u}$ a.s. for any $u > t > s \geq 0$. We call $\{\check{\Phi}_{s,t}\}$ a backward stochastic flow of C^∞-maps defined by the backward continuous symmetric SDE with coefficients $V_k(x,t), k = 0, \ldots, d'$.

We will next study the backward SDE with jumps. Let $V_k(x,t), k = 0, \dots, d'$ be time-dependent vector fields of $C^{\infty,1}$-class and let $g(x,t,z)$ be a jump-map of $C^{\infty,1,2}$-class, which are given in the definition of the forward SDE (3.10). Given $t_1 > 0$ and $x \in \mathbb{R}^d$, if a right continuous backward $\mathcal{F}_{s,t_1}$ semi-martingale $\check{X}_s, s \le t_1$ satisfies

$$\check{X}_s = x + \sum_{k=0}^{d'} \int_s^{t_1} V_k(\check{X}_r, r) \circ dW_r^k \tag{3.47}$$

$$+\lim_{\epsilon\to 0} \left\{ \int_s^{t_1}\int_{|z|\ge\epsilon} g(\check{X}_r, r, z) N(dr\, dz) - \sum_{k=1}^{d'} b_k^\epsilon \int_s^{t_1} \tilde{V}_k(\check{X}_r, r)\, dr \right\},$$

for $0 \le s < t_1$, $\check{X}_s, 0 \le s \le t_1$ is called a solution of the backward symmetric SDE with coefficients $(V_k(x,t), k = 0, \dots, d', g(x, t.z))$ starting from x at time t_1.

The existence and the uniqueness of the solution of the equation can be verified in the same way as those of the forward equation. Let $\check{X}_s^{x,t_1}, 0 \le s \le t_1$ be the solution of the equation. Then it has a modification such that it is a C^∞-function of x. We denote it by $\check{\Phi}_{s,t_1}(x)$. Then $x \to \check{\Phi}_{s,t_1}(x); \mathbb{R}^d \to \mathbb{R}^d$ are C^∞-maps a.s. It has the backward flow property. $\{\check{\Phi}_{s,t}\}$ is called the backward stochastic flow of C^∞-maps defined by the backward SDE with coefficients $(V_k(x,t), k = 0, \dots, d', g(x,t,z))$.

3.6 Forward–Backward Calculus for Continuous C^∞-Flows

Associated with the continuous symmetric SDE (3.1), we define *vector fields* (first-order differential operators) with time parameter t by

$$V_k(t) f(x) = \sum_{i=1}^{d} V_k^i(x,t) \frac{\partial f}{\partial x_i}(x), \quad k = 0, \dots, d'.$$

For a slowly increasing (or rapidly decreasing) C^∞-function f, we define a second-order differential operator $A(t)$ with time parameter $0 \le t < \infty$ by

$$A(t) f(x) = \frac{1}{2} \sum_{k=1}^{d'} V_k(t)^2 f(x) + V_0(t) f(x). \tag{3.48}$$

Then $A(t)f(x)$ is a slowly increasing (or rapidly decreasing, respectively) C^∞-function. We first discuss the forward differential calculus of the stochastic flow generated by a continuous SDE.

Proposition 3.6.1 *Let $\{\Phi_{s,t}\}$ be a continuous stochastic flow of C^∞-maps defined by a continuous symmetric SDE* (3.1). *It satisfies equation* (3.6). *Let f be a slowly increasing C^∞-function. For each s, we have the forward differential rule with respect to t:*

$$f(\Phi_{s,t})=f+\sum_{k=1}^{d'}\int_s^t V_k(r)f(\Phi_{s,r})\,dW_r^k+\int_s^t A(r)f(\Phi_{s,r})\,dr \tag{3.49}$$

$$= f+\sum_{k=0}^{d'}\int_s^t V_k(r)f(\Phi_{s,r})\circ dW_r^k. \tag{3.50}$$

Proof Set $\phi(r) = (V_1(\Phi_{s,r}, r), \dots, V_{d'}(\Phi_{s,r}, r))$ and $\upsilon(r) = V_0(\Phi_{s,r}, r)$. Then the SDE is written as $X_t = x+\int_s^t \phi(r)\circ dW_r+\int_s^r \upsilon(r)\,dr$. We will apply Theorem 2.4.1 (Itô's formula for symmetric integral). Since

$$\sum_i \upsilon^i(r)\frac{\partial f}{\partial x_i}(\Phi_{s,r}) = V_0(r)f(\Phi_{s,r}), \quad \sum_i \phi^{ik}(r)\frac{\partial f}{\partial x_i}(\Phi_{s,r}) = V_k(r)f(\Phi_{s,r}),$$

we get $f(\Phi_{s,r}) = f + \sum_{k=0}^{d'}\int_s^t V_k(r)f(\Phi_{s,r})\circ dW_r^k$ from (2.23).

Next, since the martingale part of $V_k(t)f(\Phi_{s,t})$ is

$$\sum_{i,l}\int_s^t \frac{\partial V_k(r)f}{\partial x_i}(\Phi_{s,r})V_l^i(\Phi_{s,r}, r)\,dW_r^l,$$

the symmetric integral is rewritten as

$$\begin{aligned}
&\int_s^t V_k(r)f(\Phi_{s,r})\circ dW_r^k \\
&=\int_s^t V_k(r)f(\Phi_{s,r})\,dW_r^k + \frac{1}{2}\sum_{i,l}\left\langle \int_s^t \frac{\partial V_k(r)f}{\partial x_i}(\Phi_{s,r})V_l^i(\Phi_{s,r}, r)\,dW_r^l,\, W_t^k\right\rangle_{s,t} \\
&=\int_s^t V_k(r)f(\Phi_{s,r})\,dW_r^k + \frac{1}{2}\int_s^t V_k(r)^2 f(\Phi_{s,r})\,dr.
\end{aligned}$$

Therefore we get (3.49). □

Next, we will consider the stochastic flow $\{\Phi_{s,t}\}$ as a backward process of time parameter $s \in [0, t]$ starting from x at time t. Backward Itô integrals and backward symmetric integrals are defined in Sect. 2.7.

Let f be a C^∞-function of $x \in \mathbb{R}^d$. Then for any $r < t$, the composites $f \circ \Phi_{r,t}$ are C^∞-functions of $x \in \mathbb{R}^d$, a.s. Then we may apply the differential operator $V_k(r)$ and $A(r)$ to these functions. These are denoted by $V_k(r)(f\circ\Phi_{s,t})$ and $A(r)(f\circ\Phi_{r,t})$, respectively.

Proposition 3.6.2 *Let $\{\Phi_{s,t}\}$ be the continuous stochastic flow of C^∞-maps defined by the continuous symmetric SDE* (3.1). *For a slowly increasing C^∞-function f, we have the backward differential rule with respect to s:*

$$f \circ \Phi_{s,t} = f + \sum_{k=1}^{d'} \int_s^t V_k(r)(f \circ \Phi_{r,t})\, dW_r^k + \int_s^t A(r)(f \circ \Phi_{r,t})\, dr \tag{3.51}$$

$$= f + \sum_{k=0}^{d'} \int_s^t V_k(r)(f \circ \Phi_{r,t}) \circ dW_r^k. \tag{3.52}$$

Proof Let $\Pi = \{s = r_n < \cdots < r_0 = t\}$ be a partition of the interval $[s, t]$. We have

$$\begin{aligned} & f(\Phi_{s,t}(x)) - f(x) \\ &= \sum_{m=1}^{n} \{f(\Phi_{r_m,t}(x)) - f(\Phi_{r_{m-1},t}(x))\} \\ &= \sum_i \Big\{ \sum_m \frac{\partial}{\partial x_i}(f \circ \Phi_{r_{m-1},t})(x)(\Phi^i_{r_m,r_{m-1}}(x) - x_i) \Big\} \\ &+ \frac{1}{2} \sum_{i,j} \Big\{ \sum_m \frac{\partial^2}{\partial x_i \partial x_j}(f \circ \Phi_{r_{m-1},t})(x)(\Phi^i_{r_m,r_{m-1}}(x) - x_i)(\Phi^j_{r_m,r_{m-1}}(x) - x_j) \Big\} \\ &+ O(\Pi). \end{aligned} \tag{3.53}$$

Here we used the formula $\Phi_{r_m,t}(x) = \Phi_{r_{m-1},t} \circ \Phi_{r_m,r_{m-1}}(x)$. Let $|\Pi|$ tend to 0. Then the term $O(\Pi)$ converges to 0 in probability. Note that $\Phi_{s,t}(x)$ is a solution of Itô's SDE $dX_t = \alpha(X_t, t)\, dW_t + \beta(X_t, t)\, dt$, where $\beta(x, t) = V_0(x, t) + \frac{1}{2}\sum_k V_k(t)(V_k(t))(x)$. Then we have

$$\Phi^i_{r_m,r_{m-1}}(x) - x_i = \sum_k \alpha^{ik}(x, r_m)(W^k_{r_{m-1}} - W^k_{r_m}) + \beta^i(x, r_m)(r_{m-1} - r_m) + O_m,$$

where $\sum_m O_m \to 0$ in probability as $|\Pi| \to 0$. In the above formula, we can replace $\alpha^{ik}(x, r_m)$ by $\alpha^{ik}(x, r_{m-1})$, since $\alpha^{ik}(x, r)$ is a $C^{\infty,1}$-function. Therefore, the above converges to the following backward integral:

$$\sum_{i,k} \int_s^t \frac{\partial}{\partial x_i}(f \circ \Phi_{r,t})(x)\alpha^{ik}(x, r)\, dW_r^k + \sum_i \int_s^t \frac{\partial}{\partial x_i}(f \circ \Phi_{r,t})(x)\beta^i(x, r)\, dr$$

$$+ \frac{1}{2} \sum_{i,j} \int_s^t \frac{\partial^2}{\partial x_i \partial x_j}(f \circ \Phi_{r,t})(x)\Big(\sum_k \alpha^{ik}(x, r)\alpha^{jk}(x, r)\Big)\, dr$$

$$= \sum_{k=1}^{d'} \int_s^t V_k(r)(f \circ \Phi_{r,t})(x)\, dW_r^k + \int_s^t A(r)(f \circ \Phi_{r,t})(x)\, dr.$$

Therefore we get (3.51). Further, we have for $1 \le k \le d'$

$$\int_s^t V_k(r)(f \circ \Phi_{r,t})(x) \circ dW_r^k$$
$$= \int_s^t V_k(r)(f \circ \Phi_{r,t})(x)\, dW_r^k + \frac{1}{2}\int_s^t V_k(r)^2(f \circ \Phi_{r,t})(x)\, dr,$$

in view of (2.48). From the above we get (3.52). □

Next, we consider the backward flow $\{\check{\Phi}_{s,t}\}$ defined by the backward continuous SDE (3.46).

Proposition 3.6.3 *Let $\{\check{\Phi}_{s,t}\}$ be the backward continuous stochastic flow of C^∞-maps defined by the backward continuous symmetric SDE* (3.46). *Let f be a slowly increasing C^∞-function on $\mathbb{R}^d$. Then, we have the backward differential rule with respect to s:*

$$f(\check{\Phi}_{s,t}) = f + \sum_{k=1}^{d'} \int_s^t V_k(r) f(\check{\Phi}_{r,t})\, dW_r^k + \int_s^t A(r) f(\check{\Phi}_{r,t})\, dr. \tag{3.54}$$

Further, we have the forward differential rule with respect to t.

$$f \circ \check{\Phi}_{s,t} = f + \sum_{k=1}^{d'} \int_s^t V_k(r)(f \circ \check{\Phi}_{s,r})\, dW_r^k + \int_s^t A(r)(f \circ \check{\Phi}_{s,r})\, dr. \tag{3.55}$$

Equation (3.54) is shown similarly as the proof of (3.49), using Theorem 2.7.1. Equation (3.55) is shown similarly as the proof of (3.51).

3.7 Diffeomorphic Property and Inverse Flow for Continuous SDE

Let us consider again the (forward) continuous stochastic flow $\{\Phi_{s,t}\}$ of C^∞-maps generated by the continuous SDE (3.1). In order to prove the diffeomorphic property of the flow $\Phi_{s,t}$, we want to construct a stochastic process with parameters $s, x \in \mathbb{R}^d$ denoted by $\Psi'_{s,t}(x)$, which is continuous in x, t and satisfies

$$\Phi_{s,t}(\Psi'_{s,t}(x)) = \Psi'_{s,t}(\Phi_{s,t}(x)) = x, \quad \forall x \ a.s.$$

for any $s < t$. Then $\Phi_{s,t}(x); \mathbb{R}^d \to \mathbb{R}^d$ should be diffeomorphisms a.s. Indeed, take ω for which the equality $\Psi'_{s,t}(\Phi_{s,t}(x)) = \Phi_{s,t}(\Psi'_{s,t}(x)) = x$ holds for all $x \in \mathbb{R}^d$. Then for any $x \in \mathbb{R}^d$, choose $z = \Psi'_{s,t}(x)$. Then $\Phi_{s,t}(z) = x$ holds, so that the map $\Phi_{s,t}$ is an onto map. Next suppose that $\Phi_{s,t}(x) = \Phi_{s,t}(y)$ holds for some x, y. Then $\Psi'_{s,t}(\Phi_{s,t}(x)) = \Psi'_{s,t}(\Phi_{s,t}(y))$ holds. Therefore we have $x = y$. This means that the map $\Phi_{s,t}; \mathbb{R}^d \to \mathbb{R}^d$ is one to one.

Let $\phi(x) = (\phi^1(x), \ldots, \phi^d(x))$ be a C^1-map from $\mathbb{R}^d$ into itself. The $d \times d$-matrix $\partial\phi(x) = (\frac{\partial}{\partial x_i}\phi^j(x))_{i=1,\ldots,d, j=1,\ldots,d}$ is denoted by $\nabla\phi(x)$ and is called the *Jacobian matrix*. Further, the determinant of the Jacobian matrix is denoted by $\det \nabla\phi(x)$ and is called the *Jacobian or Jacobian determinant* of ϕ. Now, if the Jacobian matrix $\nabla\Phi_{s,t}(x)$ is invertible for all x, a.s., then the inverse map $\Phi_{s,t}^{-1}$ should be C^∞ and hence $\Phi_{s,t}$ should be C^∞-diffeomorphisms.

In the next lemma, we will prove that $\nabla\Phi_{s,t}(x)$ are invertible a.s and then in Lemma 3.7.2, we will construct such a random field $\Psi'_{s,t}(x)$.

Lemma 3.7.1 *Let $\{\Phi_{s,t}\}$ be a continuous stochastic flow of C^∞-maps generated by a continuous symmetric SDE with coefficients $V_k(x,t), k = 0, \ldots, d'$. Then the family of Jacobian matrices $\{\nabla\Phi_{s,t}(x)\}$ satisfy the linear SDE*

$$\nabla\Phi_{s,t}(x) = I + \sum_{k=0}^{d'} \int_s^t \nabla V_k(\Phi_{s,r}(x), r)\nabla\Phi_{s,r}(x) \circ dW_r^k \tag{3.56}$$

for any $s < t$ and x, a.s. Further, $\nabla\Phi_{s,t}(x)$ is invertible for any $s < t, x$ almost surely, and the inverse matrix $V_{s,t}(x)$ satisfies

$$\sup_x E[|V_{s,t}(x)|^p] < \infty \tag{3.57}$$

for any $s < t$ and $p \geq 2$.

Proof Differentiating each term of equation (3.6), we get (3.56). See Proposition 2.4.3. Set $\Phi_t = \Phi_{s,t}$ and $\nabla\Phi_t = \nabla\Phi_{s,t}$. Consider a stochastic differential equation for matrix-valued process V_t with parameter x:

$$V_t = I - \sum_{k=0}^{d'} \int_s^t V_r \nabla V_k(\Phi_r(x), r) \circ dW_r^k. \tag{3.58}$$

It has a unique solution, which we denote by $V_t(x)$. Then, using Itô's formula for symmetric integrals (Theorem 2.4.1), we have

$$\begin{aligned}\circ d(V_t\nabla\Phi_t) &= \circ dV_t\nabla\Phi_t + V_t \circ d\nabla\Phi_t \\ &= -\sum_{k=0}^{d'} V_t\nabla V_k\nabla\Phi_t \circ dW_t^k + \sum_{k=0}^{d'} V_t\nabla V_k\nabla\Phi_t \circ dW_t^k = 0.\end{aligned}$$

Therefore we have $V_t(x)\nabla\Phi_t(x) = I$. This shows that the Jacobian matrix $\nabla\Phi_t(x)$ is invertible and the inverse $V_t(x) = \nabla\Phi_t(x)^{-1}$ satisfies (3.58). The equation for V_t satisfies the condition of the master equation with parameter $\lambda = x$. Then we get the inequality (3.57) in view of Lemma 3.3.2. □

Now, let us consider a forward SDE defined by

$$X_t = x - \sum_{k=0}^{d'} \int_s^t \nabla\Phi_{s,r}(X_r)^{-1} V_k(X_r, r) \circ dW_r^k. \tag{3.59}$$

It is an SDE with random coefficients $\nabla\Phi_{s,t}(x)^{-1}V_k(x,t), k = 0, \dots, d'$. We will show that it has a unique solution $X_t^{x,s}, t \geq s$. We fix s and $M > 0$. Let τ^n be the first time t such that $\sup_{|x|\leq M} |\nabla\Phi_{s,t}(x)^{-1}|$ exceeds n. These are stopping times and $\lim_{n\to\infty} \tau^n = \infty$ holds a.s. We consider

$$dX_t^n = -\sum_{k=0}^{d'} \nabla\Phi_{s,t\wedge\tau^n}(X_t^n)^{-1} V_k(X_t^n, t) \circ dW_t^k.$$

Then coefficients satisfy the Lipschitz condition. Therefore it is a master equation. Let $X_t^{n,x,s}$ be the solution. It is a C^∞-function of x. By the path-wise uniqueness of the solution, it holds that $X_t^{n,x,s} = X_t^{n+1,x,s}$ for any $|x| \leq M$ if $t < \tau^n$. Therefore there exists $X_t^{x,s}$ such that $X_t^{x,x} = X_t^{n,x,s}$ holds a.s. if $t < \tau^n$. Then $X_t^{x,s}$ is the unique solution of equation (3.59). Further, the solution $X_t^{x,s}$ of (3.59) has a modification of C^∞-maps. We denote it by $\check{\Psi}_{s,t}(x)$.

Lemma 3.7.2 *It holds that $\Phi_{s,t}(\check{\Psi}_{s,t}(x)) = \check{\Psi}_{s,t}(\Phi_{s,t}(x)) = x$ for any x a.s. for any $s < t$.*

Proof We denote $\Phi_{s,t}(x)$ and $\check{\Psi}_{s,t}(x)$ by $\Phi_t(x)$ and $\check{\Psi}_t(x)$, respectively. Then $F(x,t) \equiv \Phi_t(x)$ is a stochastic process with parameter x belonging to the class $\mathbf{L}_{\mathbb{T}}^{n+Lip,p}(\mathbb{R}^d)$ for any n, p. Further, it satisfies conditions of Theorem 2.4.2. Then, setting $F(x,t) = \Phi_t(x)$ and $X_t = \check{\Psi}_t(x)$, we have by Theorem 2.4.2,

$$\begin{aligned}
\circ d\Phi_t(\check{\Psi}_t) &= \sum_{k=0}^{d'} V_k(\Phi_t \circ \check{\Psi}_t, t) \circ dW_t^k + \nabla\Phi_t(\check{\Psi}_t) \circ d\check{\Psi}_t \\
&= \sum_{k=0}^{d'} V_k(\Phi_t \circ \check{\Psi}_t, t) \circ dW_t^k - \sum_{k=0}^{d'} V_k(\Phi_t \circ \check{\Psi}_t, t) \circ dW_t^k = 0.
\end{aligned}$$

Therefore we have $\Phi_t(\check{\Psi}_t(x)) = x$ for all x a.s. Apply Theorem 2.4.2 again for $F(x,t) = \check{\Psi}_t(x)$ and $X_t = \Phi_t(x)$. Then we have

$$\circ d\check{\Psi}_t(\Phi_t) = -\sum_{k=0}^{d'}(\nabla\Phi_t)^{-1}V_k(\check{\Psi}_t\circ\Phi_t, t)\circ dW_t^k + \nabla\check{\Psi}_t(\Phi_t)\circ d\Phi_t.$$

Since $\Phi_t\circ\check{\Psi}_t(x) = x$, we have $\nabla\Phi_t(\check{\Psi}_t)\nabla\check{\Psi}_t = I$. Therefore we have $\nabla\check{\Psi}_t = (\nabla\Phi_t)(\check{\Psi}_t)^{-1}$. Then we get $\nabla\check{\Psi}_t(\Phi_t)\circ d\Phi_t = \sum_k(\nabla\Phi_t)^{-1}V_k(\check{\Psi}_t\circ\Phi_t, t)\circ dW_t^k$. Consequently, we get $\circ d\check{\Psi}_t(\Phi_t) = 0$. This proves $\check{\Psi}_t(\Phi_t(x)) = x$ for all x a.s. □

By the above two lemmas, maps $\Phi_{s,t}$; $\mathbb{R}^d\to\mathbb{R}^d$ are diffeomorphic a.s. and $\check{\Psi}_{s,t}$ are inverse maps of $\Phi_{s,t}$, i.e., $\check{\Psi}_{s,t}(x) = \Phi_{s,t}^{-1}(x)$ holds for any x a.s. We show that $\check{X}_s := \check{\Psi}_{s,t}(x)$ satisfies the backward SDE (3.46) with coefficients $-V_k(x,t)$. We have $\Phi_{s,t}(\check{\Psi}_{s,t}(x)) = x$ for all x a.s. This implies $\nabla\Phi_{s,t}(\check{\Psi}_{s,t}(x))\nabla\check{\Psi}_{s,t}(x) = I$. Therefore we have $\nabla\Phi_{s,t}(\check{\Psi}_{s,t}(x))^{-1} = \nabla\check{\Psi}_{s,t}(x)$. Then equation (3.59) can be rewritten as

$$\check{\Psi}_{s,t}(x) = x - \sum_{k=0}^{d'}\int_s^t \nabla\check{\Psi}_{s,r}(x)V_k(\check{\Psi}_{s,r}(x), r)\circ dW_r^k. \tag{3.60}$$

Consequently, we have by Theorem 2.4.1

$$f(\check{\Psi}_{s,t}(x)) = f(x) - \sum_{k=0}^{d'}\int_s^t V_k(f\circ\check{\Psi}_{s,r})(x)\circ dW_r^k,$$

for any C^∞-function f. Then $\check{\Psi}_{s,t}(x)$ should satisfy the backward differential rule

$$f(\check{\Psi}_{s,t}(x)) = f(x) - \sum_{k=0}^{d'}\int_s^t V_k(r)f(\check{\Psi}_{r,t}(x))\circ dW_r^k,$$

in view of Proposition 3.6.3. The above equation shows that $\check{\Psi}_{s,t}(x)$ satisfies the backward symmetric SDE;

$$\check{\Psi}_{s,t}(x) = x - \sum_{k=0}^{d'}\int_s^t V_k(\check{\Psi}_{r,t}(x), r)\circ dW_r^k. \tag{3.61}$$

It satisfies for any C^∞-function f

$$f(\check{\Psi}_{s,t}) = f(x) - \sum_{k=0}^{d'}\int_s^t V_k(r)f(\check{\Psi}_{r,t})\circ dW_r^k \tag{3.62}$$

$$= f(x) - \sum_{k=1}^{d'}\int_s^t V_k(r)f(\check{\Psi}_{r,t})\, dW_r^k + \int_s^t A(r)f(\check{\Psi}_{r,t})\, dr, \tag{3.63}$$

where $\mathcal{A}(t) = \frac{1}{2}\sum_{k=1}^{d'}(-V_k(t))^2 f + (-V_0(t))f$ or simply

$$\mathcal{A}(t)f = \frac{1}{2}\sum_{k=1}^{d'} V_k(t)^2 f - V_0(t)f. \tag{3.64}$$

Summing up these facts, we get the following theorem.

Theorem 3.7.1 *Assume that coefficients $V_k(x,t), k = 0,\ldots,d'$ of a continuous symmetric SDE* (3.1) *are $C_b^{\infty,1}$-functions. Then equation defines a continuous stochastic flow of diffeomorphisms $\{\Phi_{s,t}\}$. Set $\check{\Psi}_{s,t}(x) = \Phi_{s,t}^{-1}(x)$ for any x and $s < t$. Then $\{\check{\Psi}_{s,t}\}$ is a backward continuous stochastic flow and satisfies the backward SDE* (3.61). *It satisfies* (3.62)–(3.63).

Remark It is interesting to know that the equation for the forward flow given by (3.6) and the backward equation for the inverse flow $\check{\Psi}_{s,t}$ given by (3.7) are symmetric. If we rewrite equations using Itô integrals, the symmetric property disappears. In fact, the symmetric equation (3.6) is equivalent to the Itô equation

$$\Phi_{s,t}(x) = x + \sum_{k=1}^{d'}\int_s^t V_k(\Phi_{s,r}(x),r)\,dW_r^k + \int_s^t V_0'(\Phi_{s,r}(x),r)\,dr, \tag{3.65}$$

where $V_0'(x,t) = V_0(x,t) + \frac{1}{2}\sum_{k\geq 1} V_k(t)V_k(x,t)$. Further, equation (3.61) is equivalent to the backward Itô equation

$$\check{\Psi}_{s,t}(x) = x - \sum_{k=1}^{d'}\int_s^t V_k(\check{\Psi}_{r,t}(x),r)\,dW_r^k - \int_s^t V_0''(\check{\Psi}_{r,t}(x),r)\,dr, \tag{3.66}$$

where $V_0''(x,t) = V_0(x,t) - \frac{1}{2}\sum_{k\geq 1} V_k(t)V_k(x,t)$.

Remark Consider a continuous forward SDE with coefficients $-V_j(x,t), j = 0,\ldots,d'$: The equation is written as

$$X_t = X_0 - \sum_{k=0}^{d'}\int_s^t V_k(X_r,r)\circ dW_r^j.$$

Theorem 3.7.1 tells us that the equation generates a stochastic flow of diffeomorphisms, which we denote by $\{\Psi_{s,t}\}$. Set $\Psi_{s,t}^{-1} = \check{\Phi}_{s,t}$. The theorem tells us further that $\check{X}_s^{t_1,x} := \check{\Phi}_{s,t_1}(x)$ satisfies the backward SDE (3.46). Hence it coincides with the backward stochastic flow $\{\check{\Phi}_{s,t}\}$ discussed in Sect. 3.5.

Note The study of stochastic differential equations was initiated by Itô. In [50], stochastic differential equations driven by the Wiener process and Poisson random measure are discussed. Then continuous stochastic differential equations are studied extensively.

The diffeomorphic property of maps $x \to X_t^x$ received a lot of attention around 1980. We refer Elworthy [25], Malliavin [78], Bismut [8], Le Jan [75], Harris [36], Ikeda–Watanabe [41] and Kunita [59].

A method of proving the diffeomorphic property is to approximate SDEs by a sequence of stochastic ordinary differential equations. It is described as follows. Instead of a Wiener process W_t, we consider a suitable sequence of piecewise smooth stochastic processes $W_t^n = (W_t^{1,n}, \ldots, W_t^{d',n})$ which converges to W_t, and consider a sequence of equations

$$\frac{d}{dt} X_t^n = \sum_{k=1}^{d'} V_k(x_t^n, t)\dot{W}_t^{k,n} + V_0(x_t^n, t).$$

where $\dot{W}_t^{k,n} = \frac{dW_t^{k,n}}{dt}$. Then its solution $X_t^{n,x}$ starting from x at time 0 defines a diffeomorphic map for any n. The sequence $\{X_t^{n,x}, n = 1, 2, \ldots\}$ should converge to the solution X_t^x of the SDE and it should be a diffeomorphic map. For detail, see Ikeda–Watanabe [41], Bismut [8], Malliavin [78].

Kunita [59] presented another method for proving the diffeomorphic property through a skillful use of the Kolmogorov–Totoki theorem. In this monograph, we took another method by constructing a backward flow $\check{\Psi}_{s,t}(x)$ which satisfies Lemma 3.7.2. An advantage of the new method is that it can be applied for proving the diffeomorphic property of solutions of SDE on a manifold as in Sect. 7.1.

3.8 Forward–Backward Calculus for C^∞-Flows of Jumps

We will consider stochastic flow of C^∞-maps generated by a symmetric SDE with jumps (3.10), where $V_0(x, t), \ldots, V_{d'}(x, t)$ are $C_b^{\infty,1}$-functions and $g(x, t, z)$ satisfies Condition (J.1). For a slowly increasing C^∞-function f, we define an integro-differential operator $A_J(t)$ with time parameter t by

$$A_J(t)f(x) = A(t)f + \int_{\mathbf{R}_0^{d'}} \Big\{ f(\phi_{t,z}(x)) - f(x) - 1_D(z) \sum_{k=1}^{d'} z^k \tilde{V}_k(t) f(x) \Big\} \nu(dz), \tag{3.67}$$

where $A(t)$ is the differential operator defined by (3.48) and $\phi_{t,z}(x) = g(x, t, z) + x$ and $D = \{z = (z^1, \ldots, z^{d'}) \in \mathbf{R}_0^{d'};\ |z| \le 1\}$.

Proposition 3.8.1 *Let $\{\Phi_{s,t}\}$ be a stochastic flow of C^∞-maps defined by symmetric SDE* (3.10) *with jumps. Let f be a slowly increasing C^∞-function on $\mathbb{R}^d$. Then, the stochastic flow satisfies the forward differential rule with respect to t:*

$$
\begin{aligned}
f(\Phi_{s,t}) &= f + \sum_{k=1}^{d'} \int_s^t V_k(r) f(\Phi_{s,r-})\, dW_r^k + \int_s^t A_J(r) f(\Phi_{s,r-})\, dr \\
&\quad + \int_s^t \int_{\mathbb{R}_0^{d'}} \Big\{ f(\phi_{r,z} \circ \Phi_{s,r-}) - f(\Phi_{s,r-}) \Big\} \tilde{N}(dr\, dz) \qquad (3.68) \\
&= f + \sum_{k=0}^{d'} \int_s^t V_k(r) f(\Phi_{s,r}) \circ dW_r^k \\
&\quad + \lim_{\epsilon \to 0} \Big\{ \int_s^t \int_{|z| \ge \epsilon} \Big\{ f(\phi_{r,z} \circ \Phi_{s,r-}) - f(\Phi_{s,r-}) \Big\} N(dr\, dz) \\
&\quad - \sum_{k=1}^{d'} b_\epsilon^k \int_s^t \tilde{V}_k(r) f(\Phi_{s,r-})\, dr \Big\}. \qquad (3.69)
\end{aligned}
$$

Proof The process $X_t = \Phi_{s,t}(x)$ is a solution of equation (3.10). Then it is an Itô process satisfying

$$
X_t = x + \int_s^t \alpha(X_r, r)\, dW_r + \int_s^t \beta(X_r, r)\, dr + \int_s^t \int \chi(X_{r-}, r, z) \tilde{N}_D(dr\, dz),
$$

where

$$
\alpha(x, r) = (V_1(x, r), \ldots, V_{d'}(x, r)), \quad \chi(x, r, z) = g(x, t, z),
$$

$$
\beta(x, r) = V_0(x, r) + \frac{1}{2} \sum_{k \ge 1} V_k(r) V_k(x, r) + \int_D \Big\{ g(x, r, z) - \sum_{k \ge 1} z^k \tilde{V}_k(x, r) \Big\} \nu(dz).
$$

We apply Itô's formula (Theorem 2.6.1) for the C^3-function f. Then the diffusion part of the stochastic process $f(X_t)$ is equal to

$$
\sum_{i=1}^{d} \sum_{k=1}^{d'} \int_s^t \alpha^{ik}(r) \frac{\partial f}{\partial x_i}(\Phi_{s,r-})\, dW_r^k = \sum_{k=1}^{d'} \int_s^t V_k(r) f(\Phi_{s,r-})\, dW_r^k.
$$

The jump part is equal to

$$\begin{aligned}
&\int_s^t\int\Big\{f(\phi_{r,z}(\Phi_{s,r-}))-f(\Phi_{s,r-})\Big\}\tilde{N}_D(dr\,dz)\\
&=\int_s^t\int\Big\{f(\phi_{r,z}(\Phi_{s,r-}))-f(\Phi_{s,r-})\Big\}\tilde{N}(dr\,dz)\\
&\quad+\int_s^t\int_{|z|>1}\Big\{f(\phi_{r,z}(\Phi_{s,r-}))-f(\Phi_{s,r-})\Big\}n(dr\,dz).
\end{aligned}$$

The drift part is equal to

$$\begin{aligned}
&\frac{1}{2}\sum_{i,j=1}^{d}\int_s^t\Big(\sum_{k=1}^{d'}\alpha^{ik}(r)\alpha^{jk}(r)\Big)\frac{\partial^2 f}{\partial x_i\partial x_j}(\Phi_{s,r-})\,dr+\sum_{i=1}^{d}\int_s^t\beta^i(r)\frac{\partial f}{\partial x_i}(\Phi_{s,r-})\,dr\\
&\quad+\int_s^t\int_{|z|<1}\Big\{f(\phi_{r,z}(\Phi_{s,r-}))-f(\Phi_{s,r-})-\sum_i g^i\frac{\partial f}{\partial x_i}(\Phi_{s,r-})\Big\}n(dr\,dz)\\
&=\int_s^t\Big(\frac{1}{2}\sum_{k=1}^{d'}V_k(r)^2f(\Phi_{s,r-})+V_0(r)f(\Phi_{s,r-})\Big)\,dr\\
&\quad+\int_s^t\int_{|z|<1}\Big\{f(\phi_{r,z}(\Phi_{s,r-}))-f(\Phi_{s,r-})-\sum_{k=1}^{d'}z^k\tilde{V}_k(r)f(\Phi_{s,r-})\Big\}n(dr\,dz).
\end{aligned}$$

Summing these, we get the formula (3.68).

Equation (3.69) follows from (3.68), rewriting Itô integrals by symmetric integrals. □

Proposition 3.8.2 *Let $\{\Phi_{s,t}\}$ be a stochastic flow of C^∞-maps defined by the symmetric SDE* (3.10) *with jumps. Let f be a slowly increasing C^∞-function. Then for each t, the flow $\{\Phi_{s,t}\}$ has a modification which is right continuous with respect to s a.s. and the modification satisfies the backward differential rule with respect to s:*

$$\begin{aligned}
f\circ\Phi_{s,t}&=f+\sum_{k=1}^{d'}\int_s^t V_k(r)(f\circ\Phi_{r,t})\,dW_r^k+\int_s^t A_J(r)(f\circ\Phi_{r,t})\,dr\\
&\quad+\int_s^t\int\Big\{f\circ\Phi_{r,t}\circ\phi_{r,z}-f\circ\Phi_{r,t}\Big\}\tilde{N}(dr\,dz)
\end{aligned}\qquad(3.70)$$

$$
\begin{aligned}
&= f + \sum_{k=0}^{d'} \int_s^t V_k(r)(f \circ \Phi_{r,t}) \circ dW_r^k \\
&\quad + \lim_{\epsilon \to 0} \bigg\{ \int_s^t \int_{|z| \ge \epsilon} \Big\{ f \circ \Phi_{r,t} \circ \phi_{r,z} - f \circ \Phi_{r,t} \Big\} N(dr\, dz) \\
&\quad - \int_s^t \sum_{k=1}^{d'} b_\epsilon^k \tilde{V}_k(r)(f \circ \Phi_{r,t})\, dr \bigg\}.
\end{aligned} \tag{3.71}
$$

In order to prove the proposition, we will approximate $\Phi_{s,t}(x)$ by a family of stochastic flows $\{\Phi_{s,t}^\epsilon(x), 0 < \epsilon < 1\}$ generated by SDEs:

$$
\begin{aligned}
X_t &= x + \sum_{k=1}^{d'} \int_s^t V_k(X_r, r) \circ dW_r^k + \int_s^t V_0^\epsilon(X_r, r)\, dr \\
&\quad + \int_s^t \int_{|z| \ge \epsilon} g(X_{r-}, r, z) N(dr\, dz),
\end{aligned} \tag{3.72}
$$

where $V_0^\epsilon = V_0 - \sum_{k=1}^{d'} b_\epsilon^k \tilde{V}_k$. The equation cuts off all jumps $g(X_{r-}, r, z)$, $|z| \le \epsilon$ from equation (3.10). Its Lévy measure $\nu_\epsilon(dz) = 1_{(\epsilon,\infty)}(z)\nu(dz)$ has finite mass. Let $\Phi_{s,t}^\epsilon(x)$ be stochastic flows generated by the above SDE. Let $\Phi_{s,t}^0(x)$ be the stochastic flow generated by the continuous SDE

$$
X_t = x + \sum_{k=1}^{d'} \int_s^t V_k(X_r, r) \circ dW_r^i + \int_s^t V_0^\epsilon(X_r, r)\, dr. \tag{3.73}
$$

We saw in (3.18) that, taking $s = t_0$, $\Phi_{s,t}^\epsilon(x)$ is written as

$$
\Phi_{s,t}^\epsilon(x) = \begin{cases} \Phi_{s,t}^0(x), & \text{if } t < \tau_1, \\ \Phi_{\tau_n,t}^0 \circ \phi_{\tau_n,S_n} \circ \cdots \circ \phi_{\tau_1,S_1} \circ \Phi_{s,\tau_1}^0(x), & \text{if } \tau_n \le t < \tau_{n+1}, \end{cases} \tag{3.74}
$$

where τ_m, S_m are random variables defined in Sect. 3.2. It is a right continuous stochastic flow of diffeomorphisms. Further, it is a backward semi-martingale with respect to s. Then we have

$$
\begin{aligned}
&f(\Phi_{s,t}^\epsilon(x)) - f(x) \\
&= \sum_m \big\{ f(\Phi_{\tau_m-,t}^\epsilon(x)) - f(\Phi_{\tau_{m-1},t}^\epsilon(x)) \big\} + \sum_m \big\{ f(\Phi_{\tau_m-,t}^\epsilon(x)) - f(\Phi_{\tau_m,t}^\epsilon(x) \big\} \\
&= I_1^\epsilon + I_2^\epsilon.
\end{aligned}
$$

Since $\Phi_{s,t}^\epsilon(x)$ does not have jumps at $s \in (\tau_{m-1}, \tau_m)$, we have by Proposition 3.6.2:

$$I_1^\epsilon = \sum_{k=1}^{d'} \int_s^t V_k(r)(f \circ \Phi_{r,t}^\epsilon)(x)\, dW_r^k + \int_s^t A^\epsilon(r)(f \circ \Phi_{r,t}^\epsilon)(x)\, dr,$$

where $A^\epsilon(r) = \frac{1}{2}\sum_{k\geq 1} V_k(r)^2 + V_0^\epsilon(r)$. Further, we have

$$I_2^\epsilon = \int_s^t \int_{|z|\geq\epsilon} \Big\{ f \circ \Phi_{r,t}^\epsilon \circ \phi_{r,z}(x) - f \circ \Phi_{r,t}^\epsilon(x) \Big\} N(dr\, dz).$$

Therefore $\Phi_{s,t}^\epsilon(x)$ satisfies the equality (3.70) replacing $A_J(r)$ by

$$A_J^\epsilon(r) = A^\epsilon(r) + \int_{|z|>\epsilon} \{ f \circ \phi_{r,z} - f\} \nu(dz)$$

and $\tilde{N}(dr\, dz)$ by $\tilde{N}_\epsilon(dr\, dz) = 1_{(\epsilon,\infty)}(|z|)\tilde{N}(dr\, dz)$.
We will study the convergence of $\Phi_{s,t}^\epsilon(x)$ as $\epsilon \to 0$.

Lemma 3.8.1 *For any $p \geq 2$, $0 \leq s < t \leq T$ and $\mathbf{i}$, we have*

$$\lim_{\epsilon\to 0} \sup_{x'\in\mathbb{R}^d} E[|\partial^{\mathbf{i}}\Phi_{s,t}^\epsilon(x') - \partial^{\mathbf{i}}\Phi_{s,t}(x')|^p] = 0. \tag{3.75}$$

Proof We will consider the case where $|\mathbf{i}| = 0, 1$. The case $|\mathbf{i}| \geq 2$ will be shown similarly. We shall consider pair processes $Y_{s,t}(x') = (\Phi_{s,t}(x'), \partial\Phi_{s,t}(x'))$ and $Y_{s,t}^\epsilon(x') = (\Phi_{s,t}^\epsilon(x'), \partial\Phi_{s,t}^\epsilon(x'))$, regarding $0 < \epsilon < 1$ as a parameter. The latter can be regarded as a solution of the master equation (3.21) on $\mathbb{R}^d \times \mathbb{R}^{d^2}$ with coefficients

$$\alpha_\epsilon(x, y, r) = (\alpha(x, r), \partial\alpha(x, r)y), \quad \beta_\epsilon(x, y, r) = (\beta_\epsilon(x, r), \partial\beta_\epsilon(x, r)y),$$
$$\chi_\epsilon(x, y, r, z) = (\chi(x, r, z)1_{|z|\geq\epsilon}, \partial\chi(x, r, z)y1_{|z|\geq\epsilon}),$$

and $x = \Phi_{s,t}^\epsilon(x')$, $y = \partial\Phi_{s,t}^\epsilon(x')$, where

$$\beta_\epsilon(x, r) = V_0(x, r) + \frac{1}{2}\sum_{k=1}^{d'} V_k(r)V_k(x, r)$$
$$+ \int_{|z|\geq\epsilon} \Big\{ g(x, r, z) - 1_D(\sum_{k=1}^{d'} z_k\tilde{V}_k(x, r)) \Big\} \nu(dz).$$

It holds that $|\beta_\epsilon(x, y, r) - \beta(x, y, r)| \leq c\varphi_0(\epsilon)(1 + |y|)$ and

$$\int_{\mathbb{R}_0^{d'}} \left| \frac{\chi}{\gamma(z)} - \frac{\chi_\epsilon}{\gamma(z)} \right|^p \mu(dz) = \int_{|z|<\epsilon} \left| \frac{g}{|z|} \right|^p \mu(dz) \leq c\varphi_0(\epsilon)(1 + |y|)^p,$$

where $\varphi_0(\epsilon) = \mu(\{|z| < \epsilon\}) = \int_{|z|<\epsilon} |z|^2\nu(dz)$.

We shall estimate the integral $E[|Y^\epsilon_{s,t}(x') - Y_{s,t}(x')|^p]$ for $p > 2$. Set $Y^\epsilon_t = Y^\epsilon_{s,t}(x')$ and $Y_t = Y_{s,t}(x')$. The drift term of $Y^\epsilon_{s,t}(x') - Y_{s,t}(x')$ is estimated as

$$\begin{aligned}
&\Big|\int_s^t (\beta_\epsilon(Y^\epsilon_r, r) - \beta(Y_r, r))\, dr\Big| \\
&\le \int_s^t |\beta_\epsilon(Y^\epsilon_r, r) - \beta(Y^\epsilon_r, r)|\, dr + \int_s^t |\beta(Y^\epsilon_r, r) - \beta(Y_r, r)|\, dr \\
&\le c\varphi_0(\epsilon)\int_s^t (1 + |Y^\epsilon_r|)\, dr + c\int_s^t |Y^\epsilon_r - Y_r|\, dr.
\end{aligned}$$

Since $E[|Y^\epsilon_r|^p]$ are uniformly bounded with respect to $0 < s < t < T$ and x, ϵ, we have

$$E\Big[\Big|\int_s^t (\beta_\epsilon(Y^\epsilon_r, r) - \beta(Y_r, r))\, dr\Big|^p\Big] \le c_1\varphi_0(\epsilon)^p + c_2\int_s^t E[|Y^\epsilon_r - Y_r|^p]\, dr.$$

Similar estimates are valid for the diffusion term and jump term of $Y^\epsilon_{s,t}(x') - Y_{s,t}(x')$. We have

$$\begin{aligned}
&E\Big[\Big|\int_s^t \big(\alpha_\epsilon(Y^\epsilon_r, r) - \alpha(Y_r, r)\big)\, dW_r\Big|^p\Big] \le c_1'\varphi_0(\epsilon)^p + c_2'\int_s^t E[|Y^\epsilon_r - Y_r|^p]\, dr, \\
&E\Big[\Big|\int_s^t \int \{\chi_\epsilon(Y^\epsilon_r, r, z) - \chi(Y_r, r, z)\}\tilde{N}(dr\, dz)\Big|^p\Big] \\
&\le c_1''\varphi_0(\epsilon) + c_2''\int_s^t E[|Y^\epsilon_r - Y_r|^p]\, dr.
\end{aligned}$$

Consequently, for any $p \ge 2$ there exist $c_1 > 0, c_2 > 0$ such that

$$E[|Y^\epsilon_t - Y_t|^p] \le c_1(\varphi_0(\epsilon)^p + \varphi_0(\epsilon)) + c_2\int_s^t E[|Y^\epsilon_r - Y_r|^p]\, dr.$$

Then Gronwall's inequality implies

$$E[|Y^\epsilon_t - Y_t|^p] \le C_1(\varphi_0(\epsilon)^p + \varphi_0(\epsilon)).$$

Since this is valid for any x', s, we get (3.75) in the case $|\mathbf{i}| = 0, 1$. □

We are interested in the uniform convergence of $\partial^{\mathbf{i}}\Phi^\epsilon_{s,t}(x)$ with respect to x, s. For this problem, we need Sobolev's inequality in $L^p(D, dx)$-space, where D is a bounded domain in $\mathbb{R}^d$. Let $f(x)$ be a real function on D. We denote the weak derivative of f by $\partial^{\mathbf{i}}$ by $\partial^{\mathbf{i}} f$. We denote by $\mathbb{D}_{k,p}$ the set of all f such that its weak derivatives $\partial^{\mathbf{i}} f$, $|\mathbf{i}| \le k$ belong to $L^p(D, dx)$. We quote Morrey's Sobolev inequality without proof.

Lemma 3.8.2 (Brezis, [14]) *Let D be a bounded subdomain in $\mathbb{R}^d$ with smooth boundary. Let $p > d$. Then elements of $\mathbb{D}_{k,p}$ are C^{k-1}-functions. Further, there exists a positive constant $c_{k,p}$ such that*

$$\sum_{|\mathbf{i}|\le k-1} \sup_{x\in D} |\partial^{\mathbf{i}} f(x)| \le c_{k,p}\Big(\sum_{|\mathbf{i}|\le k} \int_D |\partial^{\mathbf{i}} f(x')|^p \, dx'\Big)^{\frac{1}{p}} \tag{3.76}$$

holds for any $f \in \mathbb{D}_{k,p}$.

Lemma 3.8.3 *For any $s < t$, there exists a sequence $\{\epsilon_n\}$ converging to 0 such that solutions $\{\Phi^{\epsilon_n}_{s,t}(x)\}$ and their derivatives $\{\partial^{\mathbf{i}}\Phi^{\epsilon_n}_{s,t}(x)\}$ converge to $\Phi_{s,t}(x)$ and $\partial^{\mathbf{i}}\Phi_{s,t}(x)$, uniformly on compact sets with respect to x a.s.*

Proof By Lemma 3.8.2, we have

$$\begin{aligned} &E[\sup_{|x|\le M} |\Phi_{s,t}(x) - \Phi^{\epsilon}_{s,t}(x)|^p] \\ &\le CE\Big[\int_{|x'|\le M} |\Phi_{s,t}(x') - \Phi^{\epsilon}_{s,t}(x')|^p \, dx' + \int_{|x'|\le M} |\partial\Phi_{s,t}(x') - \partial\Phi^{\epsilon}_{s,t}(x')|^p \, dx'\Big]. \end{aligned} \tag{3.77}$$

It converges to 0 as $\epsilon \to 0$ in view of Lemma 3.8.1. Therefore a subsequence $\{\Phi^{\epsilon_n}_{s,t}(x)\}$ converges to $\Phi_{s,t}(x)$ uniformly on compact sets with respect to x, t a.s. The convergence of $\{\partial^{\mathbf{i}}\Phi^{\epsilon_n}_{s,t}(x)\}$ can be shown similarly. □

Proof of Proposition 3.8.2 Let us consider the sequence of $\{\Phi^{\epsilon_n}_{s,t}(x)\}$ obtained in Lemma 3.8.3. We saw that each $\Phi^{\epsilon_n}_{s,t}(x)$ is right continuous in s and satisfies (3.70) replacing $\tilde{N}(dr\,dz)$ by $\tilde{N}_{\epsilon_n}(dr\,dz)$. Therefore it is a right continuous backward semi-martingale with respect to s. Let n tend to ∞. Then terms

$$\sum_{k=1}^{d'} \int_s^t V_k(r)(f\circ\Phi^{\epsilon_n}_{r,t})(x)\, dW^k_r, \quad \int_s^t A^{\epsilon_n}_J(r)(f\circ\Phi^{\epsilon_n}_{r,t})(x)\, dr, \quad \text{etc.}$$

should converge a.s. Therefore the limit $\Phi_{s,t}(x)$ is a backward semi-martingale with respect to s for any t, x, and it has a right continuous modification. We denote it by $\Phi_{s,t}(x)$, again. Then the above integrals should converge to

$$\sum_{k=1}^{d'} \int_s^t V_k(r)(f\circ\Phi_{r,t})(x)\, dW^k_r, \quad \int_s^t A_J(r)(f\circ\Phi_{r,t})(x)\, dr, \quad \text{etc.}$$

Therefore $\{\Phi_{s,t}\}$ satisfies (3.70).

Finally, the equality (3.71) follows from (3.70), if we rewrite backward symmetric integrals using backward Itô integrals. □

We shall next consider differential rules for the backward flow $\{\check{\Phi}_{s,t}\}$ defined by the backward symmetric SDE (3.47). The following is the counterpart of Proposition 3.8.1, whose proof is omitted.

Proposition 3.8.3 *Let $\{\check{\Phi}_{s,t}\}$ be the backward stochastic flow of C^∞-maps defined by the backward symmetric SDE (3.47). Let f be a slowly increasing C^∞-function on $\mathbb{R}^d$. Then, we have the backward differential rule with respect to s:*

$$
\begin{aligned}
f(\check{\Phi}_{s,t}) &= f + \sum_{k=1}^{d'} \int_s^t V_k(r) f(\check{\Phi}_{r,t})\, dW_r^k + \int_s^t A_J(r) f(\check{\Phi}_{r,t})\, dr \qquad (3.78)\\
&\quad + \lim_{\epsilon\to 0} \int_s^t \int_{|z|\ge\epsilon} \Big\{ f(\phi_{r,z}\circ\check{\Phi}_{r,t}) - f(\check{\Phi}_{r,t}) \Big\} \tilde{N}(dr\,dz)\\
&= f + \sum_{k=0}^{d'} \int_s^t V_k(r) f(\check{\Phi}_{r,t}) \circ dW_r^k \qquad (3.79)\\
&\quad + \lim_{\epsilon\to 0} \Big\{ \int_s^t \int_{|z|\ge\epsilon} \{ f(\phi_{r,z}\circ\check{\Phi}_{r,t}) - f(\check{\Phi}_{r,t})\} N(dr\,dz)\\
&\quad - \sum_{k=1}^{d'} b_\epsilon^k \int_s^t \tilde{V}_k(r) f(\check{\Phi}_{r,t})\, dr \Big\}.
\end{aligned}
$$

3.9 Diffeomorphic Property and Inverse Flow for SDE with Jumps

In this section, we assume further that the jump-map $g(x,t,z)$ of the SDE (3.10) satisfies Condition (J.2) and hence $\phi_{t,z}(x) = g(x,t,z) + x$ are diffeomorphic for any t, z. We will discuss the diffeomorphic property of the solution of the SDE. Let $\epsilon > 0$. Associated with equation (3.10), we consider a forward SDE (3.72). The solution defines a right continuous stochastic flow $\{\Phi_{s,t}^\epsilon\}$ of diffeomorphisms.

Lemma 3.9.1 *Set $\check{\Psi}_{s,t}^\epsilon(x) = (\Phi_{s,t}^\epsilon)^{-1}(x)$. Then $\check{X}_s^\epsilon := \check{\Psi}_{s,t}^\epsilon(x), 0 \le s < t < \infty$ satisfies the following backward equation with jumps:*

$$
\begin{aligned}
\check{X}_s^\epsilon &= x - \sum_{k=0}^{d'} \int_s^t V_k(\check{X}_r^\epsilon, r) \circ dW_r^k \qquad (3.80)\\
&\quad - \Big\{ \int_s^t \int_{|z|\ge\epsilon} h(\check{X}_r^\epsilon, r, z) N(dr\,dz) - \sum_{k=1}^{d'} b_\epsilon^k \int_s^t \tilde{V}_k(\check{X}_r^\epsilon, r)\, dr \Big\},
\end{aligned}
$$

where $h(x,t,z) = x - \phi_{t,z}^{-1}(x)$ and

$$
\tilde{V}_k(x,t) = \partial_{z^k} g(x,t,z)\Big|_{z=0} = \partial_{z^k} h(x,t,z)\Big|_{z=0}.
$$

Proof Since the flow $\Phi_{s,t}^{\epsilon}$ is represented by (3.74), the inverse flow $\check{\Psi}_{s,t}^{\epsilon}(x)$ is represented by

$$\begin{cases} \check{\Psi}_{s,t}^{0}(x), & \text{if } t < \tau_1 \\ \check{\Psi}_{s,\tau_1}^{0} \circ \phi_{\tau_1,S_1}^{-1} \circ \cdots \circ \phi_{\tau_n,S_n}^{-1} \circ \check{\Psi}_{\tau_n,t}^{0}(x), & \text{if } \tau_n \le t < \tau_{n+1}. \end{cases} \tag{3.81}$$

It is cadlag with respect to s and t, since $\Phi_{s,t}^{\epsilon}(x)$ is cadlag with respect to s and t. Note that $\check{X}_s = \check{\Psi}_{s,t}^{0}$ satisfies the continuous backward SDE

$$\check{X}_s = x - \sum_{k=1}^{d'} \int_s^t V_k(\check{X}_r, r) \circ dW_r^k - \int_s^t V_0^{\epsilon}(\check{X}_r, r)\, dr.$$

Then the composite process $\check{X}_s^{\epsilon} = \check{\Psi}_{s,t}^{\epsilon}(x)$ should satisfy the backward SDE

$$\begin{aligned} \check{X}_s^{\epsilon} = x &- \sum_{k=1}^{d'} \int_s^t V_k(\check{X}_r^{\epsilon}, r) \circ dW_r^k - \int_s^t V_0^{\epsilon}(\check{X}_r^{\epsilon}, r)\, dr \\ &- \int_s^t \int_{|z|\ge\epsilon} h(\check{X}_r^{\epsilon}, r, z) N(dr\, dz). \end{aligned} \tag{3.82}$$

(We have shown a similar fact for a forward equation in Sect. 3.2.) The above equation can be rewritten as (3.80). □

Theorem 3.9.1 *Assume that diffusion and drift coefficients of symmetric SDE* (3.10) *are* $C_b^{\infty,1}$*-functions and jump coefficients satisfy Conditions (J.1) and (J.2). Then the equation defines a right continuous stochastic flow of diffeomorphisms* $\{\Phi_{s,t}\}$.

Let $\{\check{\Psi}_{s,t}\}$ *be the backward flow of* C^{∞}*-maps generated by a backward symmetric SDE* (3.47) *with coefficients* $(-V_k(x,t), k = 0, \dots, d', -h(x,t,z))$. *Then it holds that* $\check{\Psi}_{s,t}(x) = \Phi_{s,t}^{-1}(x)$ *for any* x *a.s. for any* $s < t$.

Proof We saw in Lemma 3.8.3 that for any $s < t$, there is a sequence $\{\epsilon_n\}$ converging to 0 such that $\Phi_{s,t}^{\epsilon_n}(x)$ converges to $\Phi_{s,t}(x)$ uniformly on compact sets with respect to x almost surely. Then by the same reasoning, the sequence $\check{\Psi}_{s,t}^{\epsilon_n}(x)$ converges to $\check{\Psi}_{s,t}(x)$ uniformly on compact sets almost surely. Since $\Phi_{s,t}^{\epsilon}(\check{\Psi}_{s,t}^{\epsilon}(x)) = \check{\Psi}_{s,t}^{\epsilon}(\Phi_{s,t}^{\epsilon}(x)) = x$ holds for all x almost surely, we get $\Phi_{s,t}(\check{\Psi}_{s,t}(x)) = \check{\Psi}_{s,t}(\Phi_{s,t}(x)) = x$ for all x almost surely. This implies that maps $\Phi_{s,t}; \mathbb{R}^d \to \mathbb{R}^d$ are one to one and onto a.s. for any $s < t$. Further, since maps $\check{\Psi}_{s,t}$ are smooth a.s., maps $\Phi_{s,t}$ are diffeomorphic a.s.

Now we will fix t_0. By the above argument, $\Phi_{t_0,t}^{-1}(x)$ exists for any t, a.s. Further, $\Phi_{t_0,t}^{-1}(x)$ should be cadlag with respect to t for any x. Finally, we define for any $s < t$,

$\Phi_{s,t} = \Phi_{t_0,t} \circ \Phi_{t_0,s}^{-1}$. Then it is right continuous in t and further, a diffeomorphic map for any $s < t$ a.s. Therefore $\{\Phi_{s,t}\}$ is a right continuous stochastic flow generated by the SDE (3.10).

Since $\check{X}_s^\epsilon = \check{\Psi}_{s,t}^\epsilon$ satisfies the backward SDE (3.82), the limit $\check{\Psi}_{s,t}$ satisfies the backward SDE with coefficients $(-V_k(x,t), k = 0, \ldots, d', -h(x,t,z))$. □

So far, we assumed that coefficients of the SDE are not big; we assumed that coefficients and their derivatives are bounded. Then we saw that the stochastic flow $\{\Phi_{s,t}\}$ belongs to $L^{\infty-}$ (Theorem 3.4.1). The property is needed for proving that $\Phi_{s,t}(x)$ is a smooth Wiener–Poisson functional in the Malliavin calculus. See Chap. 6.

We will relax conditions for jump-maps $\phi_{t,z}$ in SDE (3.10) for a while. Instead of Conditions (J.1) and (J.2), we introduce the following weaker conditions.

Condition $(J.1)_K$. *(i) The function* $g(x,t,z) := \phi_{t,z}(x) - x$ *is of* $C_b^{\infty,1,2}$*-class on* $\mathbb{R}^d \times \mathbb{T} \times \{|z| \leq c\}$ *for some* $c > 0$. *(ii) It is of* C_b^∞*-class on* $\mathbb{R}^d$ *for any* $t \in \mathbb{T}$, $|z| \geq c$ *and is piecewise continuous in* $(t,z) \in \mathbb{T} \times \{|z| > c\}$.

Condition $(J.2)_K$. *For any* t, z, *the map* $\phi_{r,z}; \mathbb{R}^d \to \mathbb{R}^d$ *is a diffeomorphism of* $\mathbb{R}^d$. *Further,* $h(x,t,z) := x - \phi_{t,z}^{-1}(x)$ *satisfy Condition* $(J.1)_K$.

Theorem 3.9.2 *Assume Conditions* $(J.1)_K$ *and* $(J.2)_K$ *for jump -maps* $\phi_{t,z}$. *Then the assertion of Theorem* 3.9.1 *is valid. Further, the integro-differential operator* $A_J(t)$ *of* (3.67) *is well defined for any* $f \in C_b^\infty$, *and further, the stochastic flow satisfies equations* (3.68), (3.69), (3.70) *and* (3.71) *for any* $f \in C_b^\infty$.

Proof We truncate big jumps and consider $g^c(x,t,z) = g(x,t,z)1_{|z|\leq c}$ and $\phi_{t,z}^c(x) = x + g^c(x,t,z)$ instead of $g(x,t,z), \phi_{t,z}$. Then the SDE with coefficients $(V_k(x,t), k = 0, \ldots, d', g^c(x,t,z))$ defines a stochastic flow of diffeomorphisms $\{\Phi_{s,t}^c(x)\}$ in view of Theorem 3.9.1. We define a stochastic flow $\Phi_{s,t}$ by

$$\Phi_{s,t}(x) = \begin{cases} \Phi_{s,t}^c(x), & \text{if } t < \tau_1, \\ \Phi_{\tau_n,t}^c \circ \phi_{\tau_n,S_n} \circ \cdots \circ \phi_{\tau_1,S_1} \circ \Phi_{s,\tau_1}^c(x), & \text{if } \tau_n \leq t < \tau_{n+1}. \end{cases}$$

Here $\tau_n, n = 1, 2, \ldots$ are jumping times of the Poisson process N_t with intensity $\lambda_c := \nu(\{|z| > c\})$ with initial time s, and $S_n, n = 1, 2, \ldots$ are independent random variables with the identical distribution $\nu_c = \nu 1_{|z|>c}/\lambda_c$. Then $\{\Phi_{s,t}\}$ is a right continuous stochastic flow of diffeomorphisms generated by SDE with coefficients $(V_k(x,t), g(x,t,z))$. Further, the inverse $\check{\Psi}_{s,t} = \Phi_{s,t}^{-1}$ satisfies the backward equation with coefficients $(-V_k(x,t), k = 0, \ldots, d', -h(x,t,z))$.

The operator $A_J(t)$ is well defined if we restrict f in C_b^∞. Equations (3.68)–(3.71) for $f \in C_b^\infty$ can be verified in the same way as in the proof of Propositions 3.8.1 and 3.8.2. □

The stochastic flow $\Phi_{s,t}(x)$ of the above theorem may not be L^p-bounded. (3.45) may be infinite.

We are also interested in the case where diffusion coefficients are unbounded. In such a case, the solution of the SDE may explode in finite time. In Chap. 7, we will discuss SDEs on manifolds, where the solution may explode in finite time. Results in Chap. 7 can be applied to SDEs on Euclidean spaces with unbounded coefficients.

Note The flow property of solutions for SDE with jumps was first studied by Fujiwara–Kunita [30]. Then, it was developed further by Fujiwara–Kunita [31] and Kunita [60]. Some technical conditions posed for jump coefficient $g(x, t, z)$ in [30, 31] are removed in [60]. The latter is closed to the conditions in this book. However, concerning the proof of the regularity of the solution with respect to the initial data, we took another method introducing the master equation. Further, for the proof of the diffeomorphic property, a method different from these works is taken, approximating SDE by a sequence of SDE with finite Lévy measure.

3.10 Simple Expressions of Equations; Cases of Weak Drift and Strong Drift

Let ν be a Lévy measure. If $\int_{0<|z|\leq 1} |z|\nu(dz)$ is finite, then d'-vector $b = \int_{0<|z|\leq 1} z\nu(dz)$ is well defined. It is called a *strong drift* of the Lévy measure ν. If ν is a finite measure, it has a strong drift. Suppose next that ν may not have the strong drift. For a given $0 < \epsilon < 1$, we define d'-vector $b_\epsilon = (b_\epsilon^1, \ldots, b_\epsilon^{d'})$ by $b_\epsilon^j = \int_{\epsilon<|z|\leq 1} z^j \nu(dz)$, where $z = (z^1, \ldots, z^{d'}) \in \mathbb{R}^{d'}$. If $b_0 = \lim_{\epsilon\to 0} b_\epsilon$ exists and is finite, the Lévy measure ν is said to have a *weak drift* b_0.

So far we studied SDEs with jumps in the case where Lévy measures ν for Poisson random measures are most general. However, if the Lévy measure ν has a weak drift, the equation can be written in a simple form.

To see this, we will reconsider the integral by Poisson random measure, assuming that its Lévy measure ν has a weak drift. Let $\psi(r, z)$ be a bounded measurable function on $\mathbb{U}$. The *improper integrals* of ψ with respect to the measures n and N are defined by

$$\int_s^t \int_{|z|>0+} \psi(r, z) n(dr\, dz) := \lim_{\epsilon\to 0} \int_s^t \int_{|z|>\epsilon} \psi(r, z)\, dr\, \nu(dz),$$

$$\int_s^t \int_{|z|>0+} \psi(r, z) N(dr\, dz) := \lim_{\epsilon\to 0} \int_s^t \int_{|z|>\epsilon} \psi(r, z) N(dr\, dz),$$

if the right-hand sides exist and are finite almost surely, respectively.

We introduce

$$\mathbf{L}_{\mathbb{U}}^0 = \Big\{\psi(r, z);\ \text{predictable, } C^2\text{-class in } z,\ \psi(r, 0) = 0 \text{ for any } r \in \mathbb{T},$$
$$\partial_z^k \psi(r, z), k \leq 2 \text{ are bounded in } (r, z) \text{ a.s.}\Big\}. \qquad (3.83)$$

Then $\mathbf{L}_{\mathbb{U}}^0$ is a subset of $\mathbf{L}_{\mathbb{U}}^{\infty-}$.

Proposition 3.10.1 *Suppose that the Lévy measure* ν *has a weak drift. Let* $\psi(u)$ *is an element of the functional space* $\mathbf{L}_{\mathbb{U}}^{0}$ *defined by* (3.83)*. Then the improper integrals* $\int_s^t\int_{|z|>0+}\psi(r,z)N(dr\,dz)$ *and* $\int_s^t\int_{|z|>0+}\psi(r,z)\,dr\nu(dz)$ *exist. Further, both are integrable with respect to* P *and satisfy*

$$\int_s^t\int_{\mathbb{R}_0^{d'}}\psi(r,z)\tilde{N}(dr\,dz) \tag{3.84}$$

$$=\int_s^t\int_{|z|>0+}\psi(r,z)N(dr\,dz)-\int_s^t\int_{|z|>0+}\psi(r,z)\,dr\nu(dz).$$

In particular, if ν *has a strong drift, the above improper integrals are replaced by Lebesgue integrals.*

Proof We have $\psi(r,z)=\partial_z\psi(r,0)\cdot z+\frac{1}{2}(\partial_z^2\psi(r,\theta z)z\cdot z^T$, by Taylor's expansion. Therefore,

$$\int_s^t\int_{1\geq|z|>\epsilon}\psi(r,z)\,dr\,d\nu=\int_s^t\partial_z\psi(r,0)\,dr\cdot\int_{1\geq|z|>\epsilon}z\nu(dz)$$
$$+\frac{1}{2}\int_s^t\int_{1\geq|z|>\epsilon}\partial_z^2\psi(r,\theta z)z\cdot z^T\,dr\,d\nu. \tag{3.85}$$

The first term of the right-hand side converges as $\epsilon\to 0$, since ν has a weak drift. The second term of the right-hand side converges, since the function $\partial_z^2\psi(r,\theta z)z\cdot z^T$ is integrable on $\{|z|\leq 1\}$ with respect to the measure $dr\,d\nu$. Therefore (3.85) converges as $\epsilon\to 0$. The limit is the improper integral $\int_s^t\int_{1\geq|z|>0+}\psi(r,z)\,dr\nu(dz)$. Then improper integral $\int_s^t\int_{|z|>0+}\psi(r,z)\,dr\nu(dz)$ is also well defined and it is integrable with respect to P. Further, in the equality

$$\int_s^t\int_{|z|>\epsilon}\psi(r,z)\tilde{N}(dr\,dz)=\int_s^t\int_{|z|>\epsilon}\psi(r,z)N(dr\,dz)-\int_s^t\int_{|z|>\epsilon}\psi(r,z)\,dr\nu(dz),$$

the left-hand side converges, since $\psi\in\mathbf{L}_{\mathbb{U}}^{\infty-}$. Therefore $\int_s^t\int_{|z|>\epsilon}\psi(r,z)N(dr\,dz)$ should converge to the improper integral $\int_s^t\int_{|z|>0+}\psi(r,z)N(dr\,dz)$ and (3.84) holds. Since $\int_s^t\int_{|z|>0+}\psi(r,z)\,dr\nu(dz)$ is integrable, $\int_s^t\int_{|z|>0+}\psi(r,z)N(dr\,dz)$ is also integrable with respect to P. □

Now, for the Itô process X_t of (2.38), we assume that $\phi(t)$ is a semi-martingale with bounded jumps and the symmetric integral by the Wiener process is well defined and that $\psi(r,z)\in\mathbf{L}_{\mathbb{U}}^{0}$. Then Itô process (2.38) is rewritten as

$$X_t=X_0+\int_{t_0}^t\phi(r)\circ dW_r+\int_{t_0}^t\upsilon(r)\,dr+\int_{t_0}^t\int_{|z|>0+}\psi(r,z)N(dr\,dz). \tag{3.86}$$

For this Itô process, Itô's formula is rewritten simply, which looks like a formula of usual differential-difference calculus. Formula (2.39) is rewritten as follows.

Theorem 3.10.1 *Assume that the Lévy measure has a weak drift. Let $X_t, t \geq t_0$ be an Itô process represented by* (3.86). *Let $f(x, t)$ be a function of $C^{3,1}$-class. Then we have for any $t_0 \leq s < t < \infty$,*

$$\begin{aligned} f(X_t, t) &= f(X_s, s) + \int_s^t \frac{\partial f}{\partial t}(X_r, r)\,dr \\ &+ \sum_{i=1}^d \sum_{k=1}^{d'} \int_s^t \frac{\partial f}{\partial x_i}(X_r, r)\phi^{ik}(r) \circ dW_r^k + \sum_{i=1}^d \int_s^t \frac{\partial f}{\partial x_i}(X_r, r)\upsilon^i(r)\,dr \\ &+ \int_s^t \int_{|z|>0+} \{f(X_{r-} + \psi(r, z), r) - f(X_{r-}, r)\} N(dr\,dz). \end{aligned} \tag{3.87}$$

Proof Rewrite the symmetric integral of the above using the Itô integral. Next rewrite the last integral with Poisson random measure N, using the compensated Poisson random measure $\tilde{N}$. Then we can show similarly to the proof of Theorem 2.6.1 that the right-hand side of the above is equal to the right-hand side of (2.39). □

Next, we consider the backward Itô process written as

$$\check{X}_t = \check{X}_0 + \int_t^{t_1} \check{\phi}(r) \circ dW_r + \int_t^{t_1} \check{\upsilon}(r)\,dr + \int_t^{t_1} \int_{|z|>0+} \check{\psi}(r, z) N(dr\,dz). \tag{3.88}$$

Formula (2.50) is rewritten as follows.

Theorem 3.10.2 *Assume that the Lévy measure has a weak drift. Let $\check{X}_t$ be a backward Itô process represented by* (3.88). *Let $f(x, t)$ be a function of $C^{3,1}$-class. Then we have for any $t_0 \leq s < t < \infty$,*

$$\begin{aligned} f(\check{X}_s, s) &= f(\check{X}_t, t) - \int_s^t \frac{\partial f}{\partial t}(\check{X}_r, r)\,dr \\ &+ \sum_{i=1}^d \sum_{k=1}^{d'} \int_s^t \frac{\partial f}{\partial x_i}(\check{X}_r, r)\check{\phi}^{ik}(r) \circ dW_r^k + \sum_{i=1}^d \int_s^t \frac{\partial f}{\partial x_i}(\check{X}_r, r)\check{\upsilon}^i(r)\,dr \\ &+ \int_s^t \int_{|z|>0+} \{f(\check{X}_r + \check{\psi}(r, z), r) - f(\check{X}_r, r)\} N(dr\,dz). \end{aligned} \tag{3.89}$$

Let us reconsider SDE (3.10) on $\mathbb{R}^d$. By Proposition 3.10.1, we have

$$\lim_{\epsilon\to 0}\left\{\int_{t_0}^t\int_{|z|\geq\epsilon} g(X_{r-},r,z)N(dr\,dz)-\sum_{k=1}^{d'} b_\epsilon^k\int_{t_0}^t \tilde{V}_k(X_{r-},r)\,dr\right\}$$

$$=\int_{t_0}^t\int_{|z|>0+} g(X_{r-},r,z)N(dr\,dz)-\sum_{k=1}^{d'} b_0^k\int_{t_0}^t \tilde{V}_k(X_{r-},r)\,dr.$$

Therefore, rewriting the vector $V_0(t)-\sum_k b_0^k\tilde{V}_k(t)$ by $V_0(t)$ and setting $\phi_{t,z}(x)=g(x,t,z)+x$, we can rewrite equation (3.10) simply as

$$X_t = X_0+\sum_{k=0}^{d'}\int_{t_0}^t V_k(X_r,r)\circ dW_r^k$$
$$+\int_{t_0}^t\int_{|z|>0+}\{\phi_{r,z}(X_{r-})-X_{r-}\}N(dr\,dz). \tag{3.90}$$

If the Lévy measure has a strong drift, the last improper integral coincides with the Lebesgue integral. This expression of the SDE interprets the geometric meaning of the solution more than expression (3.10). Indeed, if the Lévy measure is of finite mass, the solution is given by the stochastic flow $\Phi_{s,t}$ represented by (3.18), i.e., the flow is written as composites of stochastic flow $\Phi_{s,t}^0$ generated by the continuous SDE $dX_t=\sum_k V_k(X_t,t)\circ dW_t^k$ and diffeomorphic maps $\phi_{t,z}$. If the Lévy measure has infinite mass, the solution should be the limit of flows given by (3.18).

Further, for the operator $A_J(t)$ the last term of (3.67) is equal to

$$\int_{|z|>0+}\{f\circ\phi_{t,z}-f\}\nu(dz)-\sum_{k=1}^{d'} b_0^k\tilde{V}_k(t)f.$$

Then, rewriting $V_0(t)-\sum_k b_0^k\tilde{V}_k(t)$ as $V_0(t)$ again, the operator $A_J(t)$ given by (3.67) is rewritten in a simple form:

$$A_J(t)f=A(t)f+\int_{|z|>0+}\{f\circ\phi_{t,z}-f\}\nu(dz). \tag{3.91}$$

If the Lévy measure has a strong drift, the last improper integral coincides with the Lebesgue integral. We call the triple $(V_k(t), k=0,\dots,d',\phi_{t,z},\nu)$ the *characteristics of the SDE* (3.90).

Let $\{\Phi_{s,t}\}$ be the stochastic flow generated by the SDE (3.90). Then, rules of forward and backward calculus given in (3.69) and (3.71) are rewritten simply as

$$
\begin{aligned}
f(\Phi_{s,t}) &= f + \sum_{k=0}^{d'} \int_s^t V_k(r) f(\Phi_{s,r}) \circ dW_r^k \\
&\quad + \int_s^t \int_{|z|>0+} \{ f(\phi_{r,z} \circ \Phi_{s,r-}) - f(\Phi_{s,r-}) \} N(dr\,dz), && (3.92) \\
&= f + \sum_{k=1}^{d'} \int_s^t V_k(r) f(\Phi_{s,r-})\, dW_r^k + \int_s^t A_J(r) f(\Phi_{s,r})\, dr \\
&\quad + \int_s^t \int_{|z|>0} \{ f(\phi_{r,z} \circ \Phi_{s,r-}) - f(\Phi_{s,r-}) \} \tilde{N}(dr\,dz), && (3.93)
\end{aligned}
$$

$$
\begin{aligned}
f \circ \Phi_{s,t} &= f + \sum_{k=0}^{d'} \int_s^t V_k(r)(f \circ \Phi_{r,t}) \circ dW_r^k \\
&\quad + \int_s^t \int_{|z|>0+} \{ (f \circ \Phi_{r,t}) \circ \phi_{r,z} - f \circ \Phi_{r,t} \} N(dr\,dz) && (3.94) \\
&= f + \sum_{k=1}^{d'} \int_s^t V_k(r)(f \circ \Phi_{r,t})\, dW_r^k + \int_s^t A_J(r)(f \circ \Phi_{r,t})\, dr \\
&\quad + \int_s^t \int_{|z|>0} \{ (f \circ \Phi_{r,t}) \circ \phi_{r,z} - f \circ \Phi_{r,t} \} \tilde{N}(dr\,dz). && (3.95)
\end{aligned}
$$

Next, we will consider a backward SDE, which governs the inverse flow $\check{\Psi}_{s,t} = \Phi_{s,t}^{-1}$. Rewriting $V_0(t) - \sum_k b_0^k V_k(t)$ as $V_0(t)$ again, the backward SDE for $\check{X}_s = \check{\Psi}_{s,t_1}(x)$ can be rewritten as

$$
\check{X}_s = x - \sum_{k=0}^{d'} \int_s^{t_1} V_k(\check{X}_r, r) \circ dW_r^k - \int_s^{t_1} \int_{|z|>0+} \{ \check{X}_r - \phi_{r,z}^{-1}(\check{X}_r) \} N(dr\,dz). \tag{3.96}
$$

We define an integro-differential operator $\mathcal{A}_J(t)$ by

$$
\mathcal{A}_J(t) f = \mathcal{A}(t) f + \int_{|z|>0+} \{ f \circ \phi_{t,z}^{-1} - f \} \nu(dz), \tag{3.97}
$$

where $\mathcal{A}(t)$ is the differential operator defined by (3.64). Then the backward differential rule (3.78) for the inverse flow $\check{\Psi}_{s,t}$ is rewritten as

$$f \circ \check{\Psi}_{s,t} = f - \sum_{k=0}^{d'} \int_s^t V_k(r) f(\check{\Psi}_{r,t}) \circ dW_r^k$$

$$+ \int_s^t \int_{|z|>0+} \{ f(\phi_{r,z}^{-1} \circ \check{\Psi}_{r,t}) - f(\check{\Psi}_{r,t}) \} N(dr\,dz) \tag{3.98}$$

$$= f + \int_s^t \mathcal{A}_J(r) f(\check{\Psi}_{r,t})\, dr - \sum_{k=1}^{d'} \int_s^t V_k(r) f(\check{\Psi}_{r,t})\, dW_r^k$$

$$+ \int_s^t \int_{|z|>0} \{ f(\phi_{r,z}^{-1} \circ \check{\Psi}_{r,t}) - f(\check{\Psi}_{r,t}) \} \tilde{N}(dr\,dz). \tag{3.99}$$

Chapter 4
Diffusions, Jump-Diffusions and Heat Equations

Abstract We study diffusions and jump-diffusions on a Euclidean space determined by SDE studied in Chap. 3. We select topics which are related to the stochastic flow generated by the SDE; topics are concerned with heat equations and backward heat equations.

In Sects. 4.1, 4.2, 4.3, and 4.4, we consider diffusion processes on a Euclidean space. In Sect. 4.1, we show that a stochastic flow generated by a continuous symmetric SDE defines a diffusion process. Its generator $A(t)$ is represented explicitly as a second order linear partial differential operator with time parameter t, using coefficients of the SDE. Kolmogorov's forward and backward equation associated with the operator $A(t)$ will be derived. In Sect. 4.2, we discuss exponential transformation of the diffusion process by potentials. It will be shown that solutions of various types of backward heat equations will be obtained by exponential transformations by potentials. In Sect. 4.3, we study backward SDE and backward diffusions. We will see that Kolmogorov equations for backward diffusion will give the solution of heat equations. In Sect. 4.4, we present a new method of constructing the dual (adjoint) semigroup, making use of the geometric property of diffeomorphic maps $\Phi_{s,t}$ of stochastic flows. The method will present a clear geometric explanation of the adjoint operator $A(t)^*$ and the dual semigroup. We will see that the dual semigroup will be obtained by a certain exponential transformation (Feynman–Kac–Girsanov transformation) of a backward diffusion.

In Sects. 4.5 and 4.6, we consider jump-diffusions on a Euclidean space. We will extend results for diffusion studied in Sects. 4.2, 4.3, and 4.4 to those for jump-diffusions. We show that the dual semigroup of jump-diffusion is well defined if the jump coefficients are diffeomorphic. In Sect. 4.7, we return to a problem of the stochastic flow. We discuss the volume-preserving property of stochastic flows by applying properties of dual jump-diffusions. In Sect. 4.8, we consider processes killed outside of a subdomain of an Euclidean space and its dual processes.

H. Kunita, *Stochastic Flows and Jump-Diffusions*, Probability Theory and Stochastic Modelling 92, https://doi.org/10.1007/978-981-13-3801-4_4

4.1 Continuous Stochastic Flows, Diffusion Processes and Kolmogorov Equations

We will study diffusion processes on a Euclidean space determined by a continuous stochastic differential equations. Let us consider a continuous symmetric SDE on $\mathbb{R}^d$:

$$dX_t = \sum_{k=0}^{d'} V_k(X_t, t) \circ dW_t^k, \tag{4.1}$$

where $V_k(x, t), k = 0, \ldots, d'$ are $C_b^{\infty,1}$-functions of $(x, t) \in \mathbb{R}^d \times \mathbb{T}$ and $W_t = (W_t^1, \ldots, W_t^{d'}), t \in \mathbb{T}$ is a d'-dimensional Wiener process. Here, $\mathbb{T} = [0, \infty)$, $W_t^0 \equiv t$ and $V_0(X_t, t) \circ dW_t^0 \equiv V_0(X_t)\, dt$, conventionally. Let $\{\Phi_{s,t}\}$ be the stochastic flow of diffeomorphisms generated by the above continuous SDE. Then $X_t = X_t^{x,s} = \Phi_{s,t}(x), t \geq s$ is a solution of equation (4.1) starting from x at time s.

Proposition 4.1.1 *Laws of $\Phi_{s,t}(x)$:*

$$P_{s,t}(x, E) = P(\Phi_{s,t}(x) \in E), \quad 0 \leq s < t < \infty, \quad E \in \mathcal{B}(\mathbb{R}^d) \tag{4.2}$$

are determined uniquely from SDE (4.1). *Further, $\{P_{s,t}(x, E)\}$ is a transition probability.*

Proof We consider the equation (4.1) with the initial condition $t_0 = s$ and $X_0 = x$. Using the Itô integral it is written as $X_t = x + \int_s^t \alpha(X_r, r)\, dW_r + \int_s^t \beta(X_r, r)\, dr$. We saw in Sect. 3.3 that the equation has a pathwise unique solution (Theorem 3.3.1). Further, the solution X_t is approximated successively; setting $X_s^0 = x$ and

$$X_t^n = X_s^{n-1} + \int_s^t \alpha(X_r^{n-1}, r)\, dW_r + \int_s^t \beta(X_r^{n-1}, r)\, dr, \quad n = 1, 2, \ldots,$$

X_t^n converges to the solution X_t in L^p-sense, in view of Lemma 3.3.1. Since the law of X_t^n is uniquely determined by coefficients α, β and x, the law of X_t is also determined uniquely from these coefficients and initial condition.

For a bounded continuous function f, we set

$$P_{s,t} f(x) := \int_{\mathbb{R}^d} P_{s,t}(x, dy) f(y) = E[f(\Phi_{s,t}(x))]. \tag{4.3}$$

Let $\{\mathcal{F}_{s,t}, 0 \leq s < t < \infty\}$ be the two-sided filtration generated by the Wiener process W_t. Then $\Phi_{t,u}(y)$ is independent of $\mathcal{F}_{s,t}$ for any $s < t < u$. Therefore we have the semigroup property

$$P_{s,u}f(x) = E\big[E\big[f(\Phi_{t,u}(\Phi_{s,t}(x)))\big|\mathcal{F}_{s,t}\big]\big] = E\big[E\big[f(\Phi_{t,u}(y))\big]\big|_{y=\Phi_{s,t}(x)}\big]$$
$$= E\big[P_{t,u}f(\Phi_{s,t}(x))\big] = P_{s,t}P_{t,u}f(x),$$

for $s < t < u$. We have further, $\lim_{t\downarrow s} P_{s,t}f(x) = f(x)$, since $\Phi_{s,t}(x) \to x$ as $t \downarrow s$. Therefore, $\{P_{s,t}(x, E)\}$ is a transition probability. □

Now, for a given $(x, t_0) \in \mathbb{R}^d \times \mathbb{T}$, let us consider the stochastic process with parameter x, t_0 given by $X_t^{x,t_0} = \Phi_{t_0,t}(x)$. In view of the flow property $\Phi_{s,t} \circ \Phi_{t_0,s} = \Phi_{t_0,t}$, the equality $X_t^{x,t_0} = \Phi_{s,t} \circ X_s^{x,t_0}$ holds a.s. for any $t_0 < s < t$. Since $\Phi_{s,t}(x)$ and $\mathcal{F}_{t_0,s}$ are independent, we have

$$P(X_t^{x,t_0} \in E|\mathcal{F}_{t_0,s}) = P_{s,t}(X_s^{x,t_0}, E) \quad a.s.$$

Therefore the stochastic process $X_t^{x,t_0}, t \in [t_0, \infty)$ is a continuous conservative Markov process on $\mathbb{R}^d$ with semigroup $\{P_{s,t}\}$. Further, if f is a bounded continuous function, $P_{s,t}f(x)$ is also a bounded continuous function of x, since the flow $\Phi_{s,t}(x)$ is continuous in x a.s. Therefore, X_t^{x,t_0} has the strong Markov property (Proposition 1.7.1). Then X_t^{x,t_0} is a diffusion process of the initial state (x, t_0) with semigroup $\{P_{s,t}\}$. Consequently, $\{X_t^{x,s}\}$ is a system of diffusion processes with semigroup $\{P_{s,t}\}$.

A continuous function f on $\mathbb{R}^d$ is called *slowly increasing* if $|f(x)|/(1+|x|)^m$ is a bounded function for some positive integer m. A C^k-function f is called *slowly increasing* if for any $|\mathbf{i}| \le k$, $|\partial^{\mathbf{i}} f(x)|/(1+|x|)^m$ are bounded for some positive integer m. A C^∞-function f is called *slowly increasing* if it is a slowly increasing C^k-function for any k. We denote by $\mathcal{O}(\mathbb{R}^d)$ the set of all real slowly increasing C^∞-functions on $\mathbb{R}^d$.

Let $f(x)$ be a slowly increasing continuous function. Let $T > 0$. Since $E[|\Phi_{s,t}(x)|^p] \le c_p(1+|x|)^p$ holds for any $0 \le s < t \le T, x \in \mathbb{R}^d$ and $p \ge 2$ (Theorem 3.4.1), the integral $P_{s,t}f(x) = E[f(\Phi_{s,t}(x))]$ is well defined for any $s < t$ and x. Further, it is a slowly increasing function of x. Suppose further that $f(x)$ is a slowly increasing C^∞-function (or C_b^∞ function, respectively). Since $E[|\partial^{\mathbf{i}}\Phi_{s,t}(x)|^p]$ is bounded with respect to x for any $p \ge 2$ (Theorem 3.4.1), we can change the order of the derivative and expectation. In fact, $E[f(\Phi_{s,t}(x))]$ is a C^∞-function of x and satisfies

$$\partial^{\mathbf{i}} E[f(\Phi_{s,t}(x))] = E[\partial^{\mathbf{i}}(f \circ \Phi_{s,t}(x))]$$

for any $\mathbf{i}$. Therefore $P_{s,t}f(x)$ is a slowly increasing C^∞-function of x (or C_b^∞-function of x, respectively) for each $s < t$. Hence $\{P_{s,t}\}$ is the family of linear transformations on $\mathcal{O}(\mathbb{R}^d)$ (or on $C_b^\infty(\mathbb{R}^d)$), satisfying the semigroup property $P_{s,t}P_{t,u}f = P_{s,u}f$ for any $s < t < u$.

Associated with SDE (4.1), we define differential operators (vector fields) $V_k(t), k = 0, \ldots, d'$ by $V_k(t)f(x) = \sum_k V_k^i(x,t)\frac{\partial f}{\partial x_i}$ and we define a second-order partial differential operator $A(t)$ with time parameter t by

$$A(t)f(x) = \frac{1}{2}\sum_{k=1}^{d'} V_k(t)^2 f(x) + V_0(t)f(x). \tag{4.4}$$

If f is a slowly increasing C^∞-function (or C_b^∞-function), then both $V_k(t)f$ and $A(t)f(x)$ are slowly increasing C^∞-functions (or C_b^∞-functions, respectively). Let $f_{t,u}(x)$, $f(x,t)$ be C^∞-functions of x with parameters t, u. Denote them by $f_\lambda(x)$, regarding $\{t, u\}, t$ as parameter λ. Apply the differential operators $V_k(t)$, $A(t)$ to the function $f_\lambda(x)$. The functions $V_k(t)f_\lambda(x)$, $A(t)f_\lambda(x)$ are denoted by $V_k(t)f_{t,u}(x)$, $A(t)f_{t,u}(x)$, $V_k(t)f(x,t)$, $A(t)f(x,t)$, respectively.

Theorem 4.1.1 *Solutions of the continuous symmetric SDE* (4.1) *define a system of conservative diffusion process with semigroup* $\{P_{s,t}\}$. *If f is a slowly increasing C^∞-function, $P_{s,t}f(x)$ is also a slowly increasing C^∞-function for any $s < t$. Further, it is continuously differentiable with respect to s, t for $s < t$ and it satisfies the following forward and backward partial differential equations, respectively:*

$$\frac{\partial}{\partial t}P_{s,t}f(x) = P_{s,t}A(t)f(x), \quad \text{if } t > s, \tag{4.5}$$

$$\frac{\partial}{\partial s}P_{s,t}f(x) = -A(s)P_{s,t}f(x), \quad \text{if } t > s. \tag{4.6}$$

Equation (4.5) is called *Kolmogorov's forward equation* and (4.6) is called *Kolmogorov's backward equation*. Equation (4.6) shows that the function $v(x,s) := P_{s,t}f(x)$, $s < t$ is a solution of the *backward heat equation*

$$\frac{\partial}{\partial s}v(x,s) = -A(s)v(x.s)$$

with the *final condition* $\lim_{s\uparrow t} v(x,s) = f(x)$.

Proof Let f be a slowly increasing C^∞-function. Then for any $s < t$ and x, we have the forward differential calculus (Proposition 3.6.1):

$$f(\Phi_{s,t}(x)) = f(x) + \sum_{k=1}^{d'}\int_s^t V_k(r)f(\Phi_{s,r}(x))\,dW_r^k + \int_s^t A(r)f(\Phi_{s,r}(x))\,dr. \tag{4.7}$$

In equation (4.7), terms written in the stochastic integrals as dW_r^k are martingales with mean 0. Therefore, taking the expectation of each term of (4.7), we have

$$\begin{aligned} P_{s,t}f(x) &= f(x) + \int_s^t E[A(r)f(\Phi_{s,r}(x))]\,dr \\ &= f(x) + \int_s^t P_{s,r}A(r)f(x)\,dr. \end{aligned} \tag{4.8}$$

Therefore, the function $P_{s,t}f(x)$ is continuously differentiable with respect to t. Differentiating both sides, we get the equality (4.5).

We saw in Proposition 3.6.2 the rule of the backward differential calculus for the flow $\{\Phi_{s,t}\}$;

$$f(\Phi_{s,t}(x)) = f(x) + \sum_{k=1}^{d'} \int_s^t V_k(r)(f \circ \Phi_{r,t})(x)\, dW_r^k + \int_s^t A(r)(f \circ \Phi_{r,t})(x)\, dr. \tag{4.9}$$

Take the expectation of each term of (4.9). Since the second term of the right-hand side is a backward martingale, its expectation is 0. We shall consider the last term. We have

$$\begin{aligned} E\Big[\int_s^t A(r)(f \circ \Phi_{r,t})(x)\, dr\Big] &= \int_s^t E\big[A(r)(f \circ \Phi_{r,t})(x)\big]\, dr \\ &= \int_s^t A(r) E\big[f(\Phi_{r,t}(x))\big]\, dr. \end{aligned}$$

In the last equality, we used $\partial^{\mathbf{i}} E[f(\Phi_{s,t}(x))] = E[\partial^{\mathbf{i}}(f \circ \Phi_{s,t}(x))]$. Then we get

$$P_{s,t}f(x) = f(x) + \int_s^t A(r) P_{r,t} f(x)\, dr. \tag{4.10}$$

Differentiate both sides with respect to s $(< t)$ we get (4.6). □

4.2 Exponential Transformation and Backward Heat Equation

We are interested in partial differential equations (4.5) and (4.6) for more general types of operator. Let $A(t)$ be the differential operator defined by (4.4). Let $c_k(x,t), k = 0, \ldots, d'$ be $C_b^{\infty,1}$-functions on $\mathbb{R}^d \times \mathbb{T}$. We will transform the operator $A(t)$ by the functions $\mathbf{c} \equiv \{c_0, \ldots, c_{d'}\}$. We define another differential operator $A^{\mathbf{c}}(t)$ by

$$A^{\mathbf{c}}(t)f = \frac{1}{2}\sum_{k=1}^{d'} (V_k(t) + c_k(t))^2 f + (V_0(t) + c_0(t)) f. \tag{4.11}$$

Here, $(V_k(t) + c_k(t)), k = 0, \ldots, d'$ are linear operators defined by

$$(V_k(t) + c_k(t)) f(x) = \sum_i V_k^i(x,t) \frac{\partial f}{\partial x_i}(x) + c_k(x,t) f(x)$$

and $(V_k(t)+c_k(t))^2 f(x) = (V_k(t)+c_k(t))\{(V_k(t)+c_k(t))f\}(x)$. Then $A^{\mathbf{c}}(t)f(x)$ is rewritten as

$$A^{\mathbf{c}}(t)f(x) = A(t)f(x) + \sum_{k=1}^{d'} c_k(x,t)V_k(t)f(x) + \tilde{c}(x,t)f(x),$$

where

$$\tilde{c}(x,t) = c_0(x,t) + \frac{1}{2}\sum_{k=1}^{d'}\Big(V_k(t)c_k(x,t) + c_k(x,t)^2\Big). \tag{4.12}$$

We consider an *exponential functional* with coefficients $\mathbf{c} = \{c_0, \ldots, c_{d'}\}$:

$$G_{s,t}(x) = G^{\mathbf{c}}_{s,t}(x) = \exp\Big\{\sum_{k=0}^{d'}\int_s^t c_k(\Phi_{s,r}(x), r)\circ dW_r^k\Big\}. \tag{4.13}$$

Lemma 4.2.1 *For any* $\mathbf{i}$, $p \ge 2$ *and* $T > 0$, $E[|\partial^{\mathbf{i}} G_{s,t}(x)|^p]$ *is bounded with respect to* $0 \le s < t \le T$ *and* $x \in \mathbb{R}^d$.

Proof Apply Itô's formula (Theorem 2.4.1) for the function $f(x) = e^x$ and $X_t = \sum_k \int_s^t c_k'(r)\circ dW_r^k$, where $c_k'(r) := c_k(\Phi_{s,r}(x), r)$. Then we have

$$\begin{aligned} G_{s,t} &= 1 + \sum_{k=0}^{d'}\int_s^t G_{s,r}c_k'(r)\circ dW_r^k \\ &= 1 + \sum_{k=0}^{d'}\int_s^t G_{s,r}c_k'(r)\,dW_r^k + \frac{1}{2}\sum_{k=1}^{d'}\langle G_{s,t}c_k'(t), W_t^k - W_s^k\rangle_{s,t}. \end{aligned}$$

For the computation of the last term, note that the martingale part of the continuous semi-martingale $G_{s,t}c_k'(t)$ is

$$\sum_{k\ge 1}\int_s^t G_{s,r}V_k(r)c_k(\Phi_{s,r}, r)\,dW_r^k + \sum_{k\ge 1}\int_s^t G_{s,r}c_k'(r)c_k'(r)\,dW_r^k$$

in view of Itô's formula. Then we have for $k = 1, \ldots, d'$

$$\langle G_{s,t}c_k'(t), W_t^k - W_s^k\rangle = \int_s^t G_{s,r}\{V_k(r)c_k(\Phi_{s,r}, r) + c_k(\Phi_{s,r}, r)^2\}\,dr.$$

Therefore,

$$G_{s,t} = 1 + \sum_{k=1}^{d'} \int_s^t G_{s,r} c_k(\Phi_{s,r}, r)\, dW_r^k + \frac{1}{2} \int_s^t \tilde{c}(\Phi_{s,r}, r)\, dr.$$

Define

$$\alpha_x(\hat{y}, r) = \hat{y} c(\Phi_{s,r}(x), r), \quad \beta_x(\hat{y}, r) = \hat{y} \tilde{c}(\Phi_{s,r}(x), r), \quad \gamma_x(\hat{x}, r, z) = 0,$$

where $c(x, r) = (c_1(x, r), \ldots, c_{d'}(x, r))$ and $\tilde{c}(x, r)$ is given by (4.12). These functionals satisfy Conditions 1–3 of the master equation with parameter x. Further, $Y_t = Y_t^{x,s} := G_{s,t}(x)$ satisfies the equation

$$Y_t^{x,s} = 1 + \int_s^t \alpha_x(Y_r^{x,s}, r)\, dW_r + \int_s^t \beta_x(Y_r^{x,s}, r)\, dr.$$

Then in view of Theorem 3.3.1, $E[|Y_t^{x,s}|^p]$ is bounded with respect to x, s for any $p \geq 2$.

Since $Y_t^{x,s} = G_{s,t}(x)$ satisfies the above equation, the derivative $\partial G_{s,t}(x) = (\partial_{x_1} G_{s,t}(x), \ldots, \partial_{x_d} G_{s,t}(x))$ satisfies

$$\begin{aligned}\partial G_{s,t} = &\int_s^t (\partial G_{s,r} c(\Phi_{s,r}, r) + G_{s,r} \partial c(\Phi_{s,r}, r))\, dW_r \\ &+ \int_s^t (\partial G_{s,r} \tilde{c}(\Phi_{s,r}, r) + G_{s,r} \partial \tilde{c}(\Phi_{s,r}, r))\, dr.\end{aligned}$$

Since $c(\Phi_{s,r}(x), r)$, $\tilde{c}(\Phi_{s,r}(x), r)$ and $\partial c(\Phi_{s,r}(x), r)$, $\partial \tilde{c}(\Phi_{s,r}(x), r)$ are bounded with respect to r and x (x is a parameter), coefficients of the equation satisfy Conditions 1–3 of the master equation. Therefore $E[|\partial G_{s,t}(x)|^p]$ should be bounded with respect to $0 \leq s < t \leq T$ and $x \in \mathbb{R}^d$ by Theorem 3.3.1. Repeating this argument inductively, we arrive at the assertion of the lemma. ⊔

Now, $Y_t = G_{s,t}$ satisfies the linear SDE:

$$dY_t = \sum_{k=0}^{d'} c_k(X_t, t) Y_t \circ dW_t^k. \tag{4.14}$$

Let $\mathbb{R}^+$ be the open half line $(0, \infty)$. We shall consider an SDE on the product space $\mathbb{R}^d \times \mathbb{R}^+$ for the process $(X_t, Y_t) \in \mathbb{R}^d \times \mathbb{R}^+$ defined by the pair of equations (4.1) and (4.14). Then the solution starting from $(x, y) \in \mathbb{R}^d \times \mathbb{R}^+$ at time s is given by $(\Phi_{s,t}(x), G_{s,t}(x)y)$. For any $p \geq 2$, there exists $c_p > 0$ such that

$$E[|\Phi_{s,t}(x)|^p + |G_{s,t}(x)y|^p] \leq c_p(1 + |x| + |y|)^p, \quad \forall 0 \leq s < t \leq T.$$

For a slowly increasing C^∞-function $\bar{f}$ on $\mathbb{R}^d \times \mathbb{R}^+$, we can define the integral

$$\bar{P}_{s,t}\bar{f}(x,y) = E[\bar{f}(\Phi_{s,t}(x), G_{s,t}(x)y)]$$

for any s, t, x and y. It is a slowly increasing C^∞-function. We can apply the arguments of Sect. 4.1 to the diffusion process $(X_t, Y_t) = (\Phi_{s,t}(x), G_{s,t}(x)y)$. Then its generator $\bar{A}(t)$ is written by

$$\bar{A}(t)\bar{f} = \frac{1}{2}\sum_{k=1}^{d'}\Big(V_k(t) + c_k(t)y\frac{\partial}{\partial y}\Big)^2\bar{f} + \Big(V_0(t) + c_0(t)y\frac{\partial}{\partial y}\Big)\bar{f}.$$

Now take $\bar{f}(x,y) = f(x)y$, where $f(x)$ is a slowly increasing C^∞ function on $\mathbb{R}^d$. Apply it to the operator $\bar{A}(t)$. Then $A^{\mathbf{c}}(t)f(x) = \bar{A}(t)\bar{f}(x,1)$ holds.

We will define another transition function on $\mathbb{R}^d$ and its transformation by

$$P^{\mathbf{c}}_{s,t}(x,E) := E[1_E(\Phi_{s,t}(x))G^{\mathbf{c}}_{s,t}(x)],$$

$$P^{\mathbf{c}}_{s,t}f(x) := \int P^{\mathbf{c}}_{s,t}(x,dx')f(x') = E[f(\Phi_{s,t}(x))G^{\mathbf{c}}_{s,t}(x)]. \tag{4.15}$$

It is the transformation of $P_{s,t}f(x)$ by the exponential functional $G^{\mathbf{c}}_{s,t}(x)$. Since $\{\bar{P}_{s,t}\}$ satisfies the semigroup property $\bar{P}_{s,t}\bar{P}_{t,u} = \bar{P}_{s,u}$ for any $s < t < u$, $\{P^{\mathbf{c}}_{s,t}\}$ satisfies the semigroup property $P^{\mathbf{c}}_{s,t}P^{\mathbf{c}}_{t,u} = P^{\mathbf{c}}_{s,u}$. Further, for any bounded continuous function f, $P^{\mathbf{c}}_{s,t}f$ converges to f as $t \to s$. Hence $\{P^{\mathbf{c}}_{s,t}(x,E)\}$ is a transition function. It is called the *transition function of* $\Phi_{s,t}(x)$ *weighted by* $\mathbf{c}$ or simply a *weighted transition function*.

Since Kolmogorov's forward and backward equations are valid for the diffusion process $(X_t, Y_t) = (\Phi_{s,t}(x), G_{s,t}(x)y)$ on $\mathbb{R}^d \times \mathbb{R}^+$, we have the following assertion in view of Theorem 4.1.1.

Theorem 4.2.1 *If f is a slowly increasing C^∞-function of x, $P^{\mathbf{c}}_{s,t}f(x)$ is also a slowly increasing C^∞-function of x. Further, it is continuously differentiable with respect to s, t for $s < t$ and satisfies the following forward and backward partial differential equations, respectively:*

$$\frac{\partial}{\partial t}P^{\mathbf{c}}_{s,t}f(x) = P^{\mathbf{c}}_{s,t}A^{\mathbf{c}}(t)f(x), \tag{4.16}$$

$$\frac{\partial}{\partial s}P^{\mathbf{c}}_{s,t}f(x) = -A^{\mathbf{c}}(s)P^{\mathbf{c}}_{s,t}f(x). \tag{4.17}$$

Equation (4.16) is called *Kolmogorov's forward equation* and (4.17), *Kolmogorov's backward equation*.

As an application of the above theorem, we will consider the final value problem for a backward heat equation. Let $0 < t_1 < \infty$ be a fixed time, called the final time. Let $f_1(x)$ be a slowly increasing continuous function on $\mathbb{R}^d$. We want to find a function $v(x, s)$ of $C^{2,1}$-class defined on $\mathbb{R}^d \times (0, t_1)$ satisfying the following backward equation:

$$\begin{cases} \dfrac{\partial}{\partial s} v(x, s) = -A^{\mathbf{c}}(s)v(x, s), & \forall 0 < s < t_1, \\ \lim_{s \uparrow t_1} v(x, s) = f_1(x), & \forall x \in \mathbb{R}^d. \end{cases} \tag{4.18}$$

If the above function $v(x, s)$ exists, it is called a solution of the *final value problem for the backward heat equation* associated with the operator $A^{\mathbf{c}}(t)$.

Theorem 4.2.2 *If f_1 is a slowly increasing C^∞-function, the final value problem for the backward heat equation* (4.18) *has a unique slowly increasing solution $v(x, s)$. The solution is of $C^{\infty,1}$-class. It is represented by*

$$v(x, s) = E\Big[f_1(\Phi_{s,t_1}(x)) \exp\Big\{\sum_{k=0}^{d'} \int_s^{t_1} c_k(\Phi_{s,r}(x), r) \circ dW_r^k\Big\}\Big]. \tag{4.19}$$

Proof The function $v(x, s)$ given by (4.19) coincides with $P^{\mathbf{c}}_{s,t_1} f_1(x)$. Since $P^{\mathbf{c}}_{s,t_1} f_1(x)$ satisfies (4.17), v should be a solution of equation (4.18). It is a C^∞-function of x, since $v(x, s) = E[f_1(\Phi_{s,t_1}(x))G^{\mathbf{c}}_{s,t_1}(x)]$ is infinitely differentiable with respect to x. Further, we have $\partial_x^{\mathbf{i}} v(x, s) = \partial_x^{\mathbf{i}}\big(f_1(x) + \int_s^{t_1} A^{\mathbf{c}}(r)P^{\mathbf{c}}_{r,t_1} f_1(x)\, dr\big)$. Therefore $\partial_x^{\mathbf{i}} v(x, s)$ is continuously differentiable with respect to s.

We give the uniqueness of the solution. Let $v'(x, s)$ be any slowly increasing solution of equations (4.18) of $C^{2,1}$-class. Set $v_0(x, s) = v(x, s) - v'(x, s)$. Then we have by Itô's formula,

$$\begin{aligned} v_0(\Phi_{s,t}, t)G^{\mathbf{c}}_{s,t} &= v_0(x, s) + \sum_{k=1}^{d'} \int_s^t (V_k(r) + c_k(r))v_0(\Phi_{s,r}, r)G^{\mathbf{c}}_{s,r}\, dW_r^k \\ &\quad + \int_s^t \Big\{A^{\mathbf{c}}(r)v_0(\Phi_{s,r}, r) + \frac{\partial v_0}{\partial r}(\Phi_{s,r}, r)\Big\}G^{\mathbf{c}}_{s,r}\, dr. \end{aligned}$$

Since $v_0(x, s)$ is a solution of (4.18), $A^{\mathbf{c}}(s)v_0(x, s) + \frac{\partial}{\partial s} v_0(x, s) = 0$ holds. Therefore $M_t = v_0(\Phi_{s,t}(x), t)G^{\mathbf{c}}_{s,t}(x)$ is a local martingale. Further, since v_0 is of polynomial growth, $E[|M_t|^p] < \infty$. Therefore M_t is actually a martingale. Then we have $M_t = E[M_T|\mathcal{F}_t] = 0$, because $M_T = 0$ holds a.s. Therefore $v_0(\Phi_{s,t}(x), t) = 0$ holds for all x a.s. Let $\check{\Psi}_{s,t}(x)$ be the inverse map of $\Phi_{s,t}(x)$. Then we have $v_0(\Phi_{s,t} \circ \Psi_{s,t}(x), t) = 0$ a.s. for any x. Therefore we get $v_0(x, t) = 0$ for any x. Since this is valid for any t, $v_0(x, t)$ should be identically 0. □

A semigroup $\{P_{s,t}\}$ is said to be of $C^{2,1}$-class if for any C_b^2-function f, $P_{s,t} f(x)$ is a $C^{2,1}$ function of x, s for any t. The above theorem shows that for a given partial differential operator $A^{\mathbf{c}}(t)$, the semigroup of $C^{2,1}$-class satisfying Kolmogorov's backward equation exists uniquely. The semigroup is said to be generated by the operator $A^{\mathbf{c}}(t)$ and $A^{\mathbf{c}}(t)$ is called the *generator* of the semigroup. In particular, when $\mathbf{c} = 0$, a system of diffusion processes $\{X_t^{x,s} = \Phi_{s,t}(x); (x,s) \in \mathbb{R}^d \times \mathbb{T}\}$ with the semigroup $\{P_{s,t}\}$ is called a diffusion process with the generator $A(t)(= A^0(t))$ or a system of diffusion processes generated by the operator $A(t)$. Further, since the law of $\Phi_{s,t}(x)$ is uniquely determined by the operator $A(t)$, $\{\Phi_{s,t}\}$ is called the stochastic flow associated with the operator $A(t)$.

Remark It is not obvious whether the semigroup satisfying Kolmogorov's forward equation is unique or not. However, if the semigroup is defined by a diffusion process, is of $C^{2,1}$-class and satisfies Kolmogorov's forward equation, then it is unique. We will prove this fact in the case $\mathbf{c} = 0$. Let X_t is the diffusion process with the semigroup $P_{s,t}$ mentioned above. Then $N_s = P_{s,t}f(X_s)$ is a martingale, because for $s < u < t$,

$$E[N_u|\mathcal{F}_s] = E[P_{u,t}f(X_u)|\mathcal{F}_s] = P_{s,u}P_{u,t}f(X_s) = P_{s,t}f(X_s) = N_s.$$

Note that the operator $A(t)$ is written as

$$A(t)f(x) = \frac{1}{2}\sum_{i,j}\alpha^{ij}(x,t)\frac{\partial^2 f}{\partial x_i \partial x_j}(x) + \sum_i \beta^i(x,t)\frac{\partial f}{\partial x_i}(x), \tag{4.20}$$

where

$$\alpha^{ij}(x,t) = \sum_{k=1}^{d'} V_k^i(x,t)V_k^j(x,t), \quad \beta(x,t) = V_0(x,t) + \frac{1}{2}\sum_{k=1}^{d'} V_k(t)V_k(x,t).$$

Set $\beta_r^i = \beta^i(X_r, r)$ and $\beta_r = (\beta_r^1, \ldots, \beta_r^d)$, Since $f(X_t) - \int_{t_0}^t A(r)f(X_r)\,dr$ is a martingale, both

$$X_t - \int_{t_0}^t b_r\,dr \text{ and } X_t^iX_t^j - \int_{t_0}^t X_r^i\beta_r^j\,dr - \int_{t_0}^t X_r^j\beta_r^i\,dr - \int_{t_0}^t \alpha^{ij}(X_r,r)\,dr$$

are martingales. On the other hand, by Itô's formula (Theorem 2.2.1),

$$X_t^iX_t^j = X_{t_0}^iX_{t_0}^j + \int_{t_0}^t X_r^i\,dX_r^j + \int_{t_0}^t X_r^j\,dX_r^i + \langle X^i, X^j\rangle_{t_0,t}.$$

Comparing these equations, we find that

$$\langle X^i, X^j\rangle_{t_0,t} = \int_{t_0}^t \alpha^{ij}(X_r,r)\,dr.$$

Now, apply Itô's formula to the function $f(x,s) = P_{s,t}f(x)$ and $x = X_s$. Then we have

$$N_t = f(X_{t_0}, t_0) + \int_{t_0}^t \Big\{\frac{\partial}{\partial r} f(X_r, r) + A(r) f(X_r, r)\Big\}\, dr + \text{a local martingale}.$$

Therefore the integral $\int_{t_0}^t \{\cdots\}\, dr$ is a continuous martingale, so that it should be 0 a.s. Therefore the integrand $\{\cdots\}$ is 0 a.s. Since this is valid for any X_t, we have $\{\frac{\partial}{\partial r} + A(r)\} f(x, r) = 0$. This shows that $f(x, s) = P_{s,t} f$ satisfies Kolmogorov's backward equation. Then the semigroup is unique by Theorem 4.2.1.

Now, we will rewrite the exponential functional $G_{s,t}(x)$ using the Itô integral. Setting $c'_k(r) = c_k(\Phi_{s,r}(x), r), k = 0, \ldots, d'$ and $\tilde{c}'(r) = \tilde{c}(\Phi_{s,r}(x), r)$, we have

$$\begin{aligned}\sum_{k=0}^{d'} \int c'_k(r) \circ dW_r^k &= \sum_{k=1}^{d'} \int_s^t c'_k(r)\, dW_r^k + \frac{1}{2} \sum_{k=1}^{d'} \langle c'_k(t), W_t^k \rangle_{s,t} + \int_s^t c'_0(r)\, dr \\ &= \sum_{k=1}^{d'} \int_s^t c'_k(r)\, dW_r^k - \frac{1}{2} \int_s^t \sum_{k=1}^{d'} c'_k(r)^2\, dr + \int_s^t \tilde{c}'(r)\, dr.\end{aligned}$$

Therefore, $G^{\mathbf{c}}_{s,t}(x)$ can be decomposed as the product of $G^{(0)}_{s,t}$ and $G^{(1)}_{s,t}$, where

$$G^{(0)}_{s,t} = \exp\Big\{\int_s^t \tilde{c}(\Phi_{s,r}, r)\, dr\Big\},$$

$$G^{(1)}_{s,t} = \exp\Big\{\sum_{k=1}^{d'} \int_s^t c_k(\Phi_{s,r}, r)\, dW_r^k - \frac{1}{2} \sum_{k=1}^{d'} \int_s^t c_k(\Phi_{s,r}, r)^2\, dr\Big\}.$$

Set $P^{(0)}_{s,t} f(x) = E[f(\Phi_{s,t}(x)) G^{(0)}_{s,t}(x)]$. Then $\{P^{(0)}_{s,t}\}$ is a semigroup and satisfies

$$P^{(0)}_{s,t} f(x) = f(x) + \int_s^t P^{(0)}_{s,r} (A(r) + \tilde{c}(r)) f(x).$$

Hence by the transformation $P_{s,t} \to P^{(0)}_{s,t}$, the potential term $\tilde{c}(t)$ is added to the generator. The generator is changed from $A(t)$ to the operator $A^{(0)}(t) = A(t) + \tilde{c}(t)$. It is called the *Feynman–Kac transformation*. Next, $G^{(1)}_{s,t}$ is a positive martingale with mean 1. In fact, apply Proposition 2.2.3 to $Z_t \equiv G^{(1)}_{s,t}$. Since $\sum_{k=1}^{d'} \int_s^t c_k(\Phi_{s,r}(x), r)^2\, dr$ belongs to $L^{\infty-}$, Z_t is a positive $L^{\infty-}$-martingale with mean 1. Then we can define another probability measure $P^{(1)}$ by $dP^{(1)} = G^{(1)}_{s,t}\, dP$. With respect to $P^{(1)}$, $\tilde{W}_t^k = W_t^k - \int_s^t c_k(\Phi_{s,r}, r)\, dr, k = 1, \ldots, d'$ is a Wiener process by Girsanov's theorem (Theorem 2.2.2). Then $X_t = \Phi_{s,t}(x)$ satisfies

$$X_t^i = x_i + \sum_{k=0}^{d'} \int_s^t V_k^i(X_r, r) \circ d\tilde{W}_r^k + \sum_{k=1}^{d'} \int_s^t c_k(X_r, r) V_k^i(X_r, r)\, dr, \quad a.s.\ P^{(1)}.$$

Therefore, with respect to $P^{(1)}$, $X_t = \Phi_{s,t}(x)$ is a diffusion process with the generator $A^{(1)}(t)f = A(t)f + \sum_{k=1}^{d'} c_k(t)V_k(t)f$. Hence by the transformation $P_{s,t} \to P_{s,t}^{(1)}$, the drift term (first-order differential operator) $\sum_k c_k(t)V_k(t)$ is added to the generator. The generator is changed from $A(t)$ to the operator $A^{(1)}(t)$. It is called the *Girsanov transformation*. Consequently, our exponential transformation coincides with the product of the Feynman–Kac transformation and the Girsanov transformation. We will call it the *Feynman–Kac–Girsanov transformation*. Equation (4.19) is called the *Feynman–Kac–Girsanov formula*.

We are also interested in the initial value problem for a heat equation associated with the operator $A^{\mathbf{c}}(t)$. Let $0 \leq t_0 < \infty$ be a fixed time, called the initial time. Let $f_0(x)$ be a slowly increasing continuous function on $\mathbb{R}^d$. We want to find a function $u(x,t)$ of $C^{2,1}$-class defined on $\mathbb{R}^d \times (t_0, \infty)$ satisfying the equation

$$\begin{cases} \dfrac{\partial}{\partial t}u(x,t) = A^{\mathbf{c}}(t)u(x,t), & \forall t_0 < t < \infty, \\ \lim_{t \downarrow t_0} u(x,t) = f_0(x), & \forall x \in \mathbb{R}^d. \end{cases} \tag{4.21}$$

If such a function exists, it is called a solution of the *initial value problem for the heat equation* associated with the operator $A^{\mathbf{c}}(t)$.

The problem can be solved in time homogeneous case, using the Feynman–Kac–Girsanov formula.

Theorem 4.2.3 *For the operator $A^{\mathbf{c}}(t)$, we assume that $V_k(t), k = 1, \ldots, d'$ and $c_k(t), k = 0, \ldots, d'$ are time homogeneous (not depending on t). Then for any slowly increasing C^∞-function f_0, the initial value problem for the heat equation* (4.21) *has a unique slowly increasing C^∞-solution. It is given by the Feynman–Kac–Girsanov formula*

$$u(x,t) = E\Big[f_0(\Phi_{t_0,t}(x)) \exp\Big\{\sum_{k=0}^{d'} \int_{t_0}^{t} c_k(\Phi_{t_0,r}(x)) \circ dW_r^k\Big\}\Big]. \tag{4.22}$$

Proof In the time homogeneous case, the semigroup $P_{s,t}^{\mathbf{c}}$ depends on $t - s$ only, which we denote by $P_{t-s}^{\mathbf{c}}$. Then we have $\frac{\partial}{\partial s}P_{s,t}^{\mathbf{c}} f = \frac{\partial}{\partial s}P_{t-s}^{\mathbf{c}} f = -\frac{\partial}{\partial t}P_{t-s}^{\mathbf{c}} f$. Therefore equation (4.17) is written as $\frac{\partial}{\partial t}P_t^{\mathbf{c}} f = A^{\mathbf{c}}P_t^{\mathbf{c}} f$. Consequently, $u(x,t) = P_{t-t_0}^{\mathbf{c}} f_0(x)$ is the solution of the heat equation (4.21). Further, $u(x,t)$ is represented by (4.22). □

The above argument can be applied to complex-valued potentials $c_k, k = 0, \ldots, d'$. Suppose that these are pure imaginary, written as $c_k(x) = \sqrt{-1}\Theta_k(x)$. We define a complex-valued exponential functional by

$$G_{t_0,t}(x) = \exp\Big\{\sqrt{-1}\Big(\sum_{k=0}^{d'} \int_{t_0}^{t} \Theta_k(\Phi_{t_0,r}(x)) \circ dW_r^k\Big)\Big\} \tag{4.23}$$

and define a semigroup by $P^{\mathbf{c}}_{t_0,t} f(x) = E[f(\Phi_{t_0,t}(x))G_{t_0,t}(x)]$. Then $u(x,t) = P^{\mathbf{c}}_{t_0,t}\phi_0(x)$ is a solution of the heat equation associated with the operator A^{Θ} defined by

$$A^{\Theta} f = \frac{1}{2}\sum_{k=1}^{d'}(V_k + \sqrt{-1}\Theta_k)^2 f + (V_0 + \sqrt{-1}\Theta_0) f.$$

If A is a Laplacian, i.e., $V_k = \partial_{x_k}$ for $k = 1, \ldots, d'$ where $d' = d$, the operator A^{Θ} is called a *Schrodinger operator with magnetic field* $\Theta = (\Theta_0, \ldots, \Theta_{d'})$. (Matsumoto–Taniguchi [83]). The solution of the heat equation associated with the operator A^{Θ} is represented by Feynman–Kac–Girsanov formula (4.22) with $c_k = \sqrt{-1}\Theta_k, k = 0, \ldots, d$.

In the time-dependent case, discussions are complicated. It will be studied in the next section.

Remark Backward heat equations are used in mathematical finance. Consider the equation on $(0,\infty) \times [0, t_1]$ defined by

$$\frac{\partial}{\partial s} v(x,s) + \Big(\frac{\sigma(s)^2}{2} x^2 \frac{\partial^2}{\partial x^2} + rx\frac{\partial}{\partial x} - r\Big) v(x,s) = 0, \quad 0 < s < t_1,$$

$$\text{(final condition)} \quad \lim_{s\to t_1} v(x,s) = (x - K) \vee 0.$$

where $\sigma(t) > 0$ and r, K are positive constants, called the volatility, the interest rate and the exercise price, respectively. The equation is called the Black–Scholes partial differential equation. The solution $v(x,s)$ is the pricing function of the call option of exercise price K. For details, see Karatzas–Shreve [56] and Lamberton–Lapeyre [72].

The function $f_1(x) = (x - K) \vee 0$ is a continuous function but it is not smooth. Then it is not clear whether the possible solution $v(x,s)$ given by (4.19) is smooth or not with respect x. This problem will be solved by constructing the fundamental solution for the Black–Scholes partial differential equation, which will be done in Sect. 6.3, using the Malliavin calculus.

4.3 Backward Diffusions and Heat Equations

Let us consider a continuous backward symmetric SDE on $\mathbb{R}^d$ with coefficients $V_k(x,t), k = 0, \ldots, d'$, which are $C^{\infty,1}$-functions. For a given x, t, suppose that there exists a d-dimensional continuous backward semi-martingale $\check{X}_s, s < t$ which satisfies

$$\check{X}_s = x + \sum_{k=0}^{d'} \int_s^t V_k(\check{X}_r, r) \circ dW_r^k, \tag{4.24}$$

where the stochastic integrals $\circ dW_r^k, k = 1, 2, \ldots, d'$ are backward symmetric integrals and $\circ dW_r^0 = dr$. Then $\check{X}_s$ is called a solution of the equation starting from x at time t. The equation has a unique solution, which we denote by $\check{X}_s^{x,t}$. Then, there exists a continuous backward stochastic flow of diffeomorphisms $\{\check{\Phi}_{s,t}\}$ such that $\check{X}_s^{x,t} = \check{\Phi}_{s,t}(x)$ holds for any x, t, as we studied in Chap. 3. We set

$$\check{P}_{s,t}(x, E) = P(\check{\Phi}_{s,t}(x) \in E), \quad 0 \le s < t < \infty, \quad E \in \mathcal{B}(\mathbb{R}^d).$$

Let $\{\mathcal{F}_{s,t}, 0 \le s < t < \infty\}$ be the two-sided filtration generated by the Wiener process W_t. Then $\check{\Phi}_{t,u}(x)$ is independent of $\mathcal{F}_{s,t}$ for any $s < t < u$. Hence transformations $\check{P}_{s,t} f(x) := \int f(y)\check{P}_{s,t}(x, dy)$ have the backward semigroup property $\check{P}_{s,u} f(x) = \check{P}_{t,u}\check{P}_{s,t} f(x)$ for any $s < t < u$. The stochastic process $\check{X}_s = \check{X}_s^{x,t}, s \in [0, t]$ has the backward Markov property $P(\check{X}_s \in E|\mathcal{F}_{t,T}) = \check{P}_{s,t}(\check{X}_t, E)$. It has the (backward) strong Markov property. Hence $\check{X}_s^{x,t}$ is a backward diffusion process with the backward transition probabilities $\check{P}_{s,t}(x, \cdot)$. The system of processes $\{\check{X}_s^{x,t}, (x, t) \in \mathbb{R}^d \times \mathbb{T}\}$ is called a system of *backward diffusion process* with transition probabilities $\check{P}_{s,t}(x, E)$. By Proposition 3.6.3, the backward flow $\{\check{\Phi}_{s,t}\}$ satisfies the rules of the following backward and forward differential calculus, respectively:

$$\begin{aligned} f \circ \check{\Phi}_{s,t} &= f + \int_s^t A(r) f \circ \check{\Phi}_{r,t}\, dr + \sum_{k=1}^{d'} \int_s^t V_k(r) f \circ \check{\Phi}_{r,t}\, dW_r^k \qquad (4.25) \\ &= f + \int_s^t A(r)(f \circ \check{\Phi}_{s,r})\, dr + \sum_{k=1}^{d'} \int_s^t V_k(r)(f \circ \check{\Phi}_{s,r})\, dW_r^k. \end{aligned}$$

Expectations of the last terms of the above two equations are both equal to 0. Therefore, taking expectations for each terms of the above, we have

$$\check{P}_{s,t} f = f + \int_s^t \check{P}_{r,t} A(r) f\, dr, \quad \check{P}_{s,t} f = f + \int_s^t A(r)\check{P}_{s,r} f\, dr. \tag{4.26}$$

Consequently, the backward semigroup $\{\check{P}_{s,t}\}$ is generated by the same differential operator $A(t)$ given by (4.4).

Consider a *backward exponential functional* with coefficients $\mathbf{c} = \{c_0, \ldots, c_{d'}\}$:

$$\check{G}_{s,t}(x) = \check{G}^{\mathbf{c}}_{s,t}(x) = \exp\Big\{\sum_{k=0}^{d'} \int_s^t c_k(\check{\Phi}_{r,t}(x), r) \circ dW_r^k\Big\}. \tag{4.27}$$

For any slowly increasing continuous function f on $\mathbb{R}^d$, we can define the integral

$$\check{P}^{\mathbf{c}}_{s,t} f(x) := E\big[f(\check{\Phi}_{s,t}(x))\check{G}^{\mathbf{c}}_{s,t}(x)\big] \tag{4.28}$$

for any s, t and x. It is a slowly increasing C^∞-function. Further, $\{\check{P}^{\mathbf{c}}_{s,t}\}$ satisfies the backward semigroup property $\check{P}^{\mathbf{c}}_{t,u}\check{P}^{\mathbf{c}}_{s,t} f = \check{P}^{\mathbf{c}}_{s,u} f$ for any $u > t > s$. The following theorem corresponds to Theorem 4.2.1 for forward diffusion.

Theorem 4.3.1 *If f is a slowly increasing C^∞-function of x, $\check{P}^{\mathbf{c}}_{s,t} f(x)$ is also a slowly increasing C^∞-function of x. Further, it is continuously differentiable with respect to s, t for $s < t$ and satisfies the following backward and forward partial differential equations, respectively:*

$$\frac{\partial}{\partial s}\check{P}^{\mathbf{c}}_{s,t} f(x) = -\check{P}^{\mathbf{c}}_{s,t} A^{\mathbf{c}}(s) f(x), \tag{4.29}$$

$$\frac{\partial}{\partial t}\check{P}^{\mathbf{c}}_{s,t} f(x) = A^{\mathbf{c}}(t)\check{P}^{\mathbf{c}}_{s,t} f(x). \tag{4.30}$$

Proof We can show similarly to equations (4.26) that $\check{P}^{\mathbf{c}}_{s,t} f$ satisfies

$$\check{P}^{\mathbf{c}}_{s,t} f = f + \int_s^t \check{P}^{\mathbf{c}}_{r,t} A^{\mathbf{c}}(r) f\, dr, \quad \check{P}^{\mathbf{c}}_{s,t} f = f + \int_s^t A^{\mathbf{c}}(r)\check{P}^{\mathbf{c}}_{s,r} f\, dr.$$

Therefore $\check{P}^{\mathbf{c}}_{s,t} f$ is differentiable with respect to $t(> s)$ and $s(< t)$. Differentiating the first equation by s, we get (4.29) and differentiating the second equation by t, we get (4.30). □

Consequently, the backward semigroup $\{\check{P}^{\mathbf{c}}_{s,t}\}$ is generated by the differential operator $A^{\mathbf{c}}(t)$ given by (4.11). Equation (4.30) tells us that $u(x, t) = \check{P}^{\mathbf{c}}_{t_0,t} f_0(x)$ is a solution of the heat equation (4.21). Therefore we have the followings:

Theorem 4.3.2 *If f_0 is a slowly increasing C^∞-function, the heat equation* (4.21) *has a unique slowly increasing solution $u(x, t)$. The solution is a $C^{\infty,1}$-function of x, t. It is represented by the backward Feynman–Kac–Girsanov formula*

$$u(x, t) = E\Big[f_0(\check{\Phi}_{t_0,t}(x)) \exp\Big\{\sum_{k=0}^{d'} \int_{t_0}^t c_k(\check{\Phi}_{r,t}(x), r) \circ dW_r^k\Big\}\Big], \tag{4.31}$$

where $\{\check{\Phi}_{s,t}\}$ is the backward flow defined by a backward symmetric SDE with coefficients $V_k(x, t), k = 0, \ldots, d'$.

So far, we have discussed the existence of solutions for a heat equation with a smooth initial function f_0 and for a backward heat equation with a smooth final function f_1. If the function f_0 is not smooth, it is not easy to show that $u(x,t)$ defined by (4.31) is a solution of the heat equation, since it is not clear whether the function $u(x,t)$ may be smooth or not with respect to x. We will show in Chap. 6 that if the operator $A^{\mathbf{c}}(t)$ is 'nondegenerate', the function $u(x,t)$ given by (4.31) is smooth and satisfies the heat equation. An approach to the problem is to construct the fundamental solution for the equation (4.21). Another method is to extend smooth functions f_0 in formulas (4.31) to a generalized function f_0. It will be completed by using the Malliavin calculus, which will be studied in Chap. 5.

4.4 Dual Semigroup, Inverse Flow and Backward Diffusion

We will denote by C_0^∞ or $C_0^\infty(\mathbb{R}^d)$ the set of all real-valued C^∞-functions of compact supports. Let $A(t)$ be the differential operator defined by (4.4). It is a linear map from $C_0^\infty(\mathbb{R}^d)$ into itself. We shall study its dual (adjoint) operator $A(t)^*$ with respect to the Lebesgue measure dx. A linear map $A(t)^*; C_0^\infty(\mathbb{R}^d) \to C_0^\infty(\mathbb{R}^d)$ is called a dual of $A(t)$ if it satisfies

$$\int_{\mathbb{R}^d} A(t)f(x)\cdot g(x)\,dx = \int_{\mathbb{R}^d} f(x)\cdot A(t)^*g(x)\,dx \tag{4.32}$$

for any $f, g \in C_0^\infty(\mathbb{R}^d)$.

Proposition 4.4.1 *The dual $A(t)^*$ is well defined and is given by*

$$A(t)^*g = \frac{1}{2}\sum_{k=1}^{d'}(V_k(t) + \mathrm{div}\, V_k(t))^2 g - (V_0(t) + \mathrm{div}\, V_0(t))g, \tag{4.33}$$

where $\mathrm{div}\, V(x,t) := \sum_i \frac{\partial V^i}{\partial x_i}(x,t)$ *is a $C_b^{\infty,1}$-function.*

Proof For a vector field $V = \sum V^i \frac{\partial}{\partial x_i}$, we have a formula of the integration by parts:

$$\int Vf(x)\cdot g(x)\,dx = -\int f(x)\cdot Vg(x)\,dx - \int f(x)\cdot \mathrm{div}\, V(x)g(x)\,dx,$$

for f, g of $C_0^\infty(\mathbb{R}^d)$. Therefore we have $V^*g = -Vg - \mathrm{div}\, Vg$. Then the adjoint of $A(t)$ is given by $A(t)^*g = \sum_{k=1}^{d'} \frac{1}{2}(V_k(t)^*)^2 g + V_0(t)^*g$. □

The geometric meaning of terms $\mathrm{div}\, V_k(x,t)$ in the expression (4.33) is not clear. We want to define directly the dual semigroup of $\{P_{s,t}\}$, making use of the inverse of the stochastic flow $\{\Phi_{s,t}\}$. It is interesting to know how the expression of the dual operator $A(t)^*$ is related to the geometric property of diffeomorphic maps $\Phi_{s,t}$ of the stochastic flow.

Let $\{\Phi_{s,t}\}$ be the stochastic flow of diffeomorphisms generated by SDE (4.1) and let $\{P_{s,t}\}$ be the semigroup defined by (4.3). Given $0 \leq s < t < \infty$, a linear operator $P^*_{s,t}; C^\infty_b(\mathbb{R}^d) \to C^\infty_b(\mathbb{R}^d)$ is called the dual of $P_{s,t}$, if the equality

$$\int_{\mathbb{R}^d} P_{s,t} f(x) \cdot g(x)\, dx = \int_{\mathbb{R}^d} f(x) \cdot P^*_{s,t} g(x)\, dx$$

holds for all $f, g \in C^\infty_0(\mathbb{R}^d)$. We will show the dual $P^*_{s,t}$ exists uniquely for any s, t and the family $\{P^*_{s,t}\}$ is a backward semigroup with the generator $A(t)^*$.

Let $\check{\Psi}_{s,t}; \mathbb{R}^d \to \mathbb{R}^d$ be the inverse map of the diffeomorphism $\Phi_{s,t}$. Let $\nabla\check{\Psi}_{s,t}$ be the Jacobian matrix of the diffeomorphism $\check{\Psi}_{s,t}$ and let $\det \nabla\check{\Psi}_{s,t}$ be the Jacobian (Jacobian determinant). We will first remark that it is positive a.s. for any $s < t$. If t is fixed, $\check{\Psi}_{s,t}$ is continuous in s a.s., since it satisfies the backward SDE with coefficients $-V_k(x, t), k = 0, \ldots, d'$ by Theorem 3.7.1. Then its Jacobian $\det \nabla\check{\Psi}_{s,t}$ is also continuous with respect to s. Note that it does not take the value 0, since $\check{\Psi}_{s,t}$ is diffeomorphic for any s. Further, it holds that $\det \nabla\check{\Psi}_{t,t} = 1$, since $\check{\Psi}_{t,t}$ is the identity map. Consequently, $\det \nabla\check{\Psi}_{s,t}$ should be positive for all $s < t$ a.s.

Now we will define the dual semi-group $\{P^*_{s,t}\}$. We need the formula of the change of variables on $\mathbb{R}^d$. Let f be an element of C^∞_0 and let ϕ be a diffeomorphic map from $\mathbb{R}^d$ onto itself. Then we have the formula of the change of variables;

$$\int_{\mathbb{R}^d} f(x)\, dx = \int_{\mathbb{R}^d} f(\phi(x)) |\det \nabla\phi(x)|\, dx, \tag{4.34}$$

where $\nabla\phi$ is the Jacobian matrix of the map $\phi; \mathbb{R}^d \to \mathbb{R}^d$ and $\det \nabla\phi$ is the Jacobian of the map ϕ. Now take another $g \in C^\infty_0$ and replace f by $f \circ \phi \cdot g$ and ϕ by ϕ^{-1} in the above formula. Then we obtain the formula

$$\int_{\mathbb{R}^d} f(\phi(x)) g(x)\, dx = \int_{\mathbb{R}^d} f(x) g(\phi^{-1}(x)) |\det \nabla\phi^{-1}(x)|\, dx.$$

Setting $\phi(x) = \Phi_{s,t}(x)$ in the above formula, we have

$$\int_{\mathbb{R}^d} f(\Phi_{s,t}(x)) \cdot g(x)\, dx = \int_{\mathbb{R}^d} f(x) \cdot g(\check{\Psi}_{s,t}(x)) \det \nabla\check{\Psi}_{s,t}(x)\, dx, \tag{4.35}$$

almost surely. Now applying Theorem 3.4.2 to the backward SDE, we find that for any $p \geq 2$, $E[|\nabla\check{\Psi}_{s,t}(x)|^p]$ are bounded with respect to x and $0 \leq s < t \leq T$. Therefore we can define a bounded kernel by

$$P^*_{s,t}(x, E) := E[1_E(\check{\Psi}_{s,t}(x)) \det \nabla\check{\Psi}_{s,t}(x)],$$

and a linear transformation $C_0^\infty(\mathbb{R}^d) \to C_b^\infty(\mathbb{R}^d)$ by

$$P_{s,t}^* g(x) := \int_{\mathbb{R}^d} P_{s,t}^*(x, dy)g(y) = E[g(\check{\Psi}_{s,t}(x)) \det \nabla \check{\Psi}_{s,t}(x)]. \tag{4.36}$$

Taking the expectation of each side of (4.35), we have the formula of the duality:

$$\int_{\mathbb{R}^d} P_{s,t} f(x) \cdot g(x)\, dx = \int_{\mathbb{R}^d} f(x) \cdot P_{s,t}^* g(x)\, dx. \tag{4.37}$$

Consequently, $P_{s,t}^*$ of (4.36) is the dual of $P_{s,t}$ with respect to the Lebesgue measure dx.

Since $\{P_{s,t}\}$ satisfy $P_{s,u} = P_{s,t} P_{t,u}$, dual operators $\{P_{s,t}^*\}$ satisfy the backward semigroup property $P_{s,u}^* g = P_{t,u}^* P_{s,t}^* g$ for $u > t > s$ (Sect. 1.7). Further, the dual semigroup and the dual operator $A(t)^*$ are related by

$$\begin{aligned} P_{s,t}^* g(x) &= g(x) + \int_s^t P_{r,t}^* A(r)^* g(x)\, dr \\ &= g(x) + \int_s^t A(r)^* P_{s,r}^* g(x)\, dr \end{aligned}$$

for any $g \in C_0^\infty(\mathbb{R}^d)$. Therefore, $A(t)^*$ is the generator of the dual semigroup $\{P_{s,t}^*\}$.

We will study further the relation between the dual semigroup $\{P_{s,t}^*\}$ and the inverse flow $\{\check{\Psi}_{s,t}\}$. Recall that the generator of the diffusion defined by the stochastic flow $\{\Phi_{s,t}\}$ is given by (4.4). Recall further that $\check{X}_s = \check{\Psi}_{s,t}(x)$ is the solution of the continuous backward SDE with coefficients $-V_k(x,t), k = 0, \ldots, d'$ starting from x at time t. Hence $\check{X}_s = \check{\Psi}_{s,t}$ is a backward diffusion. We define its backward transition probabilities by

$$\check{\mathcal{P}}_{s,t}(x, E) := E[1_E(\check{\Psi}_{s,t}(x))].$$

It defines the backward semigroup $\{\check{\mathcal{P}}_{s,t}\}$.

We saw in Theorem 3.7.1 that the backward flow $\{\check{\Psi}_{s,t}\}$ satisfies

$$\begin{aligned} f(\check{\Psi}_{s,t}(x)) &= f(x) + \int_s^t \mathcal{A}(r) f(\check{\Psi}_{r,t}(x))\, dr - \sum_{k=1}^{d'} \int_s^t V_k(r) f(\check{\Psi}_{r,t}(x))\, dW_r^k \\ &= f(x) + \int_s^t \mathcal{A}(r)(f \circ \check{\Psi}_{s,r})(x)\, dr - \sum_{k=1}^{d'} \int_s^t V_k(r)(f \circ \check{\Psi}_{s,r})(x)\, dW_r^k, \end{aligned}$$

where

$$\mathcal{A}(t) = \frac{1}{2}\sum_{k\geq 1} V_k(t)^2 f - V_0(t) f. \tag{4.38}$$

Take expectations for each term of the above. Then the backward semigroup $\{\check{\mathcal{P}}_{s,t}\}$ satisfies equations

$$\check{\mathcal{P}}_{s,t} f = f(x) + \int_s^t \check{\mathcal{P}}_{r,t}\mathcal{A}(r) f\, dr, \quad \check{\mathcal{P}}_{s,t} f = f(x) + \int_s^t \mathcal{A}(r)\check{\mathcal{P}}_{s,r} f\, dr. \tag{4.39}$$

Therefore $\mathcal{A}(t)$ is the generator of the backward semigroup $\{\check{\mathcal{P}}_{s,t}\}$. We define a backward exponential functional with coefficients $\mathbf{c}$ associated with the inverse flow $\{\check{\Psi}_{s,t}\}$ by

$$\check{\mathcal{G}}^{\mathbf{c}}_{s,t}(x) = \exp\left\{\sum_{k=0}^{d'}\int_s^t c_k(\check{\Psi}_{r,t}(x), r) \circ dW_r^k\right\} \tag{4.40}$$

Lemma 4.4.1 *Jacobian* $\det \nabla\check{\Psi}_{s,t}(x)$ *is represented as* $\check{\mathcal{G}}^{-\mathbf{div\,V}}_{s,t}(x)$, *where* $-\mathbf{div\,V} = -(\mathrm{div}\, V_0(x,t), \ldots, \mathrm{div}\, V_{d'}(x,t))$.

Proof We will prove that $\det \nabla\check{\Psi}_{s,t}(x)$ satisfies the following backward linear equation for all x a.s.:

$$\det \nabla\check{\Psi}_{s,t}(x) = 1 - \sum_{k=0}^{d'}\int_s^t \mathrm{div}\, V_k(\check{\Psi}_{r,t}(x), r) \det \nabla\check{\Psi}_{r,t}(x) \circ dW_r^k. \tag{4.41}$$

Let us recall the rule of the backward differential calculus (3.52). It is written as

$$f(\Phi_{s,t}(x)) - f(x) = \sum_{k=0}^{d'}\int_s^t V_k(r)(f \circ \Phi_{r,t})(x) \circ dW_r^k,$$

using the backward symmetric integrals, where f is a function of $C_0^\infty(\mathbb{R}^d)$. Since $\int f(\Phi_{s,t}(x))\, dx = \int \det \nabla\check{\Psi}_{s,t}(x) f(x) dx$ holds by the formula of the change of variables, we have

$$\int_{\mathbb{R}^d} f(x)(\det \nabla\check{\Psi}_{s,t}(x) - 1)\, dx = \sum_{k=0}^{d'}\int_{\mathbb{R}^d}\left(\int_s^t V_k(r)(f \circ \Phi_{r,t})(x) \circ dW_r^k\right) dx.$$

We shall compute the right-hand side. Use the Fubini theorem (Proposition 2.4.2) for integrals $\circ dW_r^k$ and dx. Then apply the formula of the integration by parts and the formula of the change of variables. Then we have for any $k = 0, \ldots, d'$,

$$
\begin{aligned}
&\int_{\mathbb{R}^d}\Big(\int_s^t V_k(r)(f\circ\Phi_{r,t})(x)\circ dW_r^k\Big)\,dx\\
&=\int_s^t\Big(\int_{\mathbb{R}^d} V_k(r)(f\circ\Phi_{r,t})(x)\,dx\Big)\circ dW_r^k\\
&=-\int_s^t\Big(\int_{\mathbb{R}^d}\mathrm{div}V_k(x,r)f(\Phi_{r,t}(x))\,dx\Big)\circ dW_r^k\\
&=-\int_s^t\Big(\int_{\mathbb{R}^d} f(x)\mathrm{div}V_k(\check{\Psi}_{r,t}(x),r)\det\nabla\check{\Psi}_{r,t}(x)\,dx\Big)\circ dW_r^k\\
&=-\int_{\mathbb{R}^d} f(x)\Big(\int_s^t\mathrm{div}V_k(\check{\Psi}_{r,t}(x),r)\det\nabla\check{\Psi}_{r,t}(x)\circ dW_r^k\Big)\,dx.
\end{aligned}
$$

Since this is valid for any $f\in C_0^\infty(\mathbb{R}^d)$, we get the equality

$$
\det\nabla\check{\Psi}_{s,t}(x)-1=-\sum_{k=0}^{d'}\int_s^t\mathrm{div}\,V_k(\check{\Psi}_{r,t}(x),r)\det\nabla\check{\Psi}_{r,t}(x)\circ dW_r^k,
$$

a.e. x, almost surely. We know that each symmetric integral of the right-hand side is continuous in x (Proposition 2.4.3). Therefore the above equality holds for all x almost surely. This proves (4.41).

Apply Theorem 3.10.2 for $f(x)=e^x$ and $X_t=-\sum_k\int_s^t\mathrm{div}\,V_k(\check{\Psi}_{r,t},r)\circ dW_r^k$. Then we find that $\check{\mathcal{G}}_{s,t}^{-\mathbf{div}\,\mathbf{V}}(x)$ satisfies the following backward linear SDE:

$$
\check{\mathcal{G}}_{s,t}^{-\mathbf{div}\,\mathbf{V}}(x)=1-\sum_{k=0}^{d'}\int_s^t\mathrm{div}\,V_k(\check{\Psi}_{r,t}(x),r)\check{\mathcal{G}}_{r,t}^{-\mathbf{div}\,\mathbf{V}}\circ dW_r^k. \tag{4.42}
$$

The equation is of the same form as (4.41) replacing $\det\nabla\check{\Psi}_{s,t}$ by $\check{\mathcal{G}}_{s,t}^{-\mathbf{div}\,\mathbf{V}}$. Then we have $\check{\mathcal{G}}_{s,t}^{-\mathbf{div}\,\mathbf{V}}(x)=\det\nabla\check{\Psi}_{s,t}(x)$, by the uniqueness of the solution of the SDE. □

We transform the operator $\mathcal{A}(t)$ by $\mathbf{c}=-\mathbf{div}\,\mathbf{V}$. Then we get the relation

$$
A(t)^*g=\mathcal{A}^{-\mathbf{div}\,\mathbf{V}}(t)g \tag{4.43}
$$

Theorem 4.4.1 *Let $\{P_{s,t}\}$ be the semigroup defined by the stochastic flow $\{\Phi_{s,t}\}$ generated by a continuous SDE* (3.1) *and let $\{P_{s,t}^*\}$ be its dual semigroup. Then if g is a slowly increasing C^∞-function, $P_{s,t}^*g(x)$ is a slowly increasing C^∞-function for any $s<t$. Further, the dual semigroup is obtained from the backward semigroup $\{\mathcal{P}_{s,t}\}$ defined by the inverse flow $\{\check{\Psi}_{s,t}\}$, through the backward exponential transformation with coefficients* $-\mathbf{div}\,\mathbf{V}$, *i.e., for any $s<t$ and g, it holds that*

$$
P_{s,t}^*g(x)=E[g(\check{\Psi}_{s,t}(x))\det\nabla\check{\Psi}_{s,t}(x)]=\check{\mathcal{P}}_{s,t}^{-\mathbf{div}\,\mathbf{V}}g(x). \tag{4.44}
$$

Its generator is given by (4.43).

So far, we have studied the dual of the operator $A(t)$ and the dual of the semigroup $\{P_{s,t}\}$. We will extend these to the dual of $A^{\mathbf{c}}(t)$ and $\{P^{\mathbf{c}}_{s,t}\}$ which were defined in Sect. 4.2. Let $A^{\mathbf{c}}(t)$ be the differential operator defined by (4.11). Then its dual (adjoint) $A^{\mathbf{c}}(t)^*$ is given by

$$\begin{aligned} A^{\mathbf{c}}(t)^* g &= \frac{1}{2}\sum_{k=1}^{d'} (V_k(t) + c_k(t) + \operatorname{div} V_k(t))^2 g - (V_0(t) + c_0(t) + \operatorname{div} V_0(t))g \\ &= \mathcal{A}^{\mathbf{c}-\operatorname{div}\mathbf{V}}(t)g. \end{aligned} \tag{4.45}$$

Next we will define the dual of the semigroup $\{P^{\mathbf{c}}_{s,t}\}$. For any $f, g \in C_0^\infty(\mathbb{R}^d)$, we have the formula of the change of variables:

$$\begin{aligned} &\int_{\mathbb{R}^d} f(\Phi_{s,t}(x)) G_{s,t}(x) g(x)\, dx \\ &= \int_{\mathbb{R}^d} f(x) G_{s,t}(\check{\Psi}_{s,t}(x)) g(\check{\Psi}_{s,t}(x)) \det \nabla \check{\Psi}_{s,t}(x)\, dx, \end{aligned} \tag{4.46}$$

almost surely. Since $G_{s,t}(\check{\Psi}_{s,t}(x)) = \exp\left\{ \sum_{k=0}^{d'} \int_s^t c_k(\check{\Psi}_{r,t}(x)) \circ dW_r^k \right\}$, we have

$$G_{s,t}(\check{\Psi}_{s,t}(x)) \det \nabla \check{\Psi}_{s,t}(x) = \exp\left\{ \sum_{k=0}^{d'} \int_s^t (c_k - \operatorname{div} V_k)(\check{\Psi}_{r,t}(x), r) \circ dW_r^k \right\}.$$

It is L^p-bounded for any $p \geq 2$. Then

$$\check{\mathcal{P}}^{\mathbf{c}-\operatorname{div}\mathbf{V}}_{s,t} g = E\Big[g(\check{\Psi}_{s,t}) \exp\Big\{ \sum_{k=0}^{d'} \int_s^t (c_k - \operatorname{div} V_k)(\check{\Psi}_{r,t}, r) \circ dW_r^k \Big\}\Big] \tag{4.47}$$

is well defined. We can take the expectation for both sides of (4.46). Then we get

$$\int P^{\mathbf{c}}_{s,t} f(x) g(x)\, dx = \int f(x) \check{\mathcal{P}}^{\mathbf{c}-\operatorname{div}\mathbf{V}}_{s,t} g(x)\, dx. \tag{4.48}$$

Therefore $\check{\mathcal{P}}^{\mathbf{c}-\operatorname{div}\mathbf{V}}_{s,t}$ is the dual of $P^{\mathbf{c}}_{s,t}$.

Theorem 4.4.2 *Let $\{P^{\mathbf{c}}_{s,t}\}$ be the semigroup defined by* (4.15)*, which is an exponential transformation of the semigroup $\{P_{s,t}\}$. Then the dual of $\{P^{\mathbf{c}}_{s,t}\}$ exists and it is given by* (4.47)*, where $\{\check{\Psi}_{s,t}\}$ is the inverse flow of $\{\Phi_{s,t}\}$. Further, its generator is given by* (4.45).

Finally, we will discuss briefly the dual of time homogeneous diffusions. We assume that coefficients $V_k, k = 0, \ldots, d'$ of the SDE do not depend on t. Then its generator A is given by $Af = \frac{1}{2}\sum_{k\geq 1} V_k^2 + V_0$. Let P_t^* be the dual of P_t. Then these satisfy $P_s^* P_t^* = P_{t+s}^*$ for any $s, t > 0$. Hence the dual semigroup is again stationary. We can construct the dual diffusion as a forward process. Indeed, consider an SDE on $\mathbb{R}^d$ given by

$$dX_t = -\sum_{k=0}^{d'} V_k(X_t) \circ dW_t^k. \tag{4.49}$$

Let $\{\Phi_{s,t}\}$ be the stochastic flow generated by the above SDE. It is a diffusion process with the generator $\mathcal{A} = \frac{1}{2}\sum_{k\geq 1} V_k^2 - V_0$. In particular, if div $V_k(x) = 0$ holds for $k = 0, \ldots, d'$, A^* coincides with $\mathcal{A}$. Hence the dual P_t^* is a conservative semigroup defined by the forward diffusion $\Phi_{s,t}(x)$. Then $\Phi_{0,t}(x)$ coincides with the dual process of $\Phi_{0,t}(x)$ in Hunt's potential theory for Markov processes.

Note There are many studies of diffusion processes and their generators. For diffusion determined by SDE, we refer to Itô [50], Stroock–Varadhan [109], Dynkin [24], Ikeda–Watanabe [41] etc. Backward equations for diffusion determined by SDEs are studied in Freidlin–Wentzell [27] and Oksendal [90]. Our method using the rule of the backward differential calculus might be new.

Feynman–Kac transformation of the diffusion process is widely known. Girsanov transformation (Theorem 2.2.1) is also widely used. See Ikeda–Watanabe [41] for the application to the problem of solving SDE. See Karatzas–Shreve [56] and Kunita [62] for the application to mathematical finance. Our exponential transformation is the product of the above two transformations. Its application to the Schrodinger operator is discussed in Matsumoto–Taniguchi [83].

Feynman–Kac formula for the solution of time-homogeneous heat equations is well known. However, we should remark that the formula should not hold for solutions of non-homogeneous (non-stationary) heat equation. In such a case we have to consider a similar formula for a backward diffusion. Extensions to the Feynman–Kac–Girsanov formula might be new.

The dual semigroup is defined usually as $P_{s,t}^* g(y) = \int_{\mathbb{R}^d} p_{s,t}(x, y) g(x)\, dx$, if the transition function $P_{s,t}(x, dy)$ has a density function $p_{s,t}(x, y)$ with respect to the Lebesgue measure dy. However, in our definition, we do not assume the existence of the density $p_{s,t}(x, y)$.

4.5 Jump-Diffusion and Heat Equation; Case of Smooth Jumps

Let us consider a stochastic differential equation with jumps on a Euclidean space. To avoid complicated notations, we assume that the Lévy measure ν has a weak drift. We shall consider the equation studied in Sect. 3.10. It is given by

$$X_t = X_0 + \sum_{k=0}^{d'} \int_{t_0}^{t} V_k(X_r, r) \circ dW_r^k + \int_{t_0}^{t} \int_{|z|>0+} \{\phi_{r,z}(X_{r-}) - X_{r-}\} N(dr\, dz), \tag{4.50}$$

where the last integral by the Poisson random measure is the improper integral. For diffusion and drift coefficients, we assume that these are $C_b^{\infty,1}$-functions and for jump coefficients, we assume that $g(x,t,z) \equiv \phi_{t,z}(x) - x$ satisfies Condition (J.1) in Sect. 3.2.

Let $\{\Phi_{s,t}\}$ be the stochastic flow of C^∞-maps generated by equation (4.50). Then, similarly to the proof of Proposition 4.1.1, we can show that its law

$$P_{s,t}(x,E) = P(\Phi_{s,t}(x) \in E), \quad 0 \le s < t < \infty, \quad E \in \mathcal{B}(\mathbb{R}^d).$$

is determined uniquely from SDE (4.50) and $\{P_{s,t}(x,E)\}$ is a transition probability. Further, the solution $X_t, t \ge t_0$ is a conservative strong Markov process on $\mathbb{R}^d$ with the transition probability. The fact can be verified as shown in Sect. 4.1 for the diffusion process. It is called a *jump-diffusion process*. In particular, if there is no diffusion part and no drift part, i.e., $V_k(x,t) \equiv 0, k = 0, \ldots, d'$, the solution $X_t, t \ge t_0$ is called a *jump process*.

For a slowly increasing continuous function $f(x)$ on $\mathbb{R}^d$, we define $P_{s,t}f(x)$ as in Sect. 4.1. $P_{s,t}f(x)$ is again a slowly increasing continuous function for any $s < t$, since $E[|\Phi_{s,t}(x)|^p]^{1/p} \le c_p(1+|x|)$ holds for any x. It satisfies the semigroup property $P_{s,t}P_{t,u}f = P_{s,u}f$ for any $s < t < u$ and slowly increasing continuous function f. Next, if $f(x)$ is a slowly increasing C^∞-function (or C_b^∞-function), we can change the order of the derivative and expectation and we have the equality $\partial^{\mathbf{i}} E[f(\Phi_{s,t}(x))] = E[\partial^{\mathbf{i}}(f \circ \Phi_{s,t}(x))]$ for any $\mathbf{i}$. Therefore $P_{s,t}f(x)$ is a slowly increasing C^∞-function of x (or C_b^∞-function) for each $s < t$.

Now, consider the rule of the forward differential calculus (3.93) for the flow $\{\Phi_{s,t}\}$. Take expectations for each term of (3.93). Since expectations of the second and fourth terms of the right-hand side are 0, we get

$$P_{s,t}f(x) = f(x) + \int_s^t P_{s,r}A_J(r)f(x)\,dr, \tag{4.51}$$

for any slowly increasing C^∞-function f. Here $A_J(t)$ is an integro-differential operator defined by

$$A_J(t)f = A(t)f + \int_{|z|>0+} \{f \circ \phi_{t,z} - f\}\nu(dz), \tag{4.52}$$

where $A(t)$ is the differential operator defined by (4.4) and the last integral by the Lévy measure ν is the improper integral. If the Lévy measure has a strong drift, it coincides with the Lebesgue integral. Furthermore, using the rule of the backward differential calculus (3.95), we can show that $P_{s,t}f$ satisfies

$$P_{s,t}f(x) = f(x) + \int_s^t A_J(r)P_{r,t}f(x)\,dr, \tag{4.53}$$

for any slowly increasing C^∞-function f. Therefore the semigroup $\{P_{s,t}\}$ satisfies Kolmogorov's forward and backward equations associated with the operator $A_J(t)$ of (4.52). We will summarize the result.

Theorem 4.5.1 *The solution $X_t, t \geq t_0$ of* (4.50) *is a conservative jump-diffusion process. If $f(x)$ is a slowly increasing C^∞-function, $P_{s,t}f(x)$ is a slowly increasing C^∞-function of x. Further, it is continuously differentiable with respect to s, t for $s < t$ and satisfies the Kolmogorov forward equation* (4.5) *and the Kolmogorov backward equation* (4.6) *associated with the operator $A_J(t)$ given by* (4.52).

The integro-differential operator $A_J(t)$ of (4.52) is called the *generator* of the jump-diffusion process X_t. Further the flow $\{\Phi_{s,t}\}$ is called the stochastic flow associated with the operator $A_J(t)$.

Let us next consider an exponential transformation of the jump-diffusion. We will define an exponential Wiener–Poisson functional $G_{s,t}(x)$, which is an extension of the exponential Wiener functional defined by (4.13). Let $c_k(x,t), k = 0, \ldots, d'$ be $C_b^{\infty,1}$-functions on $\mathbb{R}^d \times \mathbb{T}$ and let $d_{t,z}(x)$ be a positive $C_b^{\infty,1,2}$-function of (x,t,z) such that $d_{t,0}(x) = 1$ holds for any x, t. Then both $d_{t,z}(\Phi_{s,r}(x)) - 1$ and $\log d_{r,z}(\Phi_{s,r}(x))$ belong to the class $\mathbf{L}_{\mathbb{U}}^0$ defined by (3.83). Then in view of Proposition 3.10.1, improper integrals

$$\int_s^t \int_{|z|>0+} \{d_{r,z}(\Phi_{s,r}(x)) - 1\}\, dr\, \nu(dz), \qquad \int_s^t \int_{|z|>0+} \log d_{r,z}(\Phi_{s,r-}(x)) N(dr\, dz)$$

etc. are well defined for any $s < t$ and x, since the Lévy measure has a weak drift. For a given x and $s < t$, we define an exponential functional with coefficients $\mathbf{d} = (d_{t,z}(x))$ by

$$G_{s,t}^{\mathbf{d}}(x) = \exp\left\{ \int_s^t \int_{|z|>0+} \log d_{r,z}(\Phi_{s,r-}(x)) N(dr\, dz) \right\}. \tag{4.54}$$

We set $G_{s,t}(x) = G_{s,t}^{\mathbf{c},\mathbf{d}}(x) = G_{s,t}^{\mathbf{c}}(x) G_{s,t}^{\mathbf{d}}(x)$, where $G_{s,t}^{\mathbf{c}}(x)$ is defined by (4.13) for the stochastic flow $\{\Phi_{s,t}\}$.

Lemma 4.5.1 *For any $\mathbf{i}$, $T > 0$ and $p \geq 2$, $E[|\partial^{\mathbf{i}} G_{s,t}(x)|^p]$ is bounded with respect to $0 \leq s < t \leq T$ and $x \in \mathbb{R}^d$.*

Proof We first show that $G_{s,t}$ satisfies a linear SDE

$$\begin{aligned} G_{s,t} = 1 &+ \sum_{k=1}^{d'} \int_s^t G_{s,r-} c_k'(r)\, dW_r^k + \int_s^t G_{s,r-} \bar{c}'(r)\, dr \\ &+ \int_s^t \int_{\mathbb{R}_0^{d'}} G_{s,r-} \{d_z'(r) - 1\} \tilde{N}(dr\, dz), \end{aligned} \tag{4.55}$$

where $c_k'(r) = c_k(\Phi_{s,r-}(x), r)$, $\bar{c}'(r) = \bar{c}(\Phi_{s,r-}(x), r)$, $d_z'(r) = d_{r,z}(\Phi_{s,r-}(x))$. Here,

$$\bar{c}(x,r) := c_0(x,r) + \frac{1}{2}\sum_{k=1}^{d'}\{V_k(r)c_k(x,r) + c_k(x,r)^2\}$$
$$+ \int_{|z|>0+} \{d_{r,z}(x) - 1\}\nu(dz). \tag{4.56}$$

For the proof, apply the rule of the forward differential calculus (3.87) (Theorem 3.10.1) to the function $f(x) = e^x$ and the Itô process $X_t := \log G_{s,t}, t > s$. Then $e^{X_t} - 1$ is equal to

$$\sum_{k=0}^{d'}\int_s^t e^{X_r} c_k'(r) \circ dW_r^k + \int_s^t\int_{|z|>0+} \{e^{X_{r-}+\log d_z'(r)} - e^{X_{r-}}\}N(dr\,dz)$$
$$= \sum_{k=1}^{d'}\int_s^t e^{X_{r-}} c_k'(r)\,dW_r^k + \int_s^t e^{X_{r-}}\bar{c}'(r)\,dr + \int_s^t\int_{|z|>0} e^{X_{r-}}\{d_z'(r) - 1\}\tilde{N}(dr\,dz).$$

This proves (4.55).

We may regard $G_{s,t}$ as a solution of the master equation (3.21). In fact, note that coefficients $c_i(x,r)$, $\bar{c}(x,r)$ and $d_{r,z}(x) - 1$ are bounded functions. Then, regarding x as a parameter, we set $c = (c_1, \ldots, c_{d'})$, $\alpha_x(y,r) = yc(\Phi_{s,r-}(x), r)$ and

$$\beta_x(y,r) = y\bar{c}(\Phi_{s,r-}(x), r), \quad \chi_x(y,r,z) = y(d_{r,z}(\Phi_{s,r-}(x), r) - 1).$$

These functionals satisfy Conditions 1–3 of the master equation and $y = G_{s,r}(x)$ is the solution of the master equation with the above coefficients. Then we find that $E[|G_{s,t}(x)|^p]$ is bounded with respect to $x \in \mathbb{R}^d$, $0 \le s < t \le T$ for any $p \ge 2$.

Next, the derivative $\partial G_{s,t}(x)$ satisfies

$$\partial G_{s,t} = \int_s^t (\partial G_{s,r-}c'(r) + G_{s,r-}\partial c'(r))\,dW_r$$
$$+ \int_s^t (\partial G_{s,r-}\bar{c}'(r) + G_{s,r-}\partial\bar{c}'(r))\,dr$$
$$+ \int_s^t\int \Big\{\partial G_{s,r-}(d_z'(r) - 1) + G_{s,r-}\partial(d_z'(r) - 1)\Big\}\tilde{N}(dr\,dz),$$

since $G_{s,t}$ satisfies (4.55). We will apply Theorem 3.3.1 again. Since

$$E[|G_{s,r-}\partial c'(r)|^p], \quad E\Big[|G_{s,r-}|^p\int|\partial(d_z'(r) - 1)|^p\nu(dz)\Big]$$

are bounded with respect to $s < r$ and x (x is a parameter), $E[|\partial G_{s,t}(x)|^p]$ should be bounded with respect to $s < t, x$. Repeating this argument inductively, we arrive at the assertion of the lemma. □

We can define

$$P_{s,t}^{\mathbf{c},\mathbf{d}} f(x) = E\big[f(\Phi_{s,t}(x))G_{s,t}^{\mathbf{c},\mathbf{d}}(x)\big] \tag{4.57}$$

for any slowly increasing continuous function f. Then $\{P_{s,t}^{\mathbf{c},\mathbf{d}}\}$ has the semigroup property. We define further an integro-differential operator with parameter $\mathbf{c}, \mathbf{d}$ by

$$A_J^{\mathbf{c},\mathbf{d}}(t)f(x) = A^{\mathbf{c}}(t)f(x) + \int_{|z|>0+} \{d_{t,z}(x)f(\phi_{t,z}(x)) - f(x)\}\nu(dz) \tag{4.58}$$

for slowly increasing C^∞-function f, where $A^{\mathbf{c}}(t)$ is the differential operator defined by (4.11). If the Lévy measure has a strong drift, the last integral coincides with the Lebesgue integral. It is rewritten by a direct calculation as

$$A_J^{\mathbf{c},\mathbf{d}}(t)f = A_J(t)f + \sum_{k=1}^{d'} c_k(t)V_k(t)f + \int_{|z|>0+} (d_{t,z}-1)\{f\circ\phi_{t,z} - f\}\nu(dz) + \bar{c}(t)f,$$

where $\bar{c}(x,t)$ is defined by (4.56).

Theorem 4.5.2 *If f is a slowly increasing C^∞-function of x, $P_{s,t}^{\mathbf{c},\mathbf{d}} f(x)$ is also a slowly increasing C^∞-function of x. Further, it is continuously differentiable with respect to s, t for $s < t$ and satisfies the following forward and backward integro-differential equations, respectively:*

$$\frac{\partial}{\partial t} P_{s,t}^{\mathbf{c},\mathbf{d}} f(x) = P_{s,t}^{\mathbf{c},\mathbf{d}} A_J^{\mathbf{c},\mathbf{d}}(t)f(x), \tag{4.59}$$

$$\frac{\partial}{\partial s} P_{s,t}^{\mathbf{c},\mathbf{d}} f(x) = -A_J^{\mathbf{c},\mathbf{d}}(s)P_{s,t}^{\mathbf{c},\mathbf{d}} f(x). \tag{4.60}$$

Proof The first assertion follows from Lemma 4.5.1, immediately. We prove the second assertion. Using a symmetric integral, we can show that equation (4.55) is rewritten as

$$G_{s,t} = 1 + \sum_{k=0}^{d'} \int_s^t G_{s,r}c_k'(r)\circ dW_r^k + \int_s^t \int_{|z|>0+} G_{s,r-}\{d_z'(r) - 1\}N(dr\,dz), \tag{4.61}$$

similarly to the diffusion case (Sect. 4.2). The pair $(\Phi_{s,t}(x), G_{s,t}(x)y)$ may be regarded as a $(d+1)$-dimensional jump-diffusion process determined by the pair of SDEs (4.50) and (4.61). Let $\bar{f}(x,y)$ be a slowly increasing C^∞-function on $\mathbb{R}^{d+1}$. Then, using the rule of the forward differential calculus (3.93) for the pair process, we find that

$$\bar{f}(\Phi_{s,t}(x), G_{s,t}(x)y) - \bar{f}(x, y)$$
$$= \int_a^t \bar{A}_J(r)\bar{f}(\Phi_{s,r}(x), G_{s,r}(x)y)\,dr + \text{a martingale with mean 0.}$$

Here $\bar{A}_J(t)$ is the operator given by

$$\bar{A}_J(t)\bar{f}(x,y) = \sum_{k=1}^{d'}\Big(V_k(t) + c_k(t)y\frac{\partial}{\partial y}\Big)^2\bar{f}(x,y) + \Big(V_0(t) + c_0(t)y\frac{\partial}{\partial y}\Big)\bar{f}(x,y)$$
$$+ \int_{|z|>0+}\big\{\bar{f}(\phi_{t,z}(x), \psi_{t,z}(x,y)) - \bar{f}(x,y)\big\}\nu(dz),$$

where $\psi_{t,z}(x,y) = d_{t,z}(x)y$. Hence $\bar{P}_{s,t}\bar{f}(x,y) := E[\bar{f}(\Phi_{s,t}(x), G_{s,t}(x)y)]$ satisfies $\bar{P}_{s,t}\bar{f}(x,y) = \bar{f}(x,y) + \int_s^t \bar{P}_{s,r}\bar{A}_J(r)\bar{f}(x,y)\,dr$. Further, from the rule of the backward differential calculus (3.95), we have $\bar{P}_{s,t}\bar{f}(x,y) = \bar{f}(x,y) + \int_s^t \bar{A}_J(r)\bar{P}_{r,t}\bar{f}(x,y)\,dr$. Therefore $\bar{A}_J(t)$ is the generator of the semigroup $\{\bar{P}_{s,t}\}$.

Now take $\bar{f}(x,y) = f(x)y$, where $f(x)$ is a slowly increasing smooth function on $\mathbb{R}^d$. Apply it to the operator $\bar{A}_J(t)$. Then we have $A^{\mathbf{c},\mathbf{d}}(t)f(x) = \bar{A}_J(t)\bar{f}(x,1)$. Consequently, we can show that $P_{s,t}^{\mathbf{c},\mathbf{d}}f(x)$ satisfies the forward and backward equations (4.59) and (4.60), similarly to the case of diffusion discussed in Sect. 4.2. □

Set $X_t = \sum_{k=0}^{d'}\int_s^t c_k'(r)\circ dW_r^k + \int_s^t\int_{|z|>0+} d_z'(r)N(dr\,dz)$. Using the Itô integral, it is rewritten as $X_t = X_t^{(0)} + X_t^{(1)} + X_t^{(2)}$, where $X_t^{(0)} = \int_s^t \bar{c}'(r)\,dr$ and

$$X_t^{(1)} = \sum_{k=1}^{d'}\int_s^t c_k'(r)\,dW_r^k - \frac{1}{2}\sum_{k=1}^{d'}\int_s^t c_k'(r)^2\,dr,$$
$$X_t^{(2)} = \int_s^t\int_{|z|>0+}\log d_z'(r)N(dr\,dz) - \int_s^t\int_{|z|>0+}\{d_z'(r) - 1\}\,dr\,\nu(dz).$$

Hence exponential functional $G_{s,t}(x) = G_{s,t}^{\mathbf{c},\mathbf{d}}(x)$ is decomposed as the product of three exponential functionals $G_{s,t}^{(0)}(x) = e^{X_t^{(0)}}$, $G_{s,t}^{(1)}(x) = e^{X_t^{(1)}}$ and $G_{s,t}^{(2)}(x) = e^{X_t^{(2)}}$. We saw in Sect. 4.2 that the transformation by $G_{s,t}^{(0)}(x)$ is the Feynman–Kac transformation. By this transformation, the potential term $\bar{c}(t)f$ is added to the operator $A_J(t)$. Further, the transformation by $G_{s,t}^{(1)}(x)$ is the Girsanov transformation. By this transformation, the drift term $\sum c_kV_k(t)f$ is added to $A_J(t)$. We shall consider the transformation by $G_{s,t}^{(2)}(x)$. We denote it by Z_t. Apply Theorem 3.10.1 by setting $F(x) = e^x$. Then we get

$$Z_t = 1 - \int_s^t \int Z_{r-}(d'_z(r) - 1)\, dr\nu(dz) + \int_s^t \int Z_{r-}(d'_z(r) - 1)N(dr\, dz)$$
$$= 1 + \int_s^t \int Z_{r-}(d'_z(r) - 1)\tilde{N}(dr\, dz). \tag{4.62}$$

Hence Z_t is a positive local martingale. Further, $E[Z_t^p] < \infty$ holds for any $p \geq 2$. Then Z_t is a positive martingale with mean 1. Then we can define another probability measure $P^{(2)}$ by $dP^{(2)} = G_{s,t}^{(2)}(x)\, dP$.

Lemma 4.5.2 *Set $\tilde{N}^{\mathbf{d}}(dr\, dz) = N(dr\, dz) - d'_z(r)\, dr\nu(dz)$ and*

$$Y_t = \int_s^t \int_{\mathbb{R}_0^{d'}} \psi(r, z)\tilde{N}^{\mathbf{d}}(dr\, dz), \tag{4.63}$$

where $\psi(r, z)$ is a bounded predictable random field. Then Y_t is a martingale with respect to $P^{(2)}$.

Proof Let $Z_t = e^{X_t^{(2)}}$. If the product $Z_t Y_t$ is a martingale with respect to the measure P, Y_t is a martingale with respect to $P^{(2)}$. In fact, if $Z_t Y_t$ is a martingale with respect to P, we have $E[Z_t Y_t; B] = E[Z_s Y_s; B]$ for any $t > s$ and $B \in \mathcal{F}_s$. Therefore,

$$E^{(2)}[Y_t; B] = E[Z_t Y_t; B] = E[Z_s Y_s; B] = E^{(2)}[Y_s; B], \quad \forall B \in \mathcal{F}_s,$$

showing that Y_t is a martingale with respect to the measure $P^{(2)}$.

Apply Theorem 3.10.1 to the product of Z_t and Y_t. Then we have

$$Z_t Y_t = \int_s^t \int (Z_{r-} + d'_z(r) - 1)(Y_{r-} + \psi(r, z) - Z_{r-}Y_{r-})N(dr\, dz)$$
$$- \int_s^t \int \{Y_{r-}(d'_z(r) - 1) + Z_{r-}\psi(r, z)\}\, dr\nu(dz)$$
$$= \int_s^t \int (Y_{r-}(d'_z(r) - 1) + Z_{r-}\psi(r, z))\tilde{N}(dr\, dz).$$

Therefore $Z_t Y_t$ is a martingale. □

The solution $X_t = \Phi_{s,t}(x)$ of equation (4.50) starting from x at time s satisfies (3.93). We can rewrite it as

$$f(\Phi_{s,t}) = f + \sum_{k=1}^{d'} \int_s^t V_k(r) f(\Phi_{s,r})\, dW_r^k + \int_s^t A_J^{\mathbf{d}}(r) f(\Phi_{s,r})\, dr$$
$$+ \int_s^t \int_{|z|>0+} \{f \circ \phi_{r,z} \circ \Phi_{s,r-} - f \circ \Phi_{s,r-}\}\tilde{N}^{\mathbf{d}}(dr\, dz),$$

where

$$A_J^{\mathbf{d}}(t)f = A_J(t)f + \int_{|z|>0+} (d_{t,z} - 1)(f \circ \phi_{t,z} - f)\nu(dz). \tag{4.64}$$

Define the semigroup $\{P_{s,t}^{(2)}\}$ by

$$P_{s,t}^{(2)} f(x) = \int f(\Phi_{s,t}(x))\, dP^{(2)} = E[f(\Phi_{s,t}(x))G_{s,t}^{(2)}(x)].$$

Then it satisfies

$$P_{s,t}^{(2)} f(x) = f(x) + \int_s^t P_{s,r}^{(2)} A_J^{\mathbf{d}}(r) f(x)\, dr.$$

Therefore the generator of the semigroup $\{P_{s,t}^{(2)}\}$ is $A_J^{\mathbf{d}}(t)f$. Consequently, by the transformation $P_{s,t} \to P_{s,t}^{(2)}$, the jump term $\int (d_{t,z} - 1)\{f \circ \phi_{t,z} - f\}\nu(dz)$ is added to $A_J(t)f$. It may be considered as a Girsanov transformation for a jump process.

The final value problem for the backward heat equation associated with the integro-differential operator $A_J^{\mathbf{c},\mathbf{d}}(t)$ is defined in the same way as the problem associated with the differential operator $A^{\mathbf{c}}(t)$ defined by (4.18).

Theorem 4.5.3 *The final value problem for the backward heat equation associated with operator $A_J^{\mathbf{c},\mathbf{d}}(t)$ has a unique slowly increasing solution for any slowly increasing C^∞-function f_1. Further, the solution v is represented by $v(x,s) = E[f_1(\Phi_{s,t_1}(x))G_{s,t_1}^{\mathbf{c},\mathbf{d}}(x)]$.*

The proof, which is carried out similarly to the proof of Theorem 4.2.2 will be omitted here. Consequently, for a given integro-differential operator $A_J^{\mathbf{c},\mathbf{d}}(t)$ of (4.58), there exists a unique semigroup of $C^{2,1}$-class satisfying (4.59) and (4.60). The operator $A_J^{\mathbf{c},\mathbf{d}}(t)$ is called the generator of the semigroup $\{P_{s,t}^{\mathbf{c},\mathbf{d}}\}$. In particular, if $\mathbf{c} = 0$ and $\mathbf{d} = 1$, the operator $A_J^{\mathbf{c},\mathbf{d}}(t)$ coincides with the operator $A_J(t)$. Then the semigroup $\{P_{s,t}\}$ satisfying Kolmogorov's backward equation associated with $A_J(t)$ is unique. The semigroup is said to be generated by the operator $A_J(t)$. The jump-diffusion process with the semigroup $\{P_{s,t}\}$ is called the jump-diffusion process with the generator $A_J(t)$.

The initial value problem for the heat equation associated with the integro-differential operator $A_J^{\mathbf{c},\mathbf{d}}(t)$ is defined in the same way as the problem associated with the differential operator $A^{\mathbf{c}}(t)$ defined by (4.11). In order to get the solution of the initial value problem, we consider the backward SDE with characteristics $(V_k(x,t), k = 0, \ldots, d', g(x,t,z), \nu)$. It is written as

$$\check{X}_s = x + \sum_{k=0}^{d'} \int_s^t V_k(\check{X}_r, r) \circ dW_r^k + \int_s^t \int g(\check{X}_r, r, z) N(dr\,dz). \tag{4.65}$$

The solution exists uniquely, which we denote by $\check{X}_s^{x,t}$. There exists a backward stochastic flow $\{\check{\Phi}_{s,t}\}$ of C^∞-maps such that $\check{X}_s^{x,t} = \check{\Phi}_{s,t}(x)$ holds a.s. for any t, s, x. Set $\check{P}_{s,t}(x, E) = P(\check{\Phi}_{s,t}(x) \in E)$. It is the transition function of the backward Markov process $\check{X}_s$. Further, similarly to the case of diffusion in Sect. 4.3, we can show that the backward semigroup $\{\check{P}_{s,t}\}$ satisfies

$$\check{P}_{s,t} f = f + \int_s^t \check{P}_{r,t} A_J(r) f\, dr, \quad \check{P}_{s,t} f = f + \int_s^t A_J(r) \check{P}_{s,r} f\, dr,$$

where $A_J(t)$ is the integro-differential operator given by (4.52). Therefore, the generator of the backward semigroup $\{\check{P}_{s,t}\}$ is the same as that of the forward semigroup $\{P_{s,t}\}$.

Let $\mathbf{c} = (c_0(x,t), \ldots, c_{d'}(x,t))$ and $\mathbf{d} = (d_{t,z})$ be functions satisfying the same conditions as those for $G_{s,t}$ of (4.54). We define backward exponential functionals by

$$\begin{aligned} \check{G}_{s,t}^{\mathbf{c}} &= \exp\Big\{\sum_{k=0}^{d'} \int_s^t c_k(\check{\Phi}_{r,t}, r) \circ dW_r^k, \\ \check{G}_{s,t}^{\mathbf{d}} &= \exp\Big\{\int_s^t \int_{|z|>0+} \log d_{r,z}(\check{\Phi}_{r,t}) N(dr\,dz)\Big\}. \end{aligned} \tag{4.66}$$

and $\check{G}_{s,t}^{\mathbf{c},\mathbf{d}}(x) = \check{G}_{s,t}^{\mathbf{c}}(x)\check{G}_{s,t}^{\mathbf{d}}(x)$. We set $\check{P}_{s,t}^{\mathbf{c},\mathbf{d}} f(x) = E[f(\check{\Phi}_{s,t}(x))\check{G}_{s,t}^{\mathbf{c},\mathbf{d}}(x)]$. Then $\{\check{P}_{s,t}^{\mathbf{c},\mathbf{d}}\}$ is a backward semigroup. Its generator is equal to $A_J^{\mathbf{c},\mathbf{d}}(t)$. Hence we have the following theorem.

Theorem 4.5.4 *If f_0 is a slowly increasing C^∞-function, the initial value problem for the heat equation associated with the operator $A_J^{\mathbf{c},\mathbf{d}}(t)$ has a unique slowly increasing solution $u(x,t)$, $(x,t) \in \mathbb{R}^d \times (t_0, \infty)$. It is of $C^{\infty,1}$-class and is represented by*

$$u(x,t) = E\big[f_0(\check{\Phi}_{t_0,t}(x))\check{G}_{t_0,t}^{\mathbf{c},\mathbf{d}}(x)\big], \tag{4.67}$$

where $\{\check{\Phi}_{s,t}\}$ is the right continuous backward stochastic flow generated by the symmetric backward SDE with characteristics $(V_k(x,t), k = 0, \ldots, d', g(x,t,z), \nu)$.

In particular, if $V_k(t), k = 1, \ldots, d'$ and $c_k(t), k = 0, \ldots, d', d_z(t)$ are time homogeneous (not depending on t), the solution is given by the Feynman–Kac–Girsanov formula $u(x,t) = E[f_0(\Phi_{t_0,t}(x))G_{t_0,t}^{\mathbf{c},\mathbf{d}}(x)]$, making use of the forward stochastic flow $\{\Phi_{s,t}\}$ with characteristics $(V_k, k = 0, \ldots, d', g(\cdot, z), \nu)$.

4.6 Dual Semigroup, Inverse Flow and Backward Jump-Diffusion; Case of Diffeomorphic Jumps

In this section, we assume that jump-maps $\phi_{t,z}; \mathbb{R}^d \to \mathbb{R}^d$ defined by $\phi_{t,z}(x) = g(x,t,z) + x$ are diffeomorphic i.e., they satisfy Condition (J.2) in Chap. 3. Then the stochastic flow $\{\Phi_{s,t}\}$ is a flow of diffeomorphisms by Theorem 3.9.1.

Proposition 4.6.1 *Let $A_J(t)$ be an integro-differential operator given by* (4.52). *We assume that coefficients $V_k(x,t)$, $k = 0, \ldots, d'$ are $C_b^{\infty,1}$-functions and that jump coefficient $g(x,t,z) = \phi_{t,z}(x) - x$ satisfies Conditions (J.1) and (J.2) in* Chap. 3. *Then the dual operator $A_J(t)^*$; $C_0^\infty(\mathbb{R}^d) \to C_b^\infty(\mathbb{R}^d)$ is well defined. It is represented by*

$$A_J(t)^* g = \frac{1}{2}\sum_{k=1}^{d'}(V_k(t) + \mathrm{div}\, V_k(t))^2 g - (V_0(t) + \mathrm{div}\, V_0(t))g$$
$$+ \int_{|z|>0+} \Big\{(\det \nabla\phi_{t,z}^{-1})g \circ \phi_{t,z}^{-1} - g\Big\}\nu(dz), \tag{4.68}$$

where $\phi_{t,z}^{-1}$ is the inverse map of the diffeomrphic map $\phi_{t,z}$ and $\det\nabla\phi_{t,z}^{-1}$ is its Jacobian determinant.

Before we proceed to the proof of the proposition, let us check that the last improper integral of (4.68) is well defined and is a C_b^∞-function of x, if $g \in C_0^\infty(\mathbb{R}^d)$. Note that $\phi_{t,z}^{-1}(x)$ is written as $x - h(x,t,z)$, where $h(x,t,z)$ has the same property as that of $g(x,t,z)$. Indeed, we have $\nabla_x h(x,t,z) = O(|z|)$ uniformly in x. Therefore,

$$\det \nabla\phi_{t,z}^{-1}(x) = \det(I - \nabla h(x.t,z)) = 1 + O(|z|)$$

uniformly in x. Then $\det\nabla\phi_{t,z}^{-1} > 0$ holds for sufficiently small $|z|$. Since it is continuous in z and does not take the value 0 (because of diffeomorphic maps), it should be positive for all z. Further, we get

$$\det\nabla\phi_{t,z}^{-1} g \circ \phi_{t,z}^{-1} - g = (\det\nabla\phi_{t,z}^{-1} - 1) g \circ \phi_{t,z}^{-1} + (g \circ \phi_{t,z}^{-1} - g) = O(|z|).$$

uniformly in x. Then the improper integral of the above function with respect to the Lévy measure with weak drift is well defined, and in fact, a C_b^∞-function.

Proof of Proposition 4.6.1 The operator $A_J(t)$ is written as $A(t) + A_2(t)$, where $A_2(t)f = \int_{|z|>0+}\{f \circ \phi_{t,z} - f\}\nu(dz)$. We have computed the dual of the operator $A(t)$ in Sect. 4.4. So we will consider the operator $A_2(t)$. We want to show that the dual of the operator $A_2(t)$ is represented as

$$A_2(t)^* g = \int_{|z|>0+} \big\{ \det \nabla \phi_{t,z}^{-1} \cdot g \circ \phi_{t,z}^{-1} - g \big\} \nu(dz). \tag{4.69}$$

Let ϕ be a diffeomorphic map on $\mathbb{R}^d$ such that $\det \nabla \phi(x)$ is a bounded function. Then from formula of change of variables (4.34), we have

$$\int f(\phi^{-1}(x))\, dx = \int f(x) |\det \nabla \phi(x)|\, dx,$$

$$\int f(\phi^{-1}(x)) g(x)\, dx = \int f(x) g(\phi(x)) |\det \nabla \phi(x)|\, dx,$$

for any functions f, g of $C_0^\infty(\mathbb{R}^d)$. Therefore,

$$\begin{aligned}
&\int \Big(\int_{|z|>0+} \{ f(\phi_{t,z}(x)) - f(x) \} \nu(dz) \Big) \cdot g(x)\, dx \\
&= \int_{|z|>0+} \Big(\int \big\{ f(x) g(\phi_{t,z}^{-1}(x)) |\det \nabla \phi_{t,z}^{-1}(x)| - f(x) g(x) \big\}\, dx \Big) \nu(dz) \\
&= \int f(x) \cdot \Big(\int_{|z|>0+} \Big\{ |\det \nabla \phi_{t,z}^{-1}(x)| g(\phi_{t,z}^{-1}(x)) - g(x) \Big\} \nu(dz) \Big)\, dx.
\end{aligned}$$

Since $\det \nabla \phi_{t,z}^{-1}(x)$ is positive, $\int A_2(t) f(x) \cdot g(x)\, dx = \int f(x) \cdot A_2(t)^* g(x)\, dx$ holds. These computations yield the dual formula. □

Now we will define the dual semigroup $\{P_{s,t}^*\}$, making use of the inverse flow. Discussions are similar to the diffusion case in Sect. 4.4. Let $f, g \in C_0^\infty(\mathbb{R}^d)$ and $\check{\Psi}_{s,t} = \Phi_{s,t}^{-1}$. Then we have the formula (4.35). We saw in Sect. 3.9 that the inverse flow $\{\check{\Psi}_{s,t}\}$ is a solution of a backward SDE with characteristics $(-V_k(t), k = 0, \ldots, d', \phi_{t,z}^{-1}(x) - x, \nu)$. Applying Theorem 3.38 to the backward SDE, we find that for any $p > 1$, $E[|\partial \check{\Psi}_{s,t}(x)|^p]$ are bounded with respect to x and $0 < s < t < T$. Then $E[|\det \nabla \check{\Psi}_{s,t}(x)|^p]$ is also bounded for any $p \geq 2$. Therefore for $g \in C_0^\infty(\mathbb{R}^d)$, we can define $P_{s,t}^* g(x)$ by (4.36). It is a function of $C_b^\infty(\mathbb{R}^d)$ and satisfies (4.37). Therefore $P_{s,t}^*$ defined by (4.36) is the dual of $P_{s,t}$. The dual semigroup $\{P_{s,t}^*\}$ and the dual operator $A_J(t)^*$ are related by

$$\begin{aligned}
P_{s,t}^* g(x) &= g(x) + \int_s^t P_{r,t}^* A_J(r)^* g(x)\, dr \qquad (4.70)\\
&= g(x) + \int_s^t A_J(r)^* P_{s,r}^* g(x)\, dr
\end{aligned}$$

for any $g \in C_0^\infty(\mathbb{R}^d)$. Therefore, $A_J(t)^*$ is the generator of the dual semi-group $\{P_{s,t}^*\}$.

For a fixed x, t_1, $\check{X}_s = \check{X}_s^{x,t_1} = \check{\Psi}_{s,t_1}(x)$, $s \in [0, t_1]$ is a right continuous backward stochastic process satisfying the backward SDE (3.96). It is a backward Markov process associated with the backward transition probability $\check{\mathcal{P}}_{s,t}(x, E) = P(\check{\Psi}_{s,t}(x) \in E)$. Since $\check{\Psi}_{s,t}$ satisfies (3.99), the backward semigroup $\check{\mathcal{P}}_{s,t}$ satisfies

$$\check{\mathcal{P}}_{s,t} f = f + \int_s^t \check{\mathcal{P}}_{r,t} \mathcal{A}_J(r) f dr,$$

where

$$\mathcal{A}_J(t) f = \mathcal{A}(t) f + \int_{|z|>0+} \{f \circ \phi_{t,z}^{-1} - f\} \nu(dz), \tag{4.71}$$

where $\mathcal{A}(t)$ is the differential operator defined by (4.38).

The dual operator $A_J(t)^*$ can be regarded as a transformation of the operator $\mathcal{A}_J(t)$. It is rewritten as

$$\begin{aligned} A_J(t)^* g &= \mathcal{A}_J(t) g + \sum_{k=1}^{d'} \text{div } V_k(t) V_k(t) g \\ &+ \int_{|z|>0+} (\det \nabla \phi_{t,z}^{-1} - 1)\{g \circ \phi_{t,z}^{-1} - g\} \nu(dz) + c^*(t) g, \end{aligned} \tag{4.72}$$

where

$$\begin{aligned} c^*(t) &= -\text{div } V_0(t) + \frac{1}{2} \sum_{k=1}^{d'} \Big\{ \text{div } V_k(t) V_k(t) (\text{div } V_k(t)) + \text{div } V_k(t)^2 \Big\} \\ &+ \int_{|z|>0} (\det \nabla \phi_{t,z}^{-1} - 1) \nu(dz). \end{aligned} \tag{4.73}$$

We will next consider $|\det \nabla \check{\Psi}_{s,t}(x)|$. We set $\check{J}_{s,t}(x) = \det \nabla \check{\Psi}_{s,t}(x)$.

Lemma 4.6.1 *$|\check{J}_{s,t}|$ satisfies the following backward equation:*

$$\begin{aligned} |\check{J}_{s,t}| &= 1 - \sum_{k=0}^{d'} \int_s^t \text{div } V_k(\check{\Psi}_{r,t}, r) |\check{J}_{r,t}| \circ d W_r^k \\ &+ \int_s^t \int_{|z|>0+} \Big\{ \det \nabla \phi_{r,z}^{-1}(\check{\Psi}_{r,t}) - 1 \Big\} |\check{J}_{r,t}| N(dr\, dz). \end{aligned} \tag{4.74}$$

Proof For $f \in C_0^\infty(\mathbb{R}^d)$, we have a rule of the backward differential calculus with respect to r:

$$f(\Phi_{s,t}(x)) - f(x) = \sum_{k=0}^{d'} \int_s^t V_k(r)(f \circ \Phi_{r,t})(x) \circ dW_r^k$$
$$+ \int_s^t \int_{|z|>0+} \Big\{ f(\Phi_{r,t}(\phi_{r,z}(x))) - f(\Phi_{r,t}(x)) \Big\} N(dr\, dz).$$

See (3.94). In the proof of Lemma 4.4.1, we showed the equality

$$\int \Big(f(\Phi_{s,t}(x)) - f(x) - \sum_{k=0}^{d'} \int_s^t V_k(r)(f \circ \Phi_{r,t})(x) \circ dW_r^k \Big) dx$$
$$= \int f(x) \Big(|\check{J}_{s,t}(x)| - 1 + \sum_{k=0}^{d'} \int_s^t \operatorname{div} V_k(\check{\Psi}_{r,t}(x), r) |\check{J}_{r,t}(x)| \circ dW_r^k \Big) dx.$$

Further, we have by the Fubini theorem and formula of change of variables,

$$\int_{\mathbb{R}^d} \Big(\int_s^t \int_{|z|>0+} \Big\{ f(\Phi_{r,t} \circ \phi_{r,z}(x)) - f(\Phi_{r,t}(x)) \Big\} N(drdz) \Big) dx$$
$$= \int_s^t \int_{|z|>0+} \Big(\int_{\mathbb{R}^d} \Big\{ f(\Phi_{r,t}(x)) |\det \nabla(\phi_{r,z}^{-1}(x))| - f(\Phi_{r,t}(x)) \Big\} dx \Big) N(dr\, dz)$$
$$= \int_s^t \int_{|z|>0+} \Big(\int_{\mathbb{R}^d} f(x) (|\det \nabla \phi_{r,z}^{-1}(\check{\Psi}_{r,t}(x))| - 1) |\check{J}_{r,t}(x)|\, dx \Big) N(dr\, dz)$$
$$= \int_{\mathbb{R}^d} f(x) \Big(\int_s^t \int_{|z|>0+} (|\det \nabla \phi_{r,z}^{-1}(\check{\Psi}_{r,t}(x))| - 1) |\check{J}_{r,t}(x)| N(dr\, dz) \Big) dx.$$

Since these two equalities are valid for any $f \in C_0^\infty(\mathbb{R}^d)$, we get the equality

$$|\check{J}_{s,t}| - 1 = \sum_{k=0}^{d'} \int_s^t -\operatorname{div} V_k(\check{\Psi}_{r,t}, r) |\check{J}_{r,t}| \circ dW_r^k$$
$$+ \int_s^t \int_{|z|>0+} \Big\{ |\det \nabla \phi_{r,z}^{-1}(\check{\Psi}_{r,t})| - 1 \Big\} |\check{J}_{r,t}| N(dr\, dz).$$

This proves the equality of the lemma. □

Associated with the inverse flow $\{\check{\Psi}_{s,t}\}$, we set

$$\check{\mathcal{G}}_{s,t}^{\mathbf{d}}(x) = \exp\Big\{ \int_s^t \int_{|z|>0+} \log d_{r,z}(\check{\Psi}_{r,t}(x)) N(dr\, dz) \Big\}. \tag{4.75}$$

and $\check{\mathcal{G}}_{s,t}^{\mathbf{c},\mathbf{d}}(x) = \mathcal{G}_{s,t}^{\mathbf{c}}(x) \check{\mathcal{G}}_{s,t}^{\mathbf{d}}(x)$, where $\check{\mathcal{G}}_{s,t}^{\mathbf{c}}(x)$ is given by (4.40).

Proposition 4.6.2 *Assume the same condition as in Proposition* 4.6.1. *Then* $\det \nabla \check{\Psi}_{s,t}$ *is positive and is represented by* $\det \nabla \check{\Psi}_{s,t} = \check{\mathcal{G}}_{s,t}^{-\mathbf{div}\, V, \mathbf{det}\, \nabla \boldsymbol{\phi}^{-1}}$, *where* $-\mathbf{div}\, V = (-\mathrm{div}\, V_0(x,r), \ldots, -\mathrm{div}\, V_{d'}(x,r))$ *and* $\mathbf{det}\, \nabla \boldsymbol{\phi}^{-1} = \det \nabla \phi_{r,z}^{-1}(x)$.

Proof Set

$$\check{Y}_s = -\sum_{k=0}^{d'} \int_s^t \mathrm{div}\, V_k(\check{\Psi}_{r,t}, r) \circ dW_r^k + \int_s^t \int_{|z|>0+} \log \det \nabla \phi_{r,z}^{-1}(\check{\Psi}_{r,t}) N(dr\, dz)$$

and apply the rule of the backward differential calculus (3.89) (Theorem 3.10.2) for the exponential function $f(x) = e^x$. Then we get

$$e^{\check{Y}_s} = 1 - \sum_{k=0}^{d'} \int_s^t e^{\check{Y}_r} \mathrm{div}\, V_k(\check{\Psi}_{r,t}) \circ dW_r^k$$
$$+ \int_s^t \int_{|z|>0+} e^{\check{Y}_r} \{\det \nabla \phi_{r,z}^{-1}(\check{\Psi}_{r,t}) - 1\} N(dr\, dz),$$

similarly to the proof of Lemma 4.5.1. Hence both $e^{\check{Y}_s}$ and $|\check{J}_{s,t}|$ satisfy the same backward SDE (4.72). Then we have $e^{\check{Y}_s} = |\check{J}_{s,t}|$, by the uniqueness of the solution of the SDE.

It remains to show that $\check{J}_{s,t}(y)$ is positive a.s. Instead of $\{\check{\Psi}_{s,t}\}$, we will consider a family of flows $\{\check{\Psi}_{s,t}^{\epsilon}\}$, $\epsilon > 0$, which are obtained from $\{\check{\Psi}_{s,t}\}$ by cutting off all jumps $\phi_{t,z}^{-1}$, $|z| < \epsilon$, as in the proof of Lemma 3.8.1. It is written as $\check{\Psi}_{s,t}^{\epsilon} = \check{\Psi}_{s,t}^{0}$ if $t < \tau_1$ and

$$\check{\Psi}_{s,t}^{\epsilon} = \check{\Psi}_{s,\tau_1}^{0} \circ \phi_{\tau_1,S_1}^{-1} \circ \cdots \circ \phi_{\tau_n,S_n}^{-1} \circ \check{\Psi}_{\tau_n,t}^{0}, \quad \text{if } \tau_n \le t < \tau_{n+1},$$

where $\{\check{\Psi}_{s,t}^{0}\}$ is a continuous flow determined by a continuous backward SDE, and $\{\tau_i, S_i; i = 1, 2, \ldots\}$ is a sequence of stopping times and $\mathbb{R}_0^{d'}$-valued random variables. Then Jacobian of $\check{\Psi}_{s,t}^{\epsilon}$ satisfies $\det \check{\Psi}_{s,t}^{\epsilon} = \det \check{\Psi}_{s,t}^{0} > 0$ if $t < \tau_1$ and

$$\det \nabla \check{\Psi}_{s,t}^{\epsilon} = \det \nabla \check{\Psi}_{s,\tau_1}^{0} \det \nabla \phi_{\tau_1,S_1}^{-1} \cdots \det \nabla \phi_{\tau_n,S_n}^{-1} \det \nabla \check{\Psi}_{\tau_n,t}^{0} > 0,$$

if $\tau_n \le t < \tau_{n+1}$, since $\det \nabla \check{\Psi}_{s,\tau_i}^{0}$, $\det \nabla \phi_{\tau_i,S_i}^{-1}$ etc. are all positive. Now a subsequence $\check{\Psi}_{s,t}^{\epsilon_n}(x)$ and $\nabla \check{\Psi}_{s,t}^{\epsilon_n}(x)$ converges to $\check{\Psi}_{s,t}(x)$ and $\nabla \check{\Psi}_{s,t}(x)$ in probability. Therefore $\det \nabla \check{\Psi}_{s,t}(x)$ is nonnegative. Then it is strictly positive a.s. □

By the above proposition, the dual semigroup $P_{s,t}^* g(x)$ is obtained from the backward semigroup $\check{\mathcal{P}}_{s,t} g(x) = E[g(\check{\Psi}_{s,t}(x))]$ by the exponential transformation by (4.75). Then the operator $\mathcal{A}_J(t)$ is transformed to the operator $A_J(t)^*$ of (4.68). Consequently, we have the following theorem.

Theorem 4.6.1 *Assume the same condition as in Proposition* 4.6.1.

1. *Let* $\{\check{\Psi}_{s,t}\}$ *be the inverse flow of* $\{\Phi_{s,t}\}$. *Then* $\check{X}_s = \check{X}_s^{x,t} = \check{\Psi}_{s,t}(x)$ *is a backward Markov process with the backward semigroup* $\{\check{\mathcal{P}}_{s,t}\}$ *having the generator* $\mathcal{A}_J(t)$ *defined by* (4.71).
2. *The Jacobian of* $\check{\Psi}_{s,t}(x)$ *is represented as* $\mathcal{G}^{-\mathbf{div V}, \mathbf{det} \nabla \phi^{-1}}$
3. *The dual semigroup* $\{P^*_{s,t}\}$ *is obtained by a backward exponential transformation of the* $\check{\Psi}_{s,t}$;

$$P^*_{s,t}g(x) = E\big[g(\check{\Psi}_{s,t}(x)) \det \nabla \check{\Psi}_{s,t}(x)\big] = \check{\mathcal{P}}_{s,t}^{-\mathbf{div\,V}, \mathbf{det}\nabla\phi^{-1}} g(x). \tag{4.76}$$

Further, if g *is a slowly increasing* C^∞*-function of* x, *then* $P^*_{s,t}g(x)$ *is also a slowly increasing* C^∞*-function of* x.
4. *The generator of the dual semi-group is given by* (4.68).

So far, we have studied the dual of the operator $A(t)$ and the dual of the semi-group $\{P_{s,t}\}$. We will extend these to duals of $A_J^{\mathbf{c},\mathbf{d}}(t)$ and $P_{s,t}^{\mathbf{c},\mathbf{d}}$ which were defined in Sect. 4.5. Let $A_J^{\mathbf{c},\mathbf{d}}(t)$ be the integro-differential operator defined by (4.58). Then, by a direct computation, we find that its dual $A_J^{\mathbf{c},\mathbf{d}}(t)^*$ is expressed by

$$A_J^{\mathbf{c},\mathbf{d}}(t)^* g = \mathcal{A}_J^{\mathbf{c}-\mathbf{div\,V}, \mathbf{d}\cdot\mathbf{det}\nabla\phi^{-1}}(t) g. \tag{4.77}$$

We will define the dual semigroup $\{P_{s,t}^{\mathbf{c},\mathbf{d},*}\}$. Let $G_{s,t}(x) = G_{s,t}^{\mathbf{c},\mathbf{d}}(x)$ be an exponential functional with coefficients $\mathbf{c}, \mathbf{d}$. For any C^∞-functions of compact supports f, g, we have the formula of the change of variables:

$$\begin{aligned} &\int_{\mathbb{R}^d} f(\Phi_{s,t}(x)) G_{s,t}(x) g(x)\, dx \\ &= \int_{\mathbb{R}^d} f(x) G_{s,t}(\check{\Psi}_{s,t}(x)) g(\check{\Psi}_{s,t}(x)) \det \nabla \check{\Psi}_{s,t}(x)\, dx, \end{aligned} \tag{4.78}$$

almost surely. Note $G_{s,t}^{\mathbf{c},\mathbf{d}}(\check{\Psi}_{s,t}) = \check{\mathcal{G}}_{s,t}^{\mathbf{c},\mathbf{d}}$. Then $E[|\check{\mathcal{G}}_{s,t}(x)|^p]$ is also bounded for any p. Therefore,

$$P_{s,t}^{\mathbf{c},\mathbf{d},*} g(x) = E\big[g(\check{\Psi}_{s,t}(x)) \check{\mathcal{G}}_{s,t}^{\mathbf{c},\mathbf{d}}(x) \det \nabla \check{\Psi}_{s,t}(x)\big]$$

is well defined for any $g \in C_0^\infty(\mathbb{R}^d)$. Taking expectations for both sides of (4.78), we find that $P_{s,t}^{\mathbf{c},\mathbf{d},*}$ is the dual of $P_{s,t}^{\mathbf{c},\mathbf{d}}$. Further, it is written as

$$P_{s,t}^{\mathbf{c},\mathbf{d},*} g(x) = E\big[g(\check{\Psi}_{s,t}(x)) \check{G}^*_{s,t}(x)\big], \tag{4.79}$$

where $\check{G}^*_{s,t}(x) = \check{\mathcal{G}}_{s,t}^{\mathbf{c},\mathbf{d}}(x) \det \nabla \check{\Psi}_{s,t}(x)$.

Theorem 4.6.2 *Assume the same condition as in Proposition* 4.6.1. *Let* $\{P_{s,t}^{\mathbf{c},\mathbf{d}}\}$ *be the simigroup defined by* (4.57), *which is an exponential transformation of the semigroup* $\{P_{s,t}\}$. *Then the dual of* $P_{s,t}^{\mathbf{c},\mathbf{d}}$ *exists and it is given by* (4.79), *where* $\{\check{\Psi}_{s,t}\}$ *is the inverse flow of* $\{\Phi_{s,t}\}$ *and* $\check{G}_{s,t}^{*} = \check{\mathcal{G}}_{s,t}^{\mathbf{c},\mathbf{d}} \det \nabla \check{\Psi}_{s,t}$. *Further, its generator is given by* (4.77).

Remark The backward process $\check{X}_s = \check{\Psi}_{s,t}(x) = \Phi_{s,t}^{-1}(x)$ has the left limit $\check{X}_{s-} = \lim_{\epsilon \downarrow s} \check{X}_{s-\epsilon}$. It is a modification of $\check{X}_s$ and has the backward strong Markov property. See Sect. 1.7.

Remark For the existence of the dual operator $A_J(t)^*$ and the dual semigroup $P_{s,t}^*$, we assumed that jump-maps $\phi_{t,z}$ are diffeomorphic. If jump-maps are not diffeomorphic, it should be difficult to consider dual processes.

4.7 Volume-Preserving Flows

In this section, we will discuss a different topic. We are interested how the volume of a set is changed through the transformation by stochastic flows.

Let B be a Borel subset of $\mathbb{R}^d$. We denote its volume (=Lebesgue measure) by $|B|$. Suppose that a map $\phi : \mathbb{R}^d \to \mathbb{R}^d$ is diffeomorphic. It is called *volume-preserving* if the volume of the set $\phi^{-1}(B) = \{\phi^{-1}(x); x \in B\}$ is equal to $|B|$ for any Borel set B. Note that ϕ is volume-preserving if and only if ϕ^{-1} is volume-preserving. Next, ϕ is called *volume-gaining* (or *volume-losing*) if $|\phi^{-1}(B)| \le |B|$ (or $|\phi^{-1}(B)| \ge |B|$) holds for any Borel set B. Note that ϕ is volume-losing if and only if ϕ^{-1} is volume-gaining.

By the formula of the change of the variable (4.34) we have the equality

$$\int_{\mathbb{R}^d} f(\phi(x))\, dx = \int_{\mathbb{R}^d} f(x) |\det \nabla \phi^{-1}(x)|\, dx.$$

For a bounded Borel subset B of $\mathbb{R}^d$, we set $f = 1_B$. It holds that $1_B \circ \phi = 1_{\phi^{-1}(B)}$. Therefore we have

$$|\phi^{-1}(B)| = \int_B |\det \nabla \phi^{-1}(x)|\, dx, \quad \forall B \text{ (Borel subset of } \mathbb{R}^d). \tag{4.80}$$

Hence ϕ is volume-preserving if and only if $|\det \nabla \phi^{-1}(x)| = 1$ holds for any x. Further, ϕ is volume-gaining if and only if $|\det \nabla \phi^{-1}(x)| \le 1$ holds for any x.

Given a vector field $V(x)$ on $\mathbb{R}^d$ of C_b^∞-class, let $\phi_t(x), t \ge 0, x \in \mathbb{R}^d$ be the deterministic flow such that $\phi_0(x) = x$ and satisfies $\frac{d\phi_t}{dt}(x) = V(\phi_t(x))$ for any $t \in \mathbb{R}$. Then maps $\phi_t, t \ge 0; \mathbb{R}^d \to \mathbb{R}^d$ are diffeomorphic for all t. If maps ϕ_t are volume-preserving for all $t \ge 0$, the flow ϕ_t is called volume-preserving. Volume-losing and volume-gaining flows are defined similarly. It is known that a flow ϕ_t is volume-preserving if and only if the divergence of the associated vector field V is identically 0, and further, the flow ϕ_t is volume-gaining if and only if $\mathrm{div} V(x) \geq 0$ for any x.

We shall study the similar problem for stochastic flows. Let $\{\Phi_{s,t}\}$ be the stochastic flow generated by SDE (4.50) with jumps. It is called volume-preserving or volume-gaining if maps $\Phi_{s,t}; \mathbb{R}^d \to \mathbb{R}^d$ are volume-preserving a.s. for any $s < t$ or maps $\Phi_{s,t}$ are volume-gaining a.s. for any $s < t$.

Theorem 4.7.1 *Assume the same condition as in Proposition* 4.6.1.

1. *The flow* $\{\Phi_{s,t}\}$ *is volume-preserving if and only if either one of the following holds:*
   (a) $\det \nabla \check{\Psi}_{s,t}(x) = 1$ *holds a.s. for any* $s < t$ *and* x, *where* $\check{\Psi}_{s,t} = \Phi_{s,t}^{-1}$.
   (b) $\mathrm{div}\, V_0(x,t) = 0$, $\mathrm{div}\, V_k(x,t) = 0$, $k = 1, \ldots, d'$ *and* $\det \nabla \phi_{t,z}(x) = 1$ *hold for all* x, t, z.
   (c) $A_J(t)^* = \mathcal{A}_J(t)$, *namely, the dual semigroup* $P_{s,t}^*$ *coincides with the semigroup of the inverse flow* $\{\check{\Psi}_{s,t}\}$.
2. *The flow* $\{\Phi_{s,t}\}$ *is volume-gaining if and only if either one of the following holds:*
   (a') *For any* $s < t$ *and* x $\det \nabla \check{\Psi}_{s,t}(x)$ *is less than or equal to* 1 *a.s.*
   (b') $\mathrm{div}\, V_0(x,t) \geq 0$, $\mathrm{div}\, V_k(x,t) = 0$, $k = 1, \ldots, d'$ *and* $\det \nabla \phi_{t,z}(x) \geq 1$ *holds for all* x, t, z.
3. *The flow* $\{\Phi_{s,t}\}$ *is volume-losing if and only if either one of the following holds:*
   (a") *For any* $s < t$ *and* x, $\det \nabla \check{\Psi}_{s,t}(x)$ *is greater than or equal to 1 a.s.*
   (b") $\mathrm{div}\, V_0(x,t) \leq 0$, $\mathrm{div}\, V_k(x,t) = 0$, $k = 1, \ldots, d'$ *and* $\det \nabla \phi_{t,z}(x) \leq 1$ *holds for all* x, t, z.

Proof We give the proof of assertions 1 and 2. Assertion 3 can be verified similarly. Since maps $\Phi_{s,t}; \mathbb{R}^d \to \mathbb{R}^d$ are diffeomorphic, we have (4.80). Therefore the flow $\Phi_{s,t}$ is volume preserving (or gaining) if and only if $|\det \nabla \check{\Psi}_{s,t}(x)| = 1$ a.s. (or ≤ 1 a.s., respectively). Now, $|\det \nabla \check{\Psi}_{s,t}(x)|$ satisfies the following backward equation:

$$
\begin{aligned}
|\det \nabla \check{\Psi}_{s,t}| - 1 = & -\sum_{k=0}^{d'} \int_s^t \mathrm{div}\, V_k(\check{\Psi}_{r,t}, r) |\det \nabla \check{\Psi}_{r,t}| \circ dW_r^k \\
& + \int_s^t \int_{|z|>0+} \{|\det \nabla \phi_{r,z}^{-1}(\check{\Psi}_{r,t})| - 1\} |\det \nabla \check{\Psi}_{r,t}| N(dr\, dz).
\end{aligned}
$$

See Lemma 4.6.1. If $|\det \nabla \check{\Psi}_{s,t}(x)| = 1$ (or ≤ 1) holds, the right-hand side (semimartingale) is identically 0 (or nonpositive, respectively). This implies that the martingale part is 0 and the process of bounded variation part is 0 (or nonpositive, respectively). Note that $\det \nabla \phi_{r,z}(x) = 1/\det \nabla \phi_{r,z}^{-1}(\phi_{r,z}(x))$. Then (a) implies (b) and (a') implies (b'). Conversely if (b) (or (b')) holds, then $|\det \nabla \check{\Psi}_{s,t}(x)| = 1$ (or

≤ 1) a.s. for any s, t, x. Therefore we have proved $(b) \to (a)$ (or $(b') \to (b')$). Next, since $A_J(t)^* g = \mathcal{A}_J^{-\mathbf{div}\,\mathbf{V},\mathbf{det}\nabla\phi^{-1}}(t)g$, the equivalence of (b) and (c) is obvious. □

Corollary 4.7.1 *If the flow is volume-preserving, we have* $P^*_{s,t} = \check{\mathcal{P}}_{s,t}$ *for any* $s < t$. *Further, consider time homogeneous continuous flow. If the flow is volume-preserving and* $V_0 = 0$, *then* $P_{s,t}$ *is self adjoint, i.e.,* $P^*_{s,t} = P_{s,t}$ *holds for any* $s < t$.

We are interested in the case where the dual semigroup $\{P^*_{s,t}\}$ is conservative or Markovian. The dual semigroup is conservative if and only if the Lebesgue measure dx is an invariant measure of the transition probabilities, i.e., $\int_{\mathbb{R}^d} P_{s,t}(x, E)\,dx = |E|$ holds for any $s < t$. The dual semigroup is Markovian if and only if dx is an excessive measure of the transition probabilities, i.e., $\int_{\mathbb{R}^d} P_{s,t}(x, E)\,dx \leq |E|$ holds for any $s < t$. We will give their geometric characterization.

The stochastic flow $\{\Phi_{s,t}\}$ is said to be *volume-preserving in the mean* or *volume-gaining in the mean*, if $E[|\Phi_{s,t}^{-1}(B)|] = |B|$ or $E[|\Phi_{s,t}^{-1}(B)|] \leq |B|$ holds for any $t > s$ and Borel sets B, respectively.

Theorem 4.7.2 *Assume the same condition as in Proposition* 4.6.1.

1. *The following assertions are equivalent:*

   (a) *The flow* $\{\Phi_{s,t}\}$ *is volume-preserving in the mean.*
   (b) $E[|\det \nabla\check{\Psi}_{s,t}(x)|] = 1$ *for any* $s < t$ *and* x.
   (c) *Dual semigroup* $\{P^*_{s,t}\}$ *is conservative, i.e.,* $P^*_{s,t}1(x) = 1$ *holds for any* $x, s < t$.
   (d) *The function* $c^*(t)$ *defined by* (4.73) *satisfies* $c^*(x,t) = 0$ *for any* x, t.

2. *The following assertions are equivalent:*

   (a') *The flow* $\{\Phi_{s,t}\}$ *is volume-gaining in the mean.*
   (b') $E[|\det \nabla\check{\Psi}_{s,t}(x)|] \leq 1$ *for any* $s < t$ *and* x.
   (c') *Dual semigroup* $\{P^*_{s,t}\}$ *is Markovian, i.e.,* $P^*_{s,t}1(x) \leq 1$ *holds for any* $x, s < t$.
   (d') *The function* $c^*(t)$ *defined by* (4.73) *satisfies* $c^*(x,t) \leq 0$ *for any* x, t.

Proof We have from (4.80) the equality $|\check{\Psi}_{s,t}(B)| = \int_B |\det \nabla\check{\Psi}_{s,t}(y)|\,dy$. Taking expectations for both sides, we have

$$E[|\check{\Psi}_{s,t}(B)|] = \int_B E[|\det \nabla\check{\Psi}_{s,t}(x)|]\,dx, \quad \forall B. \tag{4.81}$$

Therefore, the flow is volume-preserving in the mean (or volume-gaining in the mean), if and only if $E[|\det \nabla\check{\Psi}_{s,t}(x)|] = 1$ (or ≤ 1, respectively) holds for any $s < t$ and x. The equivalence of assertions (b), (c) and (d) (or (b'), (c') and (d'), respectively) will be obvious. □

Finally, we will consider the case where the dual semigroup $\{P^*_{s,t}\}$ is nonconservative. Then the flow is not volume-preserving. We will study the speed of the volume-gaining (or volume-losing). It should be related to the potential function $c^*(x,t)$ of the dual semigroup defined by (4.73).

Theorem 4.7.3 *For any Borel subset B of $\mathbb{R}^d$ with finite volume, we have*

$$\exists \lim_{t\to s} \frac{1}{t-s}(E[|\check{\Psi}_{s,t}(B)|] - |B|) = \int_B c^*(x,s)\,dx. \tag{4.82}$$

Proof We have from (4.81)

$$\begin{aligned}\frac{E[|\check{\Psi}_{s,t}(B)|] - |B|}{t-s} &= \int_B \frac{E[|\det \nabla\check{\Psi}_{s,t}(x)|] - 1}{t-s}\,dx = \int_B \frac{P^*_{s,t}1(x) - 1}{t-s}\,dx \\ &= \int_B \frac{\int_s^t P^*_{r,t}A(r)^*1(x)dr}{t-s}\,dx \to \int_B c^*(x,s)\,dx,\end{aligned}$$

as $t \to s$. □

For a given $s \in \mathbb{T}$, define a subset of $\mathbb{R}^d$ by $C_+(s) = \{x \in \mathbb{R}^d; c^*(x,s) > 0\}$, $C_-(s) = \{x \in \mathbb{R}^d; c^*(x,s) < 0\}$ and $C_0(s) = \{x \in \mathbb{R}^d; c^*(x,s) = 0\}$. Then the above theorem tells us that if $t - s$ is small, $\check{\Psi}_{s,t}(B)$ is volume-gaining (or volume-losing) in the mean if $B \subset C_-(s)$ (or $B \subset C_+(s)$). Further, it is volume-preserving in the mean if $B \subset C_0(s)$.

Note We discussed the volume-preserving problem with respect to the Lebesgue measure. It may be interesting to consider the problem with respect to other measures such as the Gaussian measure. We refer to Kunita–Oh [65].

4.8 Jump-Diffusion on Subdomain of Euclidean Space

Let $\{\Phi_{s,t}(x)\}$ be the right continuous stochastic flow defined by SDE (4.50) and let $G^{\mathbf{c},\mathbf{d}}_{s,t}(x)$ be an exponential functional defined by (4.54). Let D be a bounded open subset of $\mathbb{R}^d$. Let $\tau(x,s)$ be the first leaving time of the process $X_t = X_t^{x,s} = \Phi_{s,t}(x), t \in [s,\infty)$ from the set D. We shall consider the *killed process* at time $\tau(x,s)$. It is defined by $X_t'^{x,s} = \Phi_{s,t}(x)$ if $t < \tau(x,s)$ and by $X_t'^{x,s} = \infty$ if $t \geq \tau(x,s)$. We define the weighted law of the killed process by

$$\begin{aligned}Q^{\mathbf{c},\mathbf{d}}_{s,t}(x,E) &= E\big[1_E(X_t'^{x,s})G^{\mathbf{c},\mathbf{d}}_{s,t}(x)\big] \\ &= E\Big[1_E(\Phi_{s,t}(x))G^{\mathbf{c},\mathbf{d}}_{s,t}(x)\prod 1_D(\Phi_{s,r}(x))\Big],\end{aligned} \tag{4.83}$$

where the product $\prod$ is taken for all rationals r such that $s < r < t$. We set $Q^{\mathbf{c},\mathbf{d}}_{s,t} f(x) = \int_D Q^{\mathbf{c},\mathbf{d}}_{s,t}(x,dy)f(y)$. Then it is written as

$$Q^{\mathbf{c},\mathbf{d}}_{s,t} f(x) = E\Big[f(\Phi_{s,t}(x))G^{\mathbf{c},\mathbf{d}}_{s,t}(x)\prod 1_D(\Phi_{s,r}(x))\Big]. \tag{4.84}$$

Proposition 4.8.1 *$\{Q^{\mathbf{c},\mathbf{d}}_{s,t}\}$ has the semigroup property: $Q^{\mathbf{c},\mathbf{d}}_{s,t}Q^{\mathbf{c},\mathbf{d}}_{t,u} = Q^{\mathbf{c},\mathbf{d}}_{s,u}$ holds for any $s < t < u$. Further, for any $f \in C_0^\infty(D)$, $Q^{\mathbf{c},\mathbf{d}}_{s,t} f$ is differentiable with respect to t and satisfies*

$$\frac{\partial}{\partial t} Q^{\mathbf{c},\mathbf{d}}_{s,t} f(x) = Q^{\mathbf{c},\mathbf{d}}_{s,t}(A^{\mathbf{c},\mathbf{d}}_J(t)f)(x) \tag{4.85}$$

for any $s < t$ and $x \in D$.

Proof For any $s < t < u$, we have

$$\{\tau(x,s) > u\} = \{\tau(x,s) > t\} \cap \{\tau(X_t, t) > u\}.$$

Then $Q^{\mathbf{c},\mathbf{d}}_{s,u} f(x)$ is equal to

$$\begin{aligned} &E[f(\Phi_{s,u}(x))G^{\mathbf{c},\mathbf{d}}_{s,u}(x)1_{\tau(x,s)>u}] \\ &=E\Big[E\Big[f(\Phi_{t,u}(\Phi_{s,t}(x)))G^{\mathbf{c},\mathbf{d}}_{t,u}(\Phi_{s,t}(x))G^{\mathbf{c},\mathbf{d}}_{s,t}(x)1_{\tau(\Phi_{s,t}(x),t)>u}\Big|\mathcal{F}_{s,t}\Big]1_{\tau(x,s)>t}\Big] \\ &= E\big[Q^{\mathbf{c},\mathbf{d}}_{t,u} f(\Phi_{s,t}(x))G^{\mathbf{c},\mathbf{d}}_{s,t}(x)1_{\tau(x,s)>t}\big] = Q^{\mathbf{c},\mathbf{d}}_{s,t}Q^{\mathbf{c},\mathbf{d}}_{t,u} f(x). \end{aligned}$$

Therefore $Q^{\mathbf{c},\mathbf{d}}_{s,t}$ has the semigroup property. We have by Itô's formula

$$f(\Phi_{s,t}(x))G^{\mathbf{c},\mathbf{d}}_{s,t}(x) = f(x) + \int_s^t A^{\mathbf{c},\mathbf{d}}_J(r)f(\Phi_{s,r}(x))G^{\mathbf{c},\mathbf{d}}_{s,r}(x)\,dr + M_t,$$

where M_t is a martingale with mean 0. Consider the stopped process of each term of the above by the stopping time $\tau = \tau(x,s)$. Then we have

$$\begin{aligned} f(\Phi_{s,t\wedge\tau}(x))G^{\mathbf{c},\mathbf{d}}_{s,t\wedge\tau}(x) &= f(x) + \int_s^{t\wedge\tau} A^{\mathbf{c},\mathbf{d}}_J(r)f(\Phi_{s,r}(x))G^{\mathbf{c},\mathbf{d}}_{s,r}(x)\,dr + M_{t\wedge\tau} \\ &=f(x)+\int_s^t A^{\mathbf{c},\mathbf{d}}_J(r)f(\Phi_{s,r}(x))G^{\mathbf{c},\mathbf{d}}_{s,r}(x)1_{\tau>r}\,dr + M_{t\wedge\tau}. \end{aligned}$$

It holds that $f(\Phi_{s,t\wedge\tau}) = f(\Phi_{s,t})$ if $t < \tau$ and $f(\Phi_{s,t\wedge\tau}) = 0$ if $t \geq \tau$. Therefore, taking the expectation of each term of the above equality, we get

$$E[f(\Phi_{s,t}(x))G^{\mathbf{c},\mathbf{d}}_{s,t}(x)1_{\tau>t}] = f(x) + E\Big[\int_s^t A^{\mathbf{c},\mathbf{d}}_J(r)f(\Phi_{s,r}(x))G^{\mathbf{c},\mathbf{d}}_{s,r}(x)1_{\tau>r}\,dr\Big],$$

since $M_{t\wedge\tau}$ is a martingale with mean 0 by Doob's optional sampling theorem (Theorem 1.5.1). Therefore we get the equality

$$Q_{s,t}^{\mathbf{c},\mathbf{d}} f(x) = f(x) + \int_s^t Q_{s,r}^{\mathbf{c},\mathbf{d}} (A_J^{\mathbf{c},\mathbf{d}}(r) f)(x)\, dr.$$

This proves (4.85). □

In the case $\mathbf{c} = 0$ and $\mathbf{d} = 1$, we denote $Q_{s,t}^{\mathbf{c},\mathbf{d}}(x, \cdot)$ by $Q_{s,t}(x, \cdot)$. Then the killed process $X_t^{\prime x,s}$ is a strong Markov process with transition function $Q_{s,t}(x, \cdot)$.

We are interested in the dual of the above semigroup $Q_{s,t}^{\mathbf{c},\mathbf{d}} f$ with respect to dx. We define

$$Q_{s,t}^{\mathbf{c},\mathbf{d},*} g(x) = E\Big[g(\check{\Psi}_{s,t}(x)) \check{\mathcal{G}}_{s,t}^{\mathbf{c},\mathbf{d}}(x) \prod 1_D(\check{\Psi}_{r,t}(x)) |\det \nabla \check{\Psi}_{s,t}(x)|\Big], \tag{4.86}$$

where $\check{\Psi}_{r,t}(x) = \Phi_{r,t}^{-1}(x)$ and $\det \nabla \check{\Psi}_{s,t}(x)$ is its Jacobian determinant. (We set $d_{r,z} = 1$ if X_t is a diffusion.) It satisfies the backward semigroup property $Q_{t,u}^{\mathbf{c},\mathbf{d},*} Q_{s,t}^{\mathbf{c},\mathbf{d},*} g = Q_{s,u}^{\mathbf{c},\mathbf{d},*} g$ for any $s < t < u$.

Proposition 4.8.2 *$Q_{s,t}^{\mathbf{c},\mathbf{d},*}$ is the dual of $Q_{s,t}^{\mathbf{c},\mathbf{d}}$; It holds that*

$$\int_D Q_{s,t}^{\mathbf{c},\mathbf{d}} f(x) \cdot g(x)\, dx = \int_D f(x) \cdot Q_{s,t}^{\mathbf{c},\mathbf{d},*} g(x)\, dx \tag{4.87}$$

for any continuous function f, g supported by compact subsets of D. Further, it satisfies

$$Q_{s,t}^{\mathbf{c},\mathbf{d},*} g(x) = g(x) + \int_s^t Q_{r,t}^{\mathbf{c},\mathbf{d},*} (A_J^{\mathbf{c},\mathbf{d}}(r)^* g)(x)\, dr, \tag{4.88}$$

where $A_J^{\mathbf{c},\mathbf{d}}(r)^*$ is the dual operator of $A^{\mathbf{c},\mathbf{d}}(r)$.

Proof Since $\Phi_{s,r} \circ \Phi_{s,t}^{-1} = \check{\Psi}_{r,t}$ holds for $0 < r < t$, we have

$$\begin{aligned}
&\int_D f(\Phi_{s,t}(x)) g(x) G_{s,t}^{\mathbf{c},\mathbf{d}}(x) \Big(\prod 1_D(\Phi_{s,r}(x))\Big)\, dx \\
&= \int_D f(x) g(\check{\Psi}_{s,t}(x)) \check{\mathcal{G}}_{s,t}^{\mathbf{c},\mathbf{d}}(x) \Big(\prod 1_D(\check{\Psi}_{r,t}(x))\Big) |\det \nabla \check{\Psi}_{s,t}(x)|\, dx.
\end{aligned}$$

Taking the expectation for each term of the above, we get the formula (4.87). Equation (4.88) is proved similarly to the proof of (4.85). □

The semigroup $\{Q_{s,t}^{\mathbf{c},\mathbf{d}}\}$ satisfies $\frac{\partial}{\partial t} Q_{s,t}^{\mathbf{c},\mathbf{d}} f(x) = Q_{s,t}^{\mathbf{c},\mathbf{d}} A_J^{\mathbf{c},\mathbf{d}}(t) f$ (Kolmogorov's forward equation) for any C^∞-function f satisfying $f(x) = 0$ on D^c, in view of (4.85). However, Kolmogorov's backward equation for $\{Q_{s,t}^{\mathbf{c},\mathbf{d}}\}$ is not evident, since it is not clear whether $Q_{s,t}^{\mathbf{c},\mathbf{d}} f(x)$ is a smooth function of x. In Sect. 6.10, we show that it is true if the jump-diffusion is 'pseudo-elliptic', making use of the Malliavin calculus.

Chapter 5
Malliavin Calculus

Abstract We will discuss the Malliavin calculus on the Wiener space and on the space of Poisson random measures, called the Poisson space. There are extensive works on the Malliavin calculus on the Wiener space. Also, there are some interesting works on the Malliavin calculus on the Poisson space. These two types of calculus have often been discussed separately, since the Wiener space and that of the Poisson space are quite different. Even so, we are interested in the Malliavin calculus on the product of these two spaces; we want to develop these two theories in a compatible way.

In Sects. 5.1, 5.2, and 5.3, we study the Malliavin calculus on the Wiener space. We define 'H-derivative' operator D_t and its adjoint δ (Skorohod integral by Wiener process) on the Wiener space. Then, after introducing Sobolev norms for Wiener functionals, we get a useful estimate of the adjoint operator with respect to Sobolev norms (Theorem 5.2.1). In Sect. 5.3, we give Malliavin's criterion for which the law of a given Wiener functional has a smooth density. It will be stated in terms of the Malliavin covariance.

In Sects. 5.4, 5.5, 5.6, and 5.7, we study the Malliavin calculus on the space of Poisson random measure. Difference operator $\tilde{D}_u$ and its adjoint $\tilde{\delta}$ (Skorohod integral by Poisson random measure) are defined for Poisson functionals following Picard [92]. We will define a family of Sobolev norms conditioned to a family of star-shaped neighborhoods $\{A(\rho), 0 < \rho < 1\}$. Then a criterion for the smooth density of the law of a Poisson functional will be given using this family of Sobolev norms.

In Sects. 5.8, 5.9, and 5.10, we study the Malliavin calculus on the product of the Wiener space and Poisson space. Sobolev norms for Wiener–Poisson functionals are studied in Sect. 5.9. In Sect. 5.10, we study criteria for the smooth density. Finally in Sect. 5.11, we discuss the composition of a 'nondegenerate' Wiener–Poisson functional and Schwartz's distribution.

Results of this chapter will be applied in the next chapter for getting the fundamental solution of heat equations discussed in Chap. 4.

H. Kunita, *Stochastic Flows and Jump-Diffusions*, Probability Theory and Stochastic Modelling 92, https://doi.org/10.1007/978-981-13-3801-4_5

5.1 Derivative and Its Adjoint on Wiener Space

The Malliavin calculus on the Wiener space has been studied extensively since the 1980s. It is not our purpose to present the theory in full. We will restrict our attention to the problem of finding smooth density of the law of a given random variable.

Let us define a Wiener space. In this chapter, it is convenient to take a finite interval $\mathbb{T} = [0, T]$ for the time parameter. Let d' be a positive integer. Let $\mathbf{W} = \mathbf{W}_0^{d'}$ be the set of all continuous maps $w; \mathbb{T} \to \mathbb{R}^{d'}$ such that $w(0) = 0$. We denote by $\mathcal{B}(\mathbf{W})$ the smallest σ-field with respect to which $w(t)$ are measurable for any $t \in \mathbb{T}$. We denote $w(t)$ by $W_t(w) = (W_t^1(w), \ldots, W_t^{d'}(w))$. A probability measure P on $(\mathbf{W}, \mathcal{B}(\mathbf{W}))$ is called a *Wiener measure* if W_t is a d'-dimensional Wiener process. The triple $(\mathbf{W}, \mathcal{B}(\mathbf{W}), P)$ is called a (classical) *Wiener space*. Let S be a complete metric space. An S-valued $\mathcal{B}(\mathbf{W})$-measurable function $F(w)$ (random variable) is called an S-valued *Wiener functional*. If S is the space of real numbers, it is called simply a *Wiener functional*.

In our discussion of the Malliavin calculus, we will often study L^p-estimates of Wiener functionals for large p instead of L^2-estimates. Let $p > 1$. We denote by L^p, the set of all real Wiener functionals X such that $E[|X|^p] < \infty$. The norm of X is defined by $\|X\|_{L^p} = E[|X|^p]^{\frac{1}{p}}$. We set $L^{\infty-} = \bigcap_{p>1} L^p$. It is a vector space of Wiener functionals, and is an algebra, i.e., if $X, Y \in L^{\infty-}$, then $XY \in L^{\infty-}$. Further, the set $L^{\infty-}$ is an F-space with respect to a countable family of norms $\| \|_n = \| \|_{L_n}$, $n \in \mathbb{N}$, where $\mathbb{N}$ is the set of all positive integers. Here a vector space L is called an *F-space* (a version of Frechet space) if it is equipped with a countable family of norms $\| \|_n$, $n \in \mathbb{N}$ and is a complete metric space by the metric

$$d(X, Y) = \sum_{n=1}^{\infty} \frac{1}{2^n} \frac{\|X - Y\|_n}{1 + \|X - Y\|_n}.$$

Hence a sequence $\{X_m\}$ of $L^{\infty-}$ converges to $X \in L^{\infty-}$ with respect to the metric d if and only if $\|X_m - X\|_n$ converges to 0 for any positive integer n. Note that if the family of norms are ordered in a different way, $\| \|_{n'}$, $n' \in \mathbb{N}$, the corresponding metric d' defines the equivalent topology for $L^{\infty-}$.

Let H be the totality of d'-dimensional real measurable functions $h(t) = (h^1(t), \ldots, h^{d'}(t))$ on $\mathbb{T}$ such that their L^2-norms $|h|_H := (\int_{\mathbb{T}} |h(t)|^2 \, dt)^{1/2}$ are finite. Then H is a real Hilbert space with the inner product

$$\langle f, g \rangle = \int_{\mathbb{T}} \Big(\sum_{k=1}^{d'} f^k(t) g^k(t) \Big) \, dt.$$

The space H is called a *Cameron–Martin space*.

Given $h \in H$, we want to define a linear transformation T_h of a Wiener functional F of $L^{\infty-}$ by

$$T_h F(w) = F(w + \bar{h}),$$

where $\bar{h}(t) := \int_0^t h(s)\,ds, t \in \mathbb{T}$ is an element of $\mathbf{W}$. In order that it is well defined, the transformation should be absolutely continuous, i.e., if $F(w) = 0$ a.s., then $F(w + \bar{h}) = 0$ holds a.s. The fact can be verified by Girsanov's theorem (Theorem 2.2.2), to be shown below. We will show that T_h is a continuous linear transformation from $L^{\infty-}$ into itself. For $h \in H$, we set $W(h) = \sum_{k=1}^{d'} \int_{\mathbb{T}} h^k(t)\,dW_t^k$. Then $\{W(h), h \in H\}$ is a Gaussian system with mean 0 and covariance $E[W(h)W(k)] = \langle h, k\rangle$. Now, set for $t \in \mathbb{T}$,

$$Z_h(t) = \exp\left\{\sum_{k=1}^{d'} \int_0^t h^k(s)\,dW_s^k - \frac{1}{2}\int_0^t |h(s)|^2\,ds\right\}.$$

It is a positive L^p-martingale with mean 1 for any $p \geq 2$ (Proposition 2.2.3). Then

$$Z_h = Z_h(T) = \exp\left\{W(h) - \frac{1}{2}\langle h, h\rangle\right\} \tag{5.1}$$

is an element of $L^{\infty-}$. Define another probability measure Q_h by $dQ_h = Z_h\,dP$. Then by Girsanov's theorem (Theorem 2.2.2), $w(t) - \bar{h}(t)$ is a Wiener process with respect to Q_h. Therefore, the law of $F(w + \bar{h})$ with respect to P coincides with the law of F with respect to Q_h. Then we have

$$E[|T_h F|^p] = E_{Q_h}[|F|^p] = E[|F|^p Z_h] \leq E[F|^{pr}]^{\frac{1}{r}} E[Z_h^{r'}]^{\frac{1}{r'}},$$

where r' is the conjugate of r. Consequently, we get $\|T_h F\|_{L^p} \leq \|Z_h\|_{L^{r'}}^{\frac{1}{p}} \|F\|_{L^{pr}}$, showing that T_h is a continuous linear transformation from L^{pr} to L^p. In particular, if $F = 0$ a.s., then $T_h F = 0$ a.s., so that T_h is an absolutely continuous transformation. Since this is valid for any $r > 1$ and $p > 1$, T_h is a linear transformation from $L^{\infty-}$ into itself and further, the transformation is continuous with respect to the metric d. Hence T_h is a continuous linear transformation on the F-space $L^{\infty-}$.

Let $\lambda \in \mathbb{R}$. For a given $h \in H$, we consider a family of linear transformations $\{T_{\lambda h}, \lambda \in \mathbb{R}\}$ on $L^{\infty-}$. These satisfy the group property $T_{\lambda h} T_{\lambda' h} = T_{(\lambda+\lambda')h}$ for any $\lambda, \lambda' \in \mathbb{R}$. We will define the derivative operator as the infinitesimal generator of the one-parameter group of linear transformations $\{T_{\lambda h}\}$. Let $F \in L^{\infty-}$. Suppose that there exists an H-valued Wiener functional F' with $|F'|_H \in L^{\infty-}$ satisfying

$$\lim_{\lambda \to 0} d\left(\frac{T_{\lambda h} F - F}{\lambda}, \langle F', h\rangle\right) = 0, \tag{5.2}$$

for any $h \in H^{(0)}$, which is a dense subset of H. Then F is said to be *H-differentiable* and F' is called the *H- derivative*. H-derivative of a given F is at most unique. In the following, we denote it as DF. We denote by $\mathcal{D}$ the set of all H-differentiable Wiener functionals F of $L^{\infty-}$.

The H-valued functional DF is often written as $D_t F$ or $(D_t^{(1)} F, \ldots, D_t^{(d')} F)$, $t \in \mathbb{T}$. It may be regarded as an $\mathbb{R}^{d'}$-valued stochastic process. However, we should note that $D_t F$ may not be well defined for any t a.s. P, but it is well defined a.e. t, w with respect to the product measure $dt\, dP$.

We will show differential rules for composite functionals.

Proposition 5.1.1 *Let $F_1, \ldots, F_d \in \mathcal{D}$. Let $f(x_1, \ldots, x_d)$ be a C^1-function on $\mathbb{R}^d$ such that f and $\partial_{x_j} f := \frac{\partial}{\partial x_j} f$, $j = 1, \ldots, d$ are of polynomial growth. Then $f(F_1, \ldots, F_d)$ belongs to $\mathcal{D}$. Further, we have*

$$Df(F_1, \ldots, F_d) = \sum_{j=1}^{d} \partial_{x_j} f(F_1, \ldots, F_d) \cdot DF_j, \quad a.s. \tag{5.3}$$

Proof We have by the mean value theorem,

$$\frac{1}{\lambda}\Big\{ f(T_{\lambda h}F_1, \ldots, T_{\lambda h}F_d) - f(F_1, \ldots, F_d)\Big\}$$
$$= \sum_j \partial_{x_j} f(F_1 + \theta(T_{\lambda h}F_1 - F_1), \ldots, F_d + \theta(T_{\lambda h}F_d - F_d)) \times \frac{T_{\lambda h}F_j - F_j}{\lambda},$$

a.s., where $|\theta| \le 1$. Let λ tend to 0. Since F_i are H-differentiable, $T_{\lambda h}F_j$ converge to F_j and $\frac{T_{\lambda h}F_j - F_j}{\lambda}$ converge to $\langle DF_j, h\rangle$ in $L^{\infty-}$. Therefore, the above converges to $\sum_j \partial_{x_j} f(F_1, \ldots, F_d) \cdot \langle DF_j, h\rangle$ in $L^{\infty-}$. Therefore $f(F_1, \ldots, F_d)$ is H-differentiable and the equality (5.3) holds. □

For a positive integer $i \ge 2$, we denote by $H^{\otimes i}$ the i-times tensor product of H. It is again a Hilbert space with norm $|\ |_{H^{\otimes i}}$ defined by $|f|_{H^{\otimes i}} = (\int_{\mathbb{T}^i} |f(\mathbf{t})|^2 d\mathbf{t})^{\frac{1}{2}}$, where $\mathbb{T}^i$ is the i-times product of the set $\mathbb{T}$ and $\mathbf{t} = (t_1, \ldots, t_i) \in \mathbb{T}^i$. Suppose that F is H-differentiable and that the functional $\langle DF, h_1\rangle$ is H-differentiable for any $h_1 \in H$. If there is an $H \otimes H$-valued functional $D^2 F$ such that $|D^2 F|_{H\otimes H} \in L^{\infty-}$ and

$$\langle D^2 F, h_1 \otimes h_2\rangle = \langle D\langle DF, h_1\rangle, h_2\rangle, \quad \forall h_1, h_2 \in H,$$

then we say that DF is *H-differentiable* and $D^2 F$ is *the second H-derivative*.

For $i \ge 3$, we can define the *i-th H-derivative* $D^i F$, successively. Suppose that the $(i-1)$-th H-derivative $D^{i-1} F$ is well defined as an $H^{\otimes(i-1)}$-valued Wiener functional such that $|D^{i-1} F|_{H^{\otimes(i-1)}} \in L^{\infty-}$ and $\langle D^{i-1} F, h_1 \otimes \cdots \otimes h_{i-1}\rangle$ is H-

differentiable. If there is an $H^{\otimes i}$-valued functional $D^i F$ such that $|D^i F|_{H^{\otimes i}} \in L^{\infty-}$ and

$$\langle D^i F, h_1 \otimes \cdots \otimes h_{i-1} \otimes h_i \rangle = \langle D \langle D^{i-1} F, h_1 \otimes \cdots \otimes h_{i-1} \rangle, h_i \rangle,$$

for all $h_1, \ldots, h_i \in H$, then $D^i F$ is called the i-th H-derivative of F. Note that for any i, the i-th H-derivative of a given F is at most unique. $D^i F$ is an element of $H^{\otimes i}$. It is written as $D_{t_1} \cdots D_{t_i} F$, $D_{\mathbf{t}} F$ or $D^i_{\mathbf{t}} F$, where $\mathbf{t} = (t_1, \ldots, t_i) \in \mathbb{T}^i$. It is well defined a.e. with respect to the product measure $d\mathbf{t}\, dP$. We may regard it as a $(d')^i$-dimensional random field with parameter $\mathbf{t} = (t_1, \ldots, t_i) \in \mathbb{T}^i$.

The totality of Wiener functionals F in $L^{\infty-}$, such that i-th H-derivative $D^i F$ exists for any $i \leq l$, is denoted by $\mathcal{D}^l$ and we set $\mathcal{D}^\infty = \bigcap_l \mathcal{D}^l$. Elements of $\mathcal{D}^l$ and those of $\mathcal{D}^\infty$ are called l-times H-differentiable and infinitely H-differentiable, respectively.

We will give examples of H-differentiable Wiener functionals. Let $f \in H$ and let $F = W(f) = \sum_{k=1}^{d'} \int_{\mathbb{T}} f^k(t)\, dW_t^k$. It is a linear Wiener functional. We have $T_{\lambda h} F = W(f) + \lambda \langle f, h \rangle$ a.s. Therefore F is H-differentiable and $\langle DF, h \rangle = \langle f, h \rangle$ holds a.s. This proves $D_t F = f(t)$ a.e. $dt\, dP$. Since it is a constant function, its higher-order derivatives are 0, i.e., $D^i_{\mathbf{t}} F = 0$ holds if $i \geq 2$.

Let $f_1, \ldots, f_d \in H$ and let $p; \mathbb{R}^d \to \mathbb{R}$ be a polynomial. A functional F given by $p(W(f_1), \ldots, W(f_d))$ is called a *polynomial Wiener functional* or simply a *polynomial.* We denote by $\mathcal{P}$ the set of all polynomials. Note that any $F \in \mathcal{P}$ has a finite moment of any order, i.e., $E[|F|^p] < \infty$ for any $p > 1$. Therefore, $\mathcal{P}$ is a dense subset of $L^{\infty-}$.

We will show that the domain $\mathcal{D}^\infty$ contains the set of polynomials $\mathcal{P}$ and therefore it is a dense subset of $L^{\infty-}$. If F is a polynomial of the form $F = p(W(f_1), \ldots, W(f_d))$, then as in Proposition 5.1.1, F is in $\mathcal{D}^\infty$ and satisfies

$$D_t F = \sum_j \partial_{x_j} p(W(f_1), \ldots, W(f_d)) f_j(t), \quad a.e.\ dt\, dP. \tag{5.4}$$

Further, the polynomial F is i-times H-differentiable for any i and we have

$$D_{t_1} \cdots D_{t_i} F = \sum_{j_1, \ldots, j_i} \partial_{x_{j_1}} \cdots \partial_{x_{j_i}} p(W(f_1), \ldots, W(f_d)) f_{j_1}(t_1) \cdots f_{j_i}(t_i),$$

a.e. $d\mathbf{t} dP$. Hence $\mathcal{P} \subset \mathcal{D}^\infty$. If F is a polynomial of degree d, then $D^d_{\mathbf{t}} F$ is a constant function and $D^{d+1}_{\mathbf{t}} F = 0$.

Another interesting nonlinear Wiener functional, which is H-differentiable, is a solution of a continuous SDE. In Sect. 6.1, we will show the fact and will derive its derivative explicitly.

Concerning the differential operator D, we have a formula of the integration by parts.

Proposition 5.1.2 *If $F \in \mathcal{D}^\infty$, it holds that*

$$E[\langle DF, h\rangle] = E[FW(h)] \tag{5.5}$$

for any $h \in H$. Further, if $F, G \in \mathcal{D}^\infty$, it holds that

$$E[F\langle DG, h\rangle] = E[-G\langle DF, h\rangle + GFW(h)] \tag{5.6}$$

for any $h \in H$.

Proof Let $\lambda \in \mathbb{R}$ and let $Z_{\lambda h}$ be the functional defined by (5.1) for $\lambda h \in H$. Then we have $E[T_{\lambda h}F] = E[FZ_{\lambda h}]$. Differentiate both sides with respect to λ at $\lambda = 0$. Since $\frac{d}{d\lambda}E[T_{\lambda h}F]|_{\lambda=0} = E[\langle DF, h\rangle]$ and

$$d\Big(\frac{Z_{\lambda h} - 1}{\lambda}, W(h)\Big) \to 0$$

as $\lambda \to 0$, we get (5.5).

Next, if F, G are in $\mathcal{D}^\infty$, then the product FG is in $\mathcal{D}^\infty$ and satisfies $D(FG) = DF \cdot G + F \cdot DG$. Hence we get the equality (5.6). □

We will define the adjoint operator of the derivative operator D. Let $F_l, l = 1, \ldots, n$ be elements of $\mathcal{D}^\infty$ and $h_l, l = 1, \ldots, n$ be elements of H. Then $\sum_l F_l h_l$ is an H-valued random variable. We may regard $\sum F_l h_l(t)$ as a d'-dimensional stochastic process with parameter $t \in \mathbb{T}$, a.e. $dt\, dP$. It is called a *simple functional with parameter $t \in \mathbb{T}$*. We denote by $\mathcal{S}_{\mathbb{T}}^\infty$ the set of all simple functionals with parameter $\mathbb{T}$. For $X_t = \sum_{l=1}^n F_l h_l(t) \in \mathcal{S}_{\mathbb{T}}^\infty$, we set

$$\delta(X) = \sum_{l=1}^n F_l W(h_l) - \sum_{l=1}^n \langle DF_l, h_l\rangle. \tag{5.7}$$

Then δ is a linear map from $\mathcal{S}_{\mathbb{T}}^\infty$ to $L^{\infty-}$. It is called the *adjoint operator* of D, or the *Skorohod integral of X_t by the Wiener process*. The followings are known as properties of Skorohod integrals. See Nualart [88].

Lemma 5.1.1

1. *For any $X = \{X_t\} \in \mathcal{S}_{\mathbb{T}}^\infty$, $\delta(X)$ satisfies the adjoint property*

$$E[G\delta(X)] = E[\langle DG, X\rangle] \tag{5.8}$$

   for any $G \in \mathcal{D}^\infty$.
2. *For any $X = \{X_t\} \in \mathcal{S}_{\mathbb{T}}^\infty$, $\delta(X)$ is H-differentiable and satisfies the following commutation relation:*

$$D_{t'}\delta(X) = \delta(D_{t'}X) + X_{t'}, \quad a.e.\ dt'\, dP. \tag{5.9}$$

3. *The isometric property:* $E[\delta(X)\delta(Y)]$ *is equal to*

$$\sum_{k=1}^{d'} E\Big[\int_{\mathbb{T}} X_t^k Y_t^k \, dt\Big] + \sum_{k,k'=1}^{d'} E\Big[\int_{\mathbb{T}}\int_{\mathbb{T}} D_t^{(k')} X_s^k D_s^{(k')} Y_t^k \, ds\, dt\Big] \tag{5.10}$$

for any $X, Y \in \mathcal{S}_{\mathbb{T}}^{\infty}$.

Equality (5.10) is called the Shigekawa–Nualart–Pardoux energy identity (see [89, 101]).

Proof Since $G\delta(X) = \sum_l GF_l W(h_l) - \sum_l G\langle DF_l, h_l\rangle$, we have

$$E[G\delta(X)] = E\Big[\sum_l GF_l W(h_l) - \sum_l G\langle DF_l, h_l\rangle\Big].$$

Apply Proposition 5.1.2 to the above. Then the right-hand side is written as

$$\sum_l E[\langle DG, h_l\rangle F_l] = E[\langle DG, X\rangle].$$

Therefore we get (5.8).

Further, $\delta(X)$ is H-differentiable, since each term of the right-hand side of (5.7) is H-differentiable. We have indeed,

$$\begin{aligned} D_t\delta(X) &= \sum_l D_t F_l W(h_l) + \sum_l F_l h_l(t) - \sum_l D_t(\langle DF_l, h_l\rangle) \\ &= \delta(D_t X) + X_t, \quad a.e.\ dt\, dP. \end{aligned}$$

Therefore the equality of the commutation relation (5.9) holds.

Suppose $X, Y \in \mathcal{S}_{\mathbb{T}}^{\infty}$. Then we have

$$\begin{aligned} E[\delta(X)\delta(Y)] &= E[\langle X, D\delta(Y)\rangle] \\ &= E[\langle X, Y\rangle] + E[\langle X, \delta(DY)\rangle] \\ &= \sum_k E\Big[\int_{\mathbb{T}} X_t^k Y_t^k \, dt\Big] + \sum_{k,k'} E\Big[\int_{\mathbb{T}}\int_{\mathbb{T}} D_s^{(k')} X_t^k D_t^{(k')} Y_s^k \, ds\, dt\Big]. \end{aligned}$$

Here we used (5.8), (5.9) and (5.8) in turn. □

Note The derivative operator D is a basic tool in stochastic analysis on the Wiener space. It has been discussed in various contexts. Cameron [15] introduced the derivative D on the Wiener space and showed a version of the adjoint formula (5.8). It was extended and applied to potential theory on the Wiener space by Gross [34, 35]. Malliavin [77] studied the derivative operator D through the Ornstein–Uhlenbeck operator on the Wiener space. Stroock [105] discussed the finite dimensional approximation of the operator D. Ikeda–Watanabe [40, 41] and Watanabe [116] developed Malliavin's theory by introducing norms of Sobolev-type through the Ornstein–Uhlenbeck operator.

Our definition of the derivative D is close to Cameron [15], Gross [34, 35] and Shigekawa [100]. It can be applied directly to solutions F of stochastic differential equations driven by Wiener processes and further we can compute the derivatives DF explicitly. It will be discussed in Sect. 6.1.

The absolute continuity of the transformation T_h on the Wiener space was shown by Cameron–Martin [16]. It was extended to a wider class of transformations by Maruyama [82] and Girsanov [33], and is called the Girsanov transformation (Theorem 2.2).

5.2 Sobolev Norms for Wiener Functionals

Let us define Sobolev norms of L^p-type for Wiener functionals, making use of derivative operator D. It is convenient to set $\mathbb{T}^0 = H^0 = \{\emptyset\}$ (empty set), $D^0F = D^0_\emptyset F = F$ and $|D^0F|_{H^0} = |F|$. Let $\bar{\mathbb{N}}$ be the set of all nonnegative integers. For $m \in \bar{\mathbb{N}}$ and $p \geq 2$, we will define a norm for F belonging to $\mathcal{D}^m$ by $\|F\|_{0,p} = E[|D^0F|^p_{H^0}]^{\frac{1}{p}} = E[|F|^p]^{\frac{1}{p}}$ if $m = 0$ and

$$\|F\|_{m,p} := E\Big[\sum_{i=0}^m |D^iF|^p_{H^{\otimes i}}\Big]^{\frac{1}{p}} = E\Big[|F|^p + \sum_{i=1}^m \Big(\int_{\mathbb{T}^i} |D^i_{\mathbf{t}}F|^2\, d\mathbf{t}\Big)^{\frac{p}{2}}\Big]^{\frac{1}{p}}, \tag{5.11}$$

if $m \geq 1$. It is straightforward to prove that $\| \ \|_{m,p}$ is a norm for any $m \in \bar{\mathbb{N}}$ and $p \geq 2$. The system of norms $\{\| \ \|_{m,p}, m \in \bar{\mathbb{N}}, p \geq 2\}$ is called *Sobolev norms for Wiener functionals*. We denote the completion of $\mathcal{D}^m$ by the above norm by $\mathbf{D}^{m,p}$. We set $\mathbf{D}^\infty = \bigcap_{m,p} \mathbf{D}^{m,p}$. Then $\mathbf{D}^\infty \subset L^{\infty-}$. It is an F-space with countable norms $\{\| \ \|_{m,p}, m \in \bar{\mathbb{N}}, p = 2, 3, \ldots\}$. Further, $\mathcal{P} \subset \mathcal{D}^\infty \subset \mathbf{D}^\infty$. Elements of $\mathbf{D}^\infty$ are called a *smooth Wiener functional*.

Next, let $X = \{X_t\}$ be an element of $\mathcal{S}^\infty_{\mathbb{T}}$. It is written as $X = \sum_l F_l h_l$ (finite sum), where $F_l \in \mathcal{D}^\infty$ and $h_l \in H$. Its H-derivatives are defined by $D^iX = \sum_l D^iF_l h_l$. It is an $H^{\otimes(i+1)}$-valued functional. Its norm is

$$|D^iX|_{H^{\otimes(i+1)}} = \Big(\int_{\mathbb{T}^{i+1}} |\sum_l D^i_{\mathbf{t}}F_l h_l(t)|^2\, d\mathbf{t}\, dt\Big)^{\frac{1}{2}}.$$

We will define Sobolev norms for $X \in \mathcal{S}^\infty_{\mathbb{T}}$ as follows. For $m \in \bar{\mathbb{N}}$ and $p \geq 2$, we set $\|X\|_{0,p} = E[|X|^p_H]^{\frac{1}{p}}$ if $m = 0$ and

$$\|X\|_{m,p} := E\Big[\sum_{i=0}^{m} |D^i X|^p_{H^{\otimes(i+1)}}\Big]^{\frac{1}{p}} \tag{5.12}$$
$$= E\Big[\Big(\int_{\mathbb{T}} |X_t|^2\,dt\Big)^{\frac{p}{2}} + \sum_{i=1}^{m}\Big(\int_{\mathbb{T}^{i+1}} |D^i_{\mathbf{t}} X_t|^2\,d\mathbf{t}\,dt\Big)^{\frac{p}{2}}\Big]^{\frac{1}{p}},$$

if $m \geq 1$. If $X \in \mathcal{S}^\infty_{\mathbb{T}}$ does not depend on t, denote it by F. Then the above norm $\|X\|_{m,p}$ coincides with $\sqrt{T}\|F\|_{m,p}$. Hence these two norms are compatible with each other.

For any $X \in \mathcal{S}^\infty_{\mathbb{T}}$, its norms $\|X\|_{m,p}$ are finite for all m, p. The completion of $\mathcal{S}^\infty_{\mathbb{T}}$ by the above norm is denoted by $\mathbf{D}^{m,p}_{\mathbb{T}}$. We set $\mathbf{D}^\infty_{\mathbb{T}} = \bigcap_{m,p} \mathbf{D}^{m,p}_{\mathbb{T}}$. $\mathbf{D}^\infty_{\mathbb{T}}$ is an F-space with respect to countable norms $\|\ \|_{m,p}$, $m \in \bar{\mathbb{N}}$, $p = 2, 3, \ldots$.

Proposition 5.2.1 *The derivative operator D is extended as a continuous linear operator from F-space $\mathbf{D}^\infty$ to F-space $\mathbf{D}^\infty_{\mathbb{T}}$. Further, for any positive integer m and $p \geq 2$, it holds that $\|DF\|_{m,p} \leq \|F\|_{m+1,p}$ for any $F \in \mathbf{D}^\infty$.*

Proof Let $\{h_l\}$ be a complete orthonormal system in H. If $F \in \mathcal{D}^\infty$, DF is expanded as $\sum_l G_l h_l$, where $G_l = \langle DF, h_l\rangle$ and $G_l \in \mathcal{D}^\infty$. Its finite sum $\sum_{l=1}^n G_l h_l$ belongs to $\mathcal{S}^\infty_{\mathbb{T}}$. Then its limit DF belongs to $\mathbf{D}^\infty_{\mathbb{T}}$. Further, it satisfies $\|DF\|_{m-1,p} \leq \|F\|_{m,p}$ for any m and $p \geq 2$. □

The extended DF is again called the H-derivative of $F \in \mathbf{D}^\infty$.

Lemma 5.2.1 (Hölder's inequality) *If $F, G \in \mathbf{D}^\infty$ and $X \in \mathbf{D}^\infty_{\mathbb{T}}$, then $FG \in \mathbf{D}^\infty$ and $FX \in \mathbf{D}^\infty_{\mathbb{T}}$. Let $m \in \bar{\mathbb{N}}$, $p \geq 2$ and $r, r' > 1$ with $1/r + 1/r' = 1$. Then there exists a positive constant c such that*

$$\|FG\|_{m,p} \leq c\|F\|_{m,pr}\|G\|_{m,pr'},$$
$$\|FX\|_{m,p} \leq c\|F\|_{m,pr}\|X\|_{m,pr'} \tag{5.13}$$

hold for any F, G, X with the above properties.

Proof We give the proof of the latter inequality only. We first consider the case $m = 0$. We have by Hölder's inequality

$$\|FX\|^p_{0,p} = E[|FX|^p_H] \leq E[|F|^{pr}]^{\frac{1}{r}} E[|X|^{pr'}_H]^{\frac{1}{r'}} \leq \|F\|^p_{0,pr}\|X\|^p_{0,pr'}.$$

We next consider the case $m = 1$. Since $D(FX) = DF \cdot X + F \cdot DX$, we have

$$|D(FX)|_{H^{\otimes 2}} \leq |DF|_H |X|_H + |F||DX|_{H^{\otimes 2}}.$$

Therefore, using Hölder's inequality, we get

$$E[|D(FX)|^p_{H^{\otimes 2}}] \le 2^p\{E[|DF|^{pr}_H]^{\frac{1}{r}} E[|X|^{pr'}_H]^{\frac{1}{r'}} + E[|F|^{pr}]^{\frac{1}{r}} E[|DX|^{pr'}_{H^{\otimes 2}}]^{\frac{1}{r'}}\}$$
$$\le 2^p\Big\{\|F\|^p_{1,pr}\|X\|^p_{0,pr'} + \|F\|^p_{0,pr}\|X\|^p_{1,pr'}\Big\}.$$

Therefore we get (5.13) in the case $m = 1$. Repeating this argument inductively, we get (5.13) for any positive integer m. □

We will consider the L^2-extension of the adjoint operator δ, making use of Sobolev norms. The adjoint operator δ is extended to a continuous linear operator from $\mathbf{D}^{1,2}_{\mathbb{T}}$ to $\mathbf{D}^{0,2}$ and satisfies the inequality $\|\delta(X)\|_{0,2} \le \|X\|_{1,2}$ for any $X \in \mathbf{D}^{1,2}_{\mathbb{T}}$, in view of the isometric property (5.10). $\delta(X)$ is called again the *Skorohod integral* of X_t (by Wiener process).

We show that if X_t is a predictable process, the Skorohod integral coincides with the Itô integral.

Proposition 5.2.2 *Let $\{\mathcal{F}_t\}$ be the filtration generated by the Wiener process W_t and let X_t be a predictable functional with respect to the filtration $\{\mathcal{F}_t\}$, belonging to $\mathbf{D}^{1,2}_{\mathbb{T}}$. Then the Skorohod integral $\delta(X)$ coincides with the Itô integral $\sum_k \int_{\mathbb{T}} X^k_s dW^k_s$. Further, the isometric equality is written for short as*

$$E[\delta(X)^2] = \sum_{k=1}^{d'} E\Big[\int_{\mathbb{T}} (X^k_t)^2\, dt\Big]. \tag{5.14}$$

Proof Suppose that $X_t := \sum_{k=1}^{d'} \sum_{m=1}^{n} F^k_m 1_{(t_{m-1},t_m]}(t) e_k$ is a simple process, where F^k_m are real $\mathcal{F}_{t_{m-1}}$-measurable for any m and e_k, $k = 1, \dots, d'$ are unit vectors in $\mathbb{R}^{d'}$. Then the k-th component of X_t is $X^k_t = \sum_m F^k_m 1_{(t_{m-1},t_m]}(t)$. It holds that $D_t F^k_m = 0$ for $t > t_{m-1}$. Therefore we have $\langle DF^k_m, 1_{(t_{m-1},t_m]}\rangle = 0$. Consequently, we have from the formula (5.7),

$$\delta(X) = \sum_k \sum_m F^k_m (W^k_{t_m} - W^k_{t_{m-1}}) = \sum_k \int_{\mathbb{T}} X^k_s\, dW^k_s.$$

The fact can be extended to any predictable element X belonging to $\mathbf{D}^{1,2}_{\mathbb{T}}$.

If X_s is predictable then we have $D_t X_s = 0$ if $t \ge s$, as we have shown above. Therefore we have $D^{(k')}_t X^k_s D^{(k')}_s X^k_t = 0$ a.e. $ds\,dt$ for any k, k'. Consequently, the last integral in (5.10) is 0, and we have the isometric property $E[\delta(X)^2] = \sum_k E[\int_{\mathbb{T}} (X^k_t)^2\, dt]$. □

If X_t is not predictable, the Skorohod integral is called the *anticipating stochastic integral*.

Remark We shall consider the anticipating stochastic differential equation, or Skorohod SDE. Let W_t be a one-dimensional Wiener process. We denote the anticipating stochastic integral of $X_s 1_{(0,t]}(s)$ by $\int_0^t X_s \delta W_s$. We consider stochastic differential equations represented by anticipating stochastic integrals;

$$X_t = X_0 + \int_0^t \sigma(X_s)\delta W_s + \int_0^t b(X_s)\,ds,$$

where $X_0 \in \mathbf{D}^{1,2}$ and σ, b are smooth functions on $\mathbb{R}$ and $X_t \in \mathbf{D}_{\mathbb{T}}^{1,2}$. The equation is called the *Skorohod equation*. If we replace the above Skorohod integral by the Itô integral $\int_0^t \sigma(X_s)\,dW_s$ in the above equation, the equation coincides with the Itô equation discussed in Chap. 3.

We are interested in the existence and the uniqueness of the solution of the Skorohod equation. If X_0 is $\mathcal{F}_0$-measurable, the Itô equation has a unique solution. Further, it is also the solution of the Skorohod equation, since the solution X_t of the Itô equation is predictable and hence $\int_0^t \sigma(X_s)\delta W_s = \int_0^t \sigma(X_s)\,dW_s$ holds (Proposition 5.2.2). Then the Skorohod equation also has a unique solution if X_0 is $\mathcal{F}_0$-measurable. If X_0 is not $\mathcal{F}_0$-measurable, Itô equation has no solution (not well defined). However, the Skorohod equation has a solution. Indeed, let $\Phi_{0,t}(x)$ be the stochastic flow generated by the Itô equation. Then $\Phi_{0,t}(x) \in \mathbf{D}_{\mathbb{T}}^{1,2}$ for any x (see Sect. 6.1) and the composite process $X_t = \Phi_{0,t}(X_0)$ belongs to $\mathbf{D}_{\mathbb{T}}^{1,2}$. Hence the Skorohod integral $\int_0^t \sigma(X_s)\delta W_s$ is well defined and X_t satisfies the Skorohod equation. The uniqueness of the solution of the Skorohod equation appears to be a difficult problem. For details, see Nualart [88].

By the isometric property of the Skorohod integral $\delta(X)$, we have the inequality $\|\delta(X)\|_{0,2} \leq \|X\|_{1,2}$. The inequality can be extended to higher derivatives.

Proposition 5.2.3 *For any $m \in \bar{\mathbb{N}}$, there exists a positive constant c_m such that the inequality $\|\delta(X)\|_{m,2} \leq c_m \|X\|_{m+1,2}$ holds for any $X \in \mathbf{D}_{\mathbb{T}}^{\infty}$.*

Proof To avoid complicated notations, we will consider the case of a one-dimensional Wiener space ($d' = 1$), where $\mathbb{T}$ is equal to the interval $[0, 1]$. We will prove the inequality of the lemma for $X \in \mathcal{S}_{\mathbb{T}}^{\infty}$. Let i be a positive integer. Since $D_s\delta(X) = X_s + \delta(D_s X)$ holds, we have by induction

$$D_{\mathbf{s}}^i \delta(X) = \delta(D_{\mathbf{s}}^i X) + \sum_{h=1}^{i} D_{\mathbf{s}_h}^{i-1} X_{s_h}, \tag{5.15}$$

where $\mathbf{s} = (s_1, \ldots, s_i) \in \mathbb{T}^i$ and $\mathbf{s}_h = \mathbf{s} - \{s_h\}$. We will compute $L^2(d\mathbf{s}\,dP)$-norm of each term of the above. Using the adjoint formula (5.8) and the commutation relation (5.9), we have

$$
\begin{aligned}
E\Big[\int_{\mathbb{T}^i} \delta(D^i_{\mathbf{s}}X)^2\, d\mathbf{s}\Big] &= \int_{\mathbb{T}^i} E[\delta(D^i_{\mathbf{s}}X)^2]\, d\mathbf{s} \\
&= \int_{\mathbb{T}^i} \Big(\int_{\mathbb{T}} E[D_t\delta(D^i_{\mathbf{s}}X) D^i_{\mathbf{s}}X_t]\, dt\Big)\, d\mathbf{s} \\
&= \int_{\mathbb{T}^i} \Big(\int_{\mathbb{T}} \{E[\delta(D_t D^i_{\mathbf{s}}X) D^i_{\mathbf{s}}X_t] + E[|D^i_{\mathbf{s}}X_t|^2]\}\, dt\Big)\, d\mathbf{s} \\
&= \int_{\mathbb{T}^i} \Big(\int_{\mathbb{T}}\int_{\mathbb{T}} E[D_t D^i_{\mathbf{s}}X_{t'} \cdot D_{t'} D^i_{\mathbf{s}}X_t]\, dt'\, dt\Big)\, d\mathbf{s} + \int_{\mathbb{T}^i}\int_{\mathbb{T}} E[|D^i_{\mathbf{s}}X_t|^2]\, dt\, d\mathbf{s} \\
&\le \|X\|^2_{i+1,2}.
\end{aligned}
$$

Further, we have

$$
E\Big[\int_{\mathbb{T}^i} |D^{i-1}_{\mathbf{s}_h} X_{s_h}|^2\, d\mathbf{s_i}\, ds_h\Big] \le \|X\|^2_{i-1,2}.
$$

Summing up these inequalities for $i = 1, \ldots, m$, we get the inequality of the proposition. □

We will extend the inequality for arbitrary Sobolev norms. The following is an analogue of the Burkholder–Davis–Gundy inequality (Proposition 2.2.2) for Skorohod integral.

Theorem 5.2.1 *Let δ be the Skorohod integral operator. For any $m \in \bar{\mathbb{N}}$ and a positive even number p, there exists a positive constant $c_{m,p}$ such that*

$$
\|\delta(X)\|_{m,p} \le c_{m,p}\|X\|_{m+p-1,p} \tag{5.16}
$$

holds for any $X \in \mathbf{D}^\infty_{\mathbb{T}}$.

Proof It is sufficient to prove the inequality for $X \in \mathcal{S}^\infty_{\mathbb{T}}$, since $\mathcal{S}^\infty_{\mathbb{T}}$ is dense in $\mathbf{D}^\infty_{\mathbb{T}}$. We will consider the case of a one-dimensional Wiener space ($d' = 1$), where $\mathbb{T}$ is equal to the interval $[0, 1]$. Let $p \ge 2$ be a positive even number. Let $X \in \mathcal{S}^\infty_{\mathbb{T}}$. Then $\delta(X) \in L^{\infty-}$. By the adjoint property of δ (Lemma 5.1.1), we get

$$
E[\delta(X)^p] = E[\delta(X)\delta(X)^{p-1}] = E\Big[\int_{\mathbb{T}} X_{t_1} D_{t_1}(\delta(X)^{p-1})\, dt_1\Big].
$$

We will decrease the power $p-1$ of $\delta(X)^{p-1}$, step by step. By the commutation relation for δ, we have

$$
D_{t_1}(\delta(X)^{p-1}) = (p-1)\delta(X)^{p-2}(X_{t_1} + \delta(D_{t_1}X)).
$$

Therefore, the above is equal to

$$(p-1)E\Big[\Big(\int_{\mathbb{T}} X_{t_1}^2\,dt_1\Big)\delta(X)^{p-2}\Big]+(p-1)E\Big[\Big(\int_{\mathbb{T}} X_{t_1}\delta(D_{t_1}X)\,dt_1\Big)\delta(X)^{p-2}\Big].$$

Apply Lemma 5.1.1 again to the last term. Then it is written as

$$\begin{aligned}
&(p-1)E\Big[\int_{\mathbb{T}}\int_{\mathbb{T}} D_{t_1}X_{t_2}D_{t_2}(X_{t_1}\delta(X)^{p-2})\,dt_1\,dt_2\Big]\\
&=(p-1)E\Big[\int_{\mathbb{T}}\int_{\mathbb{T}} D_{t_1}X_{t_2}\Big\{D_{t_2}X_{t_1}\delta(X)^{p-2}\\
&\quad+(p-2)X_{t_1}D_{t_2}\delta(X)\cdot\delta(X)^{p-3}\Big\}\,dt_1\,dt_2\Big]\\
&=(p-1)E\Big[\int_{\mathbb{T}}\int_{\mathbb{T}} D_{t_1}X_{t_2}D_{t_2}X_{t_1}\,dt_1\,dt_2\delta(X)^{p-2}\Big]\\
&\quad+(p-1)(p-2)E\Big[\int_{\mathbb{T}}\int_{\mathbb{T}} D_{t_1}X_{t_2}X_{t_1}X_{t_2}\,dt_1\,dt_2\delta(X)^{p-3}\Big]\\
&\quad+(p-1)(p-2)E\Big[\int_{\mathbb{T}}\int_{\mathbb{T}} D_{t_1}X_{t_2}X_{t_1}\delta(D_{t_2}X)\,dt_1\,dt_2\delta(X)^{p-3}\Big].
\end{aligned}$$

Further, if $p\geq 3$ the last term is rewritten as

$$(p-1)(p-2)E\Big[\int_{\mathbb{T}}\int_{\mathbb{T}}\int_{\mathbb{T}} D_{t_3}(D_{t_1}X_{t_2}X_{t_1}\delta(X)^{p-3})D_{t_2}X_{t_3}\,dt_1\,dt_2\,dt_3\Big].$$

Repeating this argument, we find that $E[\delta(X)^p]$ is equal to sums of the following terms:

$$E\Big[\int_{\mathbb{T}^q} D_{\mathbf{t}_1}^{j_1}X_{t_1}^{i_1}\cdots D_{\mathbf{t}_q}^{j_q}X_{t_q}^{i_q}d\mathbf{t}\Big],\quad \frac{p}{2}\leq q\leq p.$$

Here i_l are nonnegative integers satisfying $i_l=1,2$ and $i_1+\cdots+i_q=p$; j_l are nonnegative integers satisfying $0\leq j_l\leq p-1$ and $j_1+\cdots+j_q\leq p$. Further, $\mathbf{t}=(t_1,\ldots,t_p)$ and $\mathbf{t}_l$ are subsets of $\mathbf{t}\setminus\{t_l\}$. If $j_l=0$, $\mathbf{t}_l$ is an empty set and it holds that $D_{\mathbf{t}_l}^{j_l}X_{t_l}^{i_l}=X_{t_l}^{i_l}$. Then, by Hölder's inequality, the above integral is dominated by

$$\prod_{l;i_l\neq 0,j_l\neq 0}E\Big[\Big(\int_{\mathbb{T}^{j_l}}\int_{\mathbb{T}}|D_{\mathbf{t}_l}^{j_l}X_{t_l}^{i_l}|^{\frac{2}{i_l}}\,d\mathbf{t}_l\,dt_l\Big)^{\frac{p}{2}}\Big]^{\frac{i_l}{p}}\times\prod_{l;i_l\neq 0,j_l=0}E\Big[\Big(\int_{\mathbb{T}}|X_{t_l}^{i_l}|^{\frac{2}{i_l}}\,dt_l\Big)^{\frac{p}{2}}\Big]^{\frac{i_l}{p}}.\tag{5.17}$$

It holds that for $i_l=1,2$ and $j_l\neq 0$,

$$\begin{aligned}
E\Big[\Big(\int_{\mathbb{T}^{j_l}}\int_{\mathbb{T}}|D_{\mathbf{t}_l}^{j_l}X_{t_l}^{i_l}|^{\frac{2}{i_l}}\,d\mathbf{t}_l\,dt_l\Big)^{\frac{p}{2}}\Big]^{\frac{i_l}{p}}&\leq E\Big[\Big(\int_{\mathbb{T}^{j_l}}|D_{\mathbf{t}_l}^{j_l}X|_H^2\,d\mathbf{t}_l\Big)^{\frac{p}{2}}\Big]^{\frac{i_l}{p}}\\
&\leq\|X\|_{j_l,p}^{i_l}\leq\|X\|_{p-1,p}^{i_l}.
\end{aligned}$$

Further, we have

$$E\Big[\Big(\int_{\mathbb{T}} |X_{t_l}^{i_l}|^{\frac{2}{i_l}}\,dt_l\Big)^{\frac{p}{2}}\Big]^{\frac{i_l}{p}} \leq \|X\|_{0,p}^{i_l} \leq \|X\|_{p-1,p}^{i_l}.$$

Then (5.17) is dominated by $\prod_{l;i_l\neq 0}\|X\|_{p-1,p}^{i_l} = \|X\|_{p-1,p}^{p}$. Consequently, there exists a positive constant c'_p such that $E[\delta(X)^p] \leq c'_p\|X\|_{p-1,p}^{p}$. This proves the inequality $\|\delta(X)\|_{0,p} \leq c_{0,p}\|X\|_{p-1,p}$.

We will next consider the case where $m \geq 1$ and p is a positive even number. By the definition of Sobolev norms, we have

$$\|\delta(X)\|_{m,p}^{p} = E[\delta(X)^p] + \sum_{i=1}^{m} E\Big[\Big(\int_{\mathbb{T}^i} |D_{\mathbf{s}}^i\delta(X)|^2\,d\mathbf{s}\Big)^{\frac{p}{2}}\Big]. \tag{5.18}$$

The first term of the right-hand side has been computed. We shall consider other terms. Note the equality (5.15). Then there is a positive constant c_p such that

$$\begin{aligned} &E\Big[\Big(\int_{\mathbb{T}^i} |D_{\mathbf{s}}^i\delta(X)|^2\,d\mathbf{s}\Big)^{\frac{p}{2}}\Big] \qquad (5.19)\\ &\leq c_p\Big\{E\Big[\Big(\int_{\mathbb{T}^i} |\delta(D_{\mathbf{s}}^iX)|^2\,d\mathbf{s}\Big)^{\frac{p}{2}}\Big] + \sum_{h=1}^{i} E\Big[\Big(\int_{\mathbb{T}^{i-1}\times\mathbb{T}} |D_{\mathbf{s_h}}^{i-1}X_{s_h}|^2\,d\mathbf{s_h}\,ds_h\Big)^{\frac{p}{2}}\Big]\Big\}. \end{aligned}$$

We have

$$E\Big[\Big(\int_{\mathbb{T}^i} |\delta(D_{\mathbf{s}}^iX)|^2\,d\mathbf{s}\Big)^{\frac{p}{2}}\Big] = E\Big[\int_{\mathbb{T}^i}\cdots\int_{\mathbb{T}^i} \delta(D_{\mathbf{s}_1}^iX)^2\cdots\delta(D_{\mathbf{s}_{\frac{p}{2}}}^iX)^2\,d\mathbf{s}_1\cdots d\mathbf{s}_{\frac{p}{2}}\Big].$$

By a computation similar to that for $E[\delta(X)^p]$, we find that the above is equal to the sum of terms

$$E\Big[\int_{\mathbb{T}^{\frac{ip}{2}}}\int_{\mathbb{T}^q} D_{\mathbf{t}_1}^{j_1}(D_{\mathbf{s}_1}^iX_{t_1})^{i_1}\cdots D_{\mathbf{t}_q}^{j_q}(D_{\mathbf{s}_{\frac{p}{2}}}^iX_{t_p})^{i_q}\,d\mathbf{s}_1\cdots d\mathbf{s}_{\frac{p}{2}}d\mathbf{t}\Big],$$

where $\mathbf{t} = (t_1,\ldots,t_q)$ and $q = \frac{p}{2},\ldots,p$. By Hölder's inequality, the above is dominated by

$$\begin{aligned} &\prod_{l;i_l\neq 0,\,j_l\neq 0} E\Big[\Big(\int_{\mathbb{T}^{j_l}}\int_{\mathbb{T}^i}\int_{\mathbb{T}} |D_{\mathbf{t}_l}^{j_l}D_{\mathbf{s}_i}^iX_{t_l}|^{\frac{i_l}{2}}\,d\mathbf{t}_l\,d\mathbf{s}_i\,dt_l\Big)^{\frac{p}{2}}\Big]^{\frac{i_l}{p}}\\ &\times \prod_{l;i_l\neq 0,\,j_l=0} E\Big[\Big(\int_{\mathbb{T}^l}\int_{\mathbb{T}} |D_{\mathbf{s}_i}^iX_{t_l}|^{\frac{i_l}{2}}\,d\mathbf{s}_i\,dt_l\Big)^{\frac{p}{2}}\Big]^{\frac{i_l}{p}}. \qquad (5.20) \end{aligned}$$

It holds that

$$E\Big[\Big(\int_{\mathbb{T}^{j_l}}\int_{\mathbb{T}^i}\int_{\mathbb{T}}|D_{\mathbf{t}_l}^{j_l}D_{\mathbf{s}_i}^i X_{t_l}|^{\frac{i_l}{2}}\,d\mathbf{t}_l\,d\mathbf{s}_i\,dt_l\Big)^{\frac{p}{2}}\Big]^{\frac{i_l}{p}} \le \|X\|_{i+j_l,p}^{i_l} \le \|X\|_{i+p-1,p}^{i_l},$$

$$E\Big[\Big(\int_{\mathbb{T}^i}\int_{\mathbb{T}}|D_{\mathbf{s}_i}^i X_{t_l}|^{\frac{i_l}{2}}\,d\mathbf{s}_l\,dt_l\Big)^{\frac{p}{2}}\Big]^{\frac{i_l}{p}} \le \|X\|_{i,p}^{i_l} \le \|X|_{i+p-1,p}^{i_l}.$$

Therefore, (5.20) is dominated by $\prod_{l;i_l\neq 0}\|X\|_{i+p-1,p}^{i_l} = \|X\|_{i+p-1,p}^{p}$. Consequently there exists a positive constant c'' such that

$$E\Big[\Big(\int_{\mathbb{T}^i}|\delta(D_{\mathbf{s}}^i X)|^2\,d\mathbf{s}\Big)^{\frac{p}{2}}\Big] \le c''\|X\|_{i+p-1,p}^{p}. \tag{5.21}$$

We shall consider the last term of (5.19). It holds that

$$E\Big[\Big(\int_{\mathbb{T}^{i-1}\times\mathbb{T}}|D_{\mathbf{s}_i}^{i-1}X_{s_h}|^2\,d\mathbf{s}_i\,ds_h\Big)^{\frac{p}{2}}\Big] \le \|X\|_{i-1,p}^{p}. \tag{5.22}$$

Substitute (5.21) and (5.22) in (5.19). Then we get

$$E\Big[\Big(\int_{\mathbb{T}^i}|D_{\mathbf{s}}^i\delta(X)|^2\,d\mathbf{s}\Big)^{\frac{p}{2}}\Big] \le c\|X\|_{i+p-1,p}^{p}.$$

Then we obtain from (5.18) the inequality $\|\delta(X)\|_{m,p}^p \le c'_{m,p}\|X\|_{m+p-1,p}^p$. This proves $\|\delta(X)\|_{m,p} \le c_{m,p}\|X\|_{m+p-1,p}$. □

We will summarize properties of the adjoint operator. For a real number r, we denote by $[r]$ the largest integer which is dominated by r. If $r > 0$, $2[r/2]$ is equal to the largest even number which is dominated by r. Further, if $r > 0$, $-2[-r/2]$ is the smallest even number which is bigger than or equal to r. We set $\langle r\rangle = -2[-r/2]$.

Theorem 5.2.2

1. *Skorohod integral operator δ is extended to a continuous linear operator from F-space $\mathbf{D}_{\mathbb{T}}^{\infty}$ to F-space $\mathbf{D}^{\infty}$. Further, for any $m \in \bar{\mathbb{N}}$ and $p \ge 2$, there exists a positive constant $c_{m,p}$ such that*

$$\|\delta(X)\|_{m,p} \le c_{m,p}\|X\|_{m+\langle p\rangle-1,\langle p\rangle} \tag{5.23}$$

   holds for any $X \in \mathbf{D}_{\mathbb{T}}^{\infty}$.
2. *Adjoint formula* (5.8) *holds for any $X \in \mathbf{D}_{\mathbb{T}}^{\infty}$ and $G \in \mathbf{D}^{\infty}$.*
3. *Commutation relation* (5.9) *holds for any $X \in \mathbf{D}_{\mathbb{T}}^{\infty}$.*
4. *Isometric property* (5.10) *holds for any $X, Y \in \mathbf{D}_{\mathbb{T}}^{\infty}$.*

Proof We will prove (5.23). Since the norm $\|X\|_{m,p}$ is nondecreasing with respect to p, we have by Theorem 5.2.1,

$$\|\delta(X)\|_{m,p} \leq \|\delta(X)\|_{m,\langle p\rangle} \leq c\|X\|_{m+\langle p\rangle-1,\langle p\rangle}.$$

Assertions (2)–(4) are immediate from Lemma 5.1.1. □

Derivative operator D and its adjoint δ can be applied to complex-valued functionals. Let $F = F_1 + iF_2$, where $F_k, k = 1, 2$ are real smooth Wiener functionals and $i = \sqrt{-1}$. Then we define DF by $DF_1 + iDF_2$. Let $X = X_1 + iX_2$, where $X_k, k = 1, 2$ are real smooth H-valued functionals. Then we define $\delta(X)$ by $\delta(X_1) + i\delta(X_2)$. Norms $\|F\|_{m,p}$ and $\|X\|_{m,p}$ for these complex-valued functionals are defined in the same way. F is called smooth if F_1 and F_2 are smooth. The set of smooth functionals is denoted again as $\mathbf{D}^\infty$. Complex H-valued smooth functionals are defined similarly and we denote by $\mathbf{D}^\infty_{\mathbb{T}}$ the set of all complex H-valued smooth functionals. Then Theorems 5.2.1 and 5.2.2 are valid for complex functionals.

Note For arguments of the Malliavin calculus based on the Ornstein–Uhlenbeck semigroup, we refer to Malliavin [77], Ikeda–Watanabe [40], Watanabe [116], and further, books of Shigekawa [102], Nualart [88] and Matsumoto–Taniguchi [83]. Sobolev-type norms for Wiener functional were defined in [40] and [116] in a different context. In these works, norms of a Wiener functional F are defined for all real m and $p > 1$ by $E[|(I - L)^{\frac{m}{2}} F|^p]^{\frac{1}{p}}$, where L is the Ornstein–Uhlenbeck operator. To avoid confusion of notations, we will denote their norms as $\|F\|'_{m,p}$. They showed that if m is a positive integer, $\sum_{0\leq m'\leq m} \|F\|'_{m',p}$ (j' are nonnegative integers) is equivalent to our norm $\|F\|_{m,p}$. The fact is called *Meyer's equivalence* [102].

Further, with respect to the estimate of the adjoint operator δ, they assert the inequality $||\delta(X)\|'_{m,p} \leq c'\|X\|'_{m+1,p}$. In the case where m is a positive integer, it is equivalent to

$$||\delta(X)\|_{m,p} \leq c\|X\|_{m+1,p}, \tag{5.24}$$

which is stronger than the inequality of our Theorem 5.2.1. However, its proof is not clear to the author. In this monograph, we have given a new direct proof for a slightly weaker assertion (Theorem 5.2.1).

Studies of the Malliavin calculus for non-classical Wiener spaces (curved Wiener space etc.) may be found in Driver's survey work on Wiener space [23].

5.3 Nondegenerate Wiener Functionals

Let $F = (F^1, \ldots, F^d)$ be a d-dimensional Wiener functional. If its components $F^i, i = 1, \ldots, d$ are infinitely H-differentiable, F is called infinitely H-differentiable. We denote by $(\mathcal{D}^\infty)^d$ the set of all such d-dimensional functionals. If each component F^i belongs to $\mathbf{D}^\infty$, F is said to belong to $(\mathbf{D}^\infty)^d$. Its Sobolev norm is defined by the sum of Sobolev norms of components $F^j, j = 1, \ldots, d$. It holds that $(\mathcal{D}^\infty)^d \subset (\mathbf{D}^\infty)^d$. Elements of the latter are called smooth Wiener functional. The space $(\mathbf{D}^\infty_{\mathbb{T}})^d$ and the norm of its element $X_t = (X_t^1, \ldots, X_t^d)$ are defined similarly.

Let $F = (F^1, \ldots, F^d)$ be an element of $(\mathbf{D}^\infty)^d$. Let G be an element of $\mathbf{D}^\infty$. We consider the law of F *weighted by* G:

$$\mu_G(dx) = E[1_{dx}(F)G].$$

In this section, we study the existence of the smooth density of the law $\mu_G(dx)$ with respect to the Lebesgue measure. For this purpose, we consider its characteristic function. It is given by

$$\psi_G(v) = \int_{\mathbb{R}^d} e^{i(v,x)} \mu_G(dx) = E[e^{i(v,F)}G], \tag{5.25}$$

where $i = \sqrt{-1}$ is pure imaginary. It is a C^∞-function of v. Indeed, let $F^{\mathbf{j}} = (F^1)^{j_1} \cdots (F^d)^{j_d}$, where $\mathbf{j} = (j_1, \ldots, j_d)$ is a multi-index of nonnegative integers $j_l, l = 1, \ldots, d$. Since $|F| \in L^{\infty-}$, the functional $e^{i(v,F)} F^{\mathbf{j}} G$ is integrable for any $\mathbf{j}$ and we can change the order of the derivatives $\partial_v^{\mathbf{j}} \equiv \partial_{v_1}^{j_1} \cdots \partial_{v_d}^{j_d}$ and integral. Then we have

$$\partial_v^{\mathbf{j}} E[e^{i(v,F)}G] = i^{|\mathbf{j}|} E[e^{i(v,F)} F^{\mathbf{j}} G].$$

This shows that the characteristic function $\psi_G(v)$ is a C^∞-function and $\partial_v^{\mathbf{j}} \psi_G(v)$ coincides with the above function.

In Chap. 1, we discussed the existence of the smooth density of a weighted law μ_G. Proposition 1.1.2 tells us that if $|\psi_G(v)||v|^n$ is integrable for some positive integer n, then the weighted law $\mu_G(dx)$ has a C^n-density function $f_G(x)$. Now, in order to show the integrability of $|\psi_G(v)||v|^n$, we need to assume that F is *nondegenerate*, which will be defined using the Malliavin covariance. For a smooth Wiener functional $F = (F^1, \ldots, F^d)$ we define a matrix R^F called the *Malliavin covariance* by

$$R^F = \Big(\langle DF^i, DF^j \rangle\Big) = \int_{\mathbb{T}} D_t F (D_t F)^T \, dt. \tag{5.26}$$

Let $S_{d-1} = \{v \in \mathbb{R}^d;\ |v| = 1\}$. For $\theta \in S_{d-1}$, we set $R^F(\theta) = (\theta, R^F\theta)$. The functional F is called *nondegenerate*, if $R^F(\theta)$ are invertible a.s. for any $\theta \in S_{d-1}$ and inverses satisfy

$$\sup_\theta E[R^F(\theta)^{-p}] < \infty \tag{5.27}$$

for any p, where the supremum for θ is taken in the set S_{d-1}. Since the norm $\| \ \|_{0,p}$ coincides with L^p-norms, the above condition is equivalent to that the inequality $\sup_\theta \|R^F(\theta)^{-1}\|_{0,p} < \infty$ holds for any $p \geq 2$.

Remark We give a simple example of a nondegenerate Wiener functional. Let $F = (F^1, \ldots, F^d)$ be a linear functional of the Wiener process W_t given by

$$F^i = \sum_k \int_{\mathbb{T}} h^{ik}(s)\, dW_s^k, \quad i = 1, \ldots, d,$$

where $h^{ik}(s), i = 1, \ldots, d, k = 1, \ldots, d'$ are square integrable functions on $\mathbb{T}$. We saw in Sect. 5.1 that the above F is H-differentiable and its derivative is computed as $D_t F = \big(h^{i,k}(t)\big) = H(t)$ ($d \times d'$-matrix). Then its Malliavin covariance is equal to a constant (non-random) matrix

$$R^F = \Big(\sum_k \int_{\mathbb{T}} h^{ik}(t)h^{jk}(t)\, dt\Big) = \int_{\mathbb{T}} H(t)H(t)^T\, dt.$$

Therefore F is nondegenerate if and only if the above matrix R^F is invertible.

The above F is Gaussian distributed and its covariance matrix is computed as $\int_{\mathbb{T}} H(t)H(t)^T\, dt$. Therefore, the Malliavin covariance R^F coincides with the covariance matrix of F. It is well known that a Gaussian distribution has a C^∞-density if its covariance matrix is nondegenerate (positive definite). The density function is given by (1.8) with $A = R^F$ and $b = 0$.

Now, we want to show that the law of any nondegenerate Wiener functional has a C^∞-density. We first discuss the inverse of the Malliavin covariance. We need a lemma.

Lemma 5.3.1 *Suppose that $G \in \mathbf{D}^\infty$ is invertible and the inverse G^{-1} belongs to $L^{\infty-}$. Then G^{-1} belongs to $\mathbf{D}^\infty$. Further, for any m, p, there exists $c_{m,p} > 0$ such that the inequality*

$$\|G^{-1}\|_{m,p} \leq c_{m,p}(1 + \|G\|_{m,2mp})^m (1 + \|G^{-1}\|_{0,2(m+1)p})^{m+1} \tag{5.28}$$

holds for any G with the above property.

Proof If $G^{-1} \in L^{\infty-}$, then $T_{\lambda h}(G^{-1}) \in L^{\infty-}$. Then $T_{\lambda h}(G)$ is invertible and satisfies $T_{\lambda h}(G^{-1}) = 1/T_{\lambda h}(G)$ a.s. Hence we have

$$\frac{T_{\lambda h}(G^{-1}) - G^{-1}}{\lambda} = \frac{-1}{T_{\lambda h}(G)G}\frac{T_{\lambda h}(G) - G}{\lambda}, \quad a.s.$$

Let λ tend to 0. The right-hand side converges to $\frac{-1}{G^2}\langle DG, h\rangle$ in L^p for any $p > 1$. This means that G^{-1} is H-differentiable and $D(G^{-1}) = \frac{-1}{G^2}DG$. Then, repeating this argument inductively, we find that the i-th H-derivative of G^{-1} exists and is equal to

$$D^i_{\mathbf{t}}G^{-1} = \sum(-1)^r\frac{G^{i-r}D^{i_1}_{\mathbf{t}_1}G\cdots D^{i_r}_{\mathbf{t}_r}G}{G^{i+1}}, \quad \mathbf{t} = (\mathbf{t}_1, \ldots, \mathbf{t}_r),$$

where the sum is taken for positive integers r and i_k such that $r \leq i$ and $i_1+\cdots+i_r = i$. Using Hölder's inequality, we get

$$\begin{aligned}
&E\Big[\Big(\int\Big|\frac{G^{i-r}D^{i_1}_{\mathbf{t}_1}G\cdots D^{i_r}_{\mathbf{t}_r}G}{G^{i+1}}\Big|^2 d\mathbf{t}\Big)^{\frac{p}{2}}\Big] \qquad (5.29)\\
&\leq E\Big[\Big(\int|G^{i-r}D^{i_1}_{\mathbf{t}_1}G\cdots D^{i_r}_{\mathbf{t}_r}G|^2 d\mathbf{t}\Big)^p\Big]^{\frac{1}{2}}E\Big[|G|^{-2(i+1)p}\Big]^{\frac{1}{2}}\\
&\leq c'\|G\|^{ip}_{i,2ip}\|G^{-1}\|^{(i+1)p}_{0,2(i+1)p}\\
&\leq c'(1+\|G\|_{i,2ip})^{ip}(1+\|G^{-1}\|_{0,2(i+1)p})^{(i+1)p}.
\end{aligned}$$

Now for a given $m \in \mathbb{N}$, the above inequality is valid for any $i = 0, 1, 2, \ldots, m$. Further, the last term of the above is nondecreasing with respect to i and p. Therefore, summing up (5.29) for $i = 0, \ldots, m$, we get the inequality

$$\|G^{-1}\|^p_{m,p} \leq c_{m,p}(1+\|G\|_{m,2mp})^{mp}(1+\|G^{-1}\|_{0,2(m+1)p})^{(m+1)p}.$$

This proves the inequality of the lemma. □

Proposition 5.3.1 *If F of $(\mathbf{D}^\infty)^d$ is nondegenerate, $R^F(\theta)^{-1}$ belongs to $\mathbf{D}^\infty$ for any θ. Further, for any m, p, there exists a positive constant $c_{m,p}$ such that*

$$\sup_\theta \|R^F(\theta)^{-1}\|_{m,p} \leq c_{m,p}(1+\|DF\|^2_{m,4mp})^m(1+\sup_\theta\|R^F(\theta)^{-1}\|_{0,2(m+1)p})^{m+1} \qquad (5.30)$$

holds for any nondegenerate F.

Proof We apply Lemma 5.3.1 for $G = R^F(\theta)$. It is sufficient to show that any m, p, there exists a positive constant $c_{m,p}$ such that

$$\sup_\theta \|R^F(\theta)\|_{m,p} \le c_{m,p}\|DF\|^2_{m,2p}. \tag{5.31}$$

The inequality is immediately verified if $m = 0$. We consider the case $m \ge 1$. Let $i \le m$. Since $R^F(\theta) = \int_{\mathbb{T}} (\theta, D_t F)^2 \, dt$, and

$$D^i_{\mathbf{t}}(\theta, D_t F)^2 = \sum_{i',i'' \in \mathbb{N}, i'+i''=i} D^{i'}_{\mathbf{t}'}(\theta, D_t F) \cdot D^{i''}_{\mathbf{t}''}(\theta, D_t F),$$

we have

$$\begin{aligned} |D^i_{\mathbf{t}} R^F(\theta)| &\le \sum_{i',i''} \int_{\mathbb{T}} |D^{i'}_{\mathbf{t}'}(\theta, D_t F) \cdot D^{i''}_{\mathbf{t}''}(\theta, D_t F)| \, dt \\ &\le \sum_{i',i''} \Big(\int_{\mathbb{T}} |D^{i'}_{\mathbf{t}'}(\theta, D_t F)|^2 \, dt \Big)^{\frac{1}{2}} \Big(\int_{\mathbb{T}} |D^{i''}_{\mathbf{t}''}(\theta, D_t F)|^2 \, dt \Big)^{\frac{1}{2}}. \end{aligned}$$

Therefore we get

$$\begin{aligned} &\|R^F(\theta)\|^p_{m,p} \le \sum_{i \le m} E\Big[\Big(\int_{\mathbb{T}^i} |D^i_{\mathbf{t}} R^F(\theta)|^2 \, d\mathbf{t} \Big)^{\frac{p}{2}} \Big] \\ &\le \sum_{i',i'' \le m} E\Big[\Big(\int_{\mathbb{T}^{i+1}} |D^{i'}_{\mathbf{t}'}(\theta, D_t F)|^2 \, dt \, d\mathbf{t} \Big)^p \Big]^{\frac{1}{2}} E\Big[\Big(\int_{\mathbb{T}^{i+1}} |D^{i''}_{\mathbf{t}''}(\theta, D_t F)|^2 \, dt \, d\mathbf{t} \Big)^p \Big]^{\frac{1}{2}} \\ &\le \sum_{i',i'' \le m} \|DF\|^p_{i',2p} \|DF\|^p_{i'',2p} \le c' \|DF\|^{2p}_{m,2p}, \end{aligned}$$

for any θ. Then the assertion of the lemma follows. □

Now, we want to show that if F is nondegenerate, its weighted characteristic function $\psi_G(v)$ is rapidly decreasing.

Theorem 5.3.1 *For any $N \in \mathbb{N}$, there exist $m \in \mathbb{N}$, $p \ge 2$ and $c > 0$ such that the inequality*

$$\Big| E\big[e^{i(v,F)} G \big] \Big| \le \frac{c}{|v|^N} \Big(\|DF\|_{m,p} \sup_\theta \|R^F(\theta)^{-1}\|_{m,p} \Big)^N \|G\|_{m,p} \tag{5.32}$$

holds for all $|v| \ge 1$ for any nondegenerate Wiener functional F of $(\mathbf{D}^\infty)^d$ and $G \in \mathbf{D}^\infty$.

Proof For a nondegenerate smooth functional F, we will define a complex H-valued Wiener functional with parameter $v \in \mathbb{R}^d \setminus \{0\}$ by

$$X_t = X_t^{F,v} = \frac{-i(v, D_t F)}{|v|^2} \cdot R^F(\theta)^{-1},$$

where $\theta = v/|v|$. Then $X_t \in \mathbf{D}_{\mathbb{T}}^{\infty}$ by Proposition 5.3.1 and Lemma 5.2.1. Further, using the adjoint equality (5.8) for the Skorohod integral, we have

$$\begin{aligned} E\big[e^{i(v,F)}\delta(XG)\big] &= E\Big[\int_{\mathbb{T}} D_t(e^{i(v,F)})X_t G\,dt\Big] \\ &= E\Big[\int_{\mathbb{T}} e^{i(v,F)} \cdot i(v, D_t F)X_t G\,dt\Big] \\ &= E[e^{i(v,F)}G], \end{aligned} \tag{5.33}$$

since $\int i(v, D_t F)X_t\,dt = 1$. We will repeat this argument. Setting $\mathcal{L}(G) = \delta(XG)$, we have the iteration formula

$$E[e^{i(v,F)}G] = E[e^{i(v,F)}\mathcal{L}(G)] = \cdots = E[e^{i(v,F)}\mathcal{L}^N(G)]$$

for $N = 1, 2, \ldots$. Consequently, using Theorem 5.2.1 and Hölder's inequality, we get

$$\begin{aligned} |E[e^{i(v,F)}G]| &\le \|\mathcal{L}^N(G)\|_{0,2} \le \|\delta(X\mathcal{L}^{N-1}(G))\|_{0,2} \\ &\le c_1\|X\mathcal{L}^{N-1}(G)\|_{1,2} \le c_2\|X\|_{1,4}\|\mathcal{L}^{N-1}(G)\|_{1,4}. \end{aligned}$$

Repeat this argument for $\|\mathcal{L}^{N-1}(G)\|_{1,4}$. Then for any N, there exist increasing sequences $m_l, p_l, l = 1, \ldots, N$ and a positive constant c_N such that

$$|E[e^{i(v,F)}G]| \le c_N\|X\|_{m_1,p_1} \cdots \|X\|_{m_N,p_N}\|G\|_{m_N,p_N} \le c_N\|X\|_{m_N,p_N}^N\|G\|_{m_N,p_N}.$$

By Hölder's inequality, we have

$$\begin{aligned} \|X\|_{m,p} &\le \frac{c_{m,p}}{|v|^2}\|i(v, D)F\|_{m,2p}\|R^F(\theta)^{-1}\|_{m,2p} \\ &\le \frac{c_{m,p}}{|v|}\|DF\|_{m,2p}\sup_{\theta}\|R^F(\theta)^{-1}\|_{m,2p}. \end{aligned}$$

Therefore we get the inequality of the theorem. □

Theorem 5.3.2 *Suppose that a d-dimensional smooth Wiener functional F is nondegenerate. Then for any $G \in \mathbf{D}^{\infty}$, the law of F weighted by G has a rapidly decreasing C^{∞}-density. Further, the density function $f_G(x)$ and its derivatives are* *given by the Fourier inversion formula; for any multi-index* $\mathbf{j} = (j_1, \ldots, j_d)$ *of nonnegative integers, we have*

$$\partial^{\mathbf{j}} f_G(x) = \frac{(-i)^{|\mathbf{j}|}}{(2\pi)^d} \int_{\mathbb{R}^d} e^{-i(v,x)} \psi_G(v) v^{\mathbf{j}} \, dv, \tag{5.34}$$

where $\psi_G(v) = E[e^{i(v,F)} G]$.

Proof We saw in Theorem 5.3.1 that the function $\psi_G(v)$ is a rapidly decreasing C^∞-function of $v \in \mathbb{R}^d$. Therefore $|\psi_G(v)||v|^{n_0}$ is integrable for any $n_0 \in \mathbb{N}$. Then the weighted law $\mu_G(dx)$ has a C^{n_0}-density $f_G(x)$. Since this is valid for any n_0, $f_G(x)$ is actually a C^∞-function. It satisfies the formula (5.34) by Proposition 1.1.2.

Further, $\psi_G(v)$ is a C^∞-function of v. Then the density $f_G(x)$ is a rapidly decreasing function. Therefore $f_G(x)$ is a rapidly decreasing C^∞-function. □

We can replace the nondegenerate condition of a smooth Wiener functional F by Malliavin's condition using the determinant of the Malliavin covariance R^F. Let $\lambda_1 \geq \lambda_2 > \cdots \geq \lambda_d \geq 0$ be eigen-values of the matrix R^F. If $\lambda_d > 0$ a.s and it satisfies $E[\lambda_d^{-p}] < \infty$ for all $p > 1$, then we have $\sup_{\theta \in S_{d-1}} E[(\theta, R^F \theta)^{-p}] < \infty$ and hence F is nondegenerate. Note that the above is equivalent to that $E[(\lambda_1 \cdots \lambda_d)^{-p}] < \infty$ holds for all $p > 1$. Therefore, if $|\det R^F| > 0$ a.s. and $E[|\det R^F|^{-p}] < \infty$ holds for all $p > 1$, then F is nondegenerate. Therefore we get a cerebrated result in the Malliavin calculus for the Wiener space.

Theorem 5.3.3 (Malliavin [77], Malliavin–Thalmaier [80]) *Let* F *be a* d*-dimensional smooth Wiener functional. Assume that* $|\det R^F| > 0$ *a.s. and* $(\det R^F)^{-1} \in L^{\infty-}$. *Then* F *is nondegenerate and the law of* F *weighted by* $G \in \mathbf{D}^\infty$ *has a rapidly decreasing* C^∞*-density.*

The Malliavin criterion for the smooth density (Theorem 5.3.3) has been applied to diffusion processes whose generator is elliptic or further, hypo-elliptic. See Malliavin [77], Kusuoka–Stroock [69, 70] etc. We will discuss the problem in Chap. 6.

Remark (Formula of integration by parts) In literatures on Malliavin calculus, the formula of the integration by parts is often obtained and it is applied to the proof of the smooth density. We will derive the formula directly, using the property of adjoint operator δ stated in Theorem 5.2.2. Let $F = (F^1, \ldots, F^d)$ be a nondegenerate smooth functional. Let f be a C_b^∞ function on $\mathbb{R}^d$. Then $f(F)$ is a one-dimensional smooth functional. By Proposition 5.1.1, we have $D(f(F)) = \sum_{i=1}^d \partial_{x_i} f(F) DF^i$. Let (γ_{ij}) be the inverse matrix of the Malliavin covariance R^F. Then we have

$$\partial_{x_i} f(F) = \langle D(f(F)), \sum_j \gamma_{ij} DF^j \rangle_{\mathbb{T}}$$

for any $i = 1, \dots, d$. Let $G \in \mathbf{D}^\infty$. Then $G \sum_j \gamma_{ij} DF^j \in \mathbf{D}^\infty_{\mathbb{T}}$, since $\gamma_{ij} \in \mathbf{D}^\infty$ by Proposition 5.3.1. Then we get

$$\begin{aligned} E[\partial_{x_i} f(F) \cdot G] &= E[\langle D(f(F)), G \sum_j \gamma_{ij} DF^j \rangle] \\ &= E[f(F) \cdot \delta(G \sum_j \gamma_{ij} DF^j)]. \end{aligned}$$

Here we use the adjoint formula (5.8). Set $l_i(G) = \delta(G \sum_j \gamma_{ij} DF^j)$. It is a continuous linear operator from $\mathbf{D}^\infty$ into itself by Theorem 5.2.2 and Proposition 5.3.1. Then we get the *formula of the integration by parts*:

$$E\big[\partial_{x_i} f \circ F \cdot G\big] = E\big[f \circ F \cdot l_i(G)\big]. \tag{5.35}$$

Further, by induction, the formula is extended as

$$E\big[\partial_{x_{i_1}} \cdots \partial_{x_{i_n}} f \circ F \cdot G\big] = E\big[f \circ F \cdot l_{i_1,\dots,i_n}(G)\big], \tag{5.36}$$

where $l_{i_1,\dots,i_n}$ is a continuous linear operator from $\mathbf{D}^\infty$ into itself by Theorem 5.2.2.

We will give an alternative proof of the smooth density for the law of F, by using the above formula of the integration by parts. Set $f(x) = e^{i(v,x)}$, where $v = (v_1, \dots, v_d)$ and $x = (x_1, \dots, x_d)$. Then it holds that $\partial_{x_{i_1}} \cdots \partial_{x_{i_n}} f(x) = i^n v_{i_1} \cdots v_{i_n} e^{i(v,F)}$. Therefore, we get by (5.36)

$$i^n v_{i_1} \cdots v_{i_n} E[e^{i(v,F)} G] = E[e^{i(v,F)} l_{i_1,\dots,i_n}(G)].$$

Therefore if $|v_{i_1} \cdots v_{i_n}| \neq 0$, we have

$$\Big|E\big[e^{i(v,F)} G\big]\Big| \leq \frac{1}{|v_{i_1} \cdots v_{i_n}|} \|l_{i_1,\dots,i_n}(G)\|_{0,2}$$

for any n. Therefore the weighted characteristic function $\psi_G(v) = E[e^{i(v,F)} G]$ is a rapidly decreasing function. Further, since F has finite moments of any order, $\psi_G(v)$ is a C^∞-function. Then weighted law of F by G has a rapidly decreasing C^∞-density.

In this monograph, we did not use the formula of the integration by parts for the proof of Theorem 5.3.3, since the formula should not hold for the Poisson space, as we will discuss shortly.

5.4 Difference Operator and Adjoint on Poisson Space

We are also interested in the problem whether the law of a functional of a Poisson random measure has a smooth density. For the problem, we will study the Malliavin calculus on the space of Poisson random measures.

Let $\mathbb{T} = [0, T]$ and $\mathbb{R}_0^{d'} = \mathbb{R}^{d'} \setminus \{0\}$. Let $\mathbb{U} = \mathbb{T} \times \mathbb{R}_0^{d'}$ and $\mathcal{B}(\mathbb{U})$ be its Borel field. Let Ω be the set of all integer-valued measures ω on $\mathbb{U}$ and let $\mathcal{B}(\Omega)$ be the smallest σ-field of Ω with respect to which $\omega(E)$, $E \in \mathcal{B}(\mathbb{U})$ are measurable. Let P be a probability measure on $(\Omega, \mathcal{B}(\Omega))$ such that $N(dt\,dz) := \omega(dt\,dz)$ is a Poisson random measure with intensity measure $n(dt\,dz) = dt\nu(dz)$, where ν is a Lévy measure on $\mathbb{R}_0^{d'}$. The triple $(\Omega, \mathcal{B}(\Omega), P)$ is called a *space of Poisson random measure with the intensity n*.

We will define two transformations $\Omega \to \Omega$ following Picard [92]. For each $u = (t, z) \in \mathbb{U}$, we define two maps $\varepsilon_u^-, \varepsilon_u^+; \Omega \to \Omega$ by

$$\varepsilon_u^- \omega(E) = \omega(E \cap \{u\}^c), \quad \forall E \in \mathcal{B}(\mathbb{U}),$$
$$\varepsilon_u^+ \omega(E) = \omega(E \cap \{u\}^c) + 1_E(u), \quad \forall E \in \mathcal{B}(\mathbb{U}).$$

Then equality $\varepsilon_u^{\theta_1} \circ \varepsilon_u^{\theta_2} = \varepsilon_u^{\theta_1}$ holds for any u and $\theta_1 = \pm$ and $\theta_2 = \pm$. For any $\omega \in \Omega$, $\varepsilon_u^+ \omega = \omega$ holds for almost all u with respect to the measure N, in view of the definition of the measure N. Further, for almost all $\omega \in \Omega$, $\varepsilon_u^- \omega = \omega$ holds for almost all u with respect to the measure n, since the set of u such that $\varepsilon_u^- \omega \neq \omega$ is at most countable and countable sets are null sets with respect to the measure $n(du) = dt\nu(dz)$. Let us remark that the Poisson random measure $\omega \in \Omega$ has no point mass almost surely, i.e., $P(\omega(\{u\}) = 0) = 1$ for any $\{u\}$, since

$$E[\omega(\{u\})] = E[N(\{u\})] = n(\{u\}) = 0.$$

Then the equality $\varepsilon_u^+ \omega(E) = \omega(E) + \delta_u(E)$ holds for almost all ω.

Let S be a complete metric space. An S-valued $\mathcal{B}(\Omega)$-measurable function $F(\omega)$ is called an S-valued *Poisson functional*. If S is the space of real numbers, it is called simply a *Poisson functional*. Transformations $\varepsilon_u^\theta, \theta = \pm 1$ induce transformations for a Poisson functional $F(\omega)$. We denote $F(\varepsilon_u^\theta \omega)$ by $F \circ \varepsilon_u^\theta(\omega)$. Let $Z_u, u \in \mathbb{U}$ be a family of real valued Poisson functionals. We assume that it is measurable with respect to (u, ω) (random field). From the definitions of transformations ε_u^+ and ε_u^-, equalities

$$\int_{\mathbb{U}} Z_u \circ \varepsilon_u^+ N(du) = \int_{\mathbb{U}} Z_u N(du), \tag{5.37}$$

$$\int_{\mathbb{U}} Z_u \circ \varepsilon_u^- n(du) = \int_{\mathbb{U}} Z_u n(du) \tag{5.38}$$

hold for almost all ω, for any positive random field $Z_u, u \in \mathbb{U}$.

Lemma 5.4.1 (Picard [91]) *If a positive random field $Z_u, u \in \mathbb{U}$ satisfies $Z_u \circ \varepsilon_u^+ = Z_u \circ \varepsilon_u^-$ for any u almost surely, then*

$$E\Big[\int_{\mathbb{U}} Z_u N(du)\Big] = E\Big[\int_{\mathbb{U}} Z_u n(du)\Big]. \tag{5.39}$$

Proof Let $\mathcal{I}$ be a sub σ-field of $\mathcal{B}(\mathbb{U}) \times \mathcal{B}(\Omega)$ generated by sets $A \times B$, where $A \in \mathcal{B}(\mathbb{U})$ and $B \in \mathcal{B}_{A^c}(\Omega)$. Here $\mathcal{B}_{A^c}(\Omega)$ is the σ-field generated by $N(C), C \subset A^c$. If Z_u satisfies $Z_u \circ \varepsilon_u^+ = Z_u \circ \varepsilon_u^-$ for all u almost surely, then Z_u is $\mathcal{I}$-measurable. In fact, let Z_u be a step functional given by

$$Z_u(\omega) = \sum_{m=1}^{n} \phi_m(\omega) 1_{A_m}(u),$$

where $\{A_1, \ldots, A_n\}$ is a measurable partition of $\mathbb{U}$ and $\phi_m, m = 1, \ldots, n$ are bounded positive $\mathcal{B}(\Omega)$-measurable functionals. If $Z_u \circ \varepsilon_u^+ = Z_u \circ \varepsilon_u^-$ holds for any $u \in \mathbb{U}$, then $\phi_m \circ \varepsilon_u^+ = \phi_m \circ \varepsilon_u^-$ holds for any $u \in A_m, m = 1, \ldots, n$. Hence each ϕ_m is $\mathcal{B}_{A_m^c}(\Omega)$-measurable. Therefore, Z_u is $\mathcal{I}$-measurable. Further, since $\int Z_u N(du) = \sum_m \phi_m N(A_m)$ and, ϕ_m and $N(A_m)$ are independent for any m, we have

$$\begin{aligned} E\Big[\int_{\mathbb{U}} Z_u N(du)\Big] &= \sum_m E[\phi_m] E[N(A_m)] \\ &= E\Big[\sum_m \phi_m n(A_m)\Big] = E\Big[\int_{\mathbb{U}} Z_u n(du)\Big]. \end{aligned}$$

Now, for a given $\mathcal{I}$-measurable positive random field Z_u, there exists a sequence of positive $\mathcal{I}$-measurable step functionals Z_u^n such that $Z_u^n \uparrow Z_u$ as $n \uparrow \infty$ for all u almost surely. Since $E[\int Z_u^n N(du)] = E[\int Z_u^n n(du)]$ holds for any Z_u^n, the equality $E[\int Z_u N(du)] = E[\int Z_u n(du)]$ holds for the limit random field Z_u. □

Corollary 5.4.1 *For any positive random field Z_u, we have*

$$E\Big[\int_{\mathbb{U}} Z_u N(du)\Big] = E\Big[\int_{\mathbb{U}} Z_u \circ \varepsilon_u^+ n(du)\Big]. \tag{5.40}$$

Proof The functional $Y_u := Z_u \circ \varepsilon_u^+$ satisfies $Y_u \circ \varepsilon_u^+ = Y_u \circ \varepsilon_u^-$ for all u almost surely. Indeed, we have

$$(Z_u \circ \varepsilon_u^+) \circ \varepsilon_u^-(\omega) = Z_u(\varepsilon_u^+ \circ \varepsilon_u^- \omega) = Z_u(\varepsilon_u^+ \omega) = Z_u \circ \varepsilon_u^+(\omega),$$

$$(Z_u \circ \varepsilon_u^+) \circ \varepsilon_u^+(\omega) = Z_u(\varepsilon_u^+ \circ \varepsilon_u^+ \omega) = Z_u(\varepsilon_u^+ \omega) = Z_u \circ \varepsilon_u^+(\omega).$$

Therefore equality (5.39) holds for Y_u. Then, using (5.37) we get

$$E\Big[\int_{\mathbb{U}} Z_u N(du)\Big] = E\Big[\int_{\mathbb{U}} Z_u \circ \varepsilon_u^+ N(du)\Big] = E\Big[\int_{\mathbb{U}} Z_u \circ \varepsilon_u^+ n(du)\Big].$$

□

We should remark that transformations ε_u^+ may not be absolutely continuous, i.e., $P(A) = 0$ may not imply $P(\varepsilon_u^+ A) = 0$ for some $A \in \mathcal{F}$. Therefore, if a Poisson functional F is defined a.s. P, $F \circ \varepsilon_u^{\pm}$ may not be defined for any u a.s. However, $F \circ \varepsilon_u^+$ is well defined a.e. (u, ω) with respect to $n(du)P(d\omega)$. Indeed, if $F = 0$ holds a.s., then $E[\int I_\epsilon F N(du)] = 0$ holds, where $1_\epsilon(u) = 1_{|z|>\epsilon}(u)$. Then we have $E[\int F \circ \varepsilon_u^+ 1_\epsilon n(du)] = 0$ in view of Corollary 5.4.1. Therefore $F \circ \varepsilon_u^+ = 0$ holds a.e. $n(du)P(d\omega)$.

For a bounded Poisson functional F, we define *difference operator* $\{\tilde{D}_u, u \in \mathbb{U}\}$ by

$$\tilde{D}_u F = F \circ \varepsilon_u^+ - F. \tag{5.41}$$

It is well defined a.e. $n(du)P(d\omega)$. If $u = (t, z)$, we denote $\tilde{D}_{(t,z)}F$ by $\tilde{D}_{t,z}F$.

For $\mathbf{u} = (u_1, \ldots, u_j) \in \mathbb{U}^j$, we set $\varepsilon_{\mathbf{u}}^+ = \varepsilon_{u_1}^+ \circ \cdots \varepsilon_{u_j}^+$ and $\tilde{D}_{\mathbf{u}} = \tilde{D}_{\mathbf{u}}^j = \tilde{D}_{u_1} \cdots \tilde{D}_{u_j}$. $\tilde{D}_{\mathbf{u}}^j F$ is well defined a.e. $dn^j\, dP$, where n^j is the j-fold product of the measure n given by $n^j(du_1 \cdots du_j) = n(du_1) \times \cdots \times n(du_j)$. The functional $\tilde{D}_{\mathbf{u}}^j F$ is invariant by the permutation; let σ be a permutation of $\{1, \ldots, j\}$. For $\mathbf{u} = (u_1, \ldots, u_j)$, set $\mathbf{u}_\sigma = (u_{\sigma_1}, \ldots, u_{\sigma_j})$. Then it holds that $F \circ \varepsilon_{\mathbf{u}}^+ = F \circ \varepsilon_{\mathbf{u}_\sigma}^+$ a.e. $dn^j\, dP$. Further, $\tilde{D}_{\mathbf{u}}^j F = \tilde{D}_{\mathbf{u}_\sigma}^j F$ holds a.e. $dn^j\, dP$. Further, we should remark that if $u_i = u_j$ $(i \neq j)$ holds for some $\mathbf{u} = (u_1, \ldots, u_j)$, then $\tilde{D}_{\mathbf{u}} F = 0$ a.e.

If $j = 0$, we regard $\mathbb{U}^0$ is an empty set $\emptyset$, $\varepsilon_\emptyset^+$ is the identity transformation and we set $\tilde{D}_\emptyset F = F$.

For $u = (r, z) \in \mathbb{U}$, we set $\gamma(u) = |z| \wedge 1$. It is a positive function of $u = (r, z)$, since $|z| > 0$ holds for any u. For $\mathbf{u} = (u_1, \ldots, u_j) \in \mathbb{U}^j$, we set $\gamma(\mathbf{u}) = \gamma(u_1) \cdots \gamma(u_j)$ and $|\mathbf{u}| = \max_{1 \le i \le j} \gamma(u_i)$. A bounded Poisson functional F is called *smooth* if

$$\sup_{\mathbf{u} \in A(1)^j} \frac{E[|\tilde{D}_{\mathbf{u}}^j F|^p]}{\gamma(\mathbf{u})^p} < \infty$$

holds for any $j \in \mathbb{N}$ and $p \geq 2$. Here, 'sup' means the essential supremum with respect to the measure n^j. The set of all smooth functionals is denoted by $\tilde{\mathcal{D}}^\infty$. A d-dimensional functional $F = (F^1, \ldots, F^d)$ is called smooth if every component F^i is smooth. We denote by $(\tilde{\mathcal{D}}^\infty)^d$ the set of all d-dimensional smooth functionals.

Remark Let us compare the above $\tilde{D}_u$ with the derivative operator D_t on the Wiener space. Recall that the derivative operator D in the Wiener space is defined by (5.2). Heuristically, we may write it as

$$D_t F = \lim_{\lambda \to 0} \frac{F(w + \lambda \delta_t) - F(w)}{\lambda}, \quad a.e.\ dt\, P(dw),$$

where δ_t is the delta measure at the point $\{t\}$. Since $\varepsilon_u^+ \omega = w + \delta_u$ holds if $\omega(\{u\}) = 0$, $\tilde{D}_u$ may be regarded as a discretization of the derivative operator D_t.

Example We give an example in which $\tilde{D}F$ is computed simply. Let F be a linear functional of a Poisson random measure, which is written as $F = \int_{\mathbb{U}} h(s, z)\tilde{N}_D(ds\, dz)$, where $h(u)$ is a bounded function. We have

$$\begin{aligned} F \circ \varepsilon_{(t,z)}^+ &= \int_{\mathbb{U}} h(s, y)\{\tilde{N}_D(ds\, dy) + \delta_{t,z}(ds\, dy)\} \\ &= \int_{\mathbb{U}} h(s, y)\tilde{N}_D(ds\, dz) + h(t, z) = F + h(t, z) \end{aligned}$$

a.e. $dn\, dP$. This shows $\tilde{D}_{(t,z)} F = h(t, z)$ a.e. $dn\, dP$. Since it is a constant function, higher differences of F are 0, i.e., $\tilde{D}_{\mathbf{u}}^j F = 0$ holds a.e. $n^j(d\mathbf{u})P(d\omega)$ for $j \geq 2$.

Another interesting Poisson functional that we can compute $F \circ \varepsilon_u^+$ and $\tilde{D}_u F$ is a solution of an SDE with jumps discussed in Chap. 3. It will be studied in Sect. 6.4.

Let $Z_u, u \in \mathbb{U}$ be a real valued Poisson random field. We denote by $\tilde{\mathcal{D}}_{\mathbb{U}}^{\infty}$ the set of all Z_u such that

$$\frac{Z_u}{\gamma(u)}, \quad \frac{\tilde{D}_{\mathbf{v}}^j Z_u}{\gamma(\mathbf{v})\gamma(u)}, \quad j = 1, 2, \ldots$$

arc boundcd a.c. $n(du)P(d\omega)$ and $n^{j+1}(d\mathbf{v}\, du)P(d\omega)$, respectively. Further, $\tilde{\mathcal{D}}_{\mathbb{U}_0}^{\infty}$ denotes the set of $Z_u \in \tilde{\mathcal{D}}_{\mathbb{U}}^{\infty}$ such that $Z_u \circ \varepsilon_u^-$ are supported by a compact subset K of $\mathbb{U}$ a.e. $n(du)P(d\omega)$. For $Z \in \tilde{\mathcal{D}}_{\mathbb{U}_0}^{\infty}$, we define the *Skorohod integral of* Z_u *(by Poisson random measure)* by

$$\tilde{\delta}(Z) = \int_{\mathbb{U}} Z_u \circ \varepsilon_u^- \tilde{N}(du) = \int_{\mathbb{U}} Z_u \circ \varepsilon_u^- (N(du) - n(du)). \tag{5.42}$$

Then $\tilde{\delta}(Z) \in L^p$ for any $p \geq 2$. In fact if K is a compact subset of $\mathbb{U}$, $N(K)$ is Poisson distributed with intensity $n(K) < \infty$ and hence $E[N(K)^p] < \infty$ holds for any $p \geq 2$.

We will show that the operator $\tilde{\delta}$ is the adjoint of $\tilde{D}$ and that it has properties similar to those of δ (Skorohod integral by Wiener process).

Lemma 5.4.2

1. *(Picard [91]) For any* $Z \in \tilde{\mathcal{D}}^{\infty}_{\mathbb{U}_0}$, $\tilde{\delta}$ *satisfies the adjoint property*

$$E[G\tilde{\delta}(Z)] = E\Big[\int_{\mathbb{U}} (\tilde{D}_u G) Z_u n(du)\Big] \tag{5.43}$$

for any bounded Poisson functional G.

2. *For any* $Z \in \tilde{\mathcal{D}}^{\infty}_{\mathbb{U}_0}$, *the commutation relation holds:*

$$\tilde{D}_{u'}\tilde{\delta}(Z) = \tilde{\delta}(\tilde{D}_{u'}Z) + Z_{u'}, \quad a.e.\ n(du')P(d\omega). \tag{5.44}$$

3. *The isometric property*

$$E[\tilde{\delta}(Z)\tilde{\delta}(Y)] = E\Big[\int_{\mathbb{U}} Z_u Y_u n(du)\Big] + E\Big[\iint_{\mathbb{U}^2} \tilde{D}_u Z_v \tilde{D}_v Y_u n(du) n(dv)\Big] \tag{5.45}$$

holds for any $Z, Y \in \tilde{\mathcal{D}}^{\infty}_{\mathbb{U}_0}$.

Proof It holds that

$$E[G\tilde{\delta}(Z)] = E\Big[\int_{\mathbb{U}} G Z_u \circ \varepsilon_u^- N(du)\Big] - E\Big[\int_{\mathbb{U}} G Z_u \circ \varepsilon_u^- n(du)\Big].$$

In view of (5.40), the first term of the right-hand side is finite and is equal to

$$E\Big[\int_{\mathbb{U}} G \circ \varepsilon_u^+ Z_u \circ \varepsilon_u^- \circ \varepsilon_u^+ n(du)\Big] = E\Big[\int_{\mathbb{U}} G \circ \varepsilon_u^+ Z_u \circ \varepsilon_u^- n(du)\Big].$$

Since $Z_u \circ \varepsilon_u^- = Z_u$ holds a.e. $n(du)P(d\omega)$, we get (5.43).

We will next show the commutation relation. We have for any $u' \in \mathbb{U}$,

$$Z_u \circ \varepsilon_u^- \circ \varepsilon_{u'}^+ = \begin{cases} Z_u \circ \varepsilon_{u'}^+ \circ \varepsilon_u^-, & \text{if } u \neq u', \\ Z_{u'} \circ \varepsilon_{u'}^-, & \text{if } u = u'. \end{cases}$$

Therefore we get

$$\begin{aligned} \tilde{D}_{u'}\tilde{\delta}(Z) &= \tilde{\delta}(Z) \circ \varepsilon_{u'}^+ - \tilde{\delta}(Z) \\ &= \int Z_u \circ \varepsilon_{u'}^+ \circ \varepsilon_u^- \tilde{N}(du) + Z_{u'} \circ \varepsilon_{u'}^- - \int Z_u \circ \varepsilon_u^- \tilde{N}(du) \\ &= \tilde{\delta}(\tilde{D}_{u'}Z) + Z_{u'} \circ \varepsilon_{u'}^- \end{aligned}$$

a.e. $n(du')P(d\omega)$. Since $Z_{u'} \circ \varepsilon_{u'}^- = Z_{u'}$ holds a.e. $n(du')P(d\omega)$, we get the equality of the commutation relation.

We will show the isometric property. Using the adjoint property (5.43) and the commutation relation (5.44), we have

$$
\begin{aligned}
E[\tilde{\delta}(Z)\tilde{\delta}(Y)] &= E\Big[\int_{\mathbb{U}} Z_u \tilde{D}_u \tilde{\delta}(Y) n(du)\Big] \\
&= E\Big[\int_{\mathbb{U}} Z_u Y_u n(du)\Big] + E\Big[\int_{\mathbb{U}} Z_u \tilde{\delta}(\tilde{D}_u Y) n(du)\Big] \\
&= E\Big[\int_{\mathbb{U}} Z_u Y_u n(du)\Big] + E\Big[\int_{\mathbb{U}}\int_{\mathbb{U}} \tilde{D}_v Z_u \tilde{D}_u Y_v n(du) n(dv)\Big].
\end{aligned}
$$

□

We will consider the Skorohod integral $\tilde{\delta}(Z)$, when $Z_u = Z_{t,z}$ is predictable with respect to t. It turns out that $\tilde{\delta}(Z)$ coincides with the Itô integral. The following proposition corresponds to Proposition 5.2.2 for the Skorohod integral $\delta(X)$ by Wiener process.

Proposition 5.4.1 *Let $\{\mathcal{F}_t\}$ be the filtration generated by the Poisson random measure N. Suppose that $Z_{t,z} \in \tilde{\mathbb{D}}^{\infty}_{\mathbb{U}_0}$ is predictable with respect to t for all z. Then Skorohod integrals $\tilde{\delta}(Z)$ coincides with Itô integral $\int_{\mathbb{U}} Z_u \tilde{N}(du)$. Further, the isometric property holds:*

$$
E[\tilde{\delta}(Z)^2] = E\Big[\int_{\mathbb{U}} Z_u^2 n(du)\Big]. \tag{5.46}
$$

Proof Suppose that $Z_{t,z}$ is a simple predictable random field given by $Z_{t,z} = \sum_m \phi_z^m 1_{(t_{m-1},t_m]}(t)$, where ϕ_z^m is $\mathcal{F}_{t_{m-1}}$-measurable. Then it holds that $Z_{t,z} \circ \varepsilon^+_{(t,z)} = Z_{t,z}$. Therefore,

$$
\tilde{\delta}(Z) = \int_{\mathbb{U}} Z_u (N(du) - n(du)) = \int_{\mathbb{U}} Z_u \tilde{N}(du).
$$

The right-hand side coincides with the Itô integral.

If Z_u is predictable, we have $\tilde{D}_{t,z} Z_{s,z'} = 0$ for $t > s$. Then $\tilde{D}_v Z_u \tilde{D}_u Z_v = 0$ holds for any $u = (t, z)$, $v = (s, z')$ such that $t \neq s$ a.e. Therefore the last term in (5.45) is equal to 0 if Z_u is predictable. □

Finally, we show that $\tilde{\delta}$ has a property, similar to the definition of δ given by (5.7).

Proposition 5.4.2 *Let $Z_u = \sum_{l=1}^n F_l h_l(u)$, where $F_l \in \tilde{\mathbb{D}}^{\infty}$ and h_l are bounded measurable function on $\mathbb{U}$ with compact support. Then we have*

$$
\tilde{\delta}(Z) = \sum_{l=1}^n F_l \tilde{N}(h_l) - \sum_{l=1}^n \int \tilde{D}_u F_l h_l(u) n(du), \quad a.s. \tag{5.47}
$$

Proof We have

$$\tilde{\delta}(F_l h_l) = \int F_l \circ \varepsilon_u^- h_l(u)\tilde{N}(du) = F_l \int h_l(u)\tilde{N}(du) - \int (F_l - F_l \circ \varepsilon_u^-) h_l(u)\tilde{N}(du).$$

We shall compute the last term. It holds by (5.37) that

$$\int (F_l - F_l \circ \varepsilon_u^-) h_l(u) N(du) = \int (F_l \circ \varepsilon_u^+ - F_l \circ \varepsilon_u^- \circ \varepsilon_u^+) h_l(u) N(du) = 0.$$

Since $F_l - F_l \circ \varepsilon_u^- = F_l \circ \varepsilon_u^+ - F_l \circ \varepsilon_u^-$ holds by definitions of ε_u^+ and ε_u^-, we have

$$\begin{aligned}\int (F_l - F_l \circ \varepsilon_u^-) h_l(u)\, n(du) &= \int (F_l \circ \varepsilon_u^+ - F_l \circ \varepsilon_u^-) h_l(u) n(du) \\ &= \int (F_l \circ \varepsilon_u^+ - F_l) h_l(u) n(du) \\ &= \int \tilde{D}_u F_l h_l(u) n(du).\end{aligned}$$

Summing up these computations for $l = 1, 2, \ldots, n$, we get (5.47). □

5.5 Sobolev Norms for Poisson Functionals

In Sect. 5.2, we defined Sobolev norms for Wiener functionals, making use of differential operator $D_t, t \in \mathbb{T}$. In this section, we will define Sobolev norms for Poisson functionals, making use of the difference operator $\tilde{D}_u F, u \in \mathbb{U}$. We should be careful about the definition of Sobolev norms, since we want to apply it for proving the existence of the smooth density of the law of 'nondegenerate' Poisson functionals. Its definition will become more complicated than that of a Wiener functional, because the smooth density of a Poisson functional F will possibly be related to the asymptotic property of $\tilde{D}_u F$ as $\gamma(u)$ converges to 0. Then we should define Sobolev norms for $\tilde{D}_u F$, restricting u to subdomains A of $\mathbb{U}$, where A are open neighborhoods of the center of the intensity measure n.

Discussions in this section and those in Sects. 5.6, 5.7, and 5.8 are close to works by Ishikawa–Kunita–Tsuchiya [46], though notations for Sobolev norms are changed.

We begin with a preliminary observation. Let F be s smooth Poisson functional. For a positive integer n, an L^2-type Sobolev norm for F, restricted to a domain $A \subset \mathbb{U}$ such that $n(A) > 0$ should be given by

$$\Big\{ E[|F|^2] + \sum_{j=1}^{n} \int_{A^j} E[|\tilde{D}_{\mathbf{u}}^j F|^2] n^j(d\mathbf{u}) \Big\}^{\frac{1}{2}}. \tag{5.48}$$

Here A^j is the j-product set of A, included in the j-product set $\mathbb{U}^j$. Further, n^j is the j-product measure of the intensity measure n defined on $\mathbb{U}^j$. The norm is well defined, since functionals $\tilde{D}_{\mathbf{u}}^j F$ are invariant by the permutation of $\mathbf{u}$. We could write it as $\|F\|_{n,2;A}$, but in order to avoid the confusion with Sobolev norm $\|F\|_{n,2}$ for a Wiener functional, we will denote (5.48) by $\|F\|_{0,n,2;A}$. (The first suffix 0 indicates that the derivative operator DF is not involved. In Sect. 5.10, we will define Sobolev norms for Wiener–Poisson functionals denoted by $\|F\|_{m,n,p;A}$.) It is compatible with Sobolev norms of L^2-type $\|F\|_{m,2}$ for Wiener functionals.

In order to define Sobolev norms of L^p $(p > 2)$ type, we will transform infinite measures n^j on A^j to a probability measure m_A^j on A^j in the following way. Let m be a bounded measure on $\mathbb{U}$ defined by

$$m(E) = \int_E \gamma(u)^2 n(du).$$

We define a probability measure m_A on A and its product measure m_A^j on A^j by

$$m_A(E) = \frac{m(A \cap E)}{m(A)}, \quad m_A^j(d\mathbf{u}) = m_A(du_1) \cdots m_A(du_j),$$

respectively, where $\mathbf{u} = (u_1, \ldots, u_j) \in \mathbb{U}^j$. Further, we define positive functions on A and A^j by

$$\gamma_A(u) = \frac{\gamma(u)}{\sqrt{m(A)}}, \quad \gamma_A(\mathbf{u}) = \gamma_A(u_1) \cdots \gamma_A(u_j).$$

Note that $\frac{1}{\sqrt{m(A)}}$ is the normalizing constant for the function $\gamma(u)$, because we have $\int_A \gamma(u)^2 n(du) = m(A)$. Then we can rewrite the L^2-Sobolev norm (5.48) as

$$\|F\|_{0,n,2;A} = \Big\{ E[|F|^2] + \sum_{j=1}^{n} \int_{A^j} \frac{E[|\tilde{D}_{\mathbf{u}}^j F|^2]}{\gamma_A(\mathbf{u})^2} m_A^j(d\mathbf{u}) \Big\}^{\frac{1}{2}}.$$

Let $A(1) = \{u \in \mathbb{U}; |z| \leq 1\}$. Suppose that A is a domain in $A(1)$ which includes the center of the Lévy measure ν and satisfies $0 < m(A) \leq 1$. For given $j \in \mathbb{N}$, $p \geq 2$, we define a norm of $F \in \tilde{\mathcal{D}}^\infty$ by $\|F\|_{0,0,p;A} = E[|F|^p]^{\frac{1}{p}}$ if $n = 0$ and if $n \geq 1$,

$$\|F\|_{0,n,p;A} = \Big\{ E[|F|^p] + \sum_{j=1}^{n} \int_{A^j} \frac{E[|\tilde{D}_{\mathbf{u}}^j F|^p]}{\gamma_A(\mathbf{u})^p} m_A^j(d\mathbf{u}) \Big\}^{\frac{1}{p}}. \tag{5.49}$$

Remark The Sobolev norm for Wiener functionals given in (5.11) is of $L^{2,p}$ type; it is L^2-type with respect to t and is L^p-type with respect to w. But the Sobolev norm $\| \ \|_{0,n,p;A}$ for Poisson functionals given in (5.49) is of $L^{p,p}$-type. We defined Sobolev norms of $L^{p,p}$-type, because the difference rule of $\tilde{D}_u$ and the differential rule of D_t are distinct. In fact, a difference rule of the operator $\tilde{D}_u$ is given by

$$\tilde{D}_u(FG) = (\tilde{D}_u F)G + G\tilde{D}_u F + \tilde{D}_u F \tilde{D}_u G, \tag{5.50}$$

while the differential rule of the operator D_t is given by a simpler form $D_t(FG) = (D_t F)G + F D_t G$.

Let $\{Z_u, u \in \mathbb{U}\}$ be a smooth Poisson random field of $\tilde{\mathcal{D}}^\infty_{\mathbb{U}_0}$. We define its norms by

$$\|Z\|_{0,0,p;A} = \Big(\int_A \frac{E[|Z_u|^p]}{\gamma_A(u)^p} m_A(du)\Big)^{\frac{1}{p}}, \tag{5.51}$$

$$\|Z\|_{0,n,p;A} = \Big\{\|Z\|^p_{0,0,p;A} + \sum_{j=1}^n \int_{A^{j+1}} \frac{E[|\tilde{D}^j_{\mathbf{u}} Z_u|^p]}{\gamma_A(\mathbf{u})^p \gamma_A(u)^p} m^j_A(d\mathbf{u}) m_A(du)\Big\}^{\frac{1}{p}}.$$

When a set A is fixed in discussions, we will often drop '; A' in the definitions of Sobolev norms and write them simply as

$$\|F\|_{0,n,p} := \|F\|_{0,n,p;A}, \quad \|Z\|_{0,n,p} := \|Z\|_{0,n,p;A}.$$

Lemma 5.5.1 (Hölder's inequality) *For any $n \in \mathbb{N}$ and $p \geq 2$, there is a positive constant c not depending on the set A such that*

$$\|FG\|_{0,n,p;A} \leq c\|F\|_{0,n,pr;A}\|G\|_{0,n,pr';A}, \tag{5.52}$$

$$\|FZ\|_{0,n,p;A} \leq c\|F\|_{0,n,pr;A}\|Z\|_{0,n,pr';A} \tag{5.53}$$

hold for any $F, G \in \tilde{\mathcal{D}}^\infty$, $Z \in \tilde{\mathcal{D}}^\infty_{\mathbb{U}_0}$, and $r, r' > 1$ with $1/r + 1/r' = 1$.

Proof We will prove (5.53) only. By Hölder's inequality, we have

$$\|FZ\|^p_{0,0,p} = \int \frac{E[|FZ_u|^p]}{\gamma_A(u)^p} m_A(du) \leq E[|F|^{pr}]^{\frac{1}{r}} \int \frac{E[|Z_u|^{pr'}]^{\frac{1}{r'}}}{\gamma_A(u)^p} m_A(du)$$
$$\leq \|F\|^p_{0,0,pr}\|Z\|^p_{0,0,pr'}.$$

Note the difference rule (5.50) for $\tilde{D}$. We can show similarly to the above

$$\int \frac{E[\tilde{D}_{u'} F Z_u|^p]}{\gamma_A(u')^p \gamma_A(u)^p} m_A(du') m_A(du) \leq \|F\|^p_{0,1,pr}\|Z\|^p_{0,0,pr'},$$

$$\int \frac{E[F \tilde{D}_{u'} Z_u|^p]}{\gamma_A(u')^p \gamma_A(u)^p} m_A(du') m_A(du) \leq \|F\|^p_{0,0,pr}\|Z\|^p_{0,1,pr'}.$$

Note that $\frac{m(A)^{1/2}}{\gamma_A(u')} = \frac{1}{\gamma(u')} \geq 1$ for $u' \in A$. Then

$$\begin{aligned}
&\int \frac{E[\tilde{D}_{u'} F \tilde{D}_{u'} Z_u|^p]}{\gamma_A(u')^p \gamma_A(u)^p} m_A(du') m_A(du) \\
&\leq m(A)^{\frac{p}{2}} \int \frac{E[\tilde{D}_{u'} F \tilde{D}_{u'} Z_u|^p]}{\gamma_A(u')^p \cdot \gamma_A(u')^p \gamma_A(u)^p} m_A(du') m_A(du) \\
&\leq \|F\|_{0,1,pr}^p \|Z\|_{0,1,pr'}^p.
\end{aligned}$$

Repeating this argument for $j = 2, 3, \ldots$, we get (5.53) for any n. □

We will define another norms that are stronger than Sobolev norms defined above. For $n \in \mathbb{N}$ and $p \geq 2$, new norms are defined by

$$\|F\|_{0,n,p}^{\sharp} = \Big\{ E[|F|^p] + \sum_{j=1}^{n} \sup_{\mathbf{u} \in A(1)^j} \frac{E[|\tilde{D}_{\mathbf{u}}^j F|^p]}{\gamma(\mathbf{u})^p} \Big\}^{\frac{1}{p}}, \tag{5.54}$$

$$\|Z\|_{0,n,p}^{\sharp} = \Big\{ \sup_{u \in A(1)} \frac{E[|Z_u|^p]}{\gamma(u)^p} + \sum_{j=1}^{n} \sup_{(\mathbf{u},u) \in A(1)^{j+1}} \frac{E[|\tilde{D}_{\mathbf{u}}^j Z_u|^p]}{\gamma(\mathbf{u})^p \gamma(u)^p} \Big\}^{\frac{1}{p}}. \tag{5.55}$$

Here, $\sup_{u \in A(1)} |\varphi(u)|$ etc. means the essential supremum of the function $\varphi(u)$ on the measurable space $(A(1), n)$, which is defined by

$$\inf \big\{ c \in \mathbb{R};\ \ |\varphi(u)| \leq c \text{ for almost every } u \in A(1) \text{ with respect to measure } n \big\}.$$

Norms $\|F\|_{0,0,p}^{\sharp}$ and $\|Z\|_{0,0,p}^{\sharp}$ are defined by cutting off terms $\sum_{j=1}^{n}$ in the above definitions. We may regard them as Sobolev norms of $L^{\infty,p}$-type, i.e., the essential supremum is taken for variable $u \in \mathbb{U}$ with respect to the measure n and the L^p-norm is taken with respect to the measure P. Thus these norms should be bigger than Sobolev norms of $L^{p,p}$-type. Indeed, we have the inequalities;

$$\begin{aligned}
\|F\|_{0,n,p;A} &\leq \|F\|_{0,n,p}^{\sharp}, \quad F \in \tilde{\mathcal{D}}^{\infty}, \\
\|Z\|_{0,n,p;A} &\leq \|Z\|_{0,n,p}^{\sharp}, \quad Z \in \tilde{\mathcal{D}}_{\mathbb{U}_0}^{\infty},
\end{aligned} \tag{5.56}$$

for any n, p and A.

We denote by $\tilde{\mathbf{D}}^{0,n,p}$ the completion of $\tilde{\mathcal{D}}^{\infty}$ by the norm $\| \ \|_{0,n,p}^{\sharp}$ and we set $\tilde{\mathbf{D}}^{\infty} = \bigcap_{n,p} \tilde{\mathbf{D}}^{0,n,p}$. Elements of $\tilde{\mathbf{D}}^{\infty}$ are called smooth Poisson functionals. Let $\tilde{\mathbf{D}}_{\mathbb{U}}^{0,n,p}$ be the completion of $\tilde{\mathcal{D}}_{\mathbb{U}_0}^{\infty}$ by the norm $\| \ \|_{0,n,p}^{\sharp}$. We set $\tilde{\mathbf{D}}_{\mathbb{U}}^{\infty} = \bigcap_{n,p} \tilde{\mathbf{D}}_{\mathbb{U}}^{0,n,p}$. Then both $\tilde{\mathbf{D}}^{\infty}$ and $\tilde{\mathbf{D}}_{\mathbb{U}}^{\infty}$ are algebras and are F-spaces with respect to countable norms $\| \ \|_{0,n,p}^{\sharp}$, $n, p \in \mathbb{N}$.

Then for any $F \in \tilde{\mathbf{D}}^\infty$ and $Z \in \tilde{\mathbf{D}}^\infty_{\mathbb{U}}$, norms $\|F\|_{0,n,p;A}$ and $\|Z\|_{0,n,p;A}$ are well defined and are finite for any n, p and $A \subset A(1)$.

It is not difficult to prove the following proposition.

Proposition 5.5.1 *The operator $\tilde{D}_u$ is extended to a continuous linear operator from F-space $\tilde{\mathbf{D}}^\infty$ to F-space $\tilde{\mathbf{D}}^\infty_{\mathbb{U}}$. Further, it holds that $\|\tilde{D}F\|^\sharp_{0,n,p} \leq \|F\|^\sharp_{0,n+1,p}$ for any $F \in \tilde{\mathbf{D}}^\infty$ for any n, p.*

Later, we will use another norms for Poisson functionals. For $n \in \mathbb{N}$, $p \geq 2$, we will define norms of F by

$$|F|^\sharp_{0,n,p} = \Big(E[|F|^p] + \sum_{j=1}^n \sup_{\mathbf{u} \in A(1)^j} E[|F \circ \varepsilon^+_{\mathbf{u}}|^p]\Big)^{\frac{1}{p}}. \tag{5.57}$$

We will study relations of these norms.

Proposition 5.5.2 *For any n, p, the norm $|\ |^\sharp_{0,n,p}$ is weaker than the norm $\|\ \|^\sharp_{0,n,p}$; there is a positive constant $c_{n,p}$ such that*

$$|F|^\sharp_{0,n,p} \leq c_{n,p} \|F\|^\sharp_{0,n,p}, \quad \forall F \in \tilde{\mathbf{D}}^\infty. \tag{5.58}$$

Proof Let $\mathbf{u} \in A(1)^j$. We can rewrite $F \circ \varepsilon^+_{\mathbf{u}}$ as

$$F \circ \varepsilon^+_{\mathbf{u}} = \sum_{\mathbf{v} \subset \mathbf{u}} \tilde{D}_{\mathbf{v}} F = F + \sum_{i=1}^j \sum_{\mathbf{v} \subset \mathbf{u}, \sharp \mathbf{v} = i} \tilde{D}_{\mathbf{v}} F. \tag{5.59}$$

Therefore we have

$$\sup_{\mathbf{u}} E[|F \circ \varepsilon^+_{\mathbf{u}}|^p] \leq 2^j \Big(E[|F|^p] + \sum_{i=1}^j \sup_{\mathbf{v}} E[|\tilde{D}^i_{\mathbf{v}} F|^p]\Big).$$

Then, introducing another norms by

$$|F|^{\sharp'}_{0,n,p} = \Big\{E[|F|^p] + \sum_{i=1}^n \sup_{\mathbf{v} \in A(1)^j} E[|\tilde{D}^i_{\mathbf{v}} F|^p]\Big\}^{\frac{1}{p}},$$

we have the inequality $(|F|^\sharp_{0,n,p})^p \leq (n+1)2^n (|F|^{\sharp'}_{0,n,p})^p$. We can show similarly the inequality $(|F|^{\sharp'}_{0,n,p})^p \leq (n+1)2^n (|F|^\sharp_{0,n,p})^p$. Therefore two norms $|F|^\sharp_{0,n,p}$ and $|F|^{\sharp'}_{0,n,p}$ are equivalent. Further, since $|F|^{\sharp'}_{0,n,p} \leq \|F\|^\sharp_{0,n,p}$ holds, the norm $|F|^\sharp_{0,n,p}$ is weaker than the norm $\|F\|^\sharp_{0,n,p}$ for any n, p. □

Let $\tilde{\mathbf{L}}^{0,n,p}$ be the completion of $\tilde{\mathcal{D}}^{\infty}$ by the norm $|\ |^{\sharp}_{0,n,p}$. We set $\tilde{\mathbf{L}}^{\infty} = \bigcap_{n,p} \tilde{\mathbf{L}}^{0,n,p}$. Then $\tilde{\mathbf{L}}^{\infty}$ is an F-space and the relation $\tilde{\mathbf{D}}^{\infty} \subset \tilde{\mathbf{L}}^{\infty}$ holds.

For a Poisson functional X_t with parameter $t \in [0, 1)$, we set

$$|X|^{\sharp}_{0,n,p} = \Big(\int_{\mathbb{T}} (|X_t|^{\sharp}_{0,n,p})^p \, dt \Big)^{\frac{1}{p}}. \tag{5.60}$$

Note Sobolev norms for Poisson functionals are defined in several ways. Ishikawa et al. defined Sobolev norms for Poisson functionals in [46]. Notations of their Sobolev norms are changed in this monograph. Norms $\|\ \|_{n,p;A}$, $\|\ \|^{*}_{n,p}$ and $|||\ |||^{*}_{n,p}$ in [46] correspond to our norms $\|\ \|_{0,n,p}$, $\|\ \|^{\sharp}_{0,n,p}$ and $|\ |^{\sharp}_{0,n,p}$, respectively.

Hayashi–Ishikawa [37] defined another Sobolev norms. Their norm $\|\ \|_{n,0,p,(\rho)}$ is close to our norm $\|\ \|_{0,n,p;A(\rho)}$, but these two are not equivalent norms.

5.6 Estimations of Two Poisson Functionals by Sobolev Norms

Let $n(dt\,dz) = dt\nu(dz)$ be the intensity measure of the Poisson random measure. The set $\{u = (t, z);\ z = 0\}$ is called the *center of the intensity measure* n. For a given family of star-shaped neighborhoods $\{A_0(\rho)\}$ of $0 \in \mathbb{R}^{d'}$ (see Sect. 1.2), we set $A(\rho) = \{u = (t, z) \in \mathbb{U};\ z \in A_0(\rho)\}$ and call $\{A(\rho)\}$ the family of star-shaped neighborhoods of the center of the intensity n. We define the function $\varphi(\rho)$ by

$$\varphi(\rho) := m(A(\rho)) = T \int_{A_0(\rho)} |z|^2 \nu(dz) = T\varphi_0(\rho). \tag{5.61}$$

Now, suppose that the Lévy measure ν satisfies the order condition of exponent α, then the function $\varphi(\rho)$ satisfies the order condition of exponent α. We will fix $0 < \rho_0 \leq 1$ such that $\varphi(\rho_0) \leq 1$.

We are interested in estimations of Poisson functionals by Sobolev norms $\{\|\ \|_{0,n,p;A(\rho)};\, 0 < \rho < \rho_0\}$ and their dependence on the parameter ρ. In this section, we will give estimations of two Poisson functionals. The first one is the Skorhod integral $\tilde{\delta}(Z\chi_\rho)$, where $Z \in \tilde{\mathcal{D}}^{\infty}_{\mathbb{U}_0}$ and χ_ρ is the indicator function $I_{A(\rho)}$ of the set $A(\rho)$. We first study the L^2-estimate. If $p = 2$, the Sobolev norm is written simply as

$$\|Z\|_{0,n,2;A(\rho)} = \Big\{ \int_{A(\rho)} E[|Z_u|^2] n(du) + \sum_{j=1}^{n} \int_{A(\rho)^{j+1}} E[|\tilde{D}^j_{\mathbf{u}} Z_u|^2] n^j(d\mathbf{u}) n(du) \Big\}^{\frac{1}{2}}.$$

Then, using the isometric property (5.45), we have

$$E[\tilde{\delta}(Z\chi_\rho)^2] = E\Big[\int_{A(\rho)} Z_u^2 n(du)\Big] + E\Big[\iint_{A(\rho)^2} \tilde{D}_u Z_v \tilde{D}_v Z_u n(du)n(dv)\Big]$$
$$\leq \|Z\|_{0,1,2;A(\rho)}^2$$

for any $0 < \rho < \rho_0$. Further, using the commutation relation for $\tilde{D}_u$ and $\tilde{\delta}$, we can show easily that for any positive integer n there is a positive constant c_n such that

$$\|\tilde{\delta}(Z\chi_\rho)\|_{0,n,2;A(\rho)} \leq c_n \|Z\|_{0,n+1,2;A(\rho)}^2 \tag{5.62}$$

holds for all $0 < \rho < \rho_0$.

We want to extend the above inequality for any $p \geq 2$. The following is the counterpart to Theorem 5.2.1, where we obtained an estimation of the Skorohod integral for Wiener functionals.

Theorem 5.6.1 ([46]) *Given $n \in \mathbb{N}$ and a positive even number $p \geq 4$, there exists a positive constant $c_{n,p}$ such that the inequality*

$$\|\tilde{\delta}(Z\chi_\rho)\|_{0,n,p;A(\rho)} \leq c_{n,p} \|Z\|_{0,n+p-1,p;A(\rho)} \tag{5.63}$$

holds for any $Z \in \tilde{\mathcal{D}}_{\mathbb{U}_0}^\infty$, where the constant $c_{n,p}$ does not depend on the sets $\{A(\rho), 0 < \rho < \rho_0\}$.

The proof of the theorem is similar to that for Theorem 5.2.1, but details are more complicated. We shall first consider the estimation of $E[\tilde{\delta}(Z\chi_\rho)^p]$ for positive even number $p \geq 4$. We prepare a lemma.

Let $1 \leq q \leq p$ be a positive integer. An element $\mathbf{u} = (u_1, \ldots, u_q)$ of the product set $\mathbb{U}^q$ is written as $\{u_1, \ldots, u_q\}$. For $\boldsymbol{\varepsilon} = (\varepsilon_1, \ldots, \varepsilon_q) \in \{0,1\}^q$ and $\mathbf{u} = \{u_1, \ldots, u_q\}$, we set $\tilde{D}_{\mathbf{u}}^{\varepsilon} = \tilde{D}_{u_1}^{\varepsilon_1} \cdots \tilde{D}_{u_q}^{\varepsilon_l} (= \tilde{D}_{u_q}^{\varepsilon_l} \cdots \tilde{D}_{u_1}^{\varepsilon_1})$.

Lemma 5.6.1 *Let $p \geq 4$ be an even number. Let $Z \in \tilde{\mathcal{D}}_{\mathbb{U}_0}^\infty$. Then $E[\tilde{\delta}(Z)^p]$ is written as sums of the following multiple integrals by $n^q(d\mathbf{u})$, $q = 1, \ldots, p$, where $n^q(d\mathbf{u}) = n(du_1) \cdots n(du_q)$.*

$$\int_{\mathbb{U}^q} E\Big[\Big(\prod_{j=1}^{p_1} \tilde{D}_{\mathbf{u}_1}^{\varepsilon_{1j}} Z_{u_1}\Big) \cdots \Big(\prod_{j=1}^{p_q} \tilde{D}_{\mathbf{u}_q}^{\varepsilon_{qj}} Z_{u_q}\Big)\Big] n^q(d\mathbf{u}). \tag{5.64}$$

Here, $p_i \in \mathbb{N}$ satisfy $p_1 + \cdots + p_q = p$ and $\varepsilon_{ij}, i = 1, \ldots, q, j = 1, \ldots, p_i$ are given by $\varepsilon_{ij} := (\varepsilon_{ij}(1), \ldots, \varepsilon_{ij}(q-1)) \in \{0,1\}^{q-1}$. Further, $\mathbf{u}_i; i = 1, \ldots, q$ are given by $\mathbf{u}_i := \mathbf{u} - \{u_i\} = \{u_i(1), \ldots, u_i(q-1)\}$, which satisfy

$$\mathbf{u} = \{u_1, \ldots, u_q\} \subset \bigcup_{i=1}^{q} \bigcup_{j=1}^{p_i} \bigcup_{k=1}^{q-1} \{u_i(k); \varepsilon_{ij}(k) = 1\}. \tag{5.65}$$

Proof Our discussion is close to the proof of Theorem 5.2.1. We have by the adjoint formula (5.43)

$$E[\tilde{\delta}(Z)^p] = E\Big[\int_{\mathbb{U}} Z_{u_1}\tilde{D}_{u_1}\tilde{\delta}(Z)^{p-1}n(du_1)\Big].$$

We will decrease the power $p-1$ of $\tilde{\delta}(Z)^{p-1}$ step by step. Using the difference rule of $\tilde{D}_u$ and the commutation relation (5.44), we have

$$\begin{aligned}\tilde{D}_{u_1}\tilde{\delta}(Z)^{p-1} &= \tilde{D}_{u_1}\tilde{\delta}(Z)\tilde{\delta}(Z)^{p-2} + \tilde{\delta}(Z)\tilde{D}_{u_1}\tilde{\delta}(Z)^{p-2} + \tilde{D}_{u_1}\tilde{\delta}(Z)\tilde{D}_{u_1}\tilde{\delta}(Z)^{p-2}\\ &= (Z_{u_1} + \tilde{\delta}(\tilde{D}_{u_1}Z))(\tilde{\delta}(Z)^{p-2} + \tilde{D}_{u_1}\tilde{\delta}(Z)^{p-2}) + \tilde{\delta}(Z)\tilde{D}_{u_1}\tilde{\delta}(Z)^{p-2}.\end{aligned}$$

Substitute the above and apply the adjoint formula again. Then we see that $E[\tilde{\delta}(Z)^p]$ is written as sums of terms

$$E\Big[\int_{\mathbb{U}^2} \tilde{D}_{u_2}^{j_1}Z_{u_1}^{i_1}\tilde{D}_{u_1}^{j_2}Z_{u_2}^{i_2}\tilde{D}_{u_2}^{j_2'}\tilde{D}_{u_1}^{j_1'}\tilde{\delta}(Z)^{p-2}n(du_1)n(du_2)\Big],$$

where $i_1, i_2, j_1, j_2, j_1', j_2'$ are nonnegative integers satisfying $i_1 + i_2 = 2$ and $j_1, j_2, j_1', j_2' \le 1$. Here $\tilde{D}_u^0$ is the identity transformation.

Repeating this procedure inductively, we find that $E[\tilde{\delta}(Z)^p]$ is written as sums of terms stated in (5.64). For further details, see [46]. □

Proof of Theorem 5.6.1 We will first prove the inequality (5.63) in the case $n = 0$. By Lemma 5.6.1, $E[\tilde{\delta}(Z\chi_\rho)^p]$ is written as a sum of terms (5.64), replacing Z by $Z\chi_\rho$. In this case, (5.64) is written as

$$\begin{aligned}&\int_{A(\rho)^q} E\Big[\Big(\prod_{j=1}^{p_1}\tilde{D}_{\mathbf{u}_1}^{\varepsilon_{1j}}Z_{u_1}\Big)\cdots\Big(\prod_{j=1}^{p_q}\tilde{D}_{\mathbf{u}_q}^{\varepsilon_{qj}}Z_{u_q}\Big)\Big]n^q(d\mathbf{u})\\ &= E\Big[\int_{A(\rho)^q}\frac{(\prod_{j=1}^{p_1}\tilde{D}_{\mathbf{u}_1}^{\varepsilon_{1j}}Z_{u_1})\cdots(\prod_{j=1}^{p_q}\tilde{D}_{\mathbf{u}_q}^{\varepsilon_{qj}}Z_{u_q})}{\gamma_{A(\rho)}(u_1)^2\cdots\gamma_{A(\rho)}(u_q)^2}m_{A(\rho)}^q(d\mathbf{u})\Big],\end{aligned}$$

since $\frac{m^q_{A(\rho)}(d\mathbf{u})}{\gamma_{A(\rho)}(u_1)^2\cdots\gamma_{A(\rho)}(u_q)^2} = 1_{A(\rho)^q}n^q(d\mathbf{u})$. Let us consider the nominator in the above expression. For $\mathbf{u} = (u_1, \ldots, u_k)$ and $\varepsilon \in \{0, 1\}^k$, we set

$$\gamma_{A(\rho)}^{\varepsilon}(\mathbf{u}) = \gamma_{A(\rho)}(u_1)^{\varepsilon_1}\cdots\gamma_{A(\rho)}(u_k)^{\varepsilon_k}.$$

We will show that there is a positive constant $C = C_p$ (not depending on $0 < \rho < 1$) such that the inequality

$$\frac{1}{\gamma_{A(\rho)}(u_1)^2\cdots\gamma_{A(\rho)}(u_q)^2} \tag{5.66}$$
$$\le \frac{C}{\left(\prod_j \gamma_{A(\rho)}(\mathbf{u}_1)^{\varepsilon_{1j}}\right)\cdots\left(\prod_j \gamma_{A(\rho)}(\mathbf{u}_q)^{\varepsilon_{qj}}\right)\gamma_{A(\rho)}(u_1)^{p_1}\cdots\gamma_{A(\rho)}(u_q)^{p_q}}$$

holds for any $\mathbf{u}\in A(\rho)^q$. Note first that by the order condition of the Lévy measure, $\varphi(\rho)\ge c\rho^\alpha$ holds for $0<\rho<\rho_0$. Then, since $m(A(\rho))=\varphi(\rho)$, we have for any $|u|\le\rho<\rho_0$,

$$\gamma_{A(\rho)}(u)=\frac{\gamma(u)}{\varphi(\rho)^{\frac{1}{2}}}\le\frac{1}{c^{\frac{1}{2}}}\rho^{1-\frac{\alpha}{2}}\le c^{-\frac{1}{2}}.$$

Now, denote by $\sharp\{\varepsilon_{ij}(k)=1\}$ the total number of triples (i,j,k) such that $\varepsilon_{ij}(k)=1$. We set $r=\sharp\{\varepsilon_{ij}(k)=1\}+p-2q$. Since $\sharp\{\varepsilon_{ij}(k)=1\}\ge q$ holds by the relation (5.65), r is a nonnegative integer. Then there are $u_{i_1},\ldots,u_{i_r}\subset\mathbf{u}$ such that

$$\frac{\left(\prod_j \gamma_{A(\rho)}(\mathbf{u}_1)^{\varepsilon_{1j}}\right)\cdots\left(\prod_j \gamma_{A(\rho)}(\mathbf{u}_q)^{\varepsilon_{qj}}\right)\gamma_{A(\rho)}(u_1)^{p_1}\cdots\gamma_{A(\rho)}(u_q)^{p_q}}{\gamma_{A(\rho)}(u_1)^2\cdots\gamma_{A(\rho)}(u_q)^2}$$
$$=\gamma_{A(\rho)}(u_{i_1})\cdots\gamma_{A(\rho)}(u_{i_r}).$$

The last term of the above is less than or equal to $c^{-r/2}$ for any $\mathbf{u}\in A(\rho)^q$. Therefore, setting $C=c^{-r/2}$, we get the inequality (5.66).

The inequality (5.66) implies

$$E\left[\int_{A(\rho)^q}\frac{|\prod_{j=1}^{p_1}\tilde{D}_{\mathbf{u}_1}^{\varepsilon_{1j}}Z_{u_1}|\cdots|\prod_{j=1}^{p_q}\tilde{D}_{\mathbf{u}_q}^{\varepsilon_{qj}}Z_{u_q}|}{\gamma_{A(\rho)}(u_1)^2\cdots\gamma_{A(\rho)}(u_q)^2}m^q_{A(\rho)}(d\mathbf{u})\right] \tag{5.67}$$
$$\le CE\left[\int_{A(\rho)^q}\prod_{j=1}^{p_1}\frac{|\tilde{D}_{\mathbf{u}_1}^{\varepsilon_{1j}}Z_{u_1}|}{\gamma_{A(\rho)}(\mathbf{u}_1)^{\varepsilon_{1j}}\gamma_{A(\rho)}(u_1)}\cdots\prod_{j=1}^{p_q}\frac{|\tilde{D}_{\mathbf{u}_q}^{\varepsilon_{qj}}Z_{u_q}|}{\gamma_{A(\rho)}(\mathbf{u}_q)^{\varepsilon_{qj}}\gamma_{A(\rho)}(u_q)}m^q_{A(\rho)}(d\mathbf{u})\right]$$

for any $0<\rho<\rho_0$ and any $Z\in\tilde{\mathcal{D}}^\infty_{\mathbb{U}_0}$. In the formula (5.67), the total number of product terms is equal to p. Apply Hölder's inequality to (5.67) with respect to the product measure $dm^q_{A(\rho)}\,dP$. Then (5.67) is dominated by

$$C\prod_{i,j}E\left[\int_{A(\rho)^q}\left|\frac{\tilde{D}_{\mathbf{u}_i}^{\varepsilon_{ij}}Z_{u_i}}{\gamma_{A(\rho)}(\mathbf{u}_i)^{\varepsilon^{ij}}\gamma_{A(\rho)}(u_i)}\right|^p m^q_{A(\rho)}(d\mathbf{u})\right]^{\frac{1}{p}}\le C\|Z\|^p_{0,q-1,p;A(\rho)}.$$

Therefore, the absolute value of (5.64) is dominated by $C'\|Z\|^p_{0,q-1,p;A(\rho)}$ for all $0<\rho<\rho_0$ and $Z\in\tilde{\mathcal{D}}^\infty_{\mathbb{U}_0}$. Note that $E[\tilde{\delta}(Z\chi_\rho)^p]$ is written as a finite sum of terms (5.64) for $q = 1, \ldots, p$. Since norms $\| \ \|_{0,n,p;A(\rho)}$ are nondecreasing with respect to n, $|E[\tilde{\delta}(Z\chi_\rho)^p]|$ is dominated by $C''\|Z\|^p_{0,p-1,p;A(\rho)}$ for all $0 < \rho < \rho_0$ and $Z \in \tilde{\mathcal{D}}^\infty_{\mathbb{U}_0}$ with another constant C''. Consequently, we get

$$\|\tilde{\delta}(Z\chi_\rho)\|^p_{0,0,p} = E[\tilde{\delta}(Z\chi_\rho)^p] \le C''\|Z\|^p_{0,p-1,p;A(\rho)}. \tag{5.68}$$

Next, we will consider the case $n \ge 1$. It holds, by the definition of Sobolev norms that

$$\|\tilde{\delta}(Z\chi_\rho)\|^p_{0,n,p;A(\rho)} = E[|\tilde{\delta}(Z\chi_\rho)|^p] + \sum_{j=1}^n \int_{A(\rho)^j} \frac{E\big[\big|\tilde{D}^j_{\mathbf{v}}\tilde{\delta}(Z\chi_\rho)\big|^p\big]}{\gamma_{A(\rho)}(\mathbf{v})^p} m^j_{A(\rho)}(d\mathbf{v}). \tag{5.69}$$

The first term of the right-hand side has been computed at (5.68). We shall consider other terms. Take any integer $1 \le j \le n$. Note the commutation relation of $\tilde{\delta}$; $\tilde{D}_v\tilde{\delta}(Z\chi_\rho) = \tilde{\delta}(\tilde{D}_v Z\chi_\rho) + Z_v\chi_\rho$. Then we have

$$\tilde{D}^j_{\mathbf{v}}\tilde{\delta}(Z\chi_\rho) = \tilde{\delta}(\tilde{D}^j_{\mathbf{v}} Z\chi_\rho) + \sum_i \tilde{D}^{j-1}_{\mathbf{v_i}} Z_{v_i}\chi_\rho,$$

where $\mathbf{v} = (v_1, \ldots, v_j)$ and $\mathbf{v}_i = \mathbf{v} - \{v_i\}, i = 1, \ldots, j$. Therefore, there is a positive constant c_p such that

$$\begin{aligned}&\int_{A(\rho)^j} \frac{E\big[\big|\tilde{D}^j_{\mathbf{v}}\tilde{\delta}(Z\chi_\rho)\big|^p\big]}{\gamma_{A(\rho)}(\mathbf{v})^p} m^j_{A(\rho)}(d\mathbf{v}) \\ &\le c_p\Big\{\int \frac{E\big[\big|\tilde{\delta}(\tilde{D}^j_{\mathbf{v}} Z\chi_\rho)\big|^p\big]}{\gamma_{A(\rho)}(\mathbf{v})^p} m^j_{A(\rho)}(d\mathbf{v}) + \sum_{i=1}^j \int \frac{E[|\tilde{D}^{j-1}_{\mathbf{v}_i} Z_{v_i}\chi_\rho|^p]}{\gamma_{A(\rho)}(\mathbf{v})^p} m^j_{A(\rho)}(d\mathbf{v})\Big\}.\end{aligned} \tag{5.70}$$

It holds that

$$\begin{aligned}E[\tilde{\delta}(\tilde{D}^j_{\mathbf{v}} Z\chi_\rho)^p] &\le c^p_{0,p}\|\tilde{D}^j_{\mathbf{v}} Z\chi_\rho\|^p_{0,p-1,p;A(\rho)} \\ &= c^p_{0,p}\sum_{j' \le p-1}\int \frac{E\big[\big|\tilde{D}^{j'}_{\mathbf{u}}\tilde{D}^j_{\mathbf{v}} Z_v\chi_\rho\big|^p\big]}{\gamma_{A(\rho)}(\mathbf{u})^p\gamma_{A(\rho)}(v)^p} m^{j'}_{A(\rho)}(d\mathbf{u}) m_{A_\rho}(dv),\end{aligned}$$

in view of (5.68) for the case $n = 0$. Then we have

$$\int_{A(\rho)^j} \frac{E\left[\tilde{\delta}(\tilde{D}_{\mathbf{u}}^j Z\chi_\rho)^p\right]}{\gamma_{A(\rho)}(\mathbf{v})^p} m_{A(\rho)}^j(d\mathbf{v})$$

$$\leq c_{0,p}^p \sum_{j' \leq p-1} \int \frac{E\left[|\tilde{D}_{\mathbf{u}}^{j'} \tilde{D}_{\mathbf{v}}^j Z_v \chi_\rho|^p\right]}{\gamma_{A(\rho)}(\mathbf{u})^p \gamma_{A(\rho)}(\mathbf{v})^p \gamma_{A(\rho)}(v)^p} m_{A(\rho)}^{j'}(d\mathbf{u}) m_{A(\rho)}^j(d\mathbf{v}) m_{A(\rho)}(dv)$$

$$\leq c_{0,p}^p \sum_{j' \leq p-1} \|Z\chi_\rho\|_{0,j+j',p;A(\rho)}^p \leq c\|Z\chi_\rho\|_{0,j+p-1,p;A(\rho)}^p. \tag{5.71}$$

Therefore, the first term of the right-hand side of (5.70) is dominated by $C\|Z\chi_\rho\|_{0,j+p-1,p;A(\rho)}^p$. We have further,

$$\int_{A(\rho)^j} \frac{E\left[|\tilde{D}_{\mathbf{v}_i}^{j-1} Z_{v_i} \chi_\rho|^p\right]}{\gamma_{A(\rho)}(\mathbf{v})^p} m_{A(\rho)}^k(d\mathbf{v}) \leq \|Z\chi_\rho\|_{0,j-1,p;A(\rho)}^p. \tag{5.72}$$

Since the fact is valid for any $j = 1, \ldots, n$, we obtain from (5.70),

$$\|\tilde{\delta}(Z\chi_\rho)\|_{0,n,p;A(\rho)}^p \leq c_{n,p}^p \|Z\chi_\rho\|_{0,n+p-1,p;A(\rho)}^p.$$

This proves the assertion of the theorem. □

We will summarize properties of Skorohod integrals by Poisson random measure.

Theorem 5.6.2

1. *Skorohod integral operator* $\tilde{\delta}$ *is extended to a continuous linear operator from F-space* $\tilde{\mathbf{D}}_{\mathbb{U}}^\infty$ *to F-space* $\tilde{\mathbf{D}}^\infty$*. Further, for any* $n \in \mathbb{N}$ *and* $p \geq 2$*, there exists a positive constant* $c_{n,p}$ *such that for any* $0 < \rho < \rho_0$

$$\|\tilde{\delta}(Z\chi_\rho)\|_{0,n,p;A(\rho)} \leq c_{n,p}\|Z\|_{0,n+\langle p\rangle-1,\langle p\rangle:A(\rho)} \tag{5.73}$$

   holds for any $Z \in \tilde{\mathbf{D}}_{\mathbb{U}}^\infty$*, where* $\langle p\rangle = -2[-p/2]$.
2. *Adjoint formula* (5.43) *holds for any* $Z \in \tilde{\mathbf{D}}_{\mathbb{U}}^\infty$ *and* $G \in \tilde{\mathbf{D}}^\infty$.
3. *Commutation relation* (5.44) *holds for any* $Z \in \tilde{\mathbf{D}}_{\mathbb{U}}^\infty$.
4. *The isometric property* (5.45) *holds for any* $Y, Z \in \tilde{\mathbf{D}}_{\mathbb{U}}^\infty$.
5. *Hölder's inequalities* (5.52), (5.53) *hold for any* $F, G \in \tilde{\mathbf{D}}^\infty$ *and* $Z \in \tilde{\mathbf{D}}_{\mathbb{U}}^\infty$. *Further, similar Hölder's inequalities are valid for norms* $\| \ \|_{0,n,p}^\sharp$.

A d-dimensional Poisson functional $F = (F^1, \ldots, F^d)$ is said to belong to the space $(\tilde{\mathbf{D}}^\infty)^d$, if all components $F^j, j = 1, \ldots, d$ belong to $\tilde{\mathbf{D}}^\infty$. Sobolev norms $\|F\|_{0,n,p;A}$ and $\|F\|_{0,n,p}^\sharp$ are defined as sums of the corresponding norms for components $F^j, j = 1, \ldots, d$.

Difference operator $\tilde{D}$ and its adjoint $\tilde{\delta}$ can be extended to complex-valued functionals. The space of complex-valued smooth Poisson functionals is denoted again by $\tilde{\mathbf{D}}^\infty$ and the space of complex $\mathbb{U}$-valued smooth functionals is denoted by $\tilde{\mathbf{D}}^\infty_{\mathbb{U}}$. Proposition 5.5.1 and Theorem 5.6.1 are valid for complex-valued functionals.

We will next study the estimation of the exponential functional $e^{i(v,\tilde{D}F)} - 1$ with respect to Sobolev norms. In the next lemma, we are interested in the dependence of the estimation with respect to parameter ρ.

Lemma 5.6.2 *For any $n \in \mathbb{N}$ and $p \geq 2$ there exists a positive constant $c_{n,p}$ such that for any $F \in (\tilde{\mathbf{D}}^\infty)^d$, $v \in \mathbb{R}^d$ and $0 < \rho < \rho_0$,*

$$\begin{aligned} &\|e^{i(v,\tilde{D}F)} - 1\|_{0,n,p;A(\rho)} \\ &\leq c_{n,p}|v|\varphi(\rho)^{\frac{1}{2}}\|\tilde{D}F\|^\sharp_{0,n,(n+1)p}(1 + |v|\rho\varphi(\rho)^{\frac{1}{2}}\|\tilde{D}F\|^\sharp_{0,n,(n+1)p})^n. \end{aligned} \tag{5.74}$$

Proof The norm $\|e^{i(v,\tilde{D}F)} - 1\|_{0,n,p;A(\rho)}$ is dominated by the sum of the following terms for $j = 0, \ldots, n$:

$$E\left[\int_{A(\rho)^{j+1}} \left(\frac{|\tilde{D}^j_{\mathbf{u}}(e^{i(v,\tilde{D}_uF)} - 1)|}{\gamma_{A(\rho)}(\mathbf{u})\gamma_{A(\rho)}(u)}\right)^p m^j_{A(\rho)}(d\mathbf{u})m_{A(\rho)}(du)\right]^{\frac{1}{p}}. \tag{5.75}$$

We will compute the functional $\tilde{D}^j_{\mathbf{u}}(e^{i(v,\tilde{D}_uF)} - 1)$. It is not simple, since we have to use the difference rule (5.50). If $j = 1$, we have

$$\begin{aligned} \tilde{D}_{u'}(e^{i(v,\tilde{D}_uF)} - 1) &= e^{i(v,\tilde{D}_uF\circ\varepsilon^+_{u'})} - e^{i(v,\tilde{D}_uF)} \\ &= e^{i(v,\tilde{D}_{u'}\tilde{D}_uF)+i(v,\tilde{D}_uF)} - e^{i(v,\tilde{D}_uF)} = e^{i(v,\tilde{D}_uF)}(e^{i(v,\tilde{D}_{u'}\tilde{D}_uF)} - 1). \end{aligned}$$

If $j = 2$, we have by the difference rule (5.50)

$$\begin{aligned} \tilde{D}_{u''}\tilde{D}_{u'}(e^{i(v,\tilde{D}_uF)} - 1) &= e^{i(v,\tilde{D}_uF)}(e^{i(v,\tilde{D}_{u''}\tilde{D}_uF)} - 1)(e^{i(v,\tilde{D}_{u'}\tilde{D}_uF)} - 1) \\ &\quad + e^{i(v,\tilde{D}_uF)}e^{i(v,\tilde{D}_{u'}\tilde{D}_uF)}(e^{i(v,\tilde{D}_{u''}\tilde{D}_{u'}\tilde{D}_uF)} - 1) \\ &\quad + e^{i(v,\tilde{D}_uF)}(e^{i(v,\tilde{D}_{u''}\tilde{D}_uF)} - 1)e^{i(v,\tilde{D}_{u'}\tilde{D}_uF)} \\ &\qquad (e^{i(v,\tilde{D}_{u''}\tilde{D}_{u'}\tilde{D}_uF)} - 1). \end{aligned}$$

Repeating this argument, we have

$$\tilde{D}^j_{\mathbf{u}}(e^{i(v,\tilde{D}_uF)} - 1) = \sum e^{iZ^{j_0,\ldots,j_j}_{\mathbf{u},u}} \prod_{l=0}^{j}(e^{i(v,\tilde{D}^{j_l}_{\mathbf{u}_l}\tilde{D}_uF)} - 1),$$

where $j_l, l = 0, \ldots, j$ are nonnegative integers satisfying $j_l \leq j$ and $j_0+\cdots+j_j \geq j$. $\mathbf{u}_l, l = 0, \ldots, j$ are subsets of $\mathbf{u}$ satisfying $\bigcup_l \mathbf{u}_l = \mathbf{u}$. Further, $Z^{j_0,\ldots,j_j}_{\mathbf{u},u}$ are sums of real functionals written as $(v, \tilde{D}^{j_l'}_{\mathbf{u}_l'}\tilde{D}_u F)$, $j_l' \le j_l - 1$, $\mathbf{u}_l' \subset \mathbf{u}_l$. The summation is taken for all such $j_0, \ldots, j_j$ and $\mathbf{u}_0, \ldots, \mathbf{u}_j$. Then it holds that $|e^{iZ^{j_0,\ldots,j_j}_{\mathbf{u},u}}| = 1$. Therefore, (5.75) is dominated by a sum of terms

$$E\Big[\int_{A(\rho)^{j+1}}\Big(\frac{\prod_{l=0}^{j}|e^{i(v,\tilde{D}^{j_l}_{\mathbf{u}_l}\tilde{D}_uF)}-1|}{\gamma_{A(\rho)}(\mathbf{u})\gamma_{A(\rho)}(u)}\Big)^p m^j_{A(\rho)}(d\mathbf{u})m_{A(\rho)}(du)\Big]^{\frac{1}{p}}$$

$$\le \varphi(\rho)^{\frac{j+1}{2}} \sup_{(\mathbf{u},u)\in A(\rho)^{j+1}} E\Big[\Big(\frac{\prod_{l=0}^{j}|e^{i(v,\tilde{D}^{j_l}_{\mathbf{u}_l}\tilde{D}_uF)}-1|}{\gamma(\mathbf{u})\gamma(u)}\Big)^p\Big]^{\frac{1}{p}}. \tag{5.76}$$

In the above we used the equality $\gamma_{A_\rho}(\mathbf{u})\gamma_{A(\rho)}(u) = \varphi(\rho)^{-\frac{j+1}{2}}\gamma(\mathbf{u})\gamma(u)$. Further, we have the inequality

$$\frac{\prod_{l=0}^{j}|e^{i(v,\tilde{D}^{j_l}_{\mathbf{u}_l}\tilde{D}_uF)}-1|}{\gamma(\mathbf{u})\gamma(u)} \le \frac{|e^{i(v,\tilde{D}^{j_0}_{\mathbf{u}_0}\tilde{D}_uF)}-1|}{\gamma(\mathbf{u}_0)\gamma(u)}\prod_{l=1}^{j}\frac{\rho|e^{i(v,\tilde{D}^{j_l}_{\mathbf{u}_l}\tilde{D}_uF)}-1|}{\gamma(\mathbf{u}_l)\gamma(u)},$$

if $\gamma(u) \le \rho$. In fact, since $\mathbf{u} \subset \bigcup_{l=0}^{j}\mathbf{u}_l$, we have $\gamma(\mathbf{u}) \ge \prod_{l=0}^{j}\gamma(\mathbf{u}_l)$ and further, $\frac{\rho}{\gamma(u)} \ge 1$ holds for $u \in A(\rho)$. Therefore, applying Hölder's inequality, (5.76) is dominated by

$$\varphi(\rho)^{\frac{j+1}{2}} \sup_{(\mathbf{u},u)\in A(\rho)^{j+1}} E\Big[\Big(\frac{|e^{i(v,\tilde{D}^{j_0}_{\mathbf{u}_0}\tilde{D}_uF)}-1|}{\gamma(\mathbf{u}_0)\gamma(u)}\Big)^p\prod_{l=1}^{j}\Big(\frac{\rho|e^{i(v,\tilde{D}^{j_l}_{\mathbf{u}_l}\tilde{D}_uF)}-1|}{\gamma(\mathbf{u}_l)\gamma(u)}\Big)^p\Big]^{\frac{1}{p}}$$

$$\le \varphi(\rho)^{\frac{j+1}{2}} I_0\cdot I_1\cdots I_j, \tag{5.77}$$

where

$$I_0 := \sup_{(\mathbf{u}_0,u)\in A(\rho)^{j+1}} E\Big[\Big(\frac{|e^{i(v,\tilde{D}^{j_0}_{\mathbf{u}_0}\tilde{D}_uF)}-1|}{\gamma(\mathbf{u}_0)\gamma(u)}\Big)^{(j+1)p}\Big]^{\frac{1}{(j+1)p}},$$

$$I_l := \sup_{(\mathbf{u}_l,u)\in A(\rho)^{j+1}} E\Big[\Big(\frac{\rho|e^{i(v,\tilde{D}^{j_l}_{\mathbf{u}_l}\tilde{D}_uF)}-1|}{\gamma(\mathbf{u}_l)\gamma(u)}\Big)^{(j+1)p}\Big]^{\frac{1}{(j+1)p}}, \quad l = 1,\ldots,j.$$

Since $|e^{i(v,\tilde{D}^{j_l}_{\mathbf{u}_0}\tilde{D}_uF)}-1| \le |(v,\tilde{D}^{j_0}_{\mathbf{u}_0}\tilde{D}_uF)|$ holds, we have

$$I_0 \le |v|\|\tilde{D}F\|^\sharp_{0,j_0,(j+1)p}, \quad I_l \le \rho|v|\|\tilde{D}F\|^\sharp_{0,j_l,(j+1)p}, \quad l = 1,\ldots,j.$$

Therefore, (5.76) is dominated by

$$c|v|^{(j+1)}\rho^j\varphi(\rho)^{\frac{j+1}{2}}(\|\tilde{D}F\|^{\sharp}_{0,j,(j+1)p})^{(j+1)}. \tag{5.78}$$

Hence (5.75) is also dominated by (5.78) (with a different constant c). Finally, summing up the above inequality for $j = 0, \dots, n$, we get the inequality of the lemma. □

5.7 Nondegenerate Poisson Functionals

Let F be an element of $(\tilde{\mathbf{D}}^{\infty})^d$. We are interested in the existence of the smooth density of the law of F. Conditions needed could be related to 'Malliavin covariance'. The *Malliavin covariance* of a smooth d-dimensional Poisson functional $F \in (\tilde{\mathbf{D}}^{\infty})^d$ conditioned to the set $A(\rho)$ is defined by

$$\tilde{R}^F_\rho = \frac{1}{\varphi(\rho)}\int_{A(\rho)} \tilde{D}_u F(\tilde{D}_u F)^T n(du), \tag{5.79}$$

where $\varphi(\rho) = n(A(\rho))$. Set $\tilde{R}^F_\rho(\theta) = (\theta, \tilde{R}^F_\rho\theta)$. F is called *nondegenerate* if $\tilde{R}^F_\rho(\theta)$ are invertible a.s., inverses belong to $\tilde{\mathbf{L}}^{\infty}$ for any $\theta \in S_{d-1}$ and satisfy

$$\sup_{(\rho,\theta)} |\tilde{R}^F_\rho(\theta)^{-1}|^{\sharp}_{0,n,p} < \infty, \quad \forall n, p,$$

where the supremum is taken for all $0 < \rho < \rho_0$ and $\theta \in S_{d-1}$.

For a Wiener functional G, we saw in Lemma 5.3.1 that if $G \in \mathbf{D}^{\infty}$ is invertible and the inverse G^{-1} belongs to $L^{\infty-}$, then G^{-1} is a smooth Wiener functional belonging to $\mathbf{D}^{\infty}$. We remark the similar fact for Poisson functionals G.

Lemma 5.7.1 *Suppose that a Poisson functional $G \in \tilde{\mathbf{D}}^{\infty}$ is invertible and G^{-1} belongs to $\tilde{\mathbf{L}}^{\infty}$. Then $G^{-1} \in \tilde{\mathbf{D}}^{\infty}$. Further, for any n, p, there is a positive constant $c_{n,p}$ such that*

$$\|G^{-1}\|^{\sharp}_{0,n,p} \le c_{n,p}(1 + \|G\|^{\sharp}_{0,n,2qp})^q(1 + |G^{-1}|^{\sharp}_{0,n,2qp})^q \tag{5.80}$$

holds for any G with the above property, where $q = 2^n - 1$.

The proof is not simple. It will be given in a more general form in Theorem 5.10.2.

By the above lemma, the nondegenerate condition for a Poisson functional $F \in (\tilde{\mathbf{D}}^{\infty})^d$ is equivalent to the condition

$$\sup_{(\rho,\theta)} \|\tilde{R}^F_\rho(\theta)^{-1}\|^{\sharp}_{0,n,p} < \infty, \quad \forall n, p.$$

In applications, however, it is not easy to handle the above nondegenerate condition. We are interested in the limit of the Malliavin covariances $\tilde{R}_\rho^F$ as $\rho \to 0$. For this, let us introduce a regularity condition for a Poisson functional. It is convenient to set $\tilde{D}_{t,0}F = 0$. Then the operator $\tilde{D}_{t,z}F$ $(= \tilde{D}_{(t,z)}F)$ is defined for any (t, z). A Poisson functional F is called *regular* if it satisfies the following.

Condition (R) $\tilde{D}_{t,z}F$ *is twice continuously differentiable with respect to* $z = (z^1, \ldots, z^{d'}) \in \mathbb{R}^{d'}$ *a.s. Set*

$$\Lambda_t(F) = \sum_{i=1}^{d'} \sup_{|z|\le 1} |\partial_{z^i} \tilde{D}_{t,z}F| + \sum_{i,j=1}^{d'} \sup_{|z|\le 1} |\partial_{z^i}\partial_{z^j} \tilde{D}_{t,z}F|. \tag{5.81}$$

Then $|\Lambda(F)|_{0,n,p}^{\sharp} < \infty$ *holds for any* n *and* p.

We shall rewrite the Malliavin covariance of F in the case where $F \in (\tilde{\mathbf{D}}^\infty)^d$ is regular. Since the measure $n(dt\,dz)$ is equal to $dt\nu(dz)$, the covariance can be written as

$$\tilde{R}_\rho^F = \int_{\mathbb{T}} \Big(\int_{A_0(\rho)} \tilde{D}_{t,z}F\tilde{D}_{t,z}F^T \nu_{A_0(\rho)}(dz) \Big)\, dt,$$

where $\nu_{A_0(\rho)}$ is a measure on $A_0(\rho)$ given by $\varphi(\rho)^{-1}\nu(dz)$. Set

$$\partial\tilde{D}_{t,0}F = (\partial_{z^1}\tilde{D}_{t,z}F, \ldots, \partial_{z^{d'}}\tilde{D}_{t,z}F)|_{z=0}. \tag{5.82}$$

We assume that the family of matrices $\Gamma_\rho = \Big(\int_{A_0(\rho)} z^i z^j \nu_{A_0(\rho)}(dz) \Big)$ converges to a matrix Γ_0 as $\rho \to 0$. Then $\tilde{R}_\rho^F$ converges to K^F, where

$$K^F = \int_{\mathbb{T}} \partial\tilde{D}_{t,0}F\,\Gamma_0(\partial\tilde{D}_{t,0}F)^T\, dt. \tag{5.83}$$

The functional K^F is called the *Malliavin covariance of* F *at the center*. Further, F is called *nondegenerate at the center* if $K^F(\theta) \equiv (\theta, K^F\theta)$ are invertible for any $\theta \in S_{d-1}$ and inverses satisfy

$$\sup_\theta |K^F(\theta)^{-1}|_{0,n,p}^{\sharp} < \infty, \quad \forall n, p. \tag{5.84}$$

Here the supremum for θ is taken in the set S_{d-1}.

Example We give a simple example where the Malliavin covariance at the center is computed directly. Let $F = (F^1, \ldots, F^d)$ be a linear functional of Poisson random measure written as $F^i = \int_{\mathbb{U}} h^i(s, z)\tilde{N}_D(ds\,dz)$. We saw in Sect. 5.4 the equality $\tilde{D}_{(t,z)}F = h(t, z)$ a.e. Therefore,

$$\tilde{R}_\rho^F = \int_{\mathbb{T}} \Big(\int h(t,z) h(t,z)^T \nu_{A0(\rho)}(dz) \Big) dt.$$

If $h(s,z)$ is differentiable in z, we have $\partial \tilde{D}_{t,0} F = \partial h(t,z)|_{z=0} = (h^{ij}(t)) \equiv H(t)$ a.e. Consequently, the Malliavin covariance of F at the center is given by

$$K^F = \int_{\mathbb{T}} H(t) \Gamma_0 H(t)^T \, dt.$$

It is a non-random matrix. Then $K^F \circ \varepsilon_{\mathbf{u}}^+ = K^F$ holds a.s. for any $\mathbf{u}$. Therefore, if the matrix K^F is positive definite (invertible), then $|K^F(\theta)^{-1}|_{0,n,p}^\sharp = K^F(\theta)^{-1}$. Hence F is nondegenerate at the center

The law of F is infinitely divisible and its characteristic function is given by

$$\psi(v) = \exp\Big\{ \int_{\mathbb{T}} \Big(\int_{\mathbb{R}^{d'}} e^{i(v,h(s,z))} - i(v, h(s,z) 1_D(z)) \nu(dz) \Big) ds \Big\}.$$

We can show similarly to the proof of Lemma 1.2.1 the inequality

$$-\mathcal{R}\Big(\int_{\mathbb{T}} \Big(\int_{\mathbb{R}_0^{d'}} \Big\{ e^{i(v,h(s,z))} - 1 - i(v, h(s,z)) 1_D(z) \Big\} \nu(dz) \Big) ds \Big) \geq \sum_{i,j} \frac{v_i}{|v|^{\frac{\alpha}{2}}} \frac{v_j}{|v|^{\frac{\alpha}{2}}} K_\epsilon^{ij},$$

where $K_\epsilon = \int_{\mathbb{T}} H(s) \Gamma_\epsilon H(s)^T \, ds$. Since $\{K_\epsilon\}$ are uniformly positive definite by our assumption, inequality $|E[e^{i(v,F)}]| \leq c_2 \exp\{-c_1 |v|^{2-\alpha}\}$ holds with positive constants c_1 and c_2, as in Lemma 1.2.1. Then the law of F has a C^∞-density.

We want to show that if $F \in (\tilde{\mathbf{D}}^\infty)^d$ is regular and is nondegenerate at the center, its law has a smooth density: However, a direct proof of the fact is difficult. As a first step, we will introduce another technical nondegenerate condition. Associated with $F \in (\tilde{\mathbf{D}}^\infty)^d$, we define a family of Poisson functionals with parameter $(\rho, v) \in (0,1) \times (\mathbb{R}^d \setminus \{0\})$:

$$Q^F(\rho, v) = \frac{1}{|v|^2 \varphi(\rho)} \int_{A(\rho)} |e^{i(v, \tilde{D}_u F)} - 1|^2 n(du). \tag{5.85}$$

Let δ be a constant satisfying $1 < \delta < 2/\alpha$, where α is the exponent of the Lévy measure. We fix it and define for $(\rho, \theta) \in (0, \rho_0) \times S_{d-1}$,

$$Q_\rho^F(\theta) = Q^F(\rho, \rho^{-\frac{1}{\delta}} \theta). \tag{5.86}$$

A Poisson functional F of $(\tilde{\mathbf{D}}^\infty)^d$ is called nondegenerate with respect to Q_ρ^F or simply δ-*nondegenerate* if $\{Q_\rho^F(\theta)\}$ are invertible a.s. for any (ρ, θ) and inverses satisfy

$$\sup_{(\rho,\theta)} |Q_\rho^F(\theta)^{-1}|_{0,n,p}^\sharp < \infty, \quad \forall n, p, \tag{5.87}$$

where the supremum is taken for $0 < \rho < \rho_0$ and $\theta \in S_{d-1}$.

We have defined three types of nondegenerate condition for a d-dimensional smooth Poisson functional F. In the remainder of this section, we will show that if F is δ-nondegenerate for some $1 < \delta < 2/\alpha$, its law has a rapidly decreasing C^∞-density. In the next section, we will show that these three nondegenerate conditions are equivalent.

The next one corresponds to Proposition 5.3.1 for Wiener functionals.

Proposition 5.7.1 *For any n, p, there exists a positive constant $c_{n,p}$ such that for any δ-nondegenerate F of $(\tilde{\mathbf{D}}^\infty)^d$,*

$$\begin{aligned} &\|Q_\rho^F(\theta)^{-1}\|_{0,n,p}^\sharp \\ &\le c_{n,p}\Big(1 + \|\tilde{D}F\|_{0,n,(n+1)2^{n+2}p}^\sharp\Big)^{(n+1)2^{n+2}} \Big(1 + |Q_\rho^F(\theta)^{-1}|_{0,n,2^{n+2}p}^\sharp\Big)^{2^n} \end{aligned} \tag{5.88}$$

holds for all (ρ, θ).

Proof We will show that for any $n \in \mathbb{N}$, $p \ge 2$, there exists a positive constant c such that

$$\sup_{(\rho,\theta)} \|Q_\rho^F(\theta)\|_{0,n,p}^\sharp \le c(1 + \|\tilde{D}F\|_{0,n,2(n+1)p}^\sharp)^{2(n+1)}. \tag{5.89}$$

Note that $Q_\rho^F(\theta)$ is rewritten as

$$Q_\rho^F(\theta) = \frac{1}{|v|^2}\int_{A(\rho)} \frac{|e^{i(v,\tilde{D}_uF)} - 1|^2}{\gamma(u)^2} m_A(du), \tag{5.90}$$

where $\rho = |v|^{-\delta}$. Since $|e^{i(v,\tilde{D}_uF)} - 1|^2 \le |(v, \tilde{D}_uF)|^2$, we have

$$\|Q_\rho^F(\theta)\|_{0,0,p}^\sharp = E[Q_\rho^F(\theta)^p]^{\frac{1}{p}} \le \|\tilde{D}F\|_{0,0,p}^\sharp. \tag{5.91}$$

Further, if $1 \le j \le n$ and $p \ge 2$,

$$\frac{E\big[|\tilde{D}_{\mathbf{u}}^j Q_\rho^F(\theta)|^p\big]^{\frac{1}{p}}}{\gamma(\mathbf{u})} \le \frac{1}{|v|^2}\frac{1}{\varphi(\rho)}\int_{A(\rho)} \frac{E\big[|\tilde{D}_{\mathbf{u}}^j(|e^{i(v,\tilde{D}_uF)} - 1|^2)|^p\big]^{\frac{1}{p}}}{\gamma(\mathbf{u})\gamma(u)^2} m_A(du),$$

where $\rho = |v|^{-\delta}$. We can show similarly to the proof of Lemma 5.6.2 that there is a positive constant c' such that the above is dominated by

$$c'\frac{|v|^{2(j+1)}\rho^{2j}}{|v|^2}(1+\|\tilde{D}F\|^{\sharp}_{0,j,2(j+1)p})^{2(j+1)} \le c'(1+\|\tilde{D}F\|^{\sharp}_{0,j,2(j+1)p})^{2(j+1)},$$

since $|v|\rho \le 1$. Therefore we get (5.89).

Now we will apply (5.80) for $G = Q^F_\rho(\theta)$. Then (5.88) follows. □

We are now in the position of establishing the smooth density of the law of a δ-nondegenerate Poisson functional.

Theorem 5.7.1 ([46]) *For any $N \in \mathbb{N}$, there exist $n \ge 1$, $p \ge 2$ and a positive constant c such that for any δ-nondegenerate F of $(\tilde{\mathbf{D}}^\infty)^d$,*

$$\left|E\left[e^{i(v,F)}G\right]\right| \tag{5.92}$$

$$\le \frac{c}{|v|^{(1-\frac{\alpha\delta}{2})N}}\left((1+\|\tilde{D}F\|^{\sharp}_{0,n,2(n+1)p})^{n+1}\sup_{(\rho,\theta)}\|Q^F_\rho(\theta)^{-1}\|^{\sharp}_{0,n,2p}\right)^N\|G\|^{\sharp}_{0,n,p}$$

hold for all $|v| \ge 1$ and $G \in \tilde{\mathbf{D}}^\infty$.

Proof We define a complex valued random field $\{Z_u = Z^{F,v}_u, u \in \mathbb{U}\}$ associated with a given δ-nondegenerate functional $F = (F^1, \ldots, F^d)$ and $v \in \mathbb{R}^d$ with $|v| \ge 1$:

$$Z_u = Z^{F,v}_u = \frac{1}{|v|^2\varphi(\rho)}\cdot\chi_\rho(u)(e^{-i(v,\tilde{D}_uF)}-1)\cdot Q^F(\rho,v)^{-1}, \tag{5.93}$$

where $\rho = |v|^{-\delta}$. It is an element of $\tilde{\mathbf{D}}^\infty_{\mathbb{T}}$ for any F, v. We will apply the adjoint equation (5.43) for $Z_u = Z_uG$ and $e^{i(v,F)}$. Then we have

$$\begin{aligned} E\left[e^{i(v,F)}\tilde{\delta}(ZG)\right] &= E\left[\int \tilde{D}_u(e^{i(v,F)})Z_uGn(du)\right] \\ &= E\left[\int e^{i(v,F)}(e^{i(v,\tilde{D}_uF)}-1)Z_un(du)\cdot G\right] \\ &= E\left[e^{i(v,F)}G\int(e^{i(v,\tilde{D}_uF)}-1)Z_un(du)\right] \\ &= E[e^{i(v,F)}G], \end{aligned} \tag{5.94}$$

since $\int(e^{i(v,\tilde{D}_uF)}-1)Z_un(du) = 1$. Therefore, setting $\mathcal{L}(G) = \tilde{\delta}(ZG)$, we have the iteration formula

$$E[e^{i(v,F)}G] = E[e^{i(v,F)}\mathcal{L}(G)] = \cdots = E[e^{i(v,F)}\mathcal{L}^N(G)], \tag{5.95}$$

for $N = 1, 2, \ldots$. Consequently, we get by using Theorem 5.6.1 and Hölder's inequality

$$|E[e^{i(v,F)}G]| \le \|\mathcal{L}^N(G)\|_{0,0,2} = \|\tilde{\delta}(Z\mathcal{L}^{N-1}(G))\|_{0,0,2}$$
$$\le c\|Z\mathcal{L}^{N-1}(G)\|_{0,1,2} \le c'\|Z\|_{0,1,4}\|\mathcal{L}^{N-1}(G)\|_{0,1,4},$$

where $\| \ \|_{0,n,p} = \| \ \|_{0,n,p;A(\rho)}$. Repeating this argument, there are increasing sequences $c_N, n_N, p_N, N = 1, 2, \ldots$ such that the inequality

$$|E[e^{i(v,F)}G]| \le c_N\|Z\|_{0,n_1,p_1} \cdots \|Z\|_{0,n_N,p_N}\|G\|_{0,n_N,p_N}$$
$$\le c_N\|Z\|^N_{0,n_N,p_N}\|G\|_{0,n_N,p_N} \tag{5.96}$$

holds, since norms $\| \ \|_{0,n,p}$ are nondecreasing with respect to n, p.

We will compute $\|Z\|_{0,n,p}$. Apply Hölder's inequality to (5.93). Then we have

$$\|Z\|_{0,n,p} \le \frac{c}{|v|^2\varphi(\rho)}\|\chi_\rho(e^{-i(v,\tilde{D}F))}-1)\|_{0,n,2p}\|Q^F(\rho,v)^{-1}\|_{0,n,2p}.$$

Set $\rho = |v|^{-\delta}$. Then we have from Lemma 5.6.2

$$\|\chi_\rho(e^{-i(v,\tilde{D}F)} - 1)\|_{0,n,2p} \le c|v|\varphi(\rho)^{\frac{1}{2}}(1 + \|\tilde{D}F\|^\sharp_{0,n,2(n+1)p})^{n+1}.$$

Further, we have

$$\|Q^F(\rho,v)^{-1}\|_{0,n,2p} \le \sup_{(\rho,\theta)} \|Q^F_\rho(\theta)^{-1}\|^\sharp_{0,n,2p}.$$

Furthermore, $\frac{1}{|v|\varphi(|v|^{-\delta})^{\frac{1}{2}}} \le \frac{c'}{|v|^{(1-\frac{\alpha\delta}{2})}}$ holds for any $|v| \ge 1$, since $\varphi(|v|^{-\delta}) \ge c|v|^{-\alpha\delta}$ by the order condition. Therefore we get

$$\|Z\|_{0,n,p} \le \frac{c'}{|v|^{(1-\frac{\alpha\delta}{2})}}(1 + \|\tilde{D}F\|^\sharp_{0,n,2(n+1)p})^{n+1} \sup_{(\rho,\theta)} \|Q^F_\rho(\theta)^{-1}\|^\sharp_{0,n,2p}.$$

Substitute the above in (5.96), then we get the inequality of the theorem. □

As an immediate consequence of Theorem 5.7.1, we have the next corollary.

Corollary 5.7.1 *If F of $(\tilde{\mathbf{D}}^\infty)^d$ is δ-nondegenerate, the law of F weighted by $G \in \tilde{\mathbf{D}}^\infty$ has a rapidly decreasing C^∞-density. The density function and its derivatives are given by the Fourier inversion formula* (5.34).

5.8 Equivalence of Nondegenerate Conditions

In Sect. 5.7, we defined three types of nondegenerate Poisson functionals. In this section, we show the equivalence of these nondegenerate conditions.

Theorem 5.8.1 ([46]) *Let F be a regular functional of $(\tilde{\mathbf{D}}^{\infty})^d$.*

1. *If F is δ-nondegenerate for some $1 < \delta < 2/\alpha$, then F is nondegenerate at the center. Further, for any n, p the inequality*

$$\sup_{\theta} |K^F(\theta)^{-1}|^{\sharp}_{0,n,p} \le \sup_{(\rho,\theta)} |Q^F_\rho(\theta)^{-1}|^{\sharp}_{0,n,p} \tag{5.97}$$

holds for all regular δ-nondegenerate functionals $F \in (\tilde{\mathbf{D}}^{\infty})^d$.

2. *If F is nondegenerate at the center, F is δ-nondegenerate for any $1 < \delta < 2/\alpha$. Further, for any n, p there is a positive constant c such that the inequality*

$$|Q^F_\rho(\theta)^{-1}|^{\sharp}_{0,n,p} \le c C^{\delta'}_{n,p}(F) |K^F(\theta)^{-1}|^{\sharp}_{0,n,2p} \tag{5.98}$$

holds for any ρ, θ and F mentioned above. Here

$$C^{\delta'}_{n,p}(F) \le (1 + |\Lambda(F)|^{\sharp}_{0,n,8(2+\delta')p})^{2(2+\delta')}(1 + |K^F(\theta)^{-1}|^{\sharp}_{0,n,8p})^{\frac{1}{2}}, \tag{5.99}$$

where $\Lambda_t(F)$ is the Poisson functional given by (5.81) and δ' is the conjugate of δ.

Proof of the first half. Suppose that F is δ-nondegenerate. Consider the integrand of $Q^F_\rho(v)$. It holds that

$$|e^{i(v,\tilde{D}_u F)} - 1|^2 \le |(v, \tilde{D}_u F)|^2 = v\partial\tilde{D}_{t,0}F zz^T \partial\tilde{D}_{t,0}F^T v^T + O(|z|^3),$$

a.e. $dndP$. Here we used the Taylor expansion $\tilde{D}_{t,z}F = \partial\tilde{D}_{t,0}Fz + O(|z|^2)$. Integrating the above by $n(du)$ on the set $\mathbb{T} \times A_0(\rho)$, we get

$$\frac{1}{|v|^2\varphi(\rho)} \int_{\mathbb{T}} \int_{A_0(\rho)} |e^{i(v,\tilde{D}_{t,z}F)} - 1|^2 dt\nu(dz)$$
$$= \frac{v}{|v|} \int_{\mathbb{T}} (\partial\tilde{D}_{t,0}F)\Gamma_\rho(\partial\tilde{D}_{t,0}F)^T dt (\frac{v}{|v|})^T + O(\rho), \quad a.s.$$

Therefore setting $\theta = v/|v|$, we have

$$\limsup_{\rho\to 0} Q^F_\rho(\theta) \le K^F(\theta), \quad \text{or} \quad \liminf_{\rho\to 0} Q^F_\rho(\theta)^{-1} \ge K^F(\theta)^{-1}, \quad a.s.$$

Operating the transformation $\varepsilon^+_{\mathbf{u}}$ on both sides of the above, we get again

$$\liminf_{\rho\to 0} Q^F_\rho(\theta)^{-1} \circ \varepsilon^+_{\mathbf{u}} \ge K^F(\theta)^{-1} \circ \varepsilon^+_{\mathbf{u}}, \quad a.e.\ dn^j dP.$$

Therefore, we have $|K^F(\theta)^{-1}|^{\sharp}_{0,n,p} \leq \sup_\rho |Q^F_\rho(\theta)^{-1}|^{\sharp}_{0,n,p}$ for all $\theta \in S_{d-1}$. Then the first assertion of the theorem follows. □

The proof of the latter half of the theorem is long. Our discussion is close to [46]. It will be completed after discussing Lemmas 5.8.1, 5.8.2, 5.8.3, and 5.8.4. We will introduce two other matrix-valued Poisson functionals. We will take any δ satisfying $1 < \delta < 2/\alpha$. Let F be a regular functional of $(\tilde{\mathbf{D}}^\infty)^d$. For $0 < \rho < \rho_0$ and $v \neq 0$ we define $Q^F_\rho(\theta)$ by (5.86) and further,

$$S^F_\rho = \frac{1}{\varphi(\rho)} \int_{A(\rho)\cap\{|\tilde{D}_u F|\leq \rho^{\frac{1}{\delta}}\}} \tilde{D}_u F(\tilde{D}_u F)^T n(du). \tag{5.100}$$

$$K^F_\rho = \frac{1}{\varphi(\rho)} \int_{\mathbb{T}} \int_{A_0(\rho)} (\partial \tilde{D}_{t,0} F) z \cdot z^T (\partial \tilde{D}_{t,0} F)^T n(dt\, dz). \tag{5.101}$$

We set further,

$$S^F_\rho(\theta) = (\theta, S^F_\rho \theta), \quad K^F_\rho(\theta) = (\theta, K^F_\rho \theta).$$

Discussions will be divided into the following three steps:

1. We compare two norms $|Q^F_\rho(\theta)^{-1}|^{\sharp}_{0,n,p}$ and $|S^F_\rho(\theta)^{-1}|^{\sharp}_{0,n,p}$. See Lemma 5.8.1.
2. We compare two norms $|K^F_\rho(\theta)^{-1}|^{\sharp}_{0,n,p}$ and $|K^F(\theta)^{-1}|^{\sharp}_{0,n,p}$. See Lemma 5.8.2.
3. We compare two norms $|S^F_\rho(\theta)^{-1}|^{\sharp}_{0,n,p}$ and $|K_\rho(\theta)^{-1}|^{\sharp}_{0,n,p}$. See Lemmas 5.8.3 and 5.8.4.

Lemma 5.8.1 *Suppose that $S^F_\rho(\theta)$ are invertible a.s. for any ρ, θ. Then $Q^F_\rho(\theta)$ are also invertible a.s. Further, for any n, p there is a positive constant c_1 such that the inequality*

$$|Q^F_\rho(\theta)^{-1}|^{\sharp}_{0,n,p} \leq c_1 |S^F_\rho(\theta)^{-1}|^{\sharp}_{0,n,p} \tag{5.102}$$

holds for any ρ, θ.

Proof Note that the inequality $|e^{i(v,\tilde{D}_u F)} - 1|^2 \geq c|(v, \tilde{D}_u F)|$ holds with some constant $c > 0$ on the set $|\tilde{D}_u F| < |v|^{-1}$. Then we have inequalities

$$Q^F_\rho(\theta) \circ \varepsilon^+_{\mathbf{u}} \geq c S^F_\rho(v) \circ \varepsilon^+_{\mathbf{u}}, \quad a.e.\ dn^j\, dP.$$

Therefore, if $S^F_\rho(v) \circ \varepsilon^+_{\mathbf{u}}$ are invertible a.e., then $Q^F_\rho(\theta) \circ \varepsilon^+_{\mathbf{u}}$ are also invertible a.e. and satisfy

$$E\big[(Q^F_\rho(\theta) \circ \varepsilon^+_{\mathbf{u}})^{-p}\big] \leq c^{-p} E\big[(S^F_\rho(\theta) \circ \varepsilon^+_{\mathbf{u}})^{-p}\big], \quad a.e.\ dn^j.$$

Taking the essential supremum with respect to $\mathbf{u}$ and summing up these for $j = 0, \ldots, n$, we get the inequality of the lemma. □

Lemma 5.8.2 *Suppose that $K^F(\theta)$ are invertible a.s. for any $\theta \in S_{d-1}$. Then $K_\rho^F(\theta)$ are also invertible a.s. for any $0 < \rho < \rho_0, \theta \in S_{d-1}$. Further, for any n, p there exist two positive constants $c_2 < c_3$ such that the inequality*

$$c_2|K^F(\theta)^{-1}|_{0,n,p}^\sharp \leq |K_\rho^F(\theta)^{-1}|_{0,n,p}^\sharp \leq c_3|K^F(\theta)^{-1}|_{0,n,p}^\sharp \tag{5.103}$$

holds for any ρ, θ.

Proof Let λ_ρ and Λ_ρ be the minimum and the maximum eigenvalues of Γ_ρ, respectively. Since $\liminf_{\rho\to 0}\Gamma_\rho \geq \Gamma_0$ and Γ_0 is nondegenerate, there exists $0 < \rho_1 < \rho_0$ such that $0 < \inf_{0<\rho<\rho_1}\lambda_\rho \leq \sup_{0<\rho<\rho_1}\Lambda_\rho < \infty$. For $\rho_1 < \rho < \rho_0$, we have $\Gamma_\rho \geq \frac{\varphi(\rho_1)}{\varphi(\rho)}\Gamma_{\rho_1}$. Therefore we have $0 < c' = \inf_{0<\rho<\rho_0}\lambda_\rho \leq \sup_{0<\rho<\rho_0}\Lambda_\rho = C' < \infty$. Then for any $0 < \rho < \rho_0$ we have the inequality

$$\frac{c'}{\Lambda_0}K^F(\theta)\circ\varepsilon_{\mathbf{u}}^+ \leq K_\rho^F(\theta)\circ\varepsilon_{\mathbf{u}}^+ \leq \frac{C'}{\lambda_0}K^F(\theta)\circ\varepsilon_{\mathbf{u}}^+, \quad a.e.\ dn^j\,dP,$$

where λ_0, Λ_0 are the minimal and the maximal eigen values of Γ_0. Then the inequality of the lemma holds. □

Now, assume that F is nondegenerate at the center. Let $\mathbf{u} \in A(1)^j$. We want to show the invertibility of $\{S_\rho^F(\theta)\circ\varepsilon_{\mathbf{u}}^+\}$ and then the estimation of $E\big[(S_\rho^F(\theta)\circ\varepsilon_{\mathbf{u}}^+)^{-p}\big]$. We will define a nonnegative functional $\varrho = \varrho^F(\theta)\circ\varepsilon_{\mathbf{u}}^+$ as the infimum of $0 < \rho < \rho_0$ which satisfies

$$\big|S_\rho^F(\theta)\circ\varepsilon_{\mathbf{u}}^+ - K_\rho^F(\theta)\circ\varepsilon_{\mathbf{u}}^+\big| > \frac{1}{2}K_\rho^F(\theta)\circ\varepsilon_{\mathbf{u}}^+.$$

(We set $\varrho = 1$ if the above is an empty set.) Then on the set $\{\varrho < \epsilon\}$, the inequality

$$\sup_{0<\rho<\epsilon}\frac{|S_\rho^F(\theta)\circ\varepsilon_{\mathbf{u}}^+ - K_\rho^F(\theta)\circ\varepsilon_{\mathbf{u}}^+|}{K_\rho^F(\theta)\circ\varepsilon_{\mathbf{u}}^+} > \frac{1}{2}$$

holds.

Lemma 5.8.3 *For any $p \geq 2$, there is a positive constant c_4 such that the inequality*

$$P(\varrho^F(\theta)\circ\varepsilon_{\mathbf{u}}^+ < \epsilon) \leq c_4 A_p^{\delta'}(F)\epsilon^p, \quad \forall 0 < \epsilon < \rho_0,\ \theta \in S_{d-1},\ \ \mathbf{u} \in A(1)^j \tag{5.104}$$

holds for all F of $(\tilde{\mathbf{D}}^\infty)^d$ which are regular and nondegenerate at the center. Here

$$A_p^{\delta'}(F) = E\Big[\int_{\mathbb{T}}(1 + |\Lambda_t(F)\circ\varepsilon_{\mathbf{u}}^+|)^{2p(2+\delta')}\,dt\Big]^{\frac{1}{2}} E[K^F(\theta)^{-2p}\circ\varepsilon_{\mathbf{u}}^+]^{\frac{1}{2}}, \tag{5.105}$$

where δ' is the conjugate of δ.

Proof We will first rewrite K_ρ^F. Since Γ_ρ is given by (1.21), the matrix K_ρ^F is written as

$$K_\rho^F=\frac{1}{\varphi(\rho)}\int_{\mathbb{T}}\int_{A_0(\rho)}(\partial\tilde{D}_{t,0}F)zz^T(\partial\tilde{D}_{t,0}F)^T n(dt\,dz).$$

Therefore,

$$K_\rho^F(\theta)=\frac{1}{\varphi(\rho)}\int_{\mathbb{T}}\int_{A_0(\rho)}(\theta,\partial\tilde{D}_{t,0}Fz)^2 n(dt\,dz),$$

where $u=(t,z)$.

Now we will fix $F,\theta,\mathbf{u}$. In the following arguments, we will write $\tilde{D}_uF\circ\varepsilon_{\mathbf{u}}^+$, $S_\rho^F(\theta)\circ\varepsilon_{\mathbf{u}}^+$, $K_\rho^F(\theta)\circ\varepsilon_{\mathbf{u}}^+$, $\Lambda_t(F)\circ\varepsilon_{\mathbf{u}}^+$ etc. as $\tilde{D}_uF$, S_ρ, K_ρ, Λ_t etc., dropping F,θ and $\varepsilon_{\mathbf{u}}^+$. We define $E(\rho)=\{(t,z)\in A(\rho);\ |\tilde{D}_{t,z}F|\le\rho^{1/\delta}\}$. Then the inequality

$$\begin{aligned}\left|S_\rho-K_\rho\right|&\le\frac{1}{\varphi(\rho)}\int_{\mathbb{T}}\int_{A_0(\rho)}\left|(\theta,\tilde{D}_{t,z}F)^2-(\theta,\partial\tilde{D}_{t,0}Fz)^2\right|1_{E(\rho)}(t,z)n(dt\,dz)\\&\quad+\frac{1}{\varphi(\rho)}\int_{\mathbb{T}}\int_{A_0(\rho)}(\theta,\partial\tilde{D}_{t,0}Fz)^2 1_{E(\rho)^c}(t,z)n(dt\,dz)\\&=I_1(\rho)+I_2(\rho)\end{aligned}$$

holds. Therefore we have

$$\begin{aligned}P(\varrho<\epsilon)&\le P\Big(\sup_{\rho<\epsilon}\frac{I_1(\rho)}{K_\rho}\ge\frac{1}{4}\Big)+P\Big(\sup_{\rho<\epsilon}\frac{I_2(\rho)}{K_\rho}\ge\frac{1}{4}\Big)\\&\le P\Big(\frac{\sup_{\rho<\epsilon}I_1(\rho)}{K}\ge\frac{c}{4}\Big)+P\Big(\frac{\sup_{\rho<\epsilon}I_2(\rho)}{K}\ge\frac{c}{4}\Big),\end{aligned}\tag{5.106}$$

where c is a positive constant such that $K_\rho\ge cK$.

We want to compute the right-hand side. We first consider $I_1(\rho)$. In view of Condition (R), the integrand of $I_1(\rho)$ is estimated as

$$|(\theta,\tilde{D}_{t,z}F)^2-(\theta,\partial\tilde{D}_{t,0}Fz)^2|\le|z|^3\Lambda_t^2,$$

where $\Lambda_t=\Lambda_t(F)\circ\varepsilon_{\mathbf{u}}^+$. Therefore, we have $I_1(\rho)\le\rho\int_{\mathbb{T}}\Lambda_t^2\,dt$. Then

$$E\Big[\sup_{\rho<\epsilon}I_1(\rho)^{2p}\Big]\le c_1\epsilon^{2p}E\Big[\int_{\mathbb{T}}|\Lambda_t|^{4p}\,dt\Big].$$

Therefore, by Tchebyschev's inequality

$$P\Big(\frac{\sup_{\rho<\epsilon} I_1(\rho)}{K} \geq \frac{c}{4}\Big) \leq c_1'\epsilon^p E\Big[\int_{\mathbb{T}} |\Lambda_t|^{4p}\,dt\Big]^{\frac{1}{2}} E[K^{-2p}]^{\frac{1}{2}}.$$

We will next consider I_2. Let δ' be the conjugate of δ. Since $m(dt\,dz) = |z|^2 n(dt\,dz)$ holds for $|z| \leq \rho_0$, we have by Tchebyschev's inequality,

$$\begin{aligned} I_2(\rho) &\leq \frac{1}{\varphi(\rho)} \int_{\mathbb{T}}\int_{A_0(\rho)} |\partial \tilde{D}_{t,0}F|^2 1_{E(\rho)^c} m(dt\,dz) \\ &\leq \frac{1}{\rho^{\delta'/\delta}} \frac{1}{\varphi(\rho)} \int_{\mathbb{T}}\int_{A_0(\rho)} |\partial \tilde{D}_{t,0}F|^2 |\tilde{D}_{t,z}F|^{\delta'} m(dt\,dz) \\ &\leq \frac{1}{\rho^{\delta'/\delta}} \frac{1}{\varphi(\rho)} \int_{\mathbb{T}}\int_{A_0(\rho)} |\partial \tilde{D}_{t,0}F|^{2+\delta'} |z|^{\delta'} m(dt\,dz) \\ &\leq \rho \int_{\mathbb{T}} \Lambda_t^{2+\delta'}\,dt. \end{aligned}$$

Therefore, we have $E[\sup_{\rho<\epsilon} I_2(\rho)^{2p}] \leq c_2\epsilon^{2p} E\Big[\int_{\mathbb{T}} |\Lambda_t|^{2p(2+\delta')}\,dt\Big]$. Then,

$$P\Big(\frac{\sup_{\rho<\epsilon} I_2(\rho)}{K} > \frac{c}{4}\Big) \leq c_2'\epsilon^p E\Big[\int_{\mathbb{T}} |\Lambda_t|^{2p(2+\delta')}\,dt\Big]^{\frac{1}{2}} E\big[K^{-2p}\big]^{\frac{1}{2}}.$$

Therefore, in view of (5.106), $P(\varrho < \epsilon)$ is dominated by

$$\begin{aligned} &\epsilon^p\Big(c_1' E\Big[\int_{\mathbb{T}} |\Lambda_t|^{4p}\,dt\Big]^{1/2} + c_2' E\Big[\int_{\mathbb{T}} |\Lambda_t|^{2p(2+\delta')}\,dt\Big]^{\frac{1}{2}}\Big) E[K^{-2p}]^{\frac{1}{2}} \\ &\leq \epsilon^p c_3' E\Big[\int_{\mathbb{T}} (1 + |\Lambda_t(F)\circ\varepsilon_{\mathbf{u}}^+|)^{2p(2+\delta')}\,dt\Big]^{\frac{1}{2}} E[K^F(\theta)^{-2p}\circ\varepsilon_{\mathbf{u}}^+]^{\frac{1}{2}}. \end{aligned}$$

Then we get the assertion of the lemma. □

Lemma 5.8.4 *If $F \in (\tilde{\mathbf{D}}^\infty)^d$ is regular and nondegenerate at the center, then functionals $\{S_\rho^F(\theta), 0 < \rho < \rho_0, \theta \in S_{d-1}\}$ are invertible. Further, for any j, p there is a positive constant c_5 such that the inequality*

$$E\big[(S_\rho^F(\theta)\circ\varepsilon_{\mathbf{u}}^+)^{-p}\big] \leq c_5 B_p^{\delta'}(F) E\big[(K_\rho^F(\theta)\circ\varepsilon_{\mathbf{u}}^+)^{-2p}\big]^{\frac{1}{2}}, \quad a.e.\ dn^j, \tag{5.107}$$

holds for any ρ, θ and for all $F \in (\tilde{\mathbf{D}}^\infty)^d$ which is regular and nondegenerate at the center. Here

$$B_p^{\delta'}(F) = \Big(1+E\Big[\int_{\mathbb{T}} (1+|\Lambda_t(F)\circ\varepsilon_{\mathbf{u}}^+|)^{8p(2+\delta')}\,dt\Big]^{\frac{1}{4}}\Big)\Big(1+E\Big[(K_\rho^F(\theta)\circ\varepsilon_{\mathbf{u}}^+)^{-8p}\Big]^{\frac{1}{4}}\Big). \tag{5.108}$$

Proof We use simplified notations S_ρ etc. as in the proof of the previous lemma. Let ϱ be the nonnegative functional defined before Lemma 5.8.3. Since S_ρ, K_ρ are left continuous with respect to ρ, we have

$$|S_{\rho\wedge\varrho} - K_{\rho\wedge\varrho}| \le \frac{1}{2} S_{\rho\wedge\varrho}.$$

Therefore we have $S_{\rho\wedge\varrho} \ge \frac{1}{2} K_{\rho\wedge\varrho}$ for any $0 < \rho < \rho_0$. Further, since $\varphi(\rho) S_\rho$ is nondecreasing with respect to ρ a.s., we have $S_\rho \ge \frac{\varphi(\varrho\wedge\rho)}{\varphi(\rho)} S_{\varrho\wedge\rho}$ if $|v| \ge 1$. Therefore,

$$S_\rho \ge \frac{1}{2} \frac{\varphi(\varrho \wedge \rho)}{\varphi(\rho)} K_{\rho\wedge\varrho} > 0, \quad \forall\, 0 < \rho < 1, \ a.s.$$

(We set $\varphi(0) = 0$.) Then we have by Schwartz's inequality

$$E\big[S_\rho^{-p}\big] \le 2E\Big[K_{\rho\wedge\varrho}^{-2p}\Big]^{\frac{1}{2}} E\Big[\Big(\frac{\varphi(\rho)}{\varphi(\varrho \wedge \rho)}\Big)^{2p}\Big]^{\frac{1}{2}}. \tag{5.109}$$

It holds that $E[K_{\rho\wedge\varrho}^{-2p}]^{\frac{1}{2}} \le c^{-p} E[K^{-2p}]^{\frac{1}{2}}$. We shall compute the last term. Set $a(x) = P(\varrho \le x)$. We shall apply Lemma 5.8.3, replacing p by $4p$. Then it holds that $a(x) \le A_{4p}^{\delta'}(F) x^{4p}$ for any x. Since $\varphi(x) \ge c_1 x^\alpha$ holds by the order condition, we have

$$E\Big[\Big(\frac{1}{\varphi(\varrho \wedge \rho)}\Big)^{2p} 1_{\varrho<\rho}\Big] = \int_0^\rho \frac{1}{\varphi(x)^{2p}} a(dx) \le c_1 A_{4p}^{\delta'}(F).$$

Therefore,

$$\begin{aligned} E\Big[\Big(\frac{\varphi(\rho)}{\varphi(\varrho \wedge \rho)}\Big)^{2p}\Big]^{\frac{1}{2}} &\le P(\varrho \ge \rho) + \varphi(\rho)^p E\Big[\Big(\frac{1}{\varphi(\varrho \wedge \rho)}\Big)^{2p} 1_{\varrho<\rho}\Big]^{\frac{1}{2}} \\ &\le c_2\Big(1 + \sqrt{A_{4p}^{\delta'}(F)}\Big), \quad \forall 0 < \rho < \rho_0. \end{aligned} \tag{5.110}$$

Consequently (5.109) and (5.110) imply the inequality of the lemma. □

Proof of Theorem 5.8.1 Inequality (5.102) holds by Lemma 5.8.1. Further, from Lemma 5.8.4, we get

$$|S_\rho^F(\theta)^{-1}|_{0,n,p}^\sharp \le c_6 C_{n,p}^{\delta'}(F) |K^F(\theta)^{-1}|_{0,n,2p}^\sharp. \tag{5.111}$$

Therefore the inequality of the theorem follows. □

We shall next study the equivalence of 'nondegenerate' and 'nondegenerate at the center'.

Theorem 5.8.2 *Let F be a regular functional of $(\tilde{\mathbf{D}}^{\infty})^d$.*

1. *If F is nondegenerate, it is nondegenerate at the center. Further, we have for any n, p,*

$$\sup_{\theta} |K^F(\theta)^{-1}|^{\sharp}_{0,n,p} \leq \sup_{(\rho,\theta)} |\tilde{R}^F_{\rho}(\theta)^{-1}|^{\sharp}_{0,n,p}. \tag{5.112}$$

2. *If F is nondegenerate at the center, it is nondegenerate. Further, for any n, p there is a positive constant c such that*

$$\sup_{(\rho,\theta)} |\tilde{R}^F_{\rho}(\theta)^{-1}|^{\sharp}_{0,n,p} \leq c C^0_{n,p}(F) \sup_{\theta} |K^F(\theta)^{-1}|^{\sharp}_{0,n,2p}, \tag{5.113}$$

where $C^0_{n,p}(F)$ is given by (5.99) with $\delta' = 0$.

Proof The first assertion can be verified similarly as the first assertion of Theorem 5.8.1. Next, suppose that F is nondegenerate at the center. We will compare $\tilde{R}^F_{\rho}(\theta)$ and $K^F_{\rho}(\theta)$. Let us replace $S^F_{\rho}(\theta)$ by $\tilde{R}^F_{\rho}(\theta)$ in the argument of Lemmas 5.8.3 and 5.8.4. We will define a nonnegative functional $\varrho = \varrho^F(\theta) \circ \varepsilon^+_{\mathbf{u}}$ as the infimum of $0 < \rho < \rho_0$ which satisfies

$$\left|\tilde{R}^F_{\rho}(\theta) \circ \varepsilon^+_{\mathbf{u}} - K^F_{\rho}(\theta) \circ \varepsilon^+_{\mathbf{u}}\right| > \frac{1}{2} K^F_{\rho}(\theta) \circ \varepsilon^+_{\mathbf{u}}.$$

Rewrite $R^F_{\rho}(\theta) \circ \varepsilon^+_{\mathbf{u}}$ etc as $\tilde{R}^F_{\rho}$ etc., dropping $\varepsilon^+_{\mathbf{u}}$. Then we have

$$|\tilde{R}^F_{\rho}(\theta) - K^F_{\rho}(\theta)| \leq \frac{1}{\varphi(\rho)} \int_{A(\rho)} \left|(\theta, \tilde{D}_{t,z}F)^2 - (\theta, \partial \tilde{D}_{t,0}Fz)^2\right| n(dt\,dz).$$

Denote the right-hand side of the above as $I_1(\rho)$ and set $I_2(\rho) = 0$. We can proceed with our arguments similarly to the proof Lemma 5.8.3. Then we get the assertion of Lemma 5.8.3, by setting $\delta' = 0$. Further, repeating arguments of Lemmas 5.8.4 and Theorem 5.8.1, we find that (5.111) should be changed to the inequality

$$|\tilde{R}^F_{\rho}(\theta)^{-1}|^{\sharp}_{0,n,p} \leq c'_6 C^0_{n,p}(F) |K^F_{\rho}(\theta)^{-1}|^{\sharp}_{0,n,2p}. \tag{5.114}$$

Therefore F is nondegenerate. The inequality (5.113) follows from (5.114) and (5.103). □

From Theorems 5.8.1 and 5.8.2, we have the following.

Theorem 5.8.3 *Let F be a regular functional of $(\tilde{\mathbf{D}}^{\infty})^d$. The following statements are equivalent:*

(i) *F is δ-nondegenerate for some $1 < \delta < 2/\alpha$.*
(ii) *F is δ-nondegenerate for any $1 < \delta < 2/\alpha$.*

(iii) *F is nondegenerate at the center.*
(iv) *F is nondegenerate.*

Note The Malliavin calculus for jump processes was studied by Bismut [10], Leandre [74], Bichiteler–Gravereau–Jacod [7]. Picard [93] proposed another approach to the Malliavin calculus on a Poisson space. The difference operator $\tilde{D}_u$ and its adjoint $\tilde{\delta}$ were introduced by Picard [92, 93]. His discussion was developed further by Picard–Savona [95], Ishikawa–Kunita [45] and Hayashi–Ishikawa [37] etc.

Picard [92] and Picard–Savona [95] introduced another condition for smooth density. For the Lévy measure ν, they assumed that there exist $0 < \alpha < 2$ and $c, C > 0$ such that the inequality

$$c\rho^{\alpha}|v|^2 \le \int_{0<|z|<\rho} (v, z)^2 \nu(dz) \le C\rho^{\alpha}|v|^2$$

holds for any $v \in \mathbb{R}^{d'}$ and $0 < \rho < \rho_0$. The condition is equivalent to that the Lévy measure is nondegenerate and satisfies the order condition of exponent α at the center with respect to family of balls $B_0(\rho) = \{z \in \mathbb{R}^{d'}; |z| < \rho\}$, i.e., in the case where the star-shaped neighborhoods $A_0(\rho)$ coincide with balls $B_0(\rho)$. Hence Theorem 5.8.1 is an extension of their work.

5.9 Product of Wiener Space and Poisson Space

We will define the product of a Wiener space and a Poisson space. Let $\mathbf{W}$ be the set of all continuous maps $w; \mathbb{T} \to \mathbb{R}^{d'}$ such that $w(0) = 0$, and let $\mathcal{B}(\mathbf{W})$ be its Borel field. Let Ω be the set of all integer-valued measures on $\mathbb{U} = \mathbb{T} \times \mathbb{R}_0^{d'}$, and let $\mathcal{B}(\Omega)$ be its Borel field. We consider the product space $\bar{\Omega} = \mathbf{W} \times \Omega$ and $\mathcal{B}(\bar{\Omega}) = \mathcal{B}(\mathbf{W}) \otimes \mathcal{B}(\Omega)$. Its sub σ-fields $\mathcal{B}(\mathbf{W}) \otimes \{\emptyset, \Omega\}$ and $\{\emptyset, \mathbf{W}\} \otimes \mathcal{B}(\Omega)$ are again denoted by $\mathcal{B}(\mathbf{W})$ and $\mathcal{B}(\Omega)$, respectively.

Let P be a probability measure on the product space $(\bar{\Omega}, \mathcal{B}(\bar{\Omega}))$, such that its restriction to $\mathcal{B}(\mathbf{W})$ is a Wiener measure and its restriction to $\mathcal{B}(\Omega)$ is a Poisson random measure associated with the Lévy measure ν, and further, $\mathcal{B}(\mathbf{W})$ and $\mathcal{B}(\Omega)$ are independent. The triple $(\bar{\Omega}, \mathcal{B}(\bar{\Omega}), P)$ is called the product of the Wiener space and the Poisson space, or simply, *Wiener–Poisson space with the Lévy measure* ν. We assume that at the center 0, the Lévy measure ν is nondegenerate and the intensity measure n satisfies the order condition of exponent $0 < \alpha < 2$ with respect to a given family of star-shaped neighborhoods $\{A(\rho), 0 < \rho < \rho_0\}$, as in Sect. 5.6.

The first component w of (w, ω) is called a *Wiener variable* and the second component ω is called a *Poisson variable*. Let S be a complete metric space. An S-valued $\mathcal{B}(\bar{\Omega})$-measurable function $F(w, \omega)$ is called an *S-valued Wiener–Poisson functional*. When $S = \mathbb{R}$, it is simply called a *Wiener–Poisson functional*. A $\mathcal{B}(\mathbf{W})$-measurable function $F(w)$ is called a Wiener functional and a $\mathcal{B}(\Omega)$-measurable function $F(\omega)$ is called a Poisson functional.

Let $p > 1$. We denote by $L^p(\bar{\Omega})$ the set of all Wiener–Poisson functionals F such that $E[|F|^p] < \infty$. We set $\bar{L}^p = L^p(\bar{\Omega})$ and $\bar{L}^{\infty-} = \bigcap_{p>1} \bar{L}^p$. It is an F-space. Let $F(w, \omega)$ be a Wiener–Poisson functional in $\bar{L}^{\infty-}$. Let H be the Cameron–Martin space introduced in Sect. 5.1. Given $h \in H$, we define transformations $T_h, h \in H$ by $T_h F(w, \omega) = F(w+\bar{h}, \omega)$, where $\bar{h}(t) = \int_0^t h(s)\, ds$. Then for any $p > 1$ and $r > 1$, T_h is a continuous linear transformation from $\bar{L}^{pr}$ to $\bar{L}^p$. Further, it is a continuous linear transformation from the F-space $\bar{L}^{\infty-}$ to itself. The fact can be verified in the same way as the transformation T_h for Wiener functionals discussed in Sect. 5.1.

We will define the derivative operator DF for a Wiener–Poisson functional $F \in \bar{L}^{\infty-}$. It is similar to the derivative operator for Wiener functionals. It may be regarded as a partial derivative of $F(w, \omega)$ with respect to Wiener variable w. Let $\lambda \in \mathbb{R}$. Then $\{T_{\lambda h}, \lambda \in \mathbb{R}\}$ is a one-parameter group of transformations on $\bar{L}^{\infty-}$. Suppose that for $F \in \bar{L}^{\infty-}$ there exists an H-valued Wiener–Poisson functional F' such that $|F'|_H \in \bar{L}^{\infty-}$ and (5.2) holds for any $h \in H^{(0)}$ with respect to the metric d of $\bar{L}^{\infty-}$, where $H^{(0)}$ is a dense subset of H. Then F is said to be *H-differentiable* and F' is called its *H-derivative*. We denote F' by DF or $D_t F$. We denote by $\bar{\mathcal{D}}$ the set of all H-differentiable Wiener–Poisson functionals. If F is a Poisson functional in $\bar{L}^{\infty-}$, then $T_{\lambda h} F = F$ holds for any λ, h. Hence F is H-differentiable and the H-derivative is 0.

For a positive integer $i \geq 2$, $D^i F$ is defined as $D^i F = D(D^{i-1} F)$ by induction, similarly to that for the Wiener functional in Sect. 5.1. It is an $H^{\otimes i}$-valued Wiener–Poisson functional with $|D^i F|_{H^{\otimes i}} \in \bar{L}^{\infty-}$. We denote it by $D^i_{\mathbf{t}} F, \mathbf{t} = (t_1, \ldots, t_i) \in \mathbb{T}^i$.

For the Poisson space and Poisson variables, we use the same notations as in Sects. 5.4, 5.5, 5.6, 5.7, and 5.8. Let $u \in \mathbb{U}$. Transformations $\varepsilon_u^{\pm}; \bar{\Omega} \to \bar{\Omega}$ is defined by $\varepsilon_u^{\pm}(w, \omega) = (w, \varepsilon_u^{\pm}\omega)$. Further, the difference operator $\tilde{D}$ is defined for F in $\bar{L}^{\infty-}$ by

$$\tilde{D}_u F(w, \omega) = F(w, \varepsilon_u^+ \omega) - F(w, \omega).$$

It is well defined for almost all (u, w, ω) with respect to $dndP$. It is written as $\tilde{D}_{t,z} F$ if $u = (t, z)$. For $\mathbf{u} = (u_1, \ldots, u_j) \in \mathbb{U}^j$, we set $\tilde{D}^j_{\mathbf{u}} = \tilde{D}_{u_1} \cdots \tilde{D}_{u_j} F$.

If F is a Wiener functional in $\bar{L}^{\infty-}$, then $F \circ \varepsilon_u^+ = F$ and hence $\tilde{D}F = 0$. Hence $\tilde{D}$ may be regarded as a partial difference of $F(w, \omega)$ with respect to Poisson variable ω.

A Wiener–Poisson functional F is called *smooth* if for any $i, j \in \mathbb{N}$ and $p \geq 2$, $\tilde{D}^j_{\mathbf{u}} F$ are infinitely H-differentiable for almost all $\mathbf{u}$ and satisfy

$$\sup_{\mathbf{u} \in A(1)^j} \frac{E\big[\big(\int_{\mathbb{T}^i} |D^i_{\mathbf{t}} \tilde{D}^j_{\mathbf{u}} F|^2\, d\mathbf{t}\big)^{\frac{p}{2}}\big]}{\gamma(\mathbf{u})^p} < \infty.$$

Here, 'sup' means the essential supremum with respect to the measure n^j. The set of all smooth functionals is denoted by $\bar{\mathcal{D}}^\infty$. A d-dimensional functional $F = (F^1, \ldots, F^d)$ is called smooth if every component F^i is smooth. We denote by $(\bar{\mathcal{D}}^\infty)^d$ the set of all d-dimensional smooth functionals.

The next lemma is shown immediately.

Lemma 5.9.1 *For $F \in \bar{\mathcal{D}}^\infty$, $D^i_{\mathbf{t}}\tilde{D}^j_{\mathbf{u}}F$ is well defined a.e. $d\mathbf{t}^i\, dn^j\, dP$ for any i, j of $\mathbb{N}$. Further, we have $D^i_{\mathbf{t}}\tilde{D}^j_{\mathbf{u}}F = \tilde{D}^j_{\mathbf{u}}D^i_{\mathbf{t}}F$ a.e. $d\mathbf{t}^i\, dn^j\, dP$.*

A Wiener–Poisson functional with parameter $\mathbb{T}$ given by $X_t = \sum_{l=1}^n F_l h_l(t)$, $t \in \mathbb{T}$ with $F_l \in \bar{\mathcal{D}}^\infty$ and $h_l \in H, l = 1, 2, \ldots$, is called a *simple functional* with parameter $\mathbb{T}$. We denote by $\bar{\mathcal{S}}^\infty_{\mathbb{T}}$ the set of all simple functionals with parameter $\mathbb{T}$. For such $X_t = \sum_l F_l h_l(t) \in \bar{\mathcal{S}}^\infty_{\mathbb{T}}$, we define the *Skorohod integral by the Wiener process* denoted by $\delta(X)$ as follows:

$$\delta(X) = \sum_{l=1}^n F_l W(h_l) - \sum_{l=1}^n \langle DF_l, h_l\rangle. \tag{5.115}$$

It is an element of $\bar{L}^{\infty-}$.

Let $Z_u, u \in \mathbb{U}$ be a Wiener–Poisson random field. We denote by $\bar{\mathcal{D}}^\infty_{\mathbb{U}}$ the set of all Z_u such that

$$\frac{D^i_{\mathbf{t}}Z_u}{\gamma(u)}, \quad \frac{\tilde{D}^j_{\mathbf{v}}D^i_{\mathbf{t}}Z_u}{\gamma(\mathbf{v})\gamma(u)}, \quad i, j = 0, 1, 2, \ldots$$

are bounded a.e. $d\mathbf{t}^i\, dn\, dP$ and $d\mathbf{t}^i\, dn^{j+1}\, dP$, respectively. Further, $\bar{\mathcal{D}}^\infty_{\mathbb{U}_0}$ denotes the set of $Z_u \in \bar{\mathcal{D}}^\infty_{\mathbb{U}}$ such that $Z_u \circ \varepsilon^-_u$ are supported by a compact subset K of $\mathbb{U}$ a.e. $dn\, dP$. For $Z \in \bar{\mathcal{D}}^\infty_{\mathbb{U}_0}$, we define the *Skorohod integral of Z_u by Poisson random measure* by

$$\tilde{\delta}(Z) = \int_{\mathbb{U}} Z_u \circ \varepsilon^-_u \tilde{N}(du) = \int_{\mathbb{U}} Z_u \circ \varepsilon^-_u (N(du) - n(du)). \tag{5.116}$$

It is an element of $\bar{L}^{\infty-}$.

Lemma 5.9.2

1. *The adjoint formula*

$$E[G\delta(X)] = E\Big[\int_{\mathbb{T}} (D_t G) X_t\, dt\Big] \tag{5.117}$$

holds for any $G \in \bar{\mathcal{D}}^\infty$ and $X \in \bar{\mathcal{S}}^\infty_{\mathbb{T}}$. Further, the adjoint formula

$$E[G\tilde{\delta}(Z)] = E\Big[\int_{\mathbb{U}} (\tilde{D}_u G) Z_u n(du)\Big] \tag{5.118}$$

holds for any bounded G *and* $Z \in \bar{\mathcal{D}}^{\infty}_{\mathbb{U}_0}$.

2. *The operator* δ *satisfies the commutation relation: if* $X \in \bar{\mathcal{S}}^{\infty}_{\mathbb{T}}$, *then*

$$D_t\delta(X) = \delta(D_t X) + X_t, \quad \tilde{D}_u\delta(X) = \delta(\tilde{D}_u X), \quad a.e. \tag{5.119}$$

The operator $\tilde{\delta}$ *satisfies the commutation relation; if* $Z \in \bar{\mathcal{D}}^{\infty}_{\mathbb{U}_0}$, *then*

$$\tilde{D}_u\tilde{\delta}(Z) = \tilde{\delta}(\tilde{D}_u Z) + Z_u, \quad D_t\tilde{\delta}(Z) = \tilde{\delta}(D_t Z), \quad a.e. \tag{5.120}$$

3. $\delta(X), \delta(Y)$ *and* $\tilde{\delta}(Z), \tilde{\delta}(Y)$ *satisfy isometric properties* (5.10) *and* (5.45), *respectively.*

Proof Let us first show the adjoint formula (5.117). We can show that equation (5.5) holds for $F \in \bar{\mathcal{D}}^{\infty}$, similarly to the proof of Proposition 5.1.2. Then (5.6) holds for any $F, G \in \bar{\mathcal{D}}^{\infty}$. Then we can verify (5.117) similarly to the proof of Lemma 5.1.1. For the proof of the adjoint formula (5.118), let us remark that Lemma 5.4.1 is valid for positive Wiener–Poisson functional Z_u. Then Corollary 5.4.1 is also valid for such Z_u. Then the adjoint formula for $\tilde{\delta}$ can be verified in the same way as in the proof of Lemma 5.4.2.

The commutation relation (5.119) for δ and D and the isometric equality (5.10) can be verified in the same way as in the proof of Lemma 5.1.1. Further, the commutation relation for $\tilde{D}$ and $\tilde{\delta}$ and its isometric property (5.45) can be verified similarly to the proof of Lemma 5.4.2.

We show that $\tilde{D}$ and δ are commutative. Suppose $X = \sum_l F_l h_l$. Then we have

$$\begin{aligned}\tilde{D}_u\delta(X) &= \tilde{D}_u\Big(\sum_l F_l W(h_l) - \sum_l \langle DF_l, h_l\rangle\Big)\\ &= \sum_l \tilde{D}_u F_l W(h_l) - \sum_l \langle \tilde{D}_u DF_l, h_l\rangle = \delta(\tilde{D}_u X).\end{aligned}$$

Hence we have $\tilde{D}_u\delta = \delta\tilde{D}_u$. Next, if $Z_u = \sum F_l\tilde{h}_l(u)$, we have

$$\begin{aligned}D_t\tilde{\delta}(Z) &= D_t\sum_l \int F_l \circ \varepsilon_u^- \tilde{h}_l(u)(N(du) - n(du))\\ &= \sum_l \int D_t F_l \circ \varepsilon_u^- \tilde{h}_l(u)(N(du) - n(du)) = \tilde{\delta}(D_t Z).\end{aligned}$$

Hence we have $D_t\tilde{\delta} = \tilde{\delta} D_t$.

Finally, isometric properties (5.10) and (5.45) are verified in the same way as in Lemmas 5.1.1 and 5.4.2, respectively. □

5.10 Sobolev Norms for Wiener–Poisson Functionals

We will define *Sobolev norms for Wiener–Poisson functionals*. These norms should be a combination of Sobolev norms for Wiener functionals and those for Poisson functionals. Let A be an open subset of $A(1) = \{u = (t, z) \in \mathbb{U};\ |z| \le 1\}$ satisfying $0 < m(A) \le 1$. Let m_A be the conditional probability measure. Let m_A^j be a j-times product measure of m_A and $\gamma_A(\mathbf{u})$ be positive functions which are defined in Sect. 5.5. Let $i, j \in \mathbb{N}$ and $p \ge 2$. Set

$$I_{i,j,p;A}(F) = \int_{A^j} \frac{E\big[\big(\int_{\mathbb{T}^i} |D_{\mathbf{t}}^i \tilde{D}_{\mathbf{u}}^j F|^2\, d\mathbf{t}\big)^{\frac{p}{2}}\big]}{\gamma_A(\mathbf{u})^p} m_A^j(d\mathbf{u}), \quad F \in \bar{\mathcal{D}}^\infty,$$

$$I_{i,j,p;A}(X) = \int_{A^j} \frac{E\big[\big(\int_{\mathbb{T}^{i+1}} |D_{\mathbf{t}}^i \tilde{D}_{\mathbf{u}}^j X_t|^2\, d\mathbf{t}\, dt\big)^{\frac{p}{2}}\big]}{\gamma_A(\mathbf{u})^p} m_A^j(d\mathbf{u}), \quad X \in \bar{\mathcal{S}}_{\mathbb{T}}^\infty,$$

$$I_{i,j,p;A}(Z) = \int_{A^{j+1}} \frac{E\big[\big(\int_{\mathbb{T}^i} |D_{\mathbf{t}}^i \tilde{D}_{\mathbf{u}}^j Z_u|^2\, d\mathbf{t}\big)^{\frac{p}{2}}\big]}{\gamma_A(\mathbf{u})^p \gamma_A(u)^p} m_A^j(d\mathbf{u}) m_A(du), \ Z \in \bar{\mathcal{D}}_{\mathbb{U}_0}^\infty.$$

If $i = 0$ or $j = 0$, we define them for $F \in \bar{\mathcal{D}}^\infty$ as $I_{0,0,p;A}(F) = E[|F|^p]$ and

$$I_{i,0,p;A}(F) = E\Big[\Big(\int_{\mathbb{T}^i} |D_{\mathbf{t}}^i F|^2\, d\mathbf{t}\Big)^{\frac{p}{2}}\Big], \quad I_{0,j,p;A}(F) = \int_{A^j} \frac{E[|\tilde{D}_{\mathbf{u}}^j F|^p]}{\gamma_A(\mathbf{u})^p} m_A^j(d\mathbf{u}).$$

We can define those for $X \in \bar{\mathcal{S}}_{\mathbb{T}}^\infty$ and $Z \in \bar{\mathcal{D}}_{\mathbb{U}_0}^\infty$ in similar ways. Then we define Sobolev norms as

$$\|F\|_{m,n,p;A} = \Big(\sum_{i=0}^m \sum_{j=0}^n I_{i,j,p;A}(F)\Big)^{\frac{1}{p}}, \quad F \in \bar{\mathcal{D}}^\infty, \tag{5.121}$$

$$\|X\|_{m,n,p;A} = \Big(\sum_{i=0}^m \sum_{j=0}^n I_{i,j,p;A}(X)\Big)^{\frac{1}{p}}, \quad X \in \bar{\mathcal{S}}_{\mathbb{T}}^\infty, \tag{5.122}$$

$$\|Z\|_{m,n,p;A} = \Big(\sum_{i=0}^m \sum_{j=0}^n I_{i,j,p;A}(Z)\Big)^{\frac{1}{p}}, \quad Z \in \bar{\mathcal{D}}_{\mathbb{U}_0}^\infty. \tag{5.123}$$

When the domain A is fixed in discussions, we will often drop ; A from these norms and write

$$\|F\|_{m,n,p;A} = \|F\|_{m,n,p}, \quad \|X\|_{m,n,p;A} = \|X\|_{m,n,p}, \quad \|Z\|_{m,n,p;A} = \|Z\|_{m,n,p}.$$

Next, we set

$$I^{\sharp}_{i,j,p}(F) = \sup_{\mathbf{u}\in A(1)^j} \frac{E[(\int_{\mathbb{T}^i} |D^i_{\mathbf{t}}\tilde{D}^j_{\mathbf{u}}F|^2\,d\mathbf{t})^{\frac{p}{2}}]}{\gamma(\mathbf{u})^p}, \quad F \in \bar{\mathcal{D}}^\infty,$$

$$I^{\sharp}_{i,j,p}(X) = \sup_{\mathbf{u}\in A(1)^j} \frac{E[(\int_{\mathbb{T}^{i+1}} |D^i_{\mathbf{t}}\tilde{D}^j_{\mathbf{u}}X_t|^2\,d\mathbf{t}\,dt)^{\frac{p}{2}}]}{\gamma(\mathbf{u})^p}, \quad X \in \bar{\mathcal{S}}^\infty_{\mathbb{T}},$$

$$I^{\sharp}_{i,j,p}(Z) = \sup_{(\mathbf{u},u)\in A(1)^{j+1}} \frac{E[(\int_{\mathbb{T}^i} |D^i_{\mathbf{t}}\tilde{D}^j_{\mathbf{u}}Z_u|^2\,d\mathbf{t})^{\frac{p}{2}}]}{\gamma(\mathbf{u})^p\gamma(u)^p}, \quad Z \in \bar{\mathcal{D}}^\infty_{\mathbb{U}_0},$$

if $i, j \geq 1$. We will define $I^{\sharp}_{i,j,p}(F)$ etc. for $i = 0$ or $j = 0$ as $I^{\sharp}_{0,0,p}(F) = E[|F|^p]$ and

$$I^{\sharp}_{i,0,p}(F) = E\Big[\Big(\int_{\mathbb{T}^i} |D^i_{\mathbf{t}}F|^2\,d\mathbf{t}\Big)^{\frac{p}{2}}\Big], \quad I^{\sharp}_{0,j,p}(F) = \sup_{\mathbf{u}\in A(1)^j} \frac{E[|\tilde{D}^j_{\mathbf{u}}F|^p]}{\gamma(\mathbf{u})^p}.$$

We can define those for $X \in \bar{\mathcal{S}}^\infty_{\mathbb{T}}$ and $Z \in \bar{\mathcal{D}}^\infty_{\mathbb{U}_0}$ in similar ways. Then we define another norms by

$$\|F\|^{\sharp}_{m,n,p} = \Big(\sum_{i=0}^{m}\sum_{j=0}^{n} I^{\sharp}_{i,j,p}(F)\Big)^{\frac{1}{p}}, \quad F \in \bar{\mathcal{D}}^\infty, \tag{5.124}$$

$$\|X\|^{\sharp}_{m,n,p} = \Big(\sum_{i=0}^{m}\sum_{j=0}^{n} I^{\sharp}_{i,j,p}(X)\Big)^{\frac{1}{p}}, \quad X \in \bar{\mathcal{S}}^\infty_{\mathbb{T}}, \tag{5.125}$$

$$\|Z\|^{\sharp}_{m,n,p} = \Big(\sum_{i=0}^{m}\sum_{j=0}^{n} I^{\sharp}_{i,j,p}(Z)\Big)^{\frac{1}{p}}, \quad Z \in \bar{\mathcal{D}}^\infty_{\mathbb{U}_0}. \tag{5.126}$$

Let $\{A(\rho), 0 < \rho < \rho_0\}$ be a family of star-shaped neighborhoods of the center of the measure n. We have inequalities

$$\|F\|_{m,n,p;A(\rho)} \leq \|F\|^{\sharp}_{m,n,p}, \quad F \in \bar{\mathcal{D}}^\infty, \tag{5.127}$$

$$\|X\|_{m,n,p;A(\rho)} \leq \|X\|^{\sharp}_{m,n,p}, \quad X \in \bar{\mathcal{S}}^\infty_{\mathbb{T}},$$

$$\|Z\|_{m,n,p;A(\rho)} \leq \|Z\|^{\sharp}_{m,n,p}, \quad Z \in \bar{\mathcal{D}}^\infty_{\mathbb{U}_0}$$

for any m, n, p, ρ. Completions of $\bar{\mathcal{D}}^\infty, \bar{\mathcal{S}}^\infty_{\mathbb{T}}$ and $\bar{\mathcal{D}}^\infty_{\mathbb{U}_0}$ by norms $\|\ \|^{\sharp}_{m,n,p}$ are denoted by $\bar{\mathbf{D}}^{m,n,p}$, $\bar{\mathbf{D}}^{m,n,p}_{\mathbb{T}}$ and $\bar{\mathbf{D}}^{m,n,p}_{\mathbb{U}}$, respectively. The intersections of these with respect to m, n, p are denoted by $\bar{\mathbf{D}}^\infty, \bar{\mathbf{D}}^\infty_{\mathbb{T}}$ and $\bar{\mathbf{D}}^\infty_{\mathbb{U}}$, respectively. These are F-spaces with respect to countable norms $\{\| \ \|^{\sharp}_{m,n,p}; m, n, p \in \mathbb{N}\}$. If F, X, Z belong to the above functional spaces, respectively, then for any A, $\|F\|_{m,n,p;A}$, $\|X\|_{m,n,p;A}$, $\|Z\|_{m,n,p;A}$ are well defined. These are norms and are dominated by $\|F\|^{\sharp}_{m,n,p}$, $\|X\|^{\sharp}_{m,n,p}$ and $\|Z\|^{\sharp}_{m,n,p}$, respectively.

The next proposition is an extension of Propositions 5.2.1, 5.5.1 and Lemma 5.9.1. Proof is straightforward and it is omitted.

Proposition 5.10.1

1. *The operator D is extended to a continuous linear operator from F-space $\bar{\mathbf{D}}^{\infty}$ to F-space $\bar{\mathbf{D}}^{\infty}_{\mathbb{T}}$. It holds that $\|DF\|^{\sharp}_{m,n,p} \le \|F\|^{\sharp}_{m+1,n,p}$ for any $m, n \in \mathbb{N}$ and $p \ge 2$.*
2. *The operator $\tilde{D}$ is extended to a continuous linear operator from F-space $\bar{\mathbf{D}}^{\infty}$ to F-space $\bar{\mathbf{D}}^{\infty}_{\mathbb{U}}$. It holds that $\|\tilde{D}F\|^{\sharp}_{m,n,p} \le \|F\|^{\sharp}_{m,n+1,p}$ for any $m, n \in \mathbb{N}$ and $p \ge 2$.*
3. *It holds that $D_t \tilde{D}_u F = \tilde{D}_u D_t F$ for any $F \in \bar{\mathbf{D}}^{\infty}$ and $D_t \tilde{D}_u Y = \tilde{D}_u D_t Y$ for any $Y \in \bar{\mathbf{D}}^{\infty}_{\mathbb{T}}$ or $Y \in \bar{\mathbf{D}}^{\infty}_{\mathbb{U}}$, a.e. (t, u, w, ω) with respect to $dtdndP$.*

The next is a combination of Lemmas 5.2.1 and 5.5.1. The proof is omitted.

Lemma 5.10.1

1. *(Hölder's inequality) For any $m, n \in \mathbb{N}$ and $p \ge 2$, there exists a positive constant c such that*

$$\|FG\|_{m,n,p;A} \le c\|F\|_{m,n,pr;A}\|G\|_{m,n,pr';A}, \tag{5.128}$$

$$\|FX\|_{m,n,p;A} \le c\|F\|_{m,n,pr;A}\|X\|_{m,n,pr';A},$$

$$\|FZ\|_{m,n,p;A} \le c\|F\|_{m,n,pr;A}\|Z\|_{m,n,pr';A}$$

hold for any $F, G \in \bar{\mathbf{D}}^{\infty}$, $X \in \bar{\mathbf{D}}^{\infty}_{\mathbb{T}}$, $Z \in \bar{\mathbf{D}}^{\infty}_{\mathbb{U}}$, domain A and $r, r' > 1$ with $1/r + 1/r' = 1$.

2. *Similar Hölder's inequalities are valid with respect to norms $\| \ \|^{\sharp}_{m,n,p}$.*

Finally the norm $|F|^{\sharp}_{0,n,p}$ for a simple Wiener–Poisson functional F is defined by (5.57). Then the norm $| \ |^{\sharp}_{0,n,p}$ is weaker than the norm $\| \ \|^{\sharp}_{m,n,p}$ for any m and satisfies $|F|^{\sharp}_{0,n,p} \le c_{m,n,p}\|F\|^{\sharp}_{m,n,p}$. Let $\bar{\mathbf{L}}^{0,n,p}$ be the completion of $\bar{\mathcal{D}}^{\infty}$ by the norm $| \ |^{\sharp}_{0,n,p}$. We set $\bar{\mathbf{L}}^{\infty} = \bigcap_{n,p} \bar{\mathbf{L}}^{0,n,p}$. Then, $\bar{\mathbf{D}}^{\infty} \subset \bar{\mathbf{L}}^{\infty}$.

A Wiener–Poisson functional $F = (F^1, \ldots, F^d)$ is said to belong to $(\bar{\mathbf{D}}^{\infty})^d$, if each component F^j belongs to $\bar{\mathbf{D}}^{\infty}$. Sobolev norms $\|F\|_{m,n,p;A}$ and $\|F\|^{\sharp}_{m,n,p}$ are defined as sums of the corresponding norms for components F^j, $j = 1, \ldots, d$. Spaces $(\bar{\mathbf{D}}^{\infty}_{\mathbb{T}})^d$, $(\bar{\mathbf{D}}^{\infty}_{\mathbb{U}})^d$ and their norms are defined similarly. Lemma 5.6.2 is extended in the following manner.

Lemma 5.10.2 *For any $m, n \in \mathbb{N}$ and $p \geq 2$, there exists a positive constant $c_{m,n,p}$ such that*

$$\|(e^{i(v,\tilde{D}F)} - 1)\|_{m,n,p;A(\rho)} \tag{5.129}$$
$$\leq c_{m,n,p}|v|\varphi(\rho)^{\frac{1}{2}}\|\tilde{D}F\|^{\sharp}_{m,n,(n+1)p}\left(1 + |v|\rho\varphi(\rho)^{\frac{1}{2}}\|\tilde{D}F\|^{\sharp}_{m,n,(n+1)p}\right)^n$$

holds for all $v \in \mathbb{R}^d, 0 < \rho < \rho_0$ and $F \in (\bar{\mathbf{D}}^\infty)^d$.

The proof will be omitted, since it can be done similarly to the proof of Lemma 5.6.2.

We shall study estimations of operators δ and $\tilde{\delta}$. By the isometric properties of operators δ and $\tilde{\delta}$, these satisfy

$$E[\delta(X)^2]^{\frac{1}{2}} \leq \|X\|^{\sharp}_{1,0,2}, \quad E[\tilde{\delta}(Z)^2]^{1/2} \leq \|Z\|^{\sharp}_{0,1,2}$$

for $X \in \bar{\mathcal{S}}_{\mathbb{T}}$ and $Z \in \bar{\mathcal{D}}^\infty_{\mathbb{U}_0}$. Therefore the operator δ is extended to a continuous linear operator from $\bar{\mathbf{D}}^{1,0,2}_{\mathbb{T}}$ to $\bar{\mathbf{D}}^{0,0,2}$. Further, $\tilde{\delta}$ is extended to a continuous linear operator from $\bar{\mathbf{D}}^{0,1,2}_{\mathbb{U}}$ to $\bar{\mathbf{D}}^{0,0,2}$.

Proposition 5.10.2 *Let $\{\mathcal{F}_t\}$ be the filtration generated by the Wiener process W_t and the Poisson random measure $N(dt\,dz)$. Suppose $X \in \bar{\mathbf{D}}^{1,0,2}_{\mathbb{T}}$ is predictable with respect to the filtration. Then the Skorohod integral $\delta(X)$ coincides with the Itô integral $\sum_k \int_{\mathbb{T}} X^k_s\, dW^k_s$. Suppose that $Z_{t,z} \in \bar{\mathbf{D}}^{0,1,2}_{\mathbb{U}}$ is predictable with respect to $\{\mathcal{F}_t\}$ for any z. Then the Skorohod integral $\tilde{\delta}(Z)$ coincides with Itô integral $\int_{\mathbb{T}}\int_{\mathbb{R}^{d'}_0} Z_{t,z}\tilde{N}(dt\,dz)$.*

The proof for δ can be carried out similarly to the proof of Proposition 5.2.2. The proof for $\tilde{\delta}$ can be carried out similarly to the proof of Proposition 5.4.1. Details are omitted. □

We will summarize properties of adjoint operators δ and $\tilde{\delta}$.

Theorem 5.10.1

1. *The adjoint operator δ is extended to a continuous linear operator from F-space $\bar{\mathbf{D}}^\infty_{\mathbb{T}}$ to F-space $\bar{\mathbf{D}}^\infty$. Further, for any $m, n \in \mathbb{N}$ and $p \geq 2$, there exists a positive constant $c = c_{m,n,p}$ not depending on domains $A(\rho)$ such that*

$$\|\delta(X)\|_{m,n,p;A(\rho)} \leq c_{m,n,p}\|X\|_{m+\langle p\rangle-1,n,\langle p\rangle;A(\rho)} \tag{5.130}$$

holds for any $0 < \rho < \rho_0$ for all $X \in \bar{\mathbf{D}}^\infty_{\mathbb{T}}$, where $\langle p\rangle = -2[-p/2]$.

The adjoint operator $\tilde{\delta}$ is extended to a continuous linear operator from F-space $\bar{\mathbf{D}}^\infty_{\mathbb{U}}$ to F-space $\bar{\mathbf{D}}^\infty$. Further, for any $m, n \in \mathbb{N}$ and $p \geq 2$, there exists a positive constant $c = c_{m,n,p}$ not depending on domains $A(\rho)$ such that

$$\|\tilde{\delta}(Z\chi_\rho)\|_{m,n,p;A(\rho)} \leq c_{m,n,p}\|Z\|_{m,n+\langle p\rangle-1,\langle p\rangle;A(\rho)} \tag{5.131}$$

holds for any $0 < \rho < \rho_0$ *for all* $Z \in \bar{\mathbf{D}}_{\mathbb{U}}^{\infty}$.

2. *For* δ, *the adjoint formula* (5.117) *holds for any* $X \in \bar{\mathbf{D}}_{\mathbb{T}}^{\infty}$ *and* $G \in \bar{\mathbf{D}}^{\infty}$. *For* $\tilde{\delta}$, *the adjoint formula* (5.118) *holds for any* $Z \in \bar{\mathbf{D}}_{\mathbb{U}}^{\infty}$ *and* $G \in \bar{\mathbf{D}}^{\infty}$.
3. *For* δ, *the commutation relation* (5.119) *holds for any* $X \in \bar{\mathbf{D}}_{\mathbb{T}}^{\infty}$. *For* $\tilde{\delta}$, *the commutation relation* (5.120) *holds for any* $Z \in \bar{\mathbf{D}}_{\mathbb{U}}^{\infty}$.
4. *For* δ, *the isometric property* (5.10) *holds for any* $X, Y \in \bar{\mathbf{D}}_{\mathbb{T}}^{\infty}$. *For* $\tilde{\delta}$, *the isometric property* (5.45) *holds for any* $Y, Z \in \bar{\mathbf{D}}_{\mathbb{U}}^{\infty}$.

Proof We will prove assertion 1 for positive even number p only. We write $A(\rho)$ as A. The operator δ on the Wiener–Poisson space satisfies the adjoint formula (5.117). Then the inequality of Theorem 5.2.1 holds for this δ, i.e., there exists $c \geq 1$ such that

$$\|\delta(X)\|_{m,0,p;A} = \|\delta(X)\|_{m,p} \leq c\|X\|_{m+p-1,p} = c\|X\|_{m+p-1,0,p;A} \tag{5.132}$$

for any domain A for all $X \in \bar{\mathcal{S}}_{\mathbb{T}}^{\infty}$. We shall compute $\|\delta(X)\|_{m,n,p;A}$ for $n \geq 1$. Since the norm is defined by (5.121), we consider $I_{i,j,p;A}(\delta(X))$ for $1 \leq i \leq m$, $1 \leq j \leq n$. Note that $\tilde{D}_{\mathbf{u}}^j\delta(X) = \delta(\tilde{D}_{\mathbf{u}}^j X)$ holds a.e. for $X \in \bar{\mathcal{S}}_{\mathbb{T}}^{\infty}$. Then

$$I_{i,j,p;A}(\delta(X)) = \int_A \frac{E[(\int |D_{\mathbf{t}}^i\delta(\tilde{D}_{\mathbf{u}}^j X)|^2\, d\mathbf{t})^{\frac{p}{2}}]}{\gamma_A(\mathbf{u})^p} m_A^j(d\mathbf{u}). \tag{5.133}$$

Taking $m = i$ in (5.132), we have

$$E\Big[\Big(\int |D_{\mathbf{t}}^i\delta(\tilde{D}_{\mathbf{u}}^j X)|^2\, d\mathbf{t}\Big)^{\frac{p}{2}}\Big] \leq c^p \sum_{1\leq i'\leq p} E\Big[\Big(\int |D_{\mathbf{t}}^{i+i'-1}\tilde{D}_{\mathbf{u}}^j X_t|^2\, d\mathbf{t}\, dt\Big)^{\frac{p}{2}}\Big],$$

a.e. $\mathbf{u}$. Then we get

$$I_{i,j,p;A}(\delta(X)) \leq c^p \sum_{1\leq i'\leq p} \|X\|_{i+i'-1,j,p;A} \leq c'\|X\|_{i+p-1,j,p;A}^p.$$

Sum up the above for $i = 0, \ldots, m$, $j = 0, \ldots, n$. Then we get

$$\|\delta(X)\|_{m,n,p;A}^p \leq C\|X\|_{m+p-1,n,p;A}^p.$$

Finally the inequality (5.130) is extended to $X \in \bar{\mathbf{D}}_{\mathbb{T}}^{\infty}$, since $\bar{\mathcal{S}}_{\mathbb{T}}^{\infty}$ is dense in $\bar{\mathbf{D}}_{\mathbb{T}}^{\infty}$.

Next consider $\tilde{\delta}(Z)$ for Wiener–Poisson functional Z. Theorem 5.6.1 is valid for Wiener–Poisson functional Z. Since D and $\tilde{\delta}$ are commutative, we get (5.131) in the same way.

Finally, assertions 2–4 are verified directly from Lemma 5.9.2. □

Differential operator D, difference operator $\tilde{D}$ and their adjoints δ, $\tilde{\delta}$ can be extended for complex-valued functionals. Theorem 5.10.1 is valid for complex valued functionals.

The next theorem is an extension of Lemma 5.3.1 for Wiener functionals.

Theorem 5.10.2 (c.f. Lemma 3.3 in [46]) *Suppose that $G \in \bar{\mathbf{D}}^\infty$ is invertible and the inverse G^{-1} belongs to $\bar{\mathbf{L}}^\infty$. Then $G^{-1} \in \bar{\mathbf{D}}^\infty$. Further, for any m, n, p, there exists $c_{m,n,p} > 0$ such that the inequality*

$$\|G^{-1}\|^\sharp_{m,n,p} \le c_{m,n,p}(1 + \|G\|^\sharp_{m,n,2qp})^q(1 + |G^{-1}|^\sharp_{0,n,2qp})^q \tag{5.134}$$

holds for any G with the above property, where $q = (m+1)(2^n - 1)$.

Proof We will fix $i, j \in \mathbb{N}$ and compute $D^i_{\mathbf{t}}\tilde{D}^j_{\mathbf{u}}(G^{-1})$. For the computation, we use difference rules for $X, Y \in \bar{\mathbf{D}}^\infty$ such that Y^{-1} exists:

$$\tilde{D}_u\left(\frac{X}{Y}\right) = \frac{\tilde{D}_u X \cdot Y - X \cdot \tilde{D}_u Y}{Y \cdot Y \circ \varepsilon_u^+} \tag{5.135}$$

and for $\mathbf{v} \in A(1)^j$,

$$\tilde{D}_u(X \circ \varepsilon^+_{\mathbf{v}}) = \sum_{\mathbf{w} \subset \mathbf{v}} \tilde{D}_u \tilde{D}_{\mathbf{w}} X, \tag{5.136}$$

where $\mathbf{w}$ runs over all subsets of $\mathbf{v}$. Then, for the functional G^{-1}, we have the equality

$$\tilde{D}_{u_1}(G^{-1}) = \frac{-\tilde{D}_{u_1} G}{G \cdot G \circ \varepsilon^+_{u_1}}. \tag{5.137}$$

Repeating operations $\tilde{D}_{u_2}, \ldots, \tilde{D}_{u_j}$ and applying the formula (5.135), we have

$$\tilde{D}^j_{\mathbf{u}}(G^{-1}) = \frac{Y_{\mathbf{u}}}{(G \circ \varepsilon^+_{\mathbf{u}})!} = Y_{\mathbf{u}} \cdot (G^{-1} \circ \varepsilon^+_{\mathbf{u}})!, \tag{5.138}$$

where $(G \circ \varepsilon^+_{\mathbf{u}})! := \prod_{\mathbf{v} \subset \mathbf{u}} G \circ \varepsilon^+_{\mathbf{v}}$ and $Y_{\mathbf{u}}$ is a linear sum of functionals of the form

$$\tilde{D}_{\mathbf{u}_1} G \cdots \tilde{D}_{\mathbf{u}_{s_j}} G, \tag{5.139}$$

where $s_j = 2^j - 1$ and $\mathbf{u}_i$ $(i = 1, \ldots, s_j)$ are subsets of $\mathbf{u}$ (including empty set) satisfying $\mathbf{u}_1 \cup \cdots \cup \mathbf{u}_{s_j} \supset \mathbf{u}$. The summation is taken for all such $\{\mathbf{u}_1, \ldots, \mathbf{u}_{s_j}\}$.

Next, with respect to the derivative operator $D_{\mathbf{t}}, \mathbf{t} \in \mathbb{T}^i$, we have the formula

$$D_{\mathbf{t}}^i\Big(\frac{X}{Y}\Big) = \sum (-1)^r \frac{D_{\mathbf{t}_0}X \cdot Y^{i-r} \cdot D_{\mathbf{t}_1}Y \cdots D_{\mathbf{t}_r}Y}{Y^{i+1}}, \tag{5.140}$$

where the sum is taken for positive integers $r \le i$, $\mathbf{t}_0, \ldots, \mathbf{t}_r$ are disjoint subsets of $\mathbf{t}$ and $\sharp(\mathbf{t}_0) + \cdots + \sharp(\mathbf{t}_r) = i$. It can be proved similarly to the proof of Lemma 5.3.1. Then from (5.138) and (5.140), $D_{\mathbf{t}}^i \tilde{D}_{\mathbf{u}}^j G^{-1}$ is written as

$$D_{\mathbf{t}}^i \tilde{D}_{\mathbf{u}}^j G^{-1} = \frac{Z_{\mathbf{t},\mathbf{u}}^{ij}}{(G \circ \varepsilon_{\mathbf{u}}^+ !)^{i+1}}, \tag{5.141}$$

where $Z_{\mathbf{t},\mathbf{u}}^{ij}$ is a linear sum of functionals of the form

$$(\tilde{D}_{\mathbf{v}_1} G \cdots \tilde{D}_{\mathbf{v}_{s_j}} G)^{i-r} \times D_{\mathbf{t}_0}(\tilde{D}_{\mathbf{v}_1^{(0)}} G \cdots \tilde{D}_{\mathbf{v}_{s_j}^{(0)}} G) \cdots D_{\mathbf{t}_r}(\tilde{D}_{\mathbf{v}_1^{(r)}} G \cdots \tilde{D}_{\mathbf{v}_{s_j}^{(r)}} G), \tag{5.142}$$

where $\mathbf{v}_j, \mathbf{v}_j^{(h)} \subset \mathbf{u}$ for $h = 0, \ldots, r$, $\mathbf{t}_1, \ldots, \mathbf{t}_r$ are disjoint subsets of $\mathbf{t} = \{t_1, \ldots, t_i\}$ and $\sharp(\mathbf{t}_0) + \cdots + \sharp(\mathbf{t}_r) = i$.

Now, we will give an estimate of the functional (5.141) with respect to the norm $\| \ \|_{i,j,p}^{\sharp}$. We have

$$\frac{E\big[\big(\int_{\mathbb{T}^i} |D_{\mathbf{t}}^i \tilde{D}_{\mathbf{u}}^j G^{-1}|^2 d\mathbf{t}\big)^{\frac{p}{2}}\big]}{\gamma(\mathbf{u})^p} = \frac{E\big[\big(\int_{\mathbb{T}^i} |Z_{\mathbf{t},\mathbf{u}}^{ij}|^2 d\mathbf{t}\big)^{\frac{p}{2}} (G \circ \varepsilon_{\mathbf{u}}^+ !)^{-(i+1)p}\big]}{\gamma(\mathbf{u})^p} \le I_1 \times I_2,$$

where

$$I_1 = \frac{E\big[\big(\int |Z_{\mathbf{t},\mathbf{u}}^{ij}|^2 d\mathbf{t}\big)^p\big]^{\frac{1}{2}}}{\gamma(\mathbf{u})^p}, \quad I_2 = E\big[(G \circ \varepsilon_{\mathbf{u}}^+ !)^{-(i+1)2p}\big]^{\frac{1}{2}}.$$

Consider the term I_1. The functional $Z_{\mathbf{t},\mathbf{u}}^{ij}$ is written as sums of terms $Z'_{\mathbf{t},\mathbf{u}}$ which are represented by $q' = (i+1)s_j$-products of functionals $D_{\mathbf{t}_h} \tilde{D}_{\mathbf{v}_l^{(h)}} G$, i.e., $Z'_{\mathbf{t},\mathbf{u}} = \prod_{h,l}^{q'} D_{\mathbf{t}_i} \tilde{D}_{\mathbf{v}_l^{(h)}} G$. Further, $\prod_{h,l} \gamma(\mathbf{v}_l^{(h)}) \le \gamma(\mathbf{u})$ holds, since $\bigcup_{h,l} \mathbf{v}_l^{(h)} \supset \mathbf{u}$. Consequently, using Hölder's inequality, we have

$$\begin{aligned} I_1^2 = \frac{E\big[\big(\int |Z'_{\mathbf{t},\mathbf{u}}|^2 d\mathbf{t}\big)^p\big]}{\gamma(\mathbf{u})^{2p}} &\le \prod_{h,l}^{q'} \frac{E\big[\big(\int |D_{\mathbf{t}_h} \tilde{D}_{\mathbf{v}_l^{(h)}} G|^2 d\mathbf{t}\big)^{q'p}\big]^{\frac{1}{q'}}}{\gamma(\mathbf{v}_l^{(h)})^{\frac{2p}{q'}}} \\ &\le c_1 (1 + \|G\|_{i,j,2q'p}^{\sharp})^{2q'p}. \end{aligned} \tag{5.143}$$

Now, since $Z_{t,u}^{ij}$ is a linear sum of $Z'_{t,u}$, there is a positive constant c_2 such that

$$I_1 \le c_2(1 + \|G\|^{\sharp}_{i,j,2qp})^{qp}. \tag{5.144}$$

On the other hand,

$$\begin{aligned} I_2 &\le \prod_{k=0}^{s_j} E\big[|G^{-(i+1)2s_j p} \circ \varepsilon^+_{\mathbf{u}_k}|\big]^{\frac{1}{2s_j}} \\ &\le (1 + |G^{-1}|^{\sharp}_{0,j,2s_j p})^{s_j p} \le (1 + |G^{-1}|^{\sharp}_{0,j,2qp})^{qp}. \end{aligned} \tag{5.145}$$

The last inequality follows since $s_j < q$ and norms $|\ |^{\sharp}_{0,n,p}$ is nondecreasing with respect to p. These two estimates (5.144) and (5.145) imply

$$\frac{E\big[\big(\int_{\mathbb{T}^i} |D^i_{\mathbf{t}} \tilde{D}^j_{\mathbf{u}} G^{-1}|^2 \, d\mathbf{t}\big)^{\frac{p}{2}}\big]}{\gamma(\mathbf{u})^p} \le c_3(1 + \|G\|^{\sharp}_{i,j,2qp})^q (1 + |G^{-1}|^{\sharp}_{0,j,2qp})^q.$$

Finally, sum up the above inequalities for $i = 1, \ldots, m$ and $j = 1, \ldots, n$ and adjoin $\|G\|_{0,0,p}$. Since the right-hand sides of above inequalities are nondecreasing with respect to m, n, we find that the seminorm $\|G^{-1}\|^{\sharp}_{m,n,p}$ is also bounded by

$$c_4(1 + \|G\|^{\sharp}_{m,n,2qp})^q (1 + |G^{-1}|^{\sharp}_{0,n,2qp})^q.$$

Therefore, we get the inequality of the theorem. □

Note In [45], Sobolev norms for Wiener–Poisson functionals are defined. Norms $|\ |_{k,l,p}$, $\|\ \|_{k,l,p}$ and $\|\ \|^{\sim}_{k,l,p}$ correspond to our $\|\ \|_{l,k,p;\mathbb{U}}$.

An estimation of $\tilde{\delta}(Z\chi_\rho)$ with respect to above norms is given in [45], Theorem 3.2. However, the discussion contains a gap; here it is rectified. The assertion of Theorem 3.2 in [45] is recovered by our inequalities (5.129) and (5.131).

5.11 Nondegenerate Wiener–Poisson Functionals

We will define nondegenerate Wiener–Poisson functionals and show the existence of smooth densities of their laws. In the first step, we will give a criterion called δ-nondegenerate, which is a combination of Malliavin's criterion for Wiener functionals given in Sect. 5.3 and a criterion for Poisson functionals given in Sect. 5.7.

Let $0 < \rho < \rho_0$ and $\theta \in S_{d-1}$. Let $1 < \delta < \frac{2}{\alpha}$. Associated with $F = (F^1, \ldots, F^d) \in (\bar{\mathbf{D}}^\infty)^d$, we will consider the following Wiener–Poisson functionals:

$$\bar{Q}^F(\rho, v) = R^F(v) + Q^F(\rho, v), \quad \bar{Q}^F_\rho(\theta) = R^F(\theta) + Q^F_\rho(\theta), \tag{5.146}$$

where R^F is defined by (5.26) and $Q^F(\rho, v)$, $Q^F_\rho(\theta)$ are defined by (5.85), (5.86), respectively. A functional $F \in (\bar{\mathbf{D}}^\infty)^d$ is called *δ-nondegenerate* (*nondegenerate with respect to $\bar{Q}^F_\rho$*) if $\bar{Q}^F_\rho(\theta)$ are invertible and inverses satisfy

$$\sup_{(\rho,\theta)} |\bar{Q}^F_\rho(\theta)^{-1}|^\sharp_{0,n,p} < \infty, \quad \forall n, p, \tag{5.147}$$

where the supremum for (ρ, θ) is taken in the domain $(0, \rho_0) \times S_{d-1}$.

We want to get estimations of $\bar{Q}^F_\rho(\theta)$ and its inverse with respect to Sobolev norms.

Proposition 5.11.1 *Suppose that $F \in (\bar{\mathbf{D}}^\infty)^d$ is δ-nondegenerate. Then $\bar{Q}^F_\rho(\theta)^{-1} \in \bar{\mathbf{D}}^\infty$ for any $0 < \rho < \rho_0$ and $\theta \in S_{d-1}$. Further, for any m, n, p, there exists a positive constant $c_{m,n,p}$ such that for any δ-nondegenerate functional F the following holds for all (ρ, θ):*

$$\begin{aligned}\|\bar{Q}^F_\rho(\theta)^{-1}\|^\sharp_{m,n,p} \le c_{m,n,p}\Big(1 + \|DF\|^\sharp_{m,n,2^{n+2}p} + \|\tilde{D}F\|^\sharp_{m,n,(n+1)2^{n+2}p}\Big)^{(n+1)2^{n+2}} \\ \times \Big(1 + |\bar{Q}^F_\rho(\theta)^{-1}|^\sharp_{0,n,2^{n+2}p}\Big)^{2^n}.\end{aligned} \tag{5.148}$$

Proof We will apply Theorem 5.10.2 for the functional $G = \bar{Q}^F_\rho(\theta)$. We need the estimate of the norm of $\bar{Q}^F_\rho(\theta)$. We have

$$\begin{aligned}\|R^F(\theta)\|^\sharp_{m,n,p} &\le c(\|DF\|^\sharp_{m,n,2p})^2, \\ \|Q^F_\rho(\theta)\|^\sharp_{m,n,p} &\le c'(1 + \|\tilde{D}F\|^\sharp_{m,n,2(n+1)p})^{2(n+1)}.\end{aligned} \tag{5.149}$$

Indeed, the first inequality is immediate from the definition of the norms. The second inequality can be verified similarly to the proof of Proposition 5.7.1, noting that operators D and $\tilde{D}$ are commutative. The above two inequalities imply the following inequality:

$$\|\bar{Q}^F_\rho(\theta)\|^\sharp_{m,n,p} \le c_{m,n,p}\big\{(\|DF\|^\sharp_{m,n,2p})^2 + (1 + \|\tilde{D}F\|^\sharp_{m,n,2(n+1)p})^{2(n+1)}\big\}.$$

Then there is a positive constant $c_{m,n,p}$ such that (5.148) holds. □

The next theorem is an extension of Theorem 5.3.1 for Wiener functionals and of Theorem 5.7.1 for Poisson functionals.

Theorem 5.11.1 *Let $\gamma_0 = 1 - \frac{\alpha\delta}{2} > 0$. Then, for any $N \in \mathbb{N}$, there exist $m, n \in \mathbb{N}$, $p > 4$ and a positive constant C such that for any d-dimensional δ-nondegenerate Wiener–Poisson functional F in $(\bar{\mathbf{D}}^\infty)^d$,*

$$\left|E\left[e^{i(v,F)}G\right]\right| \le \frac{C}{|v|^{N\gamma_0}}\Big((\|DF\|^{\sharp}_{m,n,p} + (1+\|\tilde{D}F\|^{\sharp}_{m,n.(n+1)p})^{n+1})$$

$$\times \sup_{(\rho,\theta)} \|\bar{Q}^F_\rho(\theta)^{-1}\|^{\sharp}_{m,n,p}\Big)^N \|G\|^{\sharp}_{m,n,p} \tag{5.150}$$

hold for all $|v| \ge 1$ *and* $G \in \bar{\mathbf{D}}^\infty$.

Proof We will define a d'-dimensional stochastic process X_t and a random field Z_u associated with a given δ-nondegenerate smooth Wiener–Poisson functional F:

$$X_t = X_t^{F,v} = \frac{-i(v, D_t F)}{|v|^2 \bar{Q}^F(\rho, v)}, \quad Z_u = Z_u^{F,v} = \frac{(e^{-i(v,\tilde{D}_u F)} - 1)1_{A(\rho)}(u)}{|v|^2 \varphi(\rho)\bar{Q}^F(\rho, v)}, \tag{5.151}$$

where $\rho = |v|^{-\delta}$. Adjoint formulas (5.117) and (5.118) are extended to complex valued Wiener–Poisson functionals. Then we have

$$\begin{aligned}
&E\Big[e^{i(v,F)}\{\delta(XG) + \tilde{\delta}(ZG)\}\Big] \\
=&E\Big[\int D_t(e^{i(v,F)})X_t G\,dt\Big] + E\Big[\int \tilde{D}_u(e^{i(v,F)})Z_u G n(du)\Big] \\
=& E\Big[\int e^{i(v,F)} i(v, D_t F) X_t G\,dt\Big] + E\Big[\int e^{i(v,F)}(e^{i(v,\tilde{D}_u F)} - 1) Z_u G n(du)\Big] \\
=&E\Big[\frac{e^{i(v,F)}G}{|v|^2 \bar{Q}^F(\rho, v)} \int (v, D_t F)^2\,dt\Big] \\
&+ E\Big[\frac{e^{i(v,F)}G}{|v|^2\varphi(\rho)\bar{Q}^F(\rho, v)} \int_{A(\rho)} |e^{i(v,\tilde{D}_u F)} - 1|^2 n(du)\Big] \\
=&E\big[e^{i(v,F)}G\big].
\end{aligned}$$

Therefore, setting $\mathcal{L}(G) = \delta(XG) + \tilde{\delta}(ZG)$, we have the iteration formula;

$$E[e^{i(v,F)}G] = E[e^{i(v,F)}\mathcal{L}(G)] = \cdots = E[e^{i(v,F)}\mathcal{L}^N(G)], \tag{5.152}$$

for $N = 1, 2, \ldots$. Now, note $\mathcal{L}^N(G) = \delta(X\mathcal{L}^{N-1}(G)) + \tilde{\delta}(Z\mathcal{L}^{N-1}(G))$. Then, using Theorem 5.10.1 and Hölder's inequality, we have

$$\begin{aligned}
\|\mathcal{L}^N(G)\|_{0,0,2} &\le \|\delta(X\mathcal{L}^{N-1}(G))\|_{0,0,2} + \|\tilde{\delta}(Z\mathcal{L}^{N-1}(G))\|_{0,0,2} \\
&\le c_1(\|X\mathcal{L}^{N-1}(G)\|_{1,0,2} + \|Z\mathcal{L}^{N-1}(G)\|_{0,1,2}) \\
&\le c_2(\|X\|_{1,0,4} + \|Z\|_{0,1,4})\|\mathcal{L}^{N-1}(G)\|_{1,1,4},
\end{aligned}$$

where $\| \ \|_{m,n,p} = \| \ \|_{m,n,p;A(\rho)}$. Repeating this argument inductively, there exist m_N, n_N, p_N and c_N such that

$$\|\mathcal{L}^N(G)\|_{0,0,2} \le c_N(\|X\|_{m_N,n_N,p_N} + \|Z\|_{m_N,n_N,p_N})^N \|G\|_{m_N,n_N,p_N}.$$

Further, by Hölder's inequality

$$\begin{aligned} &\|X\|_{m,n,p} + \|Z\|_{m,n,p} \\ &\le \Big(\frac{1}{|v|^2}\|i(v, DF)\|_{m,n,2p} + \frac{1}{|v|^2\varphi(\rho)}\|\chi_\rho(e^{-i(v,\tilde{D}F)} - 1)\|_{m,n,2p}\Big) \\ &\quad \times \|\bar{Q}^F(\rho, v)^{-1}\|_{m,n,2p}. \end{aligned}$$

We have

$$\|i(v, DF)\|_{m,n,2p} \le c|v|\|DF\|^\sharp_{m,n,2p}, \tag{5.153}$$

$$\|e^{-i(v,\tilde{D}F)} - 1\|_{m,n,2p} \le c|v|\varphi(\rho)^{\frac{1}{2}}(1 + \|\tilde{D}F\|^\sharp_{m,n,2(n+1)p})^{n+1},$$

$$\|\bar{Q}^F(\rho, v)^{-1}\|_{m,n,2p} \le \sup_{(\rho,\theta)} \|\bar{Q}^F_\rho(\theta)^{-1}\|^\sharp_{m,n,2p}.$$

The first inequality is immediate from the definition of the norm. The second inequality follows from Lemma 5.10.2. The third one is obvious. Then we get

$$\begin{aligned} &\|X\|_{m,n,p} + \|Z\|_{m,n,p} \\ &\le c\Big(\frac{1}{|v|}\|DF\|^\sharp_{m,n,2p} + \frac{1}{|v|\varphi(\rho)^{\frac{1}{2}}}(1 + \|\tilde{D}F\|^\sharp_{m,n,2(n+1)p})^{n+1}\Big) \\ &\quad \times \sup_{(\rho,\theta)} \|\bar{Q}^F_\rho(\theta)^{-1}\|^\sharp_{m,n,2p}, \end{aligned} \tag{5.154}$$

where $\rho = |v|^{-\delta}$. Since $\frac{1}{|v|} \le \frac{c_1}{|v|\varphi(|v|^{-\delta})} \le \frac{c_2}{|v|^{\gamma_0}}$, the last term of the above is dominated by

$$\frac{c}{|v|^{\gamma_0}}(\|DF\|^\sharp_{m,n,2p} + (1 + \|\tilde{D}F\|^\sharp_{m,n,2(n+1)p})^{n+1}) \sup_{(\rho,\theta)} \|\bar{Q}^F_\rho(\theta)^{-1}\|^\sharp_{m,n,2p}.$$

Then we get the inequality of the theorem, rewriting $2p$ as p. □

Corollary 5.11.1 *Suppose that F in $(\bar{\mathbf{D}}^\infty)^d$ is δ-nondegenerate. Then for any $G \in \bar{\mathbf{D}}^\infty$, the law of F weighted by G has a rapidly decreasing C^∞-density. Further, the density function $f_G(x)$ is given by the Fourier inversion formula* (5.34).

Now we will discuss two other nondegenerate conditions. A Wiener–Poisson functional F is called *regular* if it satisfies Condition (R) introduced in Sect. 5.7. Let F be a regular functional of $(\bar{\mathbf{D}}^\infty)^d$. We will define its *Malliavin covariance at the center* by

$$\bar{K}^F = \int_{\mathbb{T}} D_t F (D_t F)^T \, dt + \int_{\mathbb{T}} (\partial \tilde{D}_{t,0} F) \Gamma_0 (\partial \tilde{D}_{t,0} F)^T \, dt, \tag{5.155}$$

where $\partial \tilde{D}_{t,0} F$ is defined by (5.82). For $\theta \in S_{d-1}$, we set $\bar{K}^F(\theta) = (\theta, \bar{K}^F \theta)$. If $\{\bar{K}^F(\theta), \theta \in S_{d-1}\}$ are invertible and inverses satisfy

$$\sup_\theta |\bar{K}^F(\theta)^{-1}|^\sharp_{0,n,p} < \infty, \quad \forall n, p, \tag{5.156}$$

then F is called *nondegenerate at the center*.

Theorem 5.11.2 *Let F be a regular functional belonging to $(\bar{\mathbf{D}}^\infty)^d$.*

1. *If F is δ-nondegenerate for some $1 < \delta < 2/\alpha$, then F is nondegenerate at the center. Further, it holds for any n, p that*

$$\sup_\theta |\bar{K}^F(\theta)^{-1}|^\sharp_{0,n,p} \le \sup_{(\rho,\theta)} |\bar{Q}^F_\rho(\theta)^{-1}|^\sharp_{0,n,p}. \tag{5.157}$$

2. *If F is nondegenerate at the center, then F is δ-nondegenerate for any $1 < \delta < 2/\alpha$. Further, for any n, p there is a positive constant c such that the inequality*

$$\sup_{(\rho,\theta)} |\bar{Q}^F_\rho(\theta)^{-1}|^\sharp_{0,n,p} \le c C^{\delta'}_{n,p}(F) \sup_\theta |\bar{K}^F(\theta)^{-1}|^\sharp_{0,n,2p}, \tag{5.158}$$

holds for any $F \in (\bar{\mathbf{D}}^\infty)^d$, which is regular and nondegenerate at the center, where $C^{\delta'}_{n,p}(F)$ is given by (5.99) replacing $K^F(\theta)$ by $\bar{K}^F(\theta)$.

Proof The first assertion and the inequality (5.157) can be verified in the same way as in the first assertion of Theorem 5.8.1. We will prove the second assertion by a method similar to the proof of Theorem 5.8.1. We set $\bar{S}^F_\rho = R^F + S^F_\rho$ and $\bar{K}^F_\rho = R^F + K^F_\rho$, where S^F_ρ and K^F_ρ are defined by (5.100) and (5.101), respectively. Then we can verify the followings:

1. Suppose that $\bar{S}^F_\rho(\theta)$ are invertible a.s. for any ρ, θ. Then $\bar{Q}^F_\rho(\theta)$ are also invertible a.s. for any ρ, θ. Further, there is a positive constant c_1 such that the inequality

$$|\bar{Q}^F_\rho(\theta)^{-1}|^\sharp_{0,n,p} \le c_1 |\bar{S}^F_\rho(\theta)^{-1}|^\sharp_{0,n,p} \tag{5.159}$$

holds for any ρ, θ and n, p (Lemma 5.8.1).

2. Suppose that $\bar{K}^F(\theta)$ are invertible a.s. for any $\theta \in S_{d-1}$. Then $\bar{K}^F_\rho(\theta)$ are also invertible a.s. for any (ρ, θ). Further, there are two positive constants c_2, c_3 such that for any (ρ, θ), the inequalities

$$c_2 |\bar{K}^F(\theta)^{-1}|^\sharp_{0,n,p} \le |\bar{K}^F_\rho(\theta)^{-1}|^\sharp_{0,n,p} \le c_3 |\bar{K}^F(\theta)^{-1}|^\sharp_{0,n,p} \tag{5.160}$$

hold for any n, p (Lemma 5.8.2).

3. If $F \in (\bar{\mathbf{D}}^\infty)^d$ is regular and nondegenerate at the center, functionals $\{\bar{S}_\rho^F(\theta)\}$ are invertible. Further, it holds for any $p > 1$ that

$$E\big[(\bar{S}_\rho^F(\theta) \circ \varepsilon_{\mathbf{u}}^+)^{-p}\big] \leq c_5 B_p^{\delta'}(F) E\big[(\bar{K}_\rho^F(\theta) \circ \varepsilon_{\mathbf{u}}^+)^{-2p}\big]^{\frac{1}{2}}, \quad \forall \rho, \theta, \mathbf{u}, \tag{5.161}$$

where $B_p^{\delta'}(F)$ is given by (5.108). Indeed, since the equality

$$\bar{S}_\rho^F(\theta) - \bar{K}_\rho^F(\theta) = S_\rho^F(\theta) - K_\rho^F(\theta)$$

holds, discussions of Lemma 5.8.3 can be applied to the present case and we get

$$P(\varrho^F(\theta) \circ \varepsilon_{\mathbf{u}}^+ < \epsilon) \leq c_4 A_p^{\delta'}(F)\epsilon^p, \quad \forall 0 < \epsilon < 1, \ \theta \in S_{d-1}, \ \mathbf{u} \in \mathbb{U}^k \tag{5.162}$$

where $A_p^{\delta'}(F)$ is given by (5.105) (replacing K^F by $\bar{K}^F$). Then we get (5.161) similarly to the proof of Lemma 5.8.4.

Now we get from (5.161),

$$|\bar{S}_\rho^F(\theta)^{-1}|_{0,n,p}^\sharp \leq c_6 C_{n,p}^{\delta'}(F) |\bar{K}_\rho^F(\theta)^{-1}|_{0,n,2p}^\sharp. \tag{5.163}$$

Combining this with (5.160), the inequality of the theorem follows. □

Let us consider another family of Malliavin covariances conditioned to $A(\rho)$.

$$\bar{R}_\rho^F = \int_{\mathbb{T}} D_t F (D_t F)^T \, dt + \frac{1}{\varphi(\rho)} \int_{A(\rho)} \tilde{D} F_u (\tilde{D}_u F)^T n(du). \tag{5.164}$$

Suppose that $\bar{R}_\rho^F(\theta)$ are invertible and inverses satisfy

$$\sup_{(\rho,\theta)} |\bar{R}_\rho^F(\theta)^{-1}|_{0,n,p}^\sharp < \infty, \quad \forall n, p. \tag{5.165}$$

Then F is said to be *nondegenerate*. The following can be verified similarly to the case of Poisson functionals discussed in Sect. 5.8, Theorem 5.8.2.

Theorem 5.11.3 *Let F be a regular functional belonging to $(\bar{\mathbf{D}}^\infty)^d$.*

1. *If F is nondegenerate, it is nondegenerate at the center. Further, we have for any n, p,*

$$\sup_\theta |\bar{K}^F(\theta)^{-1}|_{0,n,p}^\sharp \leq \sup_{(\rho,\theta)} |\bar{R}_\rho^F(\theta)^{-1}|_{0,n,p}^\sharp. \tag{5.166}$$

2. *If F is nondegenerate at the center, it is nondegenerate. Further, for any n, p there is a positive constant c such that*

$$\sup_{(\rho,\theta)} |\bar{R}^F_\rho(\theta)^{-1}|^\sharp_{0,n,p} \leq cC^0_{n,p}(F) \sup_\theta |\bar{K}^F(\theta)^{-1}|^\sharp_{0,n,2p}, \tag{5.167}$$

for all nondegenerate F, where $C^0_{n,p}(F)$ is given by (5.99) with $\delta' = 0$ replacing $K^F(\theta)$ by $\bar{K}^F(\theta)$.

From Theorems 5.11.2 and 5.11.3, we have the following,

Theorem 5.11.4 *Let F be a regular functional belonging to $(\bar{\mathbf{D}}^\infty)^d$: The following statements are equivalent.*

(i) *F is δ-nondegenerate for some $1 < \delta < 2/\alpha$.*
(ii) *F is δ-nondegenerate for any $1 < \delta < 2/\alpha$.*
(iii) *F is nondegenerate at the center.*
(iv) *F is nondegenerate.*

Remark We will discuss briefly the density problem in the case where the Lévy measure ν may not satisfy the order condition. A typical case is that the Lévy measure is of finite mass. For a Wiener–Poisson functional $F \in (\bar{\mathbf{D}}^\infty)^d$, we consider the Malliavin covariance $R^F = \int_{\mathbb{T}} D_t F \cdot DF_t^T \, dt$. The functional F is called *strongly nondegenerate* if $\sup_\theta E[R^F(\theta)^{-p}] < \infty$ holds for any $p \geq 2$. Then we can show that for any $N \in \mathbb{N}$, there exist $m \in \mathbb{N}$, $p \geq 2$ and $c > 0$ such that the inequality

$$\left| E[e^{i(v,F)}G] \right| \leq \frac{c}{|v|^N} \Big(\|DF\|_{m,0,p} \sup_\theta \|R^F(\theta)^{-1}\|_{m,0,p} \Big)^N \|G\|_{m,0,p} \tag{5.168}$$

holds for all $|v| \geq 1$ for any strongly nondegenerate Wiener–Poisson functional F of $(\bar{\mathbf{D}}^\infty)^d$ and $G \in \bar{\mathbf{D}}^\infty$. The proof can be carried out similarly to the proof of Theorem 5.3.1, replacing norms $\| \ \|_{m,p}$ by $\| \ \|_{m,0,p}$. Consequently, if F is a strongly nondegenerate Wiener–Poisson functional in the above sense, its law weighted by G has a rapidly decreasing C^∞-density.

5.12 Compositions with Generalized Functions

Let F be a smooth Wiener–Poisson functional of $(\bar{\mathbf{D}}^\infty)^d$ and let $f(x)$ be a smooth function on $\mathbb{R}^d$. Then the composite $f(F)$ is a smooth Wiener–Poisson functional. In this section, we will define the composite of a generalized function f and a nondegenerate smooth Wiener–Poisson functional F. The composite $f(F)$ will no longer be a smooth Wiener–Poisson functional. It should be a generalized Wiener–Poisson functional.

Let us recall the Schwartz distribution or generalized function on a Euclidean space. Let $\mathcal{S} = \mathcal{S}(\mathbb{R}^d)$ be the space of complex valued rapidly decreasing C^∞-functions on $\mathbb{R}^d$. It is called the *Schwartz space*. We define semi-norms by

$$|f|_n = \Big(\sum_{|\mathbf{k}|+j\le n} \int \Big((1+|x|^2)^j |\partial^{\mathbf{k}} f(x)|\Big)^2 dx\Big)^{\frac{1}{2}}. \tag{5.169}$$

Denote the completion of $\mathcal{S}$ with respect to the norm $|\cdot|_n$ by $\mathcal{S}^n$. For a negative integer $-n$, we define the norm by

$$|f|_{-n} = \sup_{g\in\mathcal{S}^n, |g|_n\le 1} |(f,g)|, \tag{5.170}$$

where $(f,g) = \int f(x)\overline{g(x)}\,dx$. The completion of $\mathcal{S}$ by the norm $|\cdot|_{-n}$ is denoted by $\mathcal{S}^{-n}$. We set $\mathcal{S}^\infty = \bigcap_{n>0}\mathcal{S}^n$ and $\mathcal{S}^{-\infty} = \bigcup_{n>0}\mathcal{S}^{-n}$. Then it holds that $\mathcal{S}^\infty = \mathcal{S}$ and $\mathcal{S}^{-\infty} = \mathcal{S}'$ is the space of tempered distributions.

Let $f(x)$ be an integrable function on $\mathbb{R}^d$. We define the Fourier transform and the inverse Fourier transform of f by

$$\mathcal{F}f(v) = \hat{f}(v) = \Big(\frac{1}{2\pi}\Big)^{\frac{d}{2}} \int_{\mathbb{R}^d} e^{-i(v,x)} f(x)\,dx,$$

$$\check{\mathcal{F}}f(v) = \check{f}(v) = \Big(\frac{1}{2\pi}\Big)^{\frac{d}{2}} \int_{\mathbb{R}^d} e^{i(v,x)} f(x)\,dx,$$

respectively.

Proposition 5.12.1 *The Fourier transform $\mathcal{F}; \mathcal{S}\to\mathcal{S}$ is a linear, one to one and onto map. Further, it satisfies the following properties:*

1. *It holds that $\mathcal{F}\check{\mathcal{F}}f = \check{\mathcal{F}}\mathcal{F}f = f$ for any $f\in\mathcal{S}$.*
2. *(Parseval's equality) We have $\int f\bar{g}\,dx = \int \mathcal{F}f\overline{\mathcal{F}g}\,dv$ for any $f,g\in\mathcal{S}$.*
3. *(Fourier inversion formula) Let $\mathbf{i}$ and $\mathbf{j}$ be multi-indexes of nonnegative integers. Then we have*

$$(ix)^{\mathbf{i}}\partial^{\mathbf{j}} f(x) = \mathcal{F}(\partial^{\mathbf{i}}(v^{\mathbf{j}}\check{\mathcal{F}}f))(x) \tag{5.171}$$

for any $f\in\mathcal{S}$.

Proof The linear property of the Fourier transform is obvious from the definition of the transform. Then it is sufficient to prove (1)–(3) for a real function f. If f is a density function of a signed measure μ, we have the inversion formula $\mathcal{F}\check{\mathcal{F}}f = f$ in view of (1.5). Next, take the complex conjugate for both sides. Then we have $f = \overline{\mathcal{F}\check{\mathcal{F}}f} = \check{\mathcal{F}}\mathcal{F}f$. Therefore the first assertion follows. Further, equalities $\mathcal{F}\check{\mathcal{F}}f = \check{\mathcal{F}}\mathcal{F}f = f$ show that the map $\mathcal{F}:\mathcal{S}\to\mathcal{S}$ is one to one and onto.

By the Fubini theorem, we have

$$\int f(x)\overline{g(x)}\,dx = \Big(\frac{1}{2\pi}\Big)^{\frac{d}{2}} \int f(x)\overline{\Big(\int e^{i(v,x)}\hat{g}(v)\,dv\Big)}\,dx$$
$$= \Big(\frac{1}{2\pi}\Big)^{\frac{d}{2}} \int \Big(\int e^{-i(v,x)} f(x)\,dx\Big)\overline{\hat{g}(v)}\,dv = \int \hat{f}(v)\overline{\hat{g}(v)}\,dv.$$

Finally equality (5.171) follows from (1.7). □

Next, we will define generalized Wiener–Poisson functionals. Let $\| \ \|^{\sharp}_{m,n,p}$ be Sobolev norms on the Wiener–Poisson space defined in Sect. 5.10. We shall define their dual norms. For $F \in \bar{L}^{\frac{p}{p-1}}$ and $G \in \bar{\mathbf{D}}^{m,n,p}$, we set $\langle X, Y\rangle = E[XY]$. The dual norm of $\| \ \|^{\sharp}_{m,n,p}$ is defined by

$$\|F\|^{*}_{m,n,p} := \sup_{G\in\bar{\mathbf{D}}^{m,n,p},\|G\|^{\sharp}_{m,n,p}\le 1} |\langle F, G\rangle|.$$

We denote by $\bar{\mathbf{D}}^{m,n,p}_{*}$ the completion of $\bar{L}^{\frac{p}{p-1}}$ by the norm $\| \ \|^{*}_{m,n,p}$ and we set $\bar{\mathbf{D}}^{\infty}_{*} = \bigcup_{m,n\in\mathbf{N},p>1} \bar{\mathbf{D}}^{m,n,p}_{*}$. We have $\bar{\mathbf{D}}^{\infty} \subset \bar{L}^{\infty-} \subset \bar{\mathbf{D}}^{\infty}_{*}$. Elements of $\bar{\mathbf{D}}^{\infty}_{*}$ are called *generalized Wiener–Poisson functionals*. The bilinear form $\langle F, G\rangle$ is extended naturally to a generalized functional F and a smooth functional G.

Let f be a tempered distribution and let F be a nondegenerate element of $(\bar{\mathbf{D}}^{\infty})^d$. We will define the composition of f and F, following the idea of Watanabe [117] and Hayashi–Ishikawa [37]. The definition is somewhat different from theirs. We prepare a lemma.

Lemma 5.12.1 *Let F be a δ-nondegenerate Wiener–Poisson functional belonging to $(\bar{\mathbf{D}}^{\infty})^d$. Then for any $N \in \mathbb{N}$ there exist $m, n \in \mathbb{N}$, $p > 2$ and a positive constant C_N such that*

$$\|f(F)\|^{*}_{m,n,p} \le C_N |f|_{-N} \tag{5.172}$$

holds for all $f \in \mathcal{S}$.

Proof For the proof of the lemma, we will apply the Fourier transform. For a given $G \in \bar{\mathbf{D}}^{\infty}$, consider the (signed) measure $\mu_G(A) = E[1_A(F)G]$. Its Fourier transform is equal to $(2\pi)^{-\frac{d}{2}}\overline{\psi_G(v)}$, where $\psi_G(v)$ is the characteristic function of the measure μ_G. It is rapidly decreasing with respect to v. Therefore the measure μ_G has a rapidly decreasing C^{∞}-density function, which we denote by $f_G(y)$.

Let $f \in \mathcal{S}$. Then using Parseval equality, we have

$$E[f(F)G] = \int_{\mathbb{R}^d} f(y) f_G(y)\,dy = \Big(\frac{1}{2\pi}\Big)^{\frac{d}{2}} \int_{\mathbb{R}^d} \hat{f}(v)\overline{\psi_G(v)}\,dv. \tag{5.173}$$

We shall compute the semi-norm $|\psi_G|_N$. For a given $N \in \mathbb{N}$, take $N' \in \mathbb{N}$ satisfying $\gamma_0 N' > d + 1 + N$, where $\gamma_0 = 1 - \alpha\delta/2$. Then by Theorem 5.11.1, there exist m, n, p and positive constants c_1 and c_2 such that for any $|\mathbf{k}| \leq N$, the inequality

$$
\begin{aligned}
|\partial_v^{\mathbf{k}} \psi_G(v)| = |E[e^{i(v,F)} F^{\mathbf{k}} G]| &\leq c_1 |v|^{-\gamma_0 N'} \|G\|_{m,n,p}^{\sharp} \\
&\leq c_2 (1 + |v|^2)^{-\gamma_0 \frac{N'}{2}} \|G\|_{m,n,p}^{\sharp}
\end{aligned}
$$

holds if $|v| \geq 2$. Then for any $j + |\mathbf{k}| \leq N$, there is a positive constant c_N such that

$$
\begin{aligned}
\Big(\int (1 + |v|^2)^j ||\partial_v^{\mathbf{k}} \psi_G(v)|^2 \, dv \Big)^{\frac{1}{2}} &\leq c_N \Big(\int (1 + |v|^2)^{-\frac{d+1}{2}} \, dv \Big)^{\frac{1}{2}} \|G\|_{m,n,p}^{\sharp} \\
&\leq c_N' \|G\|_{m,n,p}^{\sharp}.
\end{aligned}
$$

Therefore we have $|\psi_G|_N \leq c_N' \|G\|_{m,n,p}^{\sharp}$.

We will prove (5.172). It holds that

$$
\begin{aligned}
\|f(F)\|_{m,n,p}^{*} &= \sup_{G \in \tilde{\mathbf{D}}^{\infty}, \|G\|_{m,n,p}^{\sharp} \leq 1} |E[f(F)G]| \\
&\leq \Big(\frac{1}{2\pi} \Big)^{\frac{d}{2}} \sup_{G; \|G\|_{m,n,p}^{\sharp} \leq 1} \Big| \int_{\mathbb{R}^d} \overline{\psi_G(v)} \hat{f}(v) \, dv \Big| \\
&\leq \Big(\frac{1}{2\pi} \Big)^{\frac{d}{2}} \sup_{G; \|G\|_{m,n,p}^{\sharp} \leq 1} |\psi_G|_N |\hat{f}|_{-N}.
\end{aligned}
\tag{5.174}
$$

Further,

$$
|\hat{f}|_{-N} = \sup_{|\hat{g}|_N \leq 1} |\langle f, g \rangle| = \sup_{|\hat{g}|_N \leq 1} |\langle \hat{f}, \hat{g} \rangle| = \sup_{|g|_N \leq 1} |\langle f, g \rangle| = |f|_{-N}.
$$

Here, we used the fact that $\mathcal{F}; \mathcal{S} \to \mathcal{S}$ is one to one and onto, and the Perceval equality. Therefore we get the inequality (5.172). □

By the above lemma, the composition $f(F)$ can be extended to any $f \in \mathcal{S}^{-n}$. Since $\mathcal{S}' = \bigcup_{n>0} \mathcal{S}^{-n}$, it is extended to any tempered distribution f. We set conventionally

$$
E[f(F) \cdot G] := \langle f(F), G \rangle.
\tag{5.175}
$$

We summarize the composition as a theorem.

Theorem 5.12.1 *Let F be a regular and nondegenerate element of $(\bar{\mathbf{D}}^\infty)^d$. Then for any tempered distribution f, the composition $f(F)$ can be defined as an element of $\bar{\mathbf{D}}_*^\infty$. Further, for any $N \in \mathbb{N}$, there exist $m, n \in \mathbb{N}$, $p > 2$ and a positive constant C_N such that*

$$\|f(F)\|_{m,n,p}^* \leq C_N |f|_{-N}, \quad \forall f \in \mathcal{S}^{-N}. \tag{5.176}$$

Further, it satisfies

$$E[f(F) \cdot G] = \left(\frac{1}{2\pi}\right)^{\frac{d}{2}} (\mathcal{F}f, \psi_G) \tag{5.177}$$

for any $f \in \mathcal{S}^{-\infty}$ and $G \in \bar{\mathbf{D}}^\infty$, where $\mathcal{F}f$ is the Fourier transform of f.

So far we considered the composite of a tempered distribution Φ and a nondegenerate Wiener–Poisson functional F. However, if F is a Wiener functional, the statement of Theorem 5.12.1 should be changed slightly. Let $\mathbf{D}_*^{m,p}$ be the dual of the space of Wiener functionals $\mathbf{D}^{m,p}$ and let $\| \ \|_{m,p}^*$ be its norm. Let $\mathbf{D}_*^\infty = \bigcup_{m,p} \mathbf{D}_*^{m,p}$. Elements of $\mathbf{D}_*^\infty$ is called *generalized Wiener functionals*. If f is a tempered distribution on $\mathbb{R}^d$ and F is a d-dimensional nondegenerate Wiener functional, the composition $f(F)$ is well defined as an element of $\mathbf{D}_*^\infty$. Further, the assertion of Theorem 5.12.1 is valid if we replace the inequality (5.176) by

$$\|f(F)\|_{m,p}^* \leq C_N |f|_{-N}.$$

We are particularly interested in the case where f is the delta function δ_x of the point $x \in \mathbb{R}^d$.

Corollary 5.12.1 *Let F be a d-dimensional smooth nondegenerate Wiener functional or a regular nondegenerate Wiener–Poisson functional of $(\bar{\mathbf{D}}^\infty)^d$. Let G be a smooth Wiener functional or smooth Wiener–Poisson functional, respectively. Let $\psi_G(v)$ be the characteristic function of the law of F weighted by G. Then the density function $f_G(x)$ of the law of F weighted by G is represented by*

$$f_G(x) = E[\delta_x(F) \cdot G] = \left(\frac{1}{2\pi}\right)^d \int_{\mathbb{R}^d} e^{-i(v,x)} \psi_G(v)\, dv, \tag{5.178}$$

where δ_x is the delta function at the point x.

Finally, we remark that if F is nondegenerate, the conditional law $P(G|F = x)$ is well defined for all x without ambiguity of measure 0. Indeed, it is given by

$$P(G|F = x) = \frac{E[\delta_x(F) \cdot G]}{E[\delta_x(F)]}, \tag{5.179}$$

provided that $E[\delta_x(F)] \neq 0$.

Note The composite of Schwartz distribution and nondegenerate Wiener functional was first discussed by Watanabe [116, 117], using Sobolev type norm $\| \|'_{m,p}$ defined by Orenstein–Uhlenbeck generator (Note' at the end of Sect. 5.2). The composite of Schwartz distribution and nondegenerate Wiener–Poisson functional is studied in Hayashi-Ishikawa [37] slightly different context.

Chapter 6
Smooth Densities and Heat Kernels

Abstract We discuss the existence of the smooth density of 'nondegenerate' diffusions and jump-diffusions determined by SDEs. We will use the Malliavin calculus studied in the previous chapter. In Sects. 6.1, 6.2, and 6.3, we consider diffusion processes. It will be shown that any solution of a continuous SDE defined in Chap. 3 is infinitely H-differentiable. Further, if its generator $A(t)$ is elliptic (or hypo-elliptic), its transition probability $P_{s,t}(x, E)$ has a density $p_{s,t}(x, y)$, which is a rapidly decreasing C^∞-function of x and y. We will show further that the weighted transition function has also a rapidly decreasing C^∞-function of x and y, and it is the heat kernel of the backward heat equation associated with the operator $A^{\mathbf{c}}(t)$, which were defined in Chap. 4. For the proof of these facts, we will apply the Malliavin calculus on the Wiener space discussed in Sects. 5.1, 5.2, and 5.3.

In Sects. 6.4, 6.5, and 6.6, we will study jump-diffusions. We will show that if the generator of the jump-diffusion is 'pseudo-elliptic', then its weighted transition function has a smooth density. For the proof, we will apply the Malliavin calculus on the Wiener-Poisson space discussed in Sects. 5.9, 5.10, and 5.11.

In Sects. 6.7 and 6.8, we discuss short-time estimates of heat kernels, applying the Malliavin calculus again. These facts will be applied to two problems. In Sect. 6.9, we consider the solution of SDEs with big jumps, for which the Malliavin calculus cannot be applied. Instead, we take a method of perturbation and show that the perturbation preserves the smooth density. In Sect. 6.10, we show the existence of the smooth density of the laws of the killed elliptic diffusion or the killed pseudo-elliptic jump-diffusions. The density should be the heat kernel of the backward heat equation on the domain with the Dirichlet boundary condition.

H. Kunita, *Stochastic Flows and Jump-Diffusions*, Probability Theory and Stochastic Modelling 92, https://doi.org/10.1007/978-981-13-3801-4_6

6.1 H-Derivatives of Solutions of Continuous SDE

Let us consider a continuous symmetric SDE on $\mathbb{R}^d$ defined on the Wiener space:

$$X_t = X_0 + \sum_{k=0}^{d'} \int_{t_0}^{t} V_k(X_r, r) \circ dW_r^k. \tag{6.1}$$

Here $W_r^0 = r$ and $V_0(X_r, r) \circ dW_r^0 = V_0(X_r, r)\, dr$. Further, $(W_t^1, \ldots, W_t^{d'}), t \in [0, \infty)$ is a d'-dimensional Wiener process and $\circ dW^k(t), k = 1, \ldots, d'$ denote symmetric integrals. Coefficients $V_k(x, t) = (V_k^1(x, t), \ldots, V_k^d(x, t))$ are $C_b^{\infty,1}$-functions on $\mathbb{R}^d \times [0, \infty)$. The solution defines a diffusion process with the generator $A(t)$ defined by (4.4). The operator is rewritten as (4.20). If the matrix $A(x, t) := (\alpha^{ij}(x, t)) = V(x, t)V(x, t)^T$ is positive definite for any x, t, the operator $A(t)$ is called *elliptic*. Further, if the matrix $A(x, t)$ is uniformly positive definite, the operator $A(t)$ is called *uniformly elliptic*. The associated SDE (6.1) will be called an *elliptic SDE* and a *uniformly elliptic SDE*, respectively.

Let $\{\Phi_{s,t}\}$ be the stochastic flow generated by the above equation. For a while we take s, t from the time interval $\mathbb{T} = [0, T]$. Then we can note that for any $s < t$ and $x \in \mathbb{R}^d$, $\Phi_{s,t}(x)$ is a Wiener functional discussed in Sect. 5.1. It is an element of $(L^{\infty-})^d$. We want to show (in Theorem 6.1.1) that $\Phi_{s,t}(x)$ is a d-dimensional Wiener functional belonging to the Sobolev space $(\mathbf{D}^\infty)^d$.

Let H be the Cameron–Martin space defined in Sect. 5.1. For a given $h = (h^1(r), \ldots, h^{d'}(r)) \in H$, consider an SDE with one-dimensional parameter λ in $\Lambda = [-1, 1]$ on the time interval $\mathbb{T}$:

$$X_t^{\lambda h} = X_0 + \sum_{k=1}^{d'} \int_{t_0}^{t} V_k(X_r^{\lambda h}, r)(\circ dW_r^k + \lambda h^k(r)\, dr) + \int_{t_0}^{t} V_0(X_r^{\lambda h}, r)\, dr. \tag{6.2}$$

Let X_t be the solution of (6.1) starting from X_0 at time $t_0 \in \mathbb{T}$ and let $X_t^{\lambda h}$ be the solution of (6.2). Then it holds that

$$X_t^{\lambda h}(w) = X_t(w + \lambda \bar{h}), \quad a.s. \tag{6.3}$$

for any λ, where $\bar{h}(t) = \int_0^t h(r)\, dr$. Indeed, we will approximate the solutions X_t and $X_t^{\lambda h}$ as follows. Let $N \in \mathbb{N}$. Let X_t^N and $X_t^{\lambda h, N}$ be solutions of following equations, respectively:

$$X_t^N = X_0 + \sum_{k=1}^{d'} \int_{t_0}^{t} V_k(X_{\phi_N(r)}^N, r) \circ dW_r^k + \int_{t_0}^{t} V_0(X_{\phi_N(r)}^N, r)\, dr,$$

$$\begin{aligned} X_t^{\lambda h, N} &= X_0 + \sum_{k=1}^{d'} \int_{t_0}^{t} V_k(X_{\phi_N(r)}^{\lambda h, N}, r)(\circ dW_r^k + \lambda h^k(r))\, dr \\ &\quad + \int_{t_0}^{t} V_0(X_{\phi_N(r)}^{\lambda h, N}, r)\, dr, \end{aligned}$$

where $\phi_N(r) = \frac{m}{2^N}$, if $\frac{m}{2^N} \le r < \frac{m+1}{2^N}$. Then we have $X_t^{\lambda h,N}(w) = X_t^N(w + \lambda\bar{h})$ a.s. Let N tend to ∞. Then X_t^N converges to X_t in probability and $X_t^{\lambda h,N}$ converges to $X_t^{\lambda h}$ in probability for any λ. Therefore we have $X_t^{\lambda h}(w) = X_t(w + \lambda\bar{h})$ a.s.

Let $X_t^{\lambda h,x,s}$ be the solution of the equation starting from x at time s. Then the family of the solutions $\{X_t^{\lambda h,x,s}\}$ has a modification which is infinitely differentiable with respect to x (Theorem 3.4.2 in Chap. 3). We denote it by $\Phi_{s,t}^{\lambda h}(x)$. It belongs to $(L^{\infty-})^d$ for any s, t, x, λ, h. If $\lambda = 0$, it coincides with the stochastic flow $\Phi_{s,t}(x)$ a.s.

Lemma 6.1.1 *For a given $h \in H$ and s, t $(s < t)$, the family of solutions $\{\Phi_{s,t}^{\lambda h}(x), x \in \mathbb{R}^d, \lambda \in \Lambda\}$ has a modification which is continuously differentiable in λ with respect to the metric of the F-space $(L^{\infty-})^d$. Further, we have $\sup_{|\lambda|\le 1, x\in\mathbb{R}^d} E[|\frac{d}{d\lambda}\Phi_{s,t}^{\lambda h}(x)|^p] < \infty$ for any $p \ge 2$, $s < t$ and $h \in H$.*

Proof Using the Itô integral, $\Phi_{s,t}^{\lambda h}(x)$ satisfies

$$\Phi_{s,t}^{\lambda h}(x) = x + \int_s^t \alpha(\Phi_{s,r}^{\lambda h}(x), r)\, dW_r + \int_s^t \beta_\lambda(\Phi_{s,r}^{\lambda}(x), r)\, dr,$$

where $\alpha(x, r) = (V_1(x, r), \ldots, V_{d'}(x, r))$ and

$$\beta_\lambda(x, r) = V_0(x, r) + \frac{1}{2}\sum_{k=1}^{d'} V_k(r)(V_k(r))(x) + \lambda(\sum_{k=1}^{d'} h^k(r)V_k(x, r)).$$

It satisfies Conditions 1–3 of the master equation with parameter λ, x in Sect. 3.3. Indeed, $\alpha^\lambda(x, r) := \alpha(x, r)$ does not depend on λ and is Lipschitz continuous in x. $\beta_\lambda(x, r)$ is uniformly Lipschitz continuous with respect to x. With respect to λ, it satisfies

$$\int_{\mathbb{T}} \sup_x |\beta_\lambda(x, r) - \beta_{\lambda'}(x, r)|\, dr \le c|\lambda - \lambda'|\Big(\int_{\mathbb{T}} |h(r)|\, dr\Big)$$

and $\int_{\mathbb{T}} |h(r)|\, dr < \infty$. Therefore coefficients of the equation satisfy Conditions 1–3 of the master equation. Further, $\beta_\lambda(x, r)$ is differentiable with respect to x and λ, and its derivatives satisfy Conditions 1–3 of the master equation. Then, in view of Theorem 3.3.2, solutions $\{\Phi_{s,t}^{\lambda h}(x)\}$ are included in the space $(L^{\infty-})^d$ and are continuously differentiable with respect to λ in the space $(L^{\infty-})^d$. Further, we have $\sup_{|\lambda|<1, x\in\mathbb{R}^d} E[|\frac{d}{d\lambda}\Phi_{s,t}^{\lambda h}(x)|^p] < \infty$ for any $p \ge 2$. □

Lemma 6.1.2 *Set $\frac{d}{d\lambda}\Phi_{s,t}^{\lambda h}(x)\big|_{\lambda=0} = \Xi_{s,t}^h(x)$. Then it satisfies*

$$\Xi_{s,t}^h = \sum_{k=0}^{d'} \int_s^t \nabla V_k(\Phi_{s,r}, r)\Xi_{s,r}^h \circ dW_r^k + \int_s^t \Big(\sum_{k=1}^{d'} V_k(\Phi_{s,r}, r)h^k(r)\Big)\, dr \tag{6.4}$$

for any $h \in H$, where $\nabla V_k(x,r)$ is the Jacobian matrix of the function $V_k(x,r) = (V_k^1(x,r),\ldots,V_k^d(x,r))$ given by $(\partial_{x_j} V_k^i(x,r))_{i,j=1,\ldots,d}$. Further, $\Xi_{s,t}^h(x)$ is represented by

$$\Xi_{s,t}^h(x) = \int_s^t \nabla\Phi_{s,t}(x)\nabla\Phi_{s,r}(x)^{-1}\Big(\sum_{k=1}^{d'} V_k(\Phi_{s,r}(x),r)h^k(r)\Big)\,dr, \tag{6.5}$$

where $\nabla\Phi_{s,t}(x)$ is the Jacobian matrix of $\Phi_{s,t}(x) = (\Phi_{s,t}^1(x),\ldots,\Phi_{s,t}^d(x))$.

Proof $\Phi_{s,t}^{\lambda h}(x)$ satisfies

$$\Phi_{s,t}^{\lambda h}(x) = x + \sum_{k=0}^{d'}\int_s^t V_k(\Phi_{s,r}^{\lambda h}(x),r)\circ dW_r^k + \lambda\int_s^t \Big(\sum_{k=1}^{d'} V_k(\Phi_{s,r}^{\lambda h}(x),r)h^k(r)\Big)\,dr.$$

Differentiate each term of the above with respect to λ and then set $\lambda = 0$. Then we get the equality (6.4) (See Theorem 3.3.2 and Proposition 2.4.3). On the other hand, the Jacobian matrix $\nabla\Phi_{s,t}(x)$ satisfies the linear homogeneous equation

$$\nabla\Phi_{s,t}(x) = I + \sum_{k=0}^{d'}\int_s^t \nabla V_k(\Phi_{s,r}(x),r)\nabla\Phi_{s,r}(x)\circ dW_r^k \tag{6.6}$$

(see (3.56)). Now, let us compare the two equations (6.4) and (6.6). Then the solution of the linear inhomogeneous equation (6.4) is written as (6.5), making use of the solution $\nabla\Phi_{s,t}(x)$ of the linear homogeneous equation (6.6). In fact, taking the differential of (6.5) with respect to t (Itô's formula), we can prove directly that $\Xi_{s,t}(x)$ of (6.5) satisfies SDE (6.4). □

Proposition 6.1.1 *For any $s < t$ and x, $\Phi_{s,t}(x)$ is H-differentiable. The derivative $D_r\Phi_{s,t}(x) = (D_r^{(1)}\Phi_{s,t}(x),\ldots,D_r^{(d')}\Phi_{s,t}(x))$ is given a.e. $dr\,dP$ by*

$$D_r^{(k)}\Phi_{s,t}(x) = \begin{cases} \nabla\Phi_{s,t}(x)\nabla\Phi_{s,r}(x)^{-1}V_k(\Phi_{s,r}(x),r), & \text{if } s < r < t,\\ 0, & \text{if } r < s, \text{ or } r > t.\end{cases} \tag{6.7}$$

It satisfies

$$\sup_{s<t,x}\frac{1}{\sqrt{t-s}}E\Big[\Big(\int_{\mathbb{T}}|D_r\Phi_{s,t}(x)|^2\,dr\Big)^{\frac{p}{2}}\Big]^{\frac{1}{p}} < \infty \tag{6.8}$$

for any $p \geq 2$, where the supremum is taken for $0 \leq s < t \leq T$, $x \in \mathbb{R}^d$.

Proof Denote the right-hand side of (6.7) by $F'_{s,t,x}$. Its H-norm $|F'_{s,t,x}|_H$ satisfies $E[|F'_{s,t,x}|_H^p] < \infty$ for any $p \geq 2$. Further, it holds that $\langle F'_{s,t,x}, h\rangle_H = \Xi_{s,t}^h(x)$ for any $h \in H$. Therefore we have for any $p \geq 2$

$$E\Big[\Big|\frac{\Phi_{s,t}(x, w+\lambda\bar{h}) - \Phi_{s,t}(x,w)}{\lambda} - \langle F'_{s,t,x}(w), h\rangle\Big|^p\Big] \to 0, \quad \text{as } \lambda \to 0,$$

by Theorem 3.3.2, 1. This shows that $\Phi_{s,t}(x)$ is H-differentiable and $D\Phi_{s,t}(x) = F'_{s,t,x}$.

Since $E[|\nabla\Phi_{s,t}(x)|^{2p}]$ and $E[|\nabla\Phi_{s,r}(x)^{-1}|^{2p}]$ together with $V_k(\Phi_{s,r}(x))$ are all bounded with respect to s, t, r, x, there is a positive constant c such that

$$E\Big[\Big(\int_{\mathbb{T}} |D_r\Phi_{s,t}(x)|^2\, dr\Big)^{\frac{p}{2}}\Big] \le c(t-s)^{\frac{p}{2}}. \tag{6.9}$$

Therefore we get (6.8). □

Since the flow $\{\Phi_{s,t}(x)\}$ satisfies $\Phi_{s,t}(x) = \Phi_{r,t}(\Phi_{s,r}(x))$, we have $\nabla\Phi_{s,t}(x) = \nabla\Phi_{r,t}(\Phi_{s,r}(x))\nabla\Phi_{s,r}(x)$. Then we get $\nabla\Phi_{s,t}(x)\nabla\Phi_{s,r}(x)^{-1} = \nabla\Phi_{r,t}(\Phi_{s,r}(x))$. This expression is often used for (6.7).

Theorem 6.1.1 *For any multi-index* $\mathbf{i}$ *and* $0 \le s < t \le T$, $\partial^{\mathbf{i}}\Phi_{s,t}(x)$ *is infinitely* H*-differentiable. Further, for any* m, p, $\|\partial^{\mathbf{i}}\Phi_{s,t}(x)\|_{m,p}$ *is bounded with respect to* $0 \le s < t \le T$, $x \in \mathbb{R}^d$ *if* $|\mathbf{i}| \ge 1$, *and we have*

$$\sup_{0\le s<t\le T, x} \frac{1}{\sqrt{t-s}} \|D\partial^{\mathbf{i}}\Phi_{s,t}(x)\|_{m,p} < \infty \tag{6.10}$$

for any $|\mathbf{i}| \ge 0$. *Here* $\|\ \|_{m,p}$ *are Sobolev norms for Wiener functionals defined in Sect.* 5.2.

Proof Since the pair $(\Phi_{s,t}(x), \nabla\Phi_{s,t}(x))$ is a solution of SDE (6.1) and (6.6), $\nabla\Phi_{s,t}(x)$ is H-differentiable by Proposition 6.1.1. This shows that $\partial^{\mathbf{i}}\Phi_{s,t}(x)$ is H-differentiable if $|\mathbf{i}| = 1$. Consider the triple $(\Phi_{s,t}, \nabla\Phi_{s,t}, \nabla^2\Phi_{s,t})$, where $\nabla^2\Phi_{s,t}$ is the Jacobian matrix of $\nabla\Phi_{s,t}$. It is again a solution of an SDE, and we find that $\nabla^2\Phi_{s,t}(x)$ is also H-differentiable. Repeating this argument, we find that $\partial^{\mathbf{i}}\Phi_{s,t}$ is H-differentiable for any $\mathbf{i}$.

$\Phi_{s,t}(x)$ is twice H-differentiable. Indeed, since $\nabla\Phi_{s,t}, \nabla\Phi_{s,r}^{-1}$ and $V_k(\Phi_{s,r}, r)$ are H-differentiable, their product $D_r\Phi_{s,t} \equiv \nabla\Phi_{s,t}\nabla\Phi_{s,r}^{-1}V_k(\Phi_{s,r}, r)$ is also H-differentiable. This means that $\Phi_{s,t}$ is twice H-differentiable. In the same way, $\partial^{\mathbf{i}}\Phi_{s,t}(x)$ is also twice H-differentiable. Repeating this argument we find that it is m-times differentiable for any $m \in \mathbb{N}$. We can verify (6.10) similarly to the proof of (6.9). □

Note Using Lemma 6.1.1, we showed that the flow $\Phi_{s,t}(x)$ belongs to $(\mathcal{D}^\infty)^d$ for any $0 \le s < t \le T, x$. In Watanabe [116] and Nualart [88], it is shown that $\Phi_{s,t}(x)$ is an element of $(\mathbf{D}^\infty)^d$ for any $0 \le s < t \le T$ and $x \in \mathbb{R}^d$.

6.2 Nondegenerate Diffusions

Let $R = R^{\Phi_{s,t}(x)}$ be the Malliavin covariance of the stochastic flow $\Phi_{s,t}(x)$ defined by (5.26). Note that $D_r\Phi_{s,t}(x)$ satisfies (6.7). Since $D_r\Phi_{s,t}(x) = 0$ holds for $0 < r < s$ and $t < r < T$, it is written as (omitting x),

$$R^{\Phi_{s,t}(x)} = \nabla\Phi_{s,t}\Big(\int_s^t \nabla\Phi_{s,r}^{-1}V(\Phi_{s,r},r)V(\Phi_{s,r},r)^T(\nabla\Phi_{s,r}^{-1})^T\,dr\Big)\nabla\Phi_{s,t}^T, \tag{6.11}$$

where $V(x,r) = (V_k^i(x,r))$ is a $d \times d'$-matrix and $V(x,r)^T$ is its transpose. The matrix $V(x,r)V(x,r)^T$ coincides with the matrix $A(x,r) := (\alpha^{ij}(x,r))$ of coefficients of the second-order part of the differential operator $A(r)$ given by (4.20). We set $R(\theta) = (\theta, R\theta)$ for $\theta \in S_{d-1}$.

We saw in Sect. 4.1 that the stochastic process $X_t^{x,s} := \Phi_{s,t}(x)$ with parameter x, s is a diffusion process with generator $A(t)$. A system of diffusion processes $\{X_t^{x,s} = \Phi_{s,t}(x), (x,s) \in \mathbb{T} \times \mathbb{R}^d\}$ is called *nondegenerate*, if the Malliavin covariances $R^{\Phi_{s,t}(x)}$ are invertible and the inverse belongs to $L^{\infty-}$ for any $s < t, x$, and further, these inverses satisfy

$$\sup_{|x|\le M}\sup_{\theta\in S_{d-1}} E[R^{\Phi_{s,t}(x)}(\theta)^{-p}] < \infty \tag{6.12}$$

for any $s < t$, $M > 0$ and $p > 2$. Further, if

$$\sup_{x\in\mathbb{R}^d}\sup_{\theta\in S_{d-1}} E[R^{\Phi_{s,t}(x)}(\theta)^{-p}] < \infty$$

holds for any $s < t$ and $p > 2$, the system of diffusion processes is called *uniformly nondegenerate*.

Proposition 6.2.1 *If the operator $A(t)$ of* (4.4) *is uniformly elliptic, the system of duffusion processes with the generator $A(t)$ is uniformly nondegenerate. Further, we have*

$$\sup_{0\le s<t\le T,x}\sup_{\theta\in S_{d-1}} (t-s)\|R^{\Phi_{s,t}(x)}(\theta)^{-1}\|_{m,p} < \infty \tag{6.13}$$

for any $m \in \mathbb{N}$ and $p \ge 2$.

Proof For fixed s, x, set $\varphi_r = \theta\nabla\Phi_{s,t}\nabla\Phi_{s,r}^{-1}$. Then, $R^{\Phi_{s,t}(x)}(\theta)$ is written as

$$R^{\Phi_{s,t}(x)}(\theta) = \int_s^t \varphi_r A(\Phi_{s,r},r)\varphi_r^T\,dr.$$

Let λ_r be the minimal eigen-value of the matrix $A(\Phi_{s,r}, r)$. There exists $c > 0$ such that $\lambda_r \geq c$ for any r, since the matrices $A(x,t)$ are uniformly positive definite with respect to x, t. Then we have

$$R^{\Phi_{s,t}(x)}(\theta) \geq (t-s) \cdot \frac{1}{t-s} \int_s^t \lambda_r |\varphi_r|^2 \, dr.$$

It holds that $\varphi_r \neq 0$ a.s., since the matrix $\nabla\Phi_{s,t}\nabla\Phi_{s,r}^{-1}$ is nondegenerate for any x, r, s, t. Therefore, we have

$$R^{\Phi_{s,t}(x)}(\theta)^{-1} \leq \frac{1}{(t-s)^2} \int_s^t \frac{1}{\lambda_r |\varphi_r|^2} \, dr, \quad a.s. \tag{6.14}$$

Then we get the inequality

$$E\big[R^{\Phi_{s,t}(x)}(\theta)^{-p}\big] \leq (c(t-s))^{-p} \sup_r E\big[|\varphi_r|^{-2p}\big].$$

We will show that the last term of the above is finite for any p. Let $V_{r,t}(y)$ be the inverse matrix of $\nabla\Phi_{r,t}(y)$. Since $\nabla\Phi_{s,t}(x)\nabla\Phi_{s,r}(x)^{-1} = \nabla\Phi_{r,t}(\Phi_{s,r}(x))$, $V_{r,t}(\Phi_{s,r}(x))$ is the inverse of $\nabla\Phi_{s,t}(x)\nabla\Phi_{s,r}(x)^{-1}$. Then $|\varphi_r|^{-1} \leq |V_{r,t}(\Phi_{s,r})|$ holds. We know by Lemma 3.7.1 that $E[|V_{r,t}(y)|^{2p}]$ is bounded with respect to y, r, t. Now since $V_{r,t}(y)$ and $\Phi_{s,r}$ are independent, $E[|V_{r,t}(\Phi_{s,r}(x))|^{2p}]$ is bounded with respect to x, s, r, t. This proves $\sup_{x,s,r,t} E[|\varphi_r|^{-2p}] < \infty$. Consequently, we get

$$\sup_{s<t,x} \sup_\theta (t-s)^p E[R^{\Phi_{s,t}(x)}(\theta)^{-p}] < \infty.$$

This proves (6.13) for the case $m = 0$. Since $\|D\Phi_{s,t}(x)\|_{m,p}$ is bounded with respect to x, s, t by (6.10), we get (6.13) for any m, p in view of Proposition 5.3.1. □

We will next consider the case where the operator $A(t)$ is elliptic. Then the minimum eigen-values λ_r of $A(\Phi_{s,r}, r)$ may not be uniformly positive. We want to localize the discussion. We need a lemma.

Lemma 6.2.1 *Given x, s, consider the diffusion process $X_t^{x,s} = \Phi_{s,t}(x)$, t in $[s, \infty)$. Let $M > 0$ and let $\tau = \tau(x,s)$ be the first time $t(> s)$ at which $|X_t^{x,s}| \geq M + 1$ occurs. For any $N > 2$, there is $c_N > 0$ such that the inequality*

$$\sup_{|x|\leq M} P(\tau(x,s) < t) \leq c_N (t-s)^N \tag{6.15}$$

holds for any $s < t$.

Proof By Chebyschev's inequality, we have for any $p \geq 2$,

$$\sup_{|x|\leq M} P(\tau(x,s) < t) \leq \sup_{|x|\leq M} P\Big(\sup_{s<r<t} |X_r^{x,s} - x| \geq 1\Big)$$
$$\leq \sup_{|x|\leq M} E\big[\|X^{x,s} - x\|_t^{2N}\big]$$
$$\leq c_N(t-s)^N.$$

The last inequality follows from Theorem 3.4.3. □

Proposition 6.2.2 *If the differential operator $A(t)$ of* (4.4) *is elliptic, the system of diffusion processes with the generator $A(t)$ is nondegenerate. Further, for any $M > 0$ and m, p, we have*

$$\sup_{0\leq s<t\leq T, |x|\leq M} \sup_{\theta\in S_{d-1}} (t-s)\|R^{\Phi_{s,t}(x)}(\theta)^{-1}\|_{m,p} < \infty. \tag{6.16}$$

Proof For given $M > 0, x, s$, let $\tau = \tau(x,s)$ be the first time t at which $|X_t^{x,s}| \geq M+1$ occurs, where $X_t^{x,s} = \Phi_{s,t}(x)$. Let c is a positive constant such that $(u, A(x,t)u) \geq c|u|^2$ holds if $t \in \mathbb{T}$, $|x| \leq M+1$. Then the minimal eigen-value λ_r of $A(\Phi_{s,r}(x), r)$ satisfies $\lambda_{r\wedge\tau} \geq c$. Instead of the inequality (6.14), we have the inequality

$$R^{\Phi_{s,t}(x)}(\theta)^{-1} \leq \frac{1}{(t\wedge\tau - s)^2}\int_s^t \frac{1}{\lambda_{r\wedge\tau}|\varphi_{r\wedge\tau}|^2}\,dr \leq \frac{1}{c(t\wedge\tau - s)^2}\int_s^t \frac{1}{|\varphi_{r\wedge\tau}|^2}\,dr,$$

almost surely. Therefore, we get the inequality

$$(t-s)^p E\big[R^{\Phi_{s,t}(x)}(\theta)^{-p}\big]$$
$$\leq c^{-p}\Big((t-s)^{4p}E\big[(t\wedge\tau - s)^{-4p}\big]\Big)^{\frac{1}{2}} \sup_r E\big[|\varphi_{r\wedge\tau}|^{-4p}\big]^{\frac{1}{2}}. \tag{6.17}$$

For $N > 4p$, there is $c_N > 0$ such that $\sup_{|x|\leq M} P(\tau < t) \leq c_N(t-s)^N$ holds for $\tau = \tau(x,s)$ by Lemma 6.2.1. Then we have

$$(t-s)^{2p}E\big[(t\wedge\tau - s)^{-4p}\big]^{\frac{1}{2}} \leq \Big((t-s)^{4p}\int_s^t (r-s)^{-4p}P(\tau\in dr) + P(\tau\geq t)\Big)^{\frac{1}{2}}$$
$$\leq c'\Big((t-s)^{4p}\int_s^t (r-s)^{-4p+N-1}\,dr + 1\Big)^{\frac{1}{2}}$$

for any $0 \leq s < t < T$, $|x| \leq M$. The above is bounded with respect to these s, t, x since we chose $N > 4p$.

We can show $\sup_{|x|\le M,s,r,t} E[|\varphi_{r\wedge\tau}|^{-4p}] < \infty$ similarly to the proof of Proposition 6.2.1. Therefore (6.17) is bounded for $0 \le s < t < T$ and $|x| \le M$. Then we get (6.16) in view of Proposition 5.3.1. □

6.3 Density and Fundamental Solution for Nondegenerate Diffusion

Let $c_k(x,t), k = 0,\dots,d'$ be $C_b^{\infty,1}$-functions. Let $G_{s,t}(x)$ be an exponential functional defined by

$$G_{s,t}(x) = G^{\mathbf{c}}_{s,t}(x) = \exp\Big\{\sum_{k=0}^{d'} \int_s^t c_k(\Phi_{s,r}(x), r) \circ dW_r^k\Big\}. \tag{6.18}$$

We saw in Sect. 4.2 that the pair $(\Phi_{s,t}(x), G_{s,t}(x))$ is a solution of an SDE on $(d+1)$-dimensional Euclidean space. We will apply Theorem 6.1.1 to the pair process. Then we find that $\partial^{\mathbf{i}} G_{s,t}(x)$ is a smooth Wiener functional and its Sobolev norm $\|\partial^{\mathbf{i}} G_{s,t}(x)\|_{m,p}$ is bounded with respect to $s < t, x$ for any $\mathbf{i}$ and m, p.

We define the transition function of $\Phi_{s,t}(x)$ weighted by $\mathbf{c} = (c_0,\dots,c_{d'})$ by the formula

$$P^{\mathbf{c}}_{s,t}(x, E) = E[1_E(\Phi_{s,t}(x)) G^{\mathbf{c}}_{s,t}(x)].$$

Given a differential operator $A(t)$ of (4.4) and the function $\mathbf{c}$, we define an another differential operator $A^{\mathbf{c}}(t)$ by (4.11). Then $P^{\mathbf{c}}_{s,t}(x, E)$ is the transition function of the semigroup generated by the operator $A^{\mathbf{c}}(t)$.

In this section, we will show the existence of the smooth density of the above transition function, if the system of diffusions with the generator $A(t)$ is nondegenerate. We first define its characteristic function by

$$\psi^{\mathbf{c}}_{s,t,x}(v) = \int_{\mathbb{R}^d} e^{i(v,y)} P^{\mathbf{c}}_{s,t}(x, dy) = E[e^{i(v,\Phi_{s,t}(x))} G_{s,t}(x)]. \tag{6.19}$$

For multi-index $\mathbf{i} = (i_1,\dots,i_d)$ of nonnegative integers $i_1,\dots,i_d$, we set $|\mathbf{i}| = i_1 + \cdots i_d$ and $\partial_x^{\mathbf{i}} = \partial_{x_1}^{i_1} \cdots \partial_{x_d}^{i_d}$.

Lemma 6.3.1 *Suppose that the system of diffusion processes with the generator $A(t)$ is nondegenerate. Then for any fixed $s < t$, $\psi^{\mathbf{c}}_{s,t,x}(v)$ is a C^∞-function of x and is a rapidly decreasing C^∞-function of v. Further, for any $M > 0$ and $N \in \mathbb{N}$, there exists a positive constant c such that*

$$\left|\partial_v^{\mathbf{j}} \partial_x^{\mathbf{i}} \psi^{\mathbf{c}}_{s,t,x}(v)\right| \le c \frac{(1+|x|)^{|\mathbf{j}|}}{(1+|v|)^{N-|\mathbf{i}|}} \tag{6.20}$$

holds for all $|x| \le M$ and $v \in \mathbb{R}^d$.

Furthermore, if the system of diffusions is uniformly nondegenerate, the above inequality holds for all $x, v \in \mathbb{R}^d$.

Proof We will consider the differentiability of $\psi^{\mathbf{c}}_{s,t,x}(v)$ by x. Differentiating $e^{i(v,\Phi_{s,t}(x))}G_{s,t}(x)$ with respect to x, we have $\partial^{\mathbf{i}}_x(e^{i(v,\Phi_{s,t}(x))}G_{s,t}(x)) = e^{i(v,\Phi_{s,t}(x))}G^{\mathbf{i}}_{s,t,x,v}$, where $G^{\mathbf{i}}_{s,t,x,v}$ is a finite sum of terms written as

$$\partial^{\mathbf{i}_0}G_{s,t}(x)\cdot i(v, \partial^{\mathbf{i}_1}\Phi_{s,t}(x))\cdots i(v, \partial^{\mathbf{i}_k}\Phi_{s,t}(x)), \tag{6.21}$$

where $|\mathbf{i}_0| + \cdots + |\mathbf{i}_k| = |\mathbf{i}|$. We know $\partial^{\mathbf{i}'}\Phi_{s,t}(x)$ and $\partial^{\mathbf{i}'}G_{s,t}(x)$ belong to $\mathbf{D}^\infty$ and their norms $\|\partial^{\mathbf{i}'}\Phi_{s,t}(x)\|_{m,p}$ and $\|\partial^{\mathbf{i}'}G_{s,t}(x)\|_{m,p}$ are bounded with respect to s, t, x, in view of Theorem 6.1.1. Then we can change the order of the expectation and the derivative operator and we find that $\psi^{\mathbf{c}}_{s,t,x}(v)$ is infinitely differentiable with respect to x and we get

$$\partial^{\mathbf{i}}_x\psi^{\mathbf{c}}_{s,t,x}(v) = E[e^{i(v,\Phi_{s,t}(x))}G^{\mathbf{i}}_{s,t,x,v}]$$

for any $\mathbf{i}$. Similarly, we can show the differentiability of the above function with respect to v and we get the formula

$$\partial^{\mathbf{j}}_v\partial^{\mathbf{i}}_x\psi^{\mathbf{c}}_{s,t,x}(v) = i^{|\mathbf{j}|}E[e^{i(v,\Phi_{s,t}(x))}\Phi_{s,t}(x)^{\mathbf{j}}G^{\mathbf{i}}_{s,t,x,v}]$$

for any $\mathbf{j}$. Further, for any m, p there is a positive constant c such that the inequality

$$\|\Phi_{s,t}(x)^{\mathbf{j}}G^{\mathbf{i}}_{s,t,x,v}\|_{m,p} \le c|v|^{|\mathbf{i}|}(1+|x|)^{|\mathbf{j}|}$$

holds for any $s < t$ and $x, v \in \mathbb{R}^d$. Then in view of Theorem 5.3.1, for any $N \in \mathbb{N}$ and $M > 0$, there exist c, m, p such that

$$\begin{aligned}
&|\partial^{\mathbf{j}}_v\partial^{\mathbf{i}}_x\psi^{\mathbf{c}}_{s,t,x}(v)| \\
&\le \frac{c}{|v|^N}\Big(\|D\Phi_{s,t}(x)\|_{m,p}\sup_{\theta\in S_{d-1}}\|R^{\Phi_{s,t}(x)}(\theta)^{-1}\|_{m,p}\Big)^N\|\Phi_{s,t}(x)^{\mathbf{j}}G^{\mathbf{i}}_{s,t,x,v}\|_{m,p} \\
&\le \frac{c'(1+|x|)^{|\mathbf{j}|}}{|v|^{N-|\mathbf{i}|}}
\end{aligned}$$

holds for all $|x| \le M$ and $|v| \ge 1$. Further, the inequality holds for any x if $\Phi_{s,t}(x)$ is uniformly nondegenerate. □

Theorem 6.3.1 *Assume that the system of diffusion processes with the generator* $A(t)$ *is nondegenerate. Then the transition function* $P^{\mathbf{c}}_{s,t}(x, E)$ *generated by* $A^{\mathbf{c}}(t)$ *has a rapidly decreasing* C^∞*-density* $p^{\mathbf{c}}_{s,t}(x, y)$ *for any* $s < t, x$*. Further, it satisfies the following properties:*

1. *For any $s < t$, $p^{\mathbf{c}}_{s,t}(x, y)$ is a C^∞-function of x, y and satisfies*

$$\partial_x^{\mathbf{i}} \partial_y^{\mathbf{j}} p^{\mathbf{c}}_{s,t}(x, y) = (-i)^{|\mathbf{j}|} \Big(\frac{1}{2\pi}\Big)^d \int_{\mathbb{R}^d} e^{-i(v,y)} v^{\mathbf{j}} \partial_x^{\mathbf{i}} \psi^{\mathbf{c}}_{s,t,x}(v)\, dv \tag{6.22}$$

for any $\partial_x^{\mathbf{i}}$ and $\partial_y^{\mathbf{j}}$, where $\psi^{\mathbf{c}}_{s,t,x}(v)$ is defined by (6.19).

2. *For any y, t, $p^{\mathbf{c}}_{s,t}(x, y)$ is a $C^{\infty,1}$-function of $x \in \mathbb{R}^d$ and s $(< t)$. It satisfies*

$$\frac{\partial}{\partial s} p^{\mathbf{c}}_{s,t}(x, y) = -A^{\mathbf{c}}(s)_x p^{\mathbf{c}}_{s,t}(x, y). \tag{6.23}$$

3. *Let $P^{\mathbf{c},*}_{s,t}$ be the dual operator of $P^{\mathbf{c}}_{s,t}$. Then for any $g \in C_0(\mathbb{R}^d)$, we have*

$$P^{\mathbf{c},*}_{s,t} g(y) = \int_{\mathbb{R}^d} p^{\mathbf{c}}_{s,t}(x, y) g(x)\, dx, \quad \forall y.$$

Further, for any x, s, $p^{\mathbf{c}}_{s,t}(x, y)$ is a $C^{\infty,1}$-function of $y \in \mathbb{R}^d$ and t $(> s)$. It satisfies

$$\frac{\partial}{\partial t} p^{\mathbf{c}}_{s,t}(x, y) = A^{\mathbf{c}}(t)^*_y p^{\mathbf{c}}_{s,t}(x, y), \tag{6.24}$$

where $A^{\mathbf{c}}(t)^*$ is the differential operator defined by (4.45).

Proof In view of Corollary 5.12.1, the transition function has a density $p^{\mathbf{c}}_{s,t}(x, y)$, $y \in \mathbb{R}^d$ for any s, t, x and it is given by

$$p^{\mathbf{c}}_{s,t}(x, y) = E[\delta_y(\Phi_{s,t}(x)) \cdot G^{\mathbf{c}}_{s,t}(x)] = \Big(\frac{1}{2\pi}\Big)^d \int_{\mathbb{R}^d} e^{-i(v,y)} \psi^{\mathbf{c}}_{s,t,x}(v)\, dv. \tag{6.25}$$

It is a continuous function of $s < t, x, y$ and further, a rapidly decreasing C^∞-function of y, since $\psi^{\mathbf{c}}_{s,t,x}(v)$ is a rapidly decreasing C^∞-function of v, for any $s < t, x$.

We can change the order of the derivation with respect to x and the integral in equation (6.25), because of Lemma 6.3.1, We find that $p^{\mathbf{c}}_{s,t}(x, y)$ is infinitely differentiable with respect to x and satisfies

$$\begin{aligned}\Big(\frac{1}{2\pi}\Big)^d \int e^{-i(v,y)} \partial_x^{\mathbf{i}} \psi^{\mathbf{c}}_{s,t,x}(v)\, dv &= \Big(\frac{1}{2\pi}\Big)^d \partial_x^{\mathbf{i}} \int e^{-i(v,y)} \psi^{\mathbf{c}}_{s,t,x}(v)\, dv \\ &= \partial_x^{\mathbf{i}} p^{\mathbf{c}}_{s,t}(x, y).\end{aligned}$$

Next, differentiate both sides of the above equation by y. Then we get (6.22).

We will prove 2. It holds that $P^{\mathbf{c}}_{s,t} f - f = \int_s^t A^{\mathbf{c}}(r) P^{\mathbf{c}}_{r,t} f\, dr$ for any C^∞-function f with compact supports. Therefore, the density function $p^{\mathbf{c}}_{s,t}(x, y)$ satisfies

$$p^{\mathbf{c}}_{s,t}(x,y) - \delta_y = \int_s^t A^{\mathbf{c}}(r)_x p^{\mathbf{c}}_{r,t}(x,y)\,dr.$$

Then $p^{\mathbf{c}}_{s,t}(x,y)$ is continuously differentiable with respect to s and the derivative satisfies $\frac{\partial}{\partial s} p^{\mathbf{c}}_{s,t}(x,y) = -A^{\mathbf{c}}(s)_x p^{\mathbf{c}}_{s,t}(x,y)$ for $s < t$. Consequently, for any t, y, $p^{\mathbf{c}}_{s,t}(x,y)$ is a $C^{\infty,1}$-function of x and s.

We will next prove 3. We have $P^{\mathbf{c},*}_{s,t} g(y) = \int p^{\mathbf{c}}_{s,t}(x,y) g(x)\,dx$ a.e. y for any $g \in C_0(\mathbb{R}^d)$, since $P^{\mathbf{c},*}_{s,t}$ is the dual of $P^{\mathbf{c}}_{s,t}$. The equality should holds for all y, since both terms are continuous with respect to y. Further, the dual semigroup $\{P^{\mathbf{c},*}_{s,t}\}$ satisfies $P^{\mathbf{c},*}_{s,t} g = g + \int_s^t A^{\mathbf{c}}(r)^* P^{\mathbf{c},*}_{s,r} g\,dr$ for any smooth g. Therefore $p^{\mathbf{c}}_{s,t}(x,y)$ satisfies

$$p^{\mathbf{c}}_{s,t}(x,y) - \delta_x = \int_s^t A^{\mathbf{c}}(r)^*_y p^{\mathbf{c}}_{s,r}(x,y)\,dr.$$

Then both sides are differentiable with respect to t and we get $\frac{\partial}{\partial t} p^{\mathbf{c}}_{s,t}(x,y) = A^{\mathbf{c}}(t)^*_y p^{\mathbf{c}}_{s,t}(x,y)$. □

We shall consider the *final value problem for a backward heat equation* associated with the operator $A^{\mathbf{c}}(t)$. Let $p(x,s,;y,t), x,y \in \mathbb{R}^d, 0 < s < t < \infty$ be a continuous function of x, y, s, t and $C^{2,1}$-function of x, s for any y, t. It is called the *fundamental solution of the backward heat equation* (4.18) or simply the *backward heat kernel* associated with the operator $A^{\mathbf{c}}(s)$, if it satisfies

$$\frac{\partial}{\partial s} p(x,s;y,t) = -A^{\mathbf{c}}(s)_x p(x,s;y,t), \quad x,y \in \mathbb{R}^d,\ 0 < s < t < \infty, \tag{6.26}$$

and the function

$$v(x,s) := \int_{\mathbb{R}^d} p(x,s;y,t_1) f_1(y)\,dy, \quad x \in \mathbb{R}^d,\ 0 < s < t_1 \tag{6.27}$$

is a $C^{2,1}$-function of (x,s) and is a solution of the final value problem of the backward heat equation (4.18) for any slowly increasing continuous function f_1 on $\mathbb{R}^d$.

Theorem 6.3.2 *Assume that the system of diffusion processes with the generator $A(t)$ is nondegenerate. Let $p^{\mathbf{c}}_{s,t}(x,y)$ be the density function of the transition function $P^{\mathbf{c}}_{s,t}(x,E)$ generated by $A^{\mathbf{c}}(t)$. Then $p(x,s;y,t) := p^{\mathbf{c}}_{s,t}(x,y)$ is the fundamental solution of the backward heat equation associated with the operator $A^{\mathbf{c}}(t)$.*

Proof The function $p(x,s;y,t) \equiv p^{\mathbf{c}}_{s,t}(x,y)$ satisfies (6.26) by the previous theorem. We show that it is the fundamental solution for the backward heat equation. Since $p^{\mathbf{c}}_{s,t}(x,y)$ is a rapidly decreasing C^{∞}-function of y for any t, x, we can define $v(x,s)$ by (6.27) for any slowly increasing continuous function f_1.

We show that $v(x,s)$ is a C^∞-function of x and is differentiable with respect to s. For any $M > 0$ and positive integers n, N, there is a positive constant c such that for any $|\mathbf{i}| \leq n$, $\sup_{|x|\leq M} \left|\partial_x^{\mathbf{i}} p_{s,t_1}^{\mathbf{c}}(x,y)\right| \leq \frac{c}{(1+|y|)^N}$ holds for any y, since $\partial_x^{\mathbf{i}} p_{s,t_1}(x,y)$ are rapidly decreasing with respect to y. Now let p' be a positive integer such that $f_1(y)/(1+|y|)^{p'}$ is a bounded function. Take N satisfying $N \geq p + (d+1)$. Then $\partial_x^{\mathbf{i}} p_{s,t_1}^{\mathbf{c}}(x,y) f_1(y)(1+|y|)^{d+1}$ is bounded with respect to $|x| < M$ and $y \in \mathbb{R}^d$. Then, for the function $v(x,s) = \int p_{s,t_1}^{\mathbf{c}}(x,y) f_1(y)\,dy$, we can change the order of the derivative $\partial^{\mathbf{i}}$ and the integral by $(1+|v|)^{-(d+1)}\,dv$. It means that $v(x,s)$ is n-times differentiable with respect to x and the equality $\partial_x^{\mathbf{i}} v(x,s) = \int \partial_x^{\mathbf{i}} p_{s,t_1}^{\mathbf{c}}(x,y) f_1(y)\,dy$ holds.

Further, since $\frac{\partial}{\partial s} p_{s,t_1}^{\mathbf{c}}(x,y) = -A^c(s)_x p_{s,t_1}^{\mathbf{c}}(x,y)$, for the function $v(x,s)$, we can also change the order of the derivative $\frac{\partial}{\partial s}$ and the integral. It means that $v(x,s)$ is differentiable with respect to s and $\frac{\partial}{\partial s} v(x,s) = \int \frac{\partial}{\partial s} p_{s,t_1}(x,y) f_1(y)\,dy$ holds. Therefore, $v(x,s)$ is a $C^{\infty,1}$-function and satisfies the backward heat equation associated with the operator $A^{\mathbf{c}}(t)$.

Next, note the equality $v(x,s) = E[f_1(\Phi_{s,t_1}(x))G_{s,t_1}(x)]$. Given $x \in \mathbb{R}^d$, the family of random variables $\{f_1(\Phi_{s,t_1}(x))G_{s,t_1}(x); 0 \leq s < t_1\}$ are uniformly integrable and converges to $f_1(x)$ as $s \to t_1$. Then the function $v(x,s)$ converges to $f_1(x)$ for any x as $t \to 0$. We have thus shown that $v(x,s)$ of (6.27) is a solution of the final value problem for the backward heat equation. □

If the differential operator $A^{\mathbf{c}}(t)$ is nondegenerate, the associated semigroup $\{P_{s,t}^{\mathbf{c}}\}$ of linear transformations can be extended to that of linear transformations from the space of tempered distributions to the space of C^∞-functions. The fact will be proved in Theorem 6.6.3 for the the semigroup associated with the integro-differential operator $A_J^{\mathbf{c},\mathbf{d}}(t)$.

We will next consider the heat equation (4.21) associated with $A^{\mathbf{c}}(t)$. Let $\check{p}(x,t;y,s), x, y \in \mathbb{R}^d, 0 < s < t < \infty$ be a continuous function of x, y, s, t and $C^{2,1}$-function of x, t for any y, s. It is called the *fundamental solution of the heat equation* or simply *heat kernel* associated with $A^{\mathbf{c}}(t)$ if it satisfies the equation

$$\frac{\partial}{\partial t}\check{p}(x,t;y,s) = A^{\mathbf{c}}(t)_x \check{p}(x,t;y,s), \quad x, y \in \mathbb{R}^d, 0 < s < t < \infty, \tag{6.28}$$

and the function

$$u(x,t) := \int_{\mathbb{R}^d} \check{p}(x,t;y,t_0) f_0(y)\,dy, \quad x \in \mathbb{R}^d, t_0 < t < \infty \tag{6.29}$$

is a solution of the initial value problem of the heat equation (4.21) for any slowly increasing continuous function f_0 on $\mathbb{R}^d$.

In order to construct the fundamental solution for the heat equation associated with the operator $A^{\mathbf{c}}(t)$, let us consider the backward SDE defined by (4.24), where coefficients $V_k(x,r), k = 0, \ldots, d'$ are $C_b^{\infty,1}$-functions. The solution generates a backward flow $\{\check{\Phi}_{s,t}\}$. It is a backward diffusion with the generator $A(t)$ defined by (4.4). Define $\check{P}^{\mathbf{c}}_{s,t}(x,E) = E[1_E(\check{\Phi}_{s,t}(x))\check{G}^{\mathbf{c}}_{s,t}(x)]$, where $\check{G}^{\mathbf{c}}_{s,t}(x)$ is given by (4.27). It is a backward transition function of the backward semigroup generated by the operator $A^{\mathbf{c}}(t)$.

Theorem 6.3.3 *Assume that the system of backward diffusion processes with the generator $A(t)$ is nondegenerate. Then the backward transition function generated by $A^{\mathbf{c}}(t)$ has a rapidly decreasing C^∞-density $\check{p}^{\mathbf{c}}_{s,t}(x,y)$ for any t,s,x. It is a $C_b^{\infty,1}$-function of (x,t) for any $y \in \mathbb{R}^d$ and $t(> s)$. Further, $\check{p}(x,t;y,s) := \check{p}^{\mathbf{c}}_{s,t}(x,y)$ is the fundamental solution of the heat equation associated with the operator $A^{\mathbf{c}}(t)$.*

The proof is similar to that of Theorem 6.3.2. It is omitted.

If coefficients $V_k, k = 0,\ldots,d'$ are time homogeneous, we can define the fundamental solution of the heat equation associated with $A^{\mathbf{c}}$, using the forward diffusion.

The smoothness of the transition density function $p^{\mathbf{c}}_{s,t}(x,y)$ with respect to x has not been studied in the Malliavin calculus. We showed the property by using the smoothness of the stochastic flow $\Phi_{s,t}(x)$ with respect to x.

If the operator $A(t)$ is elliptic, both the forward diffusion and the backward diffusion generated by $A(t)$ are nondegenerate. In this case, density functions $p^{\mathbf{c}}_{s,t}(x,y)$ and $\check{p}^{\mathbf{c}}_{s,t}(x,y)$ are rapidly decreasing C^∞-functions of x and y. Indeed, since the operator $\mathcal{A}(t) = \frac{1}{2}\sum_{k\geq 1} V_k(t)^2 - V_0(t)$ is elliptic, the dual transition function $P^{\mathbf{c},*}_{s,t}(y,E) := \int_E p^{\mathbf{c}}_{s,t}(x,y)\,dx$ should have a rapidly decreasing C^∞ density by Theorem 6.3.1. This shows that $p^{\mathbf{c}}_{s,t}(x,y)$ is a rapidly decreasing C^∞-function of x.

Note We can relax the elliptic condition. For two vector fields V and W, we define its *Lie bracket* by

$$[V,W]f = VWf - WVf.$$

Then $[V,W]$ is again a vector field. Now let $V_0, V_1, \ldots, V_{d'}$ be the vector fields defining the SDE. Set

$$\Sigma_0 = \{V_k(t);\ k = 1,\ldots,d'\},\ \Sigma_M = \{[V_k(t), V(t)];\ k = 0,\ldots,d', V(t) \in \Sigma_{M-1}\}.$$

If there exists $N_0 \in \mathbb{N}$ such that the family of vector fields $\bigcup_{M=0}^{N_0} \Sigma_M$ span $\mathbb{R}^d$ for all x,t, then the operator $A(t)$ is called *hypo-elliptic* or is said to satisfy the *Hörmander condition*.

There are extensive studies for the existence of the smooth density for time-homogeneous hypo-elliptic operator $A(t)$. It was shown by Malliavin [77], Kusuoka–Stroock [69] that if the operator $A(t)$ satisfies the Hörmander condition, then the diffusion is nondegenerate. The proof is not simple. We refer it to Kusuoka–Stroock [69], Norris [87], Nualart [88] and Komatsu–Takeuchi [58]. Their discussions can be applied for time dependent hypo-elliptic operator.

We can apply Theorems 6.3.1, 6.3.2, and 6.3.3 to hypo-elliptic diffusion. Note that the operator $\mathcal{A}(t)$ is hypo-elliptic if $A(t)$ is hypo-elliptic. Then its transition function has a density $p_{s,t}(x, y)$, which is a rapidly decreasing C^∞-function of x and y. Further, the heat equation associated with the operator $A^{\mathfrak{c}}(t)$ satisfying the Hörmander condition has the fundamental solution.

Note The existence of the fundamental solution for a backward heat equation is known in analysis. Let $A(t)$ be an elliptic differential operator given by (4.20). We assume that coefficients α^{ij}, β^i are bounded and are Hölder continuous. Then there exists a fundamental solution $p(x, s; y, t)$ for the operator $A(t)$, which is continuous in s, t, x, y and is C^2-class with respect to x. Results may be found in I'lin-Karashnikov-Oleinik [42], Friedman [28] and Dynkin [24]. The fundamental solution $p(x, s; y, t)$ for the operator $A(t)$ is used for constructing a diffusion process with the generator $A(t)$, without using SDEs. Define $P_{s,t} f(x) = \int p(x, s; y, t) f(y)\, dy$. Then $\{P_{s,t}\}$ is a conservative semigroup. There exists a Markov process X_t with transition probability $P_{s,t}(x, E) = \int_E p(x, s; y, t)\, dy$. We can show that X_t is a continuous process and has the strong Markov property. See Dynkin [24], Chap. 5.

In this monograph, we started with SDE and proved the existence of the fundamental solution, using the Malliavin calculus. Our result is slightly stronger than those, since $p(x, s; y, t)$ is a C^∞-function of x and y, although we assumed that coefficients of the operator $A(t)$ are of C_b^∞-class.

6.4 Solutions of SDE on Wiener–Poisson Space

Let $(\bar{\Omega}, \mathcal{B}(\bar{\Omega}), P)$ be the Wiener–Poisson space associated with the Lévy measure ν discussed in Sect. 5.9. For the Lévy measure ν we assume that at the center it is nondegenerate and satisfies the order condition of exponent $0 < \alpha < 2$ with respect to a given family of star-shaped neighborhoods $\{A_0(\rho)\}$. Further, we assume that ν has a weak drift.

Let us consider a symmetric SDE with jumps defined on $\mathbb{R}^d$:

$$X_t = X_0 + \sum_{k=0}^{d'} \int_{t_0}^t V_k(X_r, r) \circ dW_r^k + \int_{t_0}^t \int_{|z|>0+} \{\phi_{r,z}(X_{r-}) - X_{r-}\} N(dr\, dz). \tag{6.30}$$

Its diffusion part is the same as equation (6.1). For the jump part, we assume that $g(x, t, z) := \phi_{t,z}(x) - x$ satisfies Conditions (J.1) and (J.2) in Sect. 3.2 in Chap. 3.

Let $\{\Phi_{s,t}\}$ be the stochastic flow generated by the above SDE. Let $\mathbb{T} = [0, T]$ be the time parameter of the Wiener–Poisson space. We will show that for any $0 \le s < t \le T$ and $x \in \mathbb{R}^d$, $\Phi_{s,t}(x)$ is a regular Wiener–Poisson functional belonging to the space $(\bar{\mathbf{D}}^\infty)^d$ defined in Sect. 5.10. Let $u = (r, z) \in \mathbb{U}$ and ε_u^+ be the transformation on the Poisson space defined in Sect. 5.4. Then the equality $\varepsilon_u^+ \omega(E) = \omega$

$(E) + \delta_u(E)$ holds for almost all ω. Therefore, for almost all $u = (r, z)$ with respect to the measure n, $X_t := \Phi_{s,t}(x) \circ \varepsilon^+_{(r,z)}$ satisfies the equation

$$X_t = x + \sum_{k=0}^{d'} \int_s^t V_k(X_{r'}, r') \circ dW^k_{r'} + \int_s^t \int_{|z|>0+} \{\phi_{r',z'}(X_{r'-}) - X_{r'-}\} N(dr'\, dz')$$
$$+ \{\phi_{r,z}(X_{r-}) - X_{r-}\} 1_{(s,t]}(r), \quad a.s.$$

If $0 \le r < s$ or $t < r \le T$, then we have $X_t = \Phi_{s,t}(x)$. If $s \le r \le t$, the solution X_t is equal to $\Phi_{r,t} \circ \phi_{r,z} \circ \Phi_{s,r-}(x)$. Since $\Phi_{s,r}(x) = \Phi_{s,r-}(x)$ holds for all x, almost surely, we have

$$\Phi_{s,t}(x) \circ \varepsilon^+_{(r,z)} = \begin{cases} \Phi_{r,t} \circ \phi_{r,z} \circ \Phi_{s,r}(x) & \text{if } s \le r \le t, \\ \Phi_{s,t}(x), & \text{if } r < s \text{ or } r > t. \end{cases} \tag{6.31}$$

Next, let $j \ge 2$ and let $\mathbf{u} = \{(r_i, z_i), i = 1, \dots, j\} \in \mathbb{U}^j$. Let $\mathbf{u}'$ be the subset of $\mathbf{u}$ such that $s \le r_i \le t$. It is written as $\{(r_1, z_1), \dots, (r_{j'}, z_{j'})\}$, where $s < r_1 < \cdots < r_{j'} < t$ and $j' \le j$. Then we have

$$\Phi_{s,t}(x) \circ \varepsilon^+_{\mathbf{u}} = \Phi_{r_{j'},t} \circ \phi_{r_{j'},z_{j'}} \circ \cdots \circ \Phi_{r_1,r_2} \circ \phi_{r_1,z_1} \circ \Phi_{s,r_1}(x), \quad a.s. \tag{6.32}$$

It is well defined and belongs to $(\bar{L}^{\infty-})^d$ for any $s < t$, x and $\mathbf{u}$ a.s. We will study the H-differentiability of $\tilde{D}^j_{\mathbf{u}} \Phi_{s,t}(x)$.

Lemma 6.4.1 *For any $x \in \mathbb{R}^d$, $0 \le s < t \le T$ and $\mathbf{u} \in \mathbb{U}^j$, $\tilde{D}^j_{\mathbf{u}} \Phi_{s,t}(x)$ is H-differentiable. Further, H-derivative $D_r \tilde{D}^j_{\mathbf{u}} \Phi_{s,t}(x)$ is given a.e. $dr\, dP$ by*

$$\begin{cases} \tilde{D}^j_{\mathbf{u}} \big(\nabla \Phi_{r,t}(\Phi_{s,r}(x)) V_k(\Phi_{s,r}(x), r)\big), & k = 1, \dots, d', s < r < t, \\ 0, & \textit{otherwise}. \end{cases} \tag{6.33}$$

Proof Given $h = (h^1(r), \dots, h^{d'}(r)) \in H$, we set $\bar{h}(t) = \int_0^t h(r)\, dr$. We consider $\Phi^{\lambda h}_{s,t}(w) := \Phi_{s,t}(x)(w + \lambda \bar{h})$. Then,

$$\Phi^{\lambda h}_{s,t} \circ \varepsilon^+_{\mathbf{u}}$$
$$= x + \sum_{k=0}^{d'} \int_s^t V_k(\Phi^{\lambda h}_{s,r} \circ \varepsilon^+_{\mathbf{u}}, r) \circ dW^k_r + \lambda \sum_{k=1}^{d'} \int_s^t V_k(\Phi^{\lambda h}_{s,r} \circ \varepsilon^+_{\mathbf{u}}, r) h^k(r)\, dr$$
$$+ \int_s^t \int_{|z|>0+} g(\Phi^{\lambda h}_{s,r-} \circ \varepsilon^+_{\mathbf{u}}, r, z) N(dr\, dz) + \sum_{i; s \le r_i \le t} g(\Phi^{\lambda h}_{s,r_i-} \circ \varepsilon^+_{\mathbf{u}}, r_i, z_i),$$

if $\mathbf{u} = ((r_1, z_1), \ldots, (r_j, z_j))$ and $r_1 < \cdots < r_j$. Then the solution $\Phi_{s,t}^{\lambda h} \circ \varepsilon_{\mathbf{u}}^+$ is continuously differentiable with respect to λ in the F-space $\bar{L}^{\infty-}$. See Remark after Theorem 3.3.2. Let us denote its derivative at $\lambda = 0$ by $\Xi_{s,t}^h$. Then similarly to Lemma 6.1.2, it satisfies an inhomogeneous linear SDE;

$$\begin{aligned}\Xi_{s,t}^h = & \sum_{k=0}^{d'} \int_s^t \nabla V_k(\Phi_{s,r} \circ \varepsilon_{\mathbf{u}}^+, r) \Xi_{s,r}^h \circ dW_r^k \\ & + \int_s^t \int_{|z|>0+} \nabla g(\Phi_{s,r-} \circ \varepsilon_{\mathbf{u}}^+, r, z) \Xi_{s,r-}^h N(dr\, dz) \\ & + \sum_{i; s \le r_i \le t} \nabla g(\Phi_{s,r_i-} \circ \varepsilon_{\mathbf{u}}^+, r_i, z_i) \Xi_{s,r_i-}^h + \sum_{k=1}^{d'} \int_s^t V_k(\Phi_{s,r} \circ \varepsilon_{\mathbf{u}}^+, r) h^k(r)\, dr.\end{aligned}$$

On the other hand, the Jacobian matrix $\nabla \Phi_{s,t}(x) \circ \varepsilon_{\mathbf{u}}^+$ satisfies a homogeneous linear SDE:

$$\begin{aligned}\nabla \Phi_{s,t} \circ \varepsilon_{\mathbf{u}}^+ = I + & \sum_{k=0}^{d'} \int_s^t \nabla V_k(\Phi_{s,r} \circ \varepsilon_{\mathbf{u}}^+, r) \nabla \Phi_{s,r} \circ \varepsilon_{\mathbf{u}}^+ \circ dW_r^k \\ & + \int_s^t \int_{|z|>0+} \nabla g(\Phi_{s,r-} \circ \varepsilon_{\mathbf{u}}^+, r, z) \nabla \Phi_{s,r-} \circ \varepsilon_{\mathbf{u}}^+ N(dr\, dz) \\ & + \sum_{i; s \le r_i \le t} \nabla g(\Phi_{s,r_j-} \circ \varepsilon_{\mathbf{u}}^+, r_i, z_i) \nabla \Phi_{s,r_i} \circ \varepsilon_{\mathbf{u}}^+.\end{aligned}$$

Then the solution $\Xi_{s,t}^h$ of the inhomogeneous equation is written, using the solution $\nabla \Phi_{s,t} \circ \varepsilon_{\mathbf{u}}^+$ of the homogeneous equations as

$$\Xi_{s,t}^h = \int_s^t \nabla \Phi_{r,t}(\Phi_{s,r}) \circ \varepsilon_{\mathbf{u}}^+ \Big(\sum_{k=1}^{d'} V_k(\Phi_{s,r} \circ \varepsilon_{\mathbf{u}}^+, r) h^k(r) \Big)\, dr.$$

This proves that $\Phi_{s,t}(x) \circ \varepsilon_{\mathbf{u}}^+$ is H-differentiable and $D_r \Phi_{s,t}(x) \circ \varepsilon_{\mathbf{u}}^+$ is given by

$$\begin{cases} \big(\nabla \Phi_{r,t}(\Phi_{s,r}(x)) V_k(\Phi_{s,r}(x), r)\big) \circ \varepsilon_{\mathbf{u}}^+, k = 1, \ldots, d', \ s < r < t, \\ 0, \qquad \text{otherwise.} \end{cases}$$

Since $\tilde{D}_{\mathbf{u}}^j \Phi_{s,t}(x)$ is written as a linear sum of $\Phi_{s,t}(x) \circ \varepsilon_{\mathbf{v}}^+$ with $\mathbf{v} \subset \mathbf{u}$, $\tilde{D}_{\mathbf{u}}^j \Phi_{s,t}(x)$ is also H-differentiable. Further, we have $D_r \tilde{D}_{\mathbf{u}}^j \Phi_{s,t}(x) = \tilde{D}_{\mathbf{u}}^j D_r \Phi_{s,t}(x)$, since D and $\tilde{D}$ are commutative. Therefore we get (6.33) in view of Proposition 6.1.1. □

For $u = (r, z)$, we set $\gamma(u) = |z| \wedge 1$ and for $\mathbf{u} = ((r_1, z_1), \ldots, (r_j, z_j))$, we set $\gamma(\mathbf{u}) = (|z_1| \wedge 1) \cdots (|z_j| \wedge 1)$. We set $A(1) = \{u = (r, z) \in \mathbb{U};\ |z| \le 1\}$.

Lemma 6.4.2 *For any* $0 \le s < t \le T$, $j \in \mathbb{N}$, $\mathbf{i}$ *and* $p \ge 2$, *we have*

$$\sup_{s<r<t,x}\ \sup_{\mathbf{u}\in A(1)^j} \frac{E\big[\big|D_r \tilde{D}_{\mathbf{u}}^j \partial^{\mathbf{i}} \Phi_{s,t}(x)\big|^p\big]}{\gamma(\mathbf{u})^p} < \infty, \tag{6.34}$$

where the sup for $s < r < t$ *is the essential supremum with respect to the Lebesgue measure and the sup for* $\mathbf{u}$ *means the essential supremum with respect to the product measure* n^j.

Proof We first consider the case $\mathbf{i} = 0$ and $j = 1$. Let $u = (r_1, z_1) \in \mathbb{U}$, where $s < r_1 < t$. If $r_1 > r$, we have by Lemma 6.4.1,

$$D_r^{(k)}(\Phi_{r_1,t} \circ \phi_{r_1,z_1} \circ \Phi_{s,r_1}) = \nabla(\Phi_{r_1,t} \circ \phi_{r_1,z_1} \circ \Phi_{r,r_1}) V_k(\Phi_{s,r}, r).$$

Therefore, we have $D_r^{(k)} \tilde{D}_{r_1,z_1} \Phi_{s,t} = \tilde{D}_{r_1,z_1} \nabla \Phi_{r,t} \cdot V_k(\Phi_{s,r}, r)$. Similarly, if $r_1 < r$, we have $D_r^{(k)} \tilde{D}_{r_1,z_1} \Phi_{s,t}(x) = \nabla \Phi_{r,t} \cdot \tilde{D}_{r_1,z_1} V_k(\Phi_{s,r}, r)$. We claim:

$$E\big[\big|\tilde{D}_{r_1,z_1} \nabla \Phi_{r,t}\big|^p\big] \le c|z_1|^p, \quad E\big[\big|\tilde{D}_{r_1,z_1} V_k(\Phi_{s,r}, r)\big|^p\big] \le c|z_1|^p. \tag{6.35}$$

We will show the second inequality. Since $\Phi_{r_1,r}$ and Φ_{s,r_1} are independent, we have for any $p \ge 2$

$$\begin{aligned} &E\big[\big|V_k(\Phi_{r_1,r} \circ \phi_{r_1,z_1} \circ \Phi_{s,r_1}, r) - V_k(\Phi_{s,r}, r)\big|^p\big] \\ &\quad = E\Big[E\big[\big|V_k(\Phi_{r_1,r} \circ \phi_{r_1,z_1}(y), r) - V_k(\Phi_{r_1,r}(y), r)\big|^p\big]\big|_{y=\Phi_{s,r_1}}\Big]. \end{aligned} \tag{6.36}$$

Further, there exists a positive constant c_p such that

$$E\big[\big|V_k(\Phi_{r_1,r} \circ \phi_{r_1,z_1}(y), r) - V_k(\Phi_{r_1,r}(y), r)\big|^p\big] \le c_p |\phi_{r_1,z_1}(y) - y|^p, \quad \forall y, r$$

in view of (3.28). Therefore (6.36) is dominated by

$$\begin{aligned} c_p E\Big[|\phi_{r_1,z_1}(y) - y|^p_{y=\Phi_{s,r_1}}\Big] &\le c_p E\big[\big|\phi_{r_1,z_1}(\Phi_{s,r_1}) - \Phi_{s,r_1}\big|^p\big] \\ &\le c_p |z_1|^p E\big[\big|\partial_z \phi_{r_1,\theta z_1}(\Phi_{s,r_1})\big|^p\big], \end{aligned}$$

where $|\theta z_1| \le |z_1|$. Then there exists a positive constant c'_p such that

$$E\big[\big|\tilde{D}_{r_1,z_1} V_k(\Phi_{s,r}(x), r)\big|^p\big] \le c'_p |z_1|^p, \quad \forall s < t, x, r, |z_1| \le 1.$$

This proves the second inequality of (6.35). We can verify the first inequality of (6.35) in the same way.

We will next consider the case $j = 2$. Let $\mathbf{u} = \{(r_1, z_1), (r_2, z_2)\} \in \mathbb{U}^2$, where $r_1 < r_2$. Then $D_r^{(k)} \tilde{D}_{\mathbf{u}}^j \Phi_{s,t}$ is a sum of terms $\tilde{D}_{\mathbf{u}'}^{j'} \nabla \Phi_{r,t} \tilde{D}_{\mathbf{u}''} V_k(\Phi_{s,r}, r)$, where $\mathbf{u}', \mathbf{u}'' \subset \mathbf{u}$. We shall consider the term $\tilde{D}_{\mathbf{u}''} V_k(\Phi_{s,r}, r)$. If $\mathbf{u}'' = \mathbf{u}$, it holds that

$$E\big[|\tilde{D}_{\mathbf{u}}^2 V_k(\Phi_{s,r}(x), r)|^p\big] = \int I(\Phi_{r_1,r_2} \circ \phi_{r_1,z_1} \circ \Phi_{s,r_1}, \Phi_{s,r_2})\, dP, \tag{6.37}$$

where

$$\begin{aligned} I(y, y') = E\Big[\Big|&\{V_k(\Phi_{r_2,r} \circ \phi_{r_2,z_2}(y), r) - V_k(\Phi_{r_2,r}(y), r)\} \\ &- \{V_k(\Phi_{r_2,r} \circ \phi_{r_2,z_2}(y'), r) - V_k(\Phi_{r_2,r}(y'), r)\}\Big|^p\Big]. \end{aligned} \tag{6.38}$$

We show that there is a positive constant c_1 such that

$$I(y, y') \le c_1 |(\phi_{r_2,z_2}(y) - y) - (\phi_{r_2,z_2}(y') - y')|^p. \tag{6.39}$$

Set $X_r^\lambda = V_k(\Phi_{r_2,r}(\lambda), r)$. Then the $2d$-dimensional process $(X_r^{\lambda_1}, -X_r^{\lambda_2})$ satisfies

$$E[|(X_r^{\lambda_1} - X_r^{\lambda_2}) - (X_r^{\lambda_1'} - X_r^{\lambda_2'})|^p] \le c_1 |(\lambda_1 - \lambda_2) - (\lambda_1' - \lambda_2')|^p$$

with some positive constant c_1, similarly to the inequality (3.28) in Lemma 3.3.3. Setting $\lambda_1 = \phi_{r_2,z_2}(y)$, $\lambda_2 = y$, $\lambda_1' = \phi_{r_2,z_2}(y')$ and $\lambda_2' = y'$. Then we get the inequality (6.39). Therefore (6.37) is dominated by

$$\begin{aligned} &c_1 \int \Big|(\phi_{r_2,z_2}(y) - y) - (\phi_{r_2,z_2}(y') - y')\Big|^p_{y=\Phi_{r_1,r_2}\circ\phi_{r_1,z_1}\circ\Phi_{s,r_1},\, y'=\Phi_{s,r_2}} dP \\ &\le c_2 |z_2|^p E\Big[\Big|\partial_z \phi_{r_2,\theta z_2}(\Phi_{r_1,r_2} \circ \phi_{r_1,z_1} \circ \Phi_{s,r_1}) - \partial_z \phi_{r_2,\theta z_2}(\Phi_{s,r_2})\Big|^p\Big] \\ &\le c_3 |z_2|^p E\Big[\Big|\Phi_{r_1,r_2} \circ \phi_{r_1,z_1} \circ \Phi_{s,r_1} - \Phi_{s,r_2}\Big|^p\Big] \\ &\le c_4 |z_2|^p |z_1|^p. \end{aligned}$$

A similar estimation is valid for $\tilde{D}_{\mathbf{u}'}^{j'} \nabla \Phi_{r,t}$. Therefore we get the inequality

$$\sup_{s<r<t,x} E\big[\big|D_r \tilde{D}_{\mathbf{u}}^j \Phi_{s,t}(x)\big|^p\big] \le c_{j,p} |z_1|^p \cdots |z_j|^p \tag{6.40}$$

in the case $j = 2$. We can verify the above inequality in the case $j \ge 3$, similarly. Consequently, (6.34) holds in the case $\mathbf{i} = 0$. Further, we have (6.34) for any $\mathbf{i}$, since $(\Phi_{s,t}, \partial_x^{\mathbf{i}'} \Phi_{s,t}, |\mathbf{i}'| \le |\mathbf{i}|)$ is a solution of an SDE on a certain high dimensional Euclidean space. □

We can show the existence of higher derivatives $D^i_{\mathbf{r}}\tilde{D}^j_{\mathbf{u}}\partial^{\mathbf{i}}\Phi_{s,t}(x), i = 2, \ldots$ by the arguments as in Lemma 6.4.2. Further, we have

$$\sup_{\mathbf{r}\in[s,t]^i, x\in\mathbb{R}^d} \sup_{\mathbf{u}\in A(1)^j} \frac{E\big[\big|D^i_{\mathbf{r}}\tilde{D}^j_{\mathbf{u}}\partial^{\mathbf{i}}\Phi_{s,t}(x)\big|^p\big]}{\gamma(\mathbf{u})^p} < \infty \tag{6.41}$$

for any $\mathbf{i}$, i, j, p. Therefore $\partial^{\mathbf{i}}\Phi_{s,t}(x)$ is an element of $(\bar{\mathbf{D}}^\infty)^d$ for any $\mathbf{i}$, $s < t$ and x.

We will check that the solution of the SDE is regular, i.e., it satisfies Condition (R) in Sect. 5.7.

Lemma 6.4.3 *$\Phi_{s,t}(x)$ satisfies Condition (R) for any $s < t, x$.*

Proof We have $\tilde{D}_{r,z}\Phi_{s,t}(x) = \Phi_{r,t}\circ\phi_{r,z}\circ\Phi_{s,r}(x) - \Phi_{s,t}(x)$, if $s < r \le t$. Further, $\tilde{D}_{r,z}\Phi_{s,t}(x) = 0$ if $r < s$ or $r > t$. Therefore, we have the equality

$$\partial_{z^k}\tilde{D}_{r,z}\Phi_{s,t}(x) = \begin{cases} \nabla\Phi_{r,t}(\phi_{r,z}(\Phi_{s,r}(x))\partial_{z^k}\phi_{r,z}(\Phi_{s,r}(x))), & \text{if } s < r < t, \\ 0, & \text{otherwise.} \end{cases} \tag{6.42}$$

In view of Lemma 6.4.1, $E[|\nabla\Phi_{r,t}(\phi_{r,z}(\Phi_{s,r}(x)))\circ\varepsilon^+_{\mathbf{u}}|^p]$ is bounded with respect to $x \in \mathbb{R}^d, s < r < t, |z| \le 1$ and $\mathbf{u} \in A(1)^j$. Since $\partial_{z^k}\phi_{r,z}(x)$ is a bounded function with respect to r, z, x, we find $\sup_{x,s,r,t,z,\mathbf{u}} E[|\partial_{z^k}\tilde{D}_{r,z}\Phi_{s,t}(x)\circ\varepsilon^+_{\mathbf{u}}|^p] < \infty$. Similarly, we can show that for any $\partial^{\mathbf{i}}_z = \partial^{i_1}_{z^1}\cdots\partial^{i_{d'}}_{z^{d'}}$, the inequality

$$\sup_{x,s,r,t,z,\mathbf{u}} E\big[\big|\partial^{\mathbf{i}}_z\tilde{D}_{r,z}\Phi_{s,t}(x)\circ\varepsilon^+_{\mathbf{u}}\big|^p\big] < \infty$$

holds. We will apply Morrey's Sobolev inequality (Lemma 3.8.2). For any positive integer $i \ge 3$, we have

$$E\Big[\sum_{|\mathbf{i}|\le 2}\sup_{|z|\le 1}\big|\partial^{\mathbf{i}}_z\tilde{D}_{r,z}\Phi_{s,t}(x)\circ\varepsilon^+_{\mathbf{u}}\big|^p\Big] \le C^p E\Big[\sum_{|\mathbf{i}|\le i}\int_{|z|\le 1}\big|\partial^{\mathbf{i}}_z\tilde{D}_{r,z}\Phi_{s,t}(x)\circ\varepsilon^+_{\mathbf{u}}\big|^p dz\Big].$$

Take the supremum with respect to x, s, r, t and $\mathbf{u} \in A(1)^j, j = 1, \ldots, n$ for the both sides of the above. The supremum of the right-hand side is finite. Therefore the supremum of the left-hand side is also finite. Therefore Condition (R) is satisfied.

□

Norms $\|\ \|^\sharp_{m,n,p}$ for Wiener–Poisson functionals were defined in Sect. 5.10. Using these norms, we have the following theorem from (6.41).

Theorem 6.4.1 *Let $\{\Phi_{s,t}\}$ be the stochastic flow generated by SDE* (6.30). *Then $\Phi_{s,t}(x)$ is a regular Wiener–Poisson functional belonging to $(\bar{\mathbf{D}}^\infty)^d$, for any $s, t \in \mathbb{T}$ and $x \in \mathbb{R}^d$. Further,*

$$\|\partial^{\mathbf{i}}_x\Phi_{s,t}(x)\|^\sharp_{m,n,p}, \quad \|D\Phi_{s,t}(x)\|^\sharp_{m,n,p}, \quad \|\tilde{D}\Phi_{s,t}(x)\|^\sharp_{m,n,p}$$

are bounded with respect to s, t, x if $|\mathbf{i}| \geq 1$, where $\| \|_{m,n.p}^{\sharp}$ are Sobolev norms for Wiener-Poisson functionals defined in Sect. 5.10.

6.5 Nondegenerate Jump-Diffusions

The solution of SDE (6.30) is a jump-diffusion on $\mathbb{R}^d$. We saw in Sect. 4.5 that it generator $A_J(t)$ is given by (4.52). Set $V(x,t) = (V_1(x,t), \ldots, V_{d'}(x,t))$. If the matrix $V(x,t)V(x,t)^T$ is positive definite for any t, x, the operator $A_J(t)$ is called *elliptic*. Further, if the matrix $V(x,t)V(x,t)^T$ is uniformly positive definite with respect to x, t, the operator $A_J(t)$ is called *uniformly elliptic*. We shall consider the possibly non-elliptic case. Let $\{\Gamma_\rho\}$ be the family of $d' \times d'$-matrices defined by (1.21) and let Γ_0 be its limit as $\rho \to 0$. For jump-maps $\phi_{t,z}$, we will define tangent vector fields at $z = 0$ by

$$\tilde{V}_k(x,t) = \partial_{z^k}\phi_{t,z}(x)\Big|_{z=0}, \quad k = 1, \ldots, d'.$$

Set $\tilde{V}(x,t) = (\tilde{V}_1(x,t), \ldots, \tilde{V}_{d'}(x,t))$ and define $d \times d$-matrix by

$$A(x,t) = V(x,t)V(x,t)^T + \tilde{V}(x,t)\Gamma_0\tilde{V}(x,t)^T.$$

If $A(x,t)$ is positive definite for any x, t, the operator (4.52) is called *pseudo-elliptic*. Further, if the matrix $A(x,t)$ is uniformly positive definite, the operator $A_J(t)$ is called *uniformly pseudo-elliptic*.

Let $\{\Phi_{s,t}\}$ be the stochastic flow generated by SDE (6.30). The Malliavin covariance of $\Phi_{s,t}(x)$ at the center was defined by (5.155). If $s < r \leq t$, we have $D_r^{(k)}\Phi_{s,t}(x) = \nabla\Phi_{r,t}(\Phi_{s,r}(x))V_k(\Phi_{s,r}(x))$ and

$$\begin{aligned}
\partial_{z^k}\tilde{D}_{r,z}\Phi_{s,t}(x)\big|_{z=0} &= \partial_{z^k}\Phi_{r,t} \circ \phi_{r,z} \circ \Phi_{s,r}(x)|_{z=0} \\
&= \nabla\Phi_{r,t}(\Phi_{s,r}(x))\partial_{z^k}\phi_{r,z}(\Phi_{s,r}(x))\big|_{z=0} = \nabla\Phi_{r,t}(\Phi_{s,r}(x))\tilde{V}_k(\Phi_{s,r}(x), r).
\end{aligned}$$

If $r > t$ or $r < s$, we have $D_r\Phi_{s,t}(x) = 0$ and $\tilde{D}_{r,z}\Phi_{s,t}(x) = 0$. Therefore the Malliavin covariance at the center is written as

$$\begin{aligned}
\bar{K}^{\Phi_{s,t}(x)} &= \int_s^t \nabla\Phi_{r,t}(\Phi_{s,r}(x))V(\Phi_{s,r}(x), r)V(\Phi_{s,r}(x), r)^T\nabla\Phi_{r,t}(\Phi_{s,r}(x))^T\, dr \\
&\quad + \int_s^t \nabla\Phi_{r,t}(\Phi_{s,r}(x))\tilde{V}(\Phi_{s,r}(x), r)\Gamma_0\tilde{V}(\Phi_{s,r}(x), r)^T\nabla\Phi_{r,t}(\Phi_{s,r}(x))^T\, dr \\
&= \int_s^t \nabla\Phi_{r,t}(\Phi_{s,r}(x))A(\Phi_{s,r}(x), r)\nabla\Phi_{r,t}(\Phi_{s,r}(x))^T\, dr.
\end{aligned}$$

A system of jump-diffusions $\{X_t^{x,s} = \Phi_{s,t}(x), (x,s) \in \mathbb{T} \times \mathbb{R}^d\}$ with the generator $A_J(t)$ is called *uniformly nondegenerate at the center*, if $\bar{K}^{\Phi_{s,t}(x)}$ are invertible and for any $n, p > 2$, $\sup_\theta |\bar{K}^{\Phi_{s,t}(x)}(\theta)^{-1}|_{0,n,p}^\sharp$ are bounded with respect to $x \in \mathbb{R}^d$. Here $|\ |_{0,n,p}^\sharp$ are norms defined by (5.57). Further, if the above is bounded with respect to x on any compact subset of $\mathbb{R}^d$, the system of jump-diffusions is called *nondegenerate at the center*.

Theorem 6.5.1 *If the intrgro-differential operator $A_J(t)$ of* (4.52) *is uniformly pseudo-elliptic, the system of jump-diffusions with the generator $A_J(t)$ is uniformly nondegenerate at the center. Further, the inequality*

$$\sup_{s<t,x} \sup_\theta (t-s)|\bar{K}^{\Phi_{s,t}(x)}(\theta)^{-1}|_{0,n,p}^\sharp < \infty \tag{6.43}$$

holds for any $n \in \mathbb{N}$, $p > 2$.

For the proof, we need a lemma.

Lemma 6.5.1 *Jacobian matrices $\nabla\Phi_{s,t}(x)$ are invertible a.s. Let $V_{s,t}(x)$ be the inverse matrix. Then for any $p \geq 2$ and $j \in \mathbb{N}$,*

$$\sup_{s<t,x} \sup_{\mathbf{u}\in A(1)^j} E[|V_{s,t}(x) \circ \varepsilon_{\mathbf{u}}^+|^p] < \infty. \tag{6.44}$$

Proof The stochastic flow $\Phi_{s,t} = \Phi_{s,t}(x)$ satisfies

$$\Phi_{s,t} = x + \sum_{k=0}^{d'} \int_s^t V_k(\Phi_{s,r}, r) \circ dW_r^k + \int_s^t \int_{|z|>0+} g(\Phi_{s,r-}, r, z) N(dr\,dz).$$

Differentiating each term of the above with respect to x, we get an SDE for the Jacobian matrix $J_t = J_{s,t} = \nabla\Phi_{s,t}(x)$. It is written as

$$J_t = I + \sum_{k=0}^{d'} \int_s^t \nabla V_k(r) J_r \circ dW^k(r) + \int_s^t \int_{|z|>0+} \nabla g(r,z) J_r N(dr\,dz),$$

where $\nabla V_k(r) := \nabla V_k(\Phi_{s,r}, r)$ and $\nabla g(r,z) := \nabla g(\Phi_{s,r-}, r, z)$. Now consider an SDE for an unknown $d \times d$-matrix-valued process $V_t = V_{s,t}$:

$$V_t = I - \sum_{k=0}^{d'} \int_s^t V_r \nabla V_k(r) \circ dW_r^k - \int_0^t \int_{|z|>0+} V_{r-} \nabla h(r,z) N(dr\,dz), \tag{6.45}$$

where $\nabla h(r,z) := \nabla h(\phi_{r,z}(\Phi_{s,r-}), r, z)$ and $h(x,r,z)$ is given by $h(x,r,z) := x - \phi_{r,z}^{-1}(x)$. We shall prove that V_t is the inverse matrix of J_t. Apply the rule of the differential calculus for jump processes (Theorem 3.10.1) to the product $V_t J_t$. Using symmetric integrals, it holds that

$$V_t J_t = V_s J_s - \sum_k \int_s^t V_r \nabla V_k(r) J_r \circ dW_r^j + \sum_k \int_0^t V_r \nabla V_k(r) J_r \circ dW_r^j$$
$$+ \int_s^t \int_{|z|>0+} \Big\{ V_{r-}(I - \nabla h(r,z))(I + \nabla g(r,z)) J_{r-} - V_{r-} J_{r-} \Big\} dN.$$

We have

$$(I - \nabla h(r,z))(I + \nabla g(r,z)) = \nabla \phi_{r,z}^{-1}(\phi_{r,z}(\Phi_{s,r-}))\nabla \phi_{r,z}(\Phi_{s,r-}) = I.$$

Consequently, we get $V_t J_t = I$, proving that $V_{s,t}$ is the inverse matrix of $J_{s,t}$.

Now, equation (6.45) is a linear equation for V_t, where coefficients $\nabla V_j(\cdots)$ and $\nabla h(\cdots)$ are bounded uniformly with respect to parameter x. Rewrite the equation using Itô integrals. Then we can apply Lemma 3.3.2 and show the inequality $\sup_{s<t,x} E[|V_{s,t}(x)|^p] < \infty$ for any $p > 2$.

Further, the above discussion is valid if we transform $J_{s,t}$ and $V_{s,t}$ by $J_{s,t} \circ \varepsilon_{\mathbf{u}}^+$ and $V_{s,t} \circ \varepsilon_{\mathbf{u}}^+$, respectively; $V_{s,t} \circ \varepsilon_{\mathbf{u}}^+$ is the inverse of $J_{s,t} \circ \varepsilon_{\mathbf{u}}^+$ and the former satisfies the inequality of the lemma. □

Proof of Theorem 6.5.1 Our discussion is close to that of Proposition 6.2.1. Set $\varphi_r = \varphi_r^{s,t,x} = \theta \nabla \Phi_{r,t}(\Phi_{s,r}(x))$. Then $\bar{K}^{\Phi_{s,t}(x)}(\theta)$ is written as

$$\bar{K}^{\Phi_{s,t}(x)}(\theta) = \int_s^t \varphi_r A(\Phi_{s,r}(x), r) \varphi_r^T \, dr. \tag{6.46}$$

For any s, t, x, it holds that $\varphi_r = \varphi_r^{(s,t,x)} \neq 0$ a.s., since the Jacobian matrix $\nabla \Phi_{r,t}(x)$ is invertible for any x, r, t. Let λ_r be the minimal eigen-value of the matrix $A(\Phi_{s,r}(x), r)$. Then we have

$$\bar{K}^{\Phi_{s,t}(x)}(\theta)^{-1} \le \frac{1}{(t-s)^2} \int_s^t \frac{1}{\lambda_r |\varphi_r|^2} \, dr$$

similarly to the proof of Proposition 6.2.1. The inequality is valid for the transformation $\varepsilon_{\mathbf{u}}^+$. Since $\lambda_r \ge c > 0$ holds for any r and ω, we get the inequality

$$E[(\bar{K}^{\Phi_{s,t}(x)}(\theta) \circ \varepsilon_{\mathbf{u}}^+)^{-p}] \le (c(t-s))^{-p} \sup_r E[|\varphi_r \circ \varepsilon_{\mathbf{u}}^+|^{-2p}].$$

We will show that the last term of the above is finite for any p. Let $V_{r,t}(x)$ be the inverse matrix of $\nabla \Phi_{r,t}(x)$. Since $\nabla \Phi_{r,t}(\Phi_{s,r}(x)) V_{r,t}(\Phi_{s,r}(x)) = I$, we have

$$|\varphi_r \circ \varepsilon_{\mathbf{u}}^+|^{-1} \le |V_{r,t}(\Phi_{s,r}(x)) \circ \varepsilon_{\mathbf{u}}^+|. \tag{6.47}$$

The above is L^{2p}-bounded for any $2p > 2$ in view of Lemma 6.5.1. Then we get $\sup_{s,t,x,r,\mathbf{u}} E[|\varphi_r \circ \varepsilon_{\mathbf{u}}^+|^{-2p}] < \infty$. Consequently, the inequality (6.43) holds. Then the jump-diffusion is uniformly nondegenerate at the center. □

We shall next consider the pseudo-elliptic case. We want to show that the jump-diffusion is nondegenerate at the center. In the case of elliptic diffusions, we proved this by introducing localizing stopping times $\tau(x, s)$ of Lemma 6.2.1. Loosely, the lemma shows that the process $\Phi_{s,t}(x)$ moves slowly, since $P(\tau(x, s) - s < \delta) = O(\delta^N)$ holds for any $N > 2$. However, the fact does not hold for jump-diffusions. Jump-diffusions could move rapidly. Even so, we will show that a hitting time $\tau(x, s)$ of $\Phi_{s,t}(x)$ to a set V satisfies $P(\tau(x, s) - s < \delta) = O(\delta^N)$, if it jumps at least N times before hitting the set V.

We will discuss it precisely. Let $\{\phi_{t,z}\}$ be jump-maps of the jump-diffusion. For a subset U of $\mathbb{R}^d$, we set

$$\bar{\phi}(U) = \overline{\bigcup \phi_{t,z}(U)}, \quad \bar{\phi}^i(U) = \bar{\phi}(\bar{\phi}^{i-1}(U)), \quad i = 2, 3, \ldots,$$

where $\cup$ is taken for all $t \in \mathbb{T}, z \in \mathrm{Supp}(\nu)$. It holds that $U \subset \bar{\phi}^i(U) \subset \bar{\phi}^{i+1}(U)$ for any set U. Similarly, we define

$$\bar{\phi}^{-1}(V) = \overline{\bigcup \phi_{t,z}^{-1}(V)}, \quad \bar{\phi}^{-i}(V) = \bar{\phi}^{-1}(\bar{\phi}^{-i-1}(V)), \quad i = 2, 3, \ldots$$

For two subsets U, V of $\mathbb{R}^d$, we set $d(U, V) = \inf_{x \in U, y \in V} d(x, y)$, where $d(x, y)$ is the metric of two points x, y.

In the following argument, we often set $\Phi_{s,\infty} = \infty$ conventionally and we use the notation $\infty - \infty = 0$.

Lemma 6.5.2 *Let N be a positive integer satisfying $d(\bar{\phi}^N(U), V) > 0$. For given x, s, let $\tau = \tau(x, s)$ be the hitting time of the jump-diffusion process $X_t^{x,s} = \Phi_{s,t}(x), t > s$ to the set V. Then there exists $c > 0$ such that*

$$\sup_{x \in U} P(\tau(x, s) < t) \leq c(t - s)^N \tag{6.48}$$

holds for any $s < t$.

Proof For $\epsilon > 0$ we define the ϵ-neighborhood of the set $U \subset \mathbb{R}^d$ by $U_\epsilon = \{x; d(x, U) \leq \epsilon\}$. We set

$$\psi_\epsilon(U) = \bar{\phi}(U_\epsilon), \quad \psi_\epsilon^n(U) = \psi_\epsilon(\psi_\epsilon^{n-1}(U)), \quad n = 2, \ldots, N$$

inductively. Then if $d(\bar{\phi}^N(U), V) > 0$ holds, $d(\psi_\epsilon^N(U), \bar{V}) > 0$ holds for sufficiently small $\epsilon > 0$. We have $U \subset \psi_\epsilon(U) \subset \cdots \subset \psi_\epsilon^N(U) \subset V$.

Let $x \in U$. Let $\sigma_1 = \sigma_1(x,s)$ be the first leaving time of $X_t = X_t^{x,s} = \Phi_{s,t}(x)$ from the ball $B_\epsilon(x) = \{y \in \mathbb{R}^d; |y-x| < \epsilon\}$. Then $X_{\sigma_1} \in \psi_\epsilon^1(U)$, if $\sigma_1 < T$ *a.s.* We define stopping times $\sigma_2 = \sigma_2(x,s), \ldots, \sigma_N = \sigma_N(x,s)$ inductively as

$$\sigma_n = \sigma_n(x,s) = \inf\{r > \sigma_{n-1};\ |X_r - X_{\sigma_{n-1}}| \geq \epsilon\},$$
$$= \infty, \quad \text{if } \{\cdots\} \text{ is empty.}$$

Then we have $X_{\sigma_n} \in \psi_\epsilon^n(U)$ if $\sigma_n < \infty$ *a.s.* for any $n = 2, \ldots, N$. Consequently, we have $X_{\sigma_N} \notin V$ if $\sigma_N < T$ *a.s.* This means $\tau \geq \sigma_N$ a.s.

Now, we have $\tau \geq \sigma_N \geq \sum_{n=1}^N (\sigma_n - \sigma_{n-1})$, where $\sigma_0 = s$. Therefore,

$$\{\tau < t\} \subset \bigcap_{n=1}^N \{|X_{\sigma_n} - X_{\sigma_{n-1}}| 1_{(\sigma_n - \sigma_{n-1}) < t-s} \geq \epsilon\}.$$

We shall consider the conditional probability of the last term: Since $\Phi_{s,u}(x) = \Phi_{s,t}(\Phi_{t,u}(x))$ holds for any $s < t < u$ and x a.s., $\sigma_N(x,s)$ coincides with the first leaving time of $X_r^{(N-1)} := \Phi_{\sigma_{N-1},r}(X_{\sigma_{N-1}})$ from the ball $B_\epsilon(X_{\sigma_{N-1}})$. Therefore we have

$$\sigma_N(x,s) = \sigma_1(X_{\sigma_{N-1}}, \sigma_{N-1}), \quad a.s.$$

By Chebyschev's inequality and the strong Markov property, we have

$$\begin{aligned}
&P(|X_{\sigma_N} - X_{\sigma_{N-1}}| 1_{\sigma_N - \sigma_{N-1} < t-s} \geq \epsilon | \mathcal{F}_{\sigma_{N-1}}) \\
&\leq \epsilon^{-2} E[|X_{\sigma_N} - X_{\sigma_{N-1}}|^2 1_{\sigma_N - \sigma_{N-1} < t-s} | \mathcal{F}_{\sigma_{N-1}}] \\
&\leq \epsilon^{-2} E[|X_{\sigma_1(y,r)} - y|^2 1_{\sigma_1(y,r) < t-s}]\big|_{y = X_{\sigma_{N-1}}, r = \sigma_{N-1}} \\
&\leq \epsilon^{-2} E[|X_{\sigma_1(y,r)\wedge t} - y|^2]\big|_{y = X_{\sigma_{N-1}}, r = \sigma_{N-1}} \\
&\leq c\epsilon^{-2}(t-s), \quad a.s.
\end{aligned}$$

for any $x \in U$. Here we used inequality $\sup_y E[|\Phi_{s,\sigma_1(y,r)\wedge t}(y) - y|^2] \leq c(t-s)$. Therefore,

$$P(\{\tau < t\}) \leq c\epsilon^{-2}(t-s) P(\bigcap_{n=1}^{N-1} \{|X_{\sigma_n} - X_{\sigma_{n-1}}| 1_{\sigma_n - \sigma_{n-1} < t-s} > \epsilon\})$$

for any $x \in U$. Repeating this argument, $P(\tau < t) \leq (c\epsilon^{-2})^N (t-s)^N$ holds for any $x \in U$. □

Lemma 6.5.3 *Let K be a positive constant satisfying $|g(x,z)| \le K$ for any x, z. For given $M > 0, \epsilon > 0$ and a positive integer N, set $M_N = M + N(K+\epsilon)$. Let $\tau = \tau(x,s)$ be the hitting time of the process $X_t^{x,s} = \Phi_{s,t}(x)$ to the set $\{y; |y| > M_N\}$. Then there is a positive constant c such that*

$$\sup_{|x|\le M} P(\tau(x,s) < t) \le c(t-s)^N \tag{6.49}$$

holds for any $s < t$.

Proof Set $U = \{y; |y| \le M\}$. Then $\phi^N(U) \subset \{y; |y| \le M + NK\}$. Hence if $V = \{y; |y| > M_N\}$, we have $d(\psi^N(U), V) > 0$. Therefore (6.49) holds by Lemma 6.5.2. □

Theorem 6.5.2 *If the operator $A_J(t)$ of (4.52) is pseudo-elliptic, the system of jump-diffusions with the generator $A_J(t)$ is nondegenerate at the center. Further, the inequality*

$$\sup_{s,t\in\mathbb{T}, s<t, |x|\le M} \sup_{\theta\in S_{d-1}} (t-s)|\bar{K}^{\Phi_{s,t}(x)}(\theta)^{-1}|_{0,n,p}^{\sharp} < \infty \tag{6.50}$$

holds for any $n \in \mathbb{N}$, $p > 2$ and $M > 0$.

Proof Consider again the formula (6.46). Let λ_r be the minimal eigen-value of $A(\Phi_{s,r}(x), r)$. It may not be uniformly positive with respect to r, ω. So we need a localization. For a given $p > 2$, we will fix $N \in \mathbb{N}$ such that $N > 2p$. Let $\tau = \tau(x,s)$ be the stopping time satisfying (6.49). Since $A(x,t)$ is positive definite, we can take $c > 0$ such that $\theta A(x,t)\theta^T \ge c$ holds for all $\theta \in S_{d-1}$, $|x| \le M + N(K+\epsilon)$. Then we have $\bar{K}^{\Phi_{s,t}(x)}(\theta) \ge \int_s^{t\wedge\tau} \lambda_r |\varphi_r|^2\, dr$ for $|x| \le M$. Then $\bar{K}^{\Phi_{s,t}(x)}(\theta)$ are invertible a.s. and inverses satisfy

$$\bar{K}^{\Phi_{s,t}(x)}(\theta)^{-1} \le c(t\wedge\tau - s)^{-2} \int_s^{t\wedge\tau} \frac{1}{\lambda_r|\varphi_r|^2}\, dr.$$

Note $\lambda_r \ge c$ for $r < \tau$. Then we get

$$(t-s)^p E[\bar{K}^{\Phi_{s,t}(x)}(\theta)^{-p}] \le c^{-p}((t-s)^{4p} E[(t\wedge\tau - s)^{-4p}])^{\frac{1}{2}} \sup_r E[|\varphi_{r\wedge\tau}|^{-4p}]^{\frac{1}{2}}.$$

We can verify $\sup_{s<t,|x|\le M}(t-s)^{4p} E[(t\wedge\tau - s)^{-4p}] < \infty$, similarly to the proof of Proposition 6.2.2. Further, we can show $\sup_{x,r} E[|\varphi_{r\wedge\hat{\tau}}|^{-4p}] < \infty$, similarly to the proof of Proposition 6.2.1. Consequently, we get

$$\sup_{s<t,|x|\le M} \sup_{\theta} (t-s)^p E[\bar{K}^{\Phi_{s,t}(x)}(\theta)^{-p}] < \infty.$$

Then we get the inequality (6.50) in the case $n = 0$ ($\varepsilon_{\mathbf{u}}^+$ is identity).

In the case where $n \geq 1$, let $\tilde{\tau}(x,s)$ be the hitting time of the process $X_t^{x,s} = \Phi_{s,t}(x)$ to the set $\{|y| > M + (N+n)(K+\epsilon)\}$. Then we have the inequality $\tilde{\tau}(x,s) \circ \varepsilon_{\mathbf{u}}^+ \geq \tau(x,s)$ a.s. if $\mathbf{u} \in A(1)^k$, where $k \leq n$. Therefore,

$$\sup_{|x|\leq M} \sup_{\mathbf{u}\in A(1)^k} P(\tilde{\tau}(x,s) \circ \varepsilon_{\mathbf{u}}^+ < t) \leq c(t-s)^N, \quad s < t.$$

Using this stopping time, we have

$$\sup_{s<t,|x|\leq M} \sup_{\mathbf{u}\in A(1)^k} \sup_{\theta} (t-s)^p E[(\bar{K}^{\Phi_{s,t}(x)}(\theta) \circ \varepsilon_{\mathbf{u}}^+)^{-p}] < \infty.$$

Then we get (6.50) for the case $n \geq 1$. □

Now, we will discuss the relation with δ-nondegenerate property. Let δ be any constant satisfying $1 < \delta < 2/\alpha$. Define

$$\begin{aligned}\bar{Q}_\rho^{\Phi_{s,t}(x)}(\theta) &= \int_{\mathbb{T}} (\theta, D_r\Phi_{s,t}(x))^2\, dr \\ &\quad + \frac{1}{|v|^2\varphi(\rho)} \int_{A(\rho)} |e^{i(v,\tilde{D}_u\Phi_{s,t}(x))} - 1|^2 n(du)\Big|_{v=\rho^{-\frac{1}{\delta}}\theta}.\end{aligned}$$

In the next proposition, we extend a result for diffusion processes stated in Propositions 6.2.1 and 6.2.2 to that for jump-diffusion processes.

Proposition 6.5.1 *Let $\{\Phi_{s,t}\}$ be the stochastic flow associated with the operator $A_J(t)$.*

1. *If $A_J(t)$ is uniformly pseudo-elliptic, $\bar{Q}_\rho^{\Phi_{s,t}(x)}(\theta)$ are invertible and satisfy the inequality*

$$\sup_{s,t\in\mathbb{T},s<t,x} \sup_{(\rho,\theta)} (t-s)\|\bar{Q}_\rho^{\Phi_{s,t}(x)}(\theta)^{-1}\|_{m,n,p}^{\sharp} < \infty \tag{6.51}$$

for any m, n, p.

2. *If $A_J(t)$ is pseudo-elliptic, $\bar{Q}_\rho^{\Phi_{s,t}(x)}(\theta)$ are invertible and for any $M > 0$ satisfy the inequality*

$$\sup_{s<t,|x|\leq M} \sup_{(\rho,\theta)} (t-s)\|\bar{Q}_\rho^{\Phi_{s,t}(x)}(\theta)^{-1}\|_{m,n,p}^{\sharp} < \infty \tag{6.52}$$

for any m, n, p.

Proof We shall consider the uniformly pseudo-elliptic case only. We will apply Theorem 5.11.2 for $\bar{K}^{\Phi_{s,t}(x)}$. Note that it satisfies (6.43) and that $C_{n,p}^{\delta'}(\Phi_{s,t}(x))$ given by (5.99) is bounded with respect to t, x. Then we have

$$\sup_{s<t,x} \sup_{(\rho,\theta)} (t-s)|\bar{Q}_{\rho}^{\Phi_{s,t}(x)}(\theta)^{-1}|_{0,n,p}^{\sharp} < \infty$$

by Theorem 5.11.2. Next, we will prove (6.51) by applying Theorem 5.10.2. Note that there is $q > 1$ such that the inequality

$$\begin{aligned}
&(t-s)\|\bar{Q}_{\rho'}^{\Phi_{s,t}(x)}(\theta)^{-1}\|_{m,n,p}^{\sharp} \\
&\leq c_{m,n,p}\Big(1+\frac{1}{t-s}\|\bar{Q}_{\rho'}^{\Phi_{s,t}(x)}(\theta)\|_{m,n,2pq}^{\sharp}\Big)^{q}\Big(1+|(t-s)\bar{Q}_{\rho'}^{\Phi_{s,t}(x)}(\theta)^{-1}|_{0,n,2pq}^{\sharp}\Big)^{q}
\end{aligned}$$

holds by Theorem 5.10.2. We shall consider the right hand side. Since

$$\begin{aligned}
\frac{1}{t-s}\bar{Q}_{\rho'}^{\Phi_{s,t}(x)}(v) &= \frac{1}{t-s}\int_0^t \frac{(v, D_r\Phi_{s,t}(x))^2}{|v|^2}\,dr \\
&\quad+\frac{1}{\varphi(\rho')}\frac{1}{t-s}\int_s^t\int_{A_0(\rho')}\frac{|e^{i(v,\tilde{D}_u\Phi_{s,t}(x))}-1|^2}{|v|^2}n(du)
\end{aligned}$$

holds, we can show that its $\| \ \|_{m,n,p}^{\sharp}$-norm is dominated by

$$c(\|D\Phi_{s,t}(x)\|_{m,n,2p}^{\sharp})^2 + c'(1+\|\tilde{D}\Phi_{s,t}(x)\|_{m,n,2(n+1)p}^{\sharp})^{2(n+1)},$$

similarly to the proof of Proposition 5.11.1. Consequently, there is a positive constant c_0 such that

$$\sup_{(\rho,\theta)}\Big(1+\frac{1}{t-s}\|\bar{Q}_{\rho'}^{\Phi_{s,t}(x)}(\theta)\|_{m,n,2pq}^{\sharp}\Big)^{q} \leq c_0.$$

Then we have (6.51). □

Note If the generator $A(t)$ is not pseudo-elliptic, we define a sequence of a family of vector fields $\bar{\Sigma}_M$; $M = 0, 1, 2, \ldots$ by

$$\begin{aligned}
\bar{\Sigma}_0 &= \{V_1, \ldots, V_{d'}, \tilde{V}_1, \ldots, \tilde{V}_{d'}\}, \\
\bar{\Sigma}_M &= \{[V_0, V], [V_k, V], [\tilde{V}_k, V]; k = 1, \ldots, d', V \in \bar{\Sigma}_{M-1}\}, \quad M = 1, 2, \ldots
\end{aligned}$$

If $\bigcup_M \bar{\Sigma}_M$ span $\mathbb{R}^d$ for any x, the generator A is said to satisfy the Hörmander condition or called hypo pseudo-elliptic. The existence of the smooth density for hypo pseudo-elliptic jump-diffusion was studied by Kunita [64]. The discussion is complicated.

6.6 Density and Fundamental Solution for Nondegenerate Jump-Diffusion

Let $\Phi_{s,t}(x)$ be the jump-diffusion (stochastic flow) with generator (4.52). For a given x and $s < t$, we consider an exponential functional:

$$G_{s,t} = \exp\Big\{\sum_{k=0}^{d'} \int_s^t c_k(\Phi_{s,r}, r) \circ dW_r^k + \int_s^t \int_{|z|>0+} \log d_{r,z}(\Phi_{s,r-}) N(dr\, dz)\Big\}, \tag{6.53}$$

where $c_k(x,t), k = 0, \ldots, d'$ are $C_b^{\infty,1}$-functions on $\mathbb{R}^d \times \mathbb{T}$, and $d_{t,z}(x)$ is a positive $C_b^{\infty,1,2}$-function on $\mathbb{R}^d \times \mathbb{T} \times \mathbb{R}^{d'}$ such that $d_{t,0}(x) = 1$ holds for any x, t. The pair $(\Phi_{s,t}(x), G_{s,t}(x)y)$ is a solution of an $\mathbb{R}^{d+1}$-dimensional SDE and we may apply Theorem 6.4.1. Then we find that $\partial^{\mathbf{i}} G_{s,t}(x)$ are in $\bar{\mathbf{D}}^\infty$ and norms $\|\partial^{\mathbf{i}} G_{s,t}(x)\|_{m,n,p}^\sharp$ are bounded with respect to $s < t, x$ for any $\mathbf{i}$ and m, n, p. We set $\mathbf{c} = (c_0(x,t), \ldots, c_{d'}(x,t))$ and $\mathbf{d} = (d_{t,z}(x))$ and denote the above $G_{s,t}$ by $G_{s,t}^{\mathbf{c},\mathbf{d}}$. We define the law of $\Phi_{s,t}(x)$ weighted by $G_{s,t}^{\mathbf{c},\mathbf{d}}(x)$ by $P_{s,t}^{\mathbf{c},\mathbf{d}}(x, E) = E[1_E(\Phi_{s,t}(x)) G_{s,t}^{\mathbf{c},\mathbf{d}}(x)]$.

Given an integro-differential operator $A_J(t)$ of (4.52) and functions $\mathbf{c}, \mathbf{d}$, we define another operator $A_J^{\mathbf{c},\mathbf{d}}(t)$ by (4.58). Then it is the generator of the semigroup $\{P_{s,t}^{\mathbf{c},\mathbf{d}}\}$ (Theorem 4.5.2).

Lemma 6.6.1 *Assume that the system of jump-diffusions with generator $A_J(t)$ is nondegenerate at the center. Then the characteristic function $\psi_{s,t,x}^{\mathbf{c},\mathbf{d}}(v)$ of the transition function $P_{s,t}^{\mathbf{c},\mathbf{d}}(x, E)$ has properties similar to those of $\psi_{s,t,x}^{\mathbf{c}}(v)$ in Lemma* 6.3.1.

Proof The characteristic function of $P_{s,t}^{\mathbf{c},\mathbf{d}}(x, E)$ is written as

$$\psi_{s,t,x}^{\mathbf{c},\mathbf{d}}(v) = E[e^{i(v, \Phi_{s,t}(x))} G_{s,t}^{\mathbf{c},\mathbf{d}}(x)].$$

It is a C^∞-function of v, because $\Phi_{s,t}(x)$ has finite moments of any order. We will show that it is a rapidly decreasing. Since $\bar{Q}_\rho^{\Phi_{s,t}(x)}(\theta)$ satisfies (6.52) by Proposition 6.5.1, for any $N \in \mathbb{N}$ and $M > 0$, there are m, n, p and a positive constant c such that

$$\begin{aligned}\Big|\psi_{s,t,x}^{\mathbf{c},\mathbf{d}}(v)\Big| \leq \frac{C}{|v|^{N\gamma_0}} \bigg(\Big\{\|D\Phi_{s,t}(x)\|_{m,n,2p}^\sharp + \Big(1 + \|\tilde{D}\Phi_{s,t}(x)\|_{m,n,2(n+1)p}^\sharp\Big)^{n+1}\Big\} \\ \times \sup_{(\rho,\theta)} \|\bar{Q}_\rho^{\Phi_{s,t}(x)}(\theta)^{-1}\|_{m,n,p}^\sharp\bigg)^N \|G_{s,t}^{\mathbf{c},\mathbf{d}}(x)\|_{m,n,p}^\sharp\end{aligned} \tag{6.54}$$

for all $|v| \geq 1$, $0 \leq s < t \leq T$ and $|x| \leq M$ in view of Theorem 5.11.1. Then we can show that $\psi^{\mathbf{c},\mathbf{d}}_{s,t,x}(v)$ is infinitely differentiable with respect to x and v and satisfies the inequality similar to (6.20), replacing N by $N\gamma_0$. □

Theorem 6.6.1 *Assume that the system of jump-diffusions with the generator $A_J(t)$ is nondegenerate at the center. Then for any $0 \leq s < t < \infty$ and $x \in \mathbb{R}^d$, the transition function $P^{\mathbf{c},\mathbf{d}}_{s,t}(x, E)$ generated by $A^{\mathbf{c},\mathbf{d}}_J(t)$ has a rapidly decreasing C^∞-density $p^{\mathbf{c},\mathbf{d}}_{s,t}(x, y)$ with respect to the Lebesgue measure. Further, it satisfies the following properties.*

1. *For any $s < t$, it is a C^∞-function of x, y and satisfies, for any $\partial^{\mathbf{i}}_x$ and $\partial^{\mathbf{k}}_y$,*

$$\partial^{\mathbf{i}}_x \partial^{\mathbf{j}}_y p^{\mathbf{c},\mathbf{d}}_{s,t}(x, y) = (-i)^{|\mathbf{j}|}\Big(\frac{1}{2\pi}\Big)^d \int_{\mathbb{R}^d} e^{-i(v,y)} v^{\mathbf{j}} \partial^{\mathbf{i}}_x \psi^{\mathbf{c},\mathbf{d}}_{s,t,x}(v)\, dv. \tag{6.55}$$

2. *For any y, t, it is a $C^{\infty,1}$-function of $x \in \mathbb{R}^d$ and s $(< t)$. It satisfies*

$$\frac{\partial}{\partial s} p^{\mathbf{c},\mathbf{d}}_{s,t}(x, y) = -A^{\mathbf{c},\mathbf{d}}_J(s)_x p^{\mathbf{c},\mathbf{d}}_{s,t}(x, y). \tag{6.56}$$

3. *For any x, s, it is a $C^{\infty,1}$-function of $y \in \mathbb{R}^d$ and t $(> s)$. It satisfies*

$$\frac{\partial}{\partial t} p^{\mathbf{c},\mathbf{d}}_{s,t}(x, y) = A^{\mathbf{c},\mathbf{d}}_J(t)^*_y p^{\mathbf{c},\mathbf{d}}_{s,t}(x, y), \tag{6.57}$$

where $A^{\mathbf{c},\mathbf{d}}_J(t)^*$ is the operator given by (4.77).

The above theorem can be verified similarly to Theorem 6.3.1 (diffusion case), making use of Lemma 6.6.1.

We consider the final value problem of the backward heat equation associated with the operator $A^{\mathbf{c},\mathbf{d}}_J(t)$. Its fundamental solution is defined in the same way as the fundamental solution of the backward heat equation associated with $A^{\mathbf{c}}(t)$ in Sect. 6.3.

Theorem 6.6.2 *Assume that the system of jump-diffusions with generator $A_J(t)$ is nondegenerate at the center. Let $p^{\mathbf{c},\mathbf{d}}_{s,t}(x, y)$ be the smooth density of the transition function $P^{\mathbf{c},\mathbf{d}}_{s,t}(x, E)$ generated by $A^{\mathbf{c},\mathbf{d}}_J(t)$. Then $p(x, s; y, t) := p^{\mathbf{c},\mathbf{d}}_{s,t}(x, y)$ is the fundamental solution of the backward heat equation associated with the operator $A^{\mathbf{c},\mathbf{d}}_J(t)$.*

We can extend the function $f_1(x)$ of the final condition for the backward heat equation to any tempered distribution f_1. Schwartz's distribution is discussed briefly in Sect. 5.12. We may assume that f_1 belongs to $\mathcal{S}^{-N}$ for sufficiently large N. Assume that the operator $A^{\mathbf{c},\mathbf{d}}_J(t)$ is uniformly nondegenerate at the center. Then Theorem 5.12.1 and its proof tells us that for any $N \in \mathbb{N}$ and fixed $s < t$, there are m, n, p and a positive constant C_N such that

$$\|f(\Phi_{s,t}(x))\|^*_{m,n,p} \le C_N |f|_{-N}, \quad f \in \mathcal{S}^{-N} \tag{6.58}$$

holds for any x, where the norm $|\ |_{-N}$ is defined by (5.170). Indeed, in Theorem 5.12.1, the constant C_N in (5.176) is given by $\sup_{\|G\|^\sharp_{m,n,p}\le 1}|\psi^x_G|_N$, where $\psi^x_G(v) = E[e^{i(v,\Phi_{s,t}(x))}G]$. The above function is a rapidly decreasing C^∞-function of v. Further, for any positive integer m and multi-index $\mathbf{i}$, $(1+|v|^2)^m \partial^{\mathbf{i}}_v \psi^x_G(v)$ is uniformly bounded with respect to x, v and G with $\|G\|^\sharp_{m,n,p} \le 1$, since ψ^x_G satisfies the inequality (6.54) and

$$\|D\Phi_{s,t}(x)\|^\sharp_{m,n,p}, \quad \|\tilde{D}\Phi_{s,t}(x)\|^\sharp_{m,n,p}, \quad \sup_{(\rho,\theta)} \|\bar{Q}^{\Phi_{s,t}(x)}_\rho(\theta)^{-1}\|^\sharp_{m,n,p}$$

are bounded with respect to x. Then $\sup_{\|G\|^\sharp_{m,n,p}\le 1}|\psi^x_G|_N$ is bounded with respect to x.

Theorem 6.6.3 *Assume that the system of jump-diffusions with the generator $A_J(t)$ is uniformly nondegenerate at the center. Then the semigroup of linear operators $P^{\mathbf{c},\mathbf{d}}_{s,t}$ is extended to that of linear operators from the space of tempered distributions to the space of C^∞-functions.*

For a tempered distribution f_1, set $v(x,s) = E[f_1(\Phi_{s,t_1}(x)) \cdot G^{\mathbf{c},\mathbf{d}}_{s,t_1}(x)]$. Then it is a $C^{\infty,1}$-function of (x,s). Further, it is differentiable with respect to s $(< t_1)$ and satisfies the backward integro-differential equation

$$\frac{\partial}{\partial s} v(x,s) = -A^{\mathbf{c},\mathbf{d}}_J(s) v(x,s) \tag{6.59}$$

and the final condition $\lim_{s\to t_1} v(x,s) = f_1$.

Proof We will approximate f_1 by a sequence of smooth functions $f_m, m = 2, \ldots$ in $\mathcal{S}$ with respect to the norm $|\ |_{-N}$. Set $v_m(x,s) = E[f_m(\Phi_{s,t}(x))G_{s,t}(x)]$, where $G_{s,t} = G^{\mathbf{c},\mathbf{d}}_{s,t}$ Then $v_m(x,t)$ converges to $v(x,s)$ uniformly in x, in view of the inequality (6.58). Therefore $v(x,s)$ is a continuous function. Further, we have

$$\begin{aligned}\partial^{\mathbf{i}} v_m(x,t) &= \partial^{\mathbf{i}} E[f_m(\Phi_{s,t}(x))G_{s,t}(x)] \\ &= \sum E\big[\partial^{\mathbf{i}_1} f_m(\Phi_{s,t}(x)) \cdot \partial^{\mathbf{i}_2}\Phi_{s,t}(x), \partial^{\mathbf{i}_3} G_{s,t}(x)\big],\end{aligned}$$

where the sum is taken for $\mathbf{i}_1, \mathbf{i}_2$ and $\mathbf{i}_3$ such that $|\mathbf{i}_1|, |\mathbf{i}_2|, |\mathbf{i}_3| \le |\mathbf{i}|$. The right-hand side of the above converges again uniformly in x in view of (6.58). Since this is valid for any derivative operator $\partial^{\mathbf{i}}$, the limit function $v(x,t)$ is a C^∞-function of x.

Now each function $v_m(x,s)$ satisfies $\frac{\partial}{\partial s} v_m(x,s) = -A^{\mathbf{c},\mathbf{d}}_J(s) v_m(x,s)$ for $s < t_1$. Then the limit function $v(x,s)$ should be differentiable with respect to $s < t_1$ and satisfy the same equation.

Let $\varphi \in \mathcal{S}_\infty$. By the change of variables, we have

$$\int v(x,s)\varphi(x)dx = \int E[f_1(\Phi_{s,t_1}(x))G_{s,t_1}(x)\varphi(x)]\,dx$$
$$= \int E[f_1(y)G_{s,t_1}(\check{\Psi}_{s,t_1}(y))\varphi(\check{\Psi}_{s,t_1}(y))|\det \nabla\check{\Psi}_{s,t_1}(y)|]\,dy.$$

Let s tend to t_1. Then the last term converges to $\langle f_1, \varphi\rangle$. Therefore $v(x,s)$ converges to f_1 as $s \to t_1$. □

Finally, we will discuss the fundamental solution for the heat equation associated with the nondgenerate operator $A_J^{\mathbf{c},\mathbf{d}}(t)$. Consider the backward SDE defined with characteristics $(V_j(x,t), j = 0,\ldots,d', g(x,t,z), \nu)$. Let $\check{\Phi}_{s,t}(x)$ be the backward flow of diffeomorphisms generated by the backward SDE and let $\check{G}_{s,t}^{\mathbf{c},\mathbf{d}}$ be the exponential function defined in Sect. 4.5. Then $\check{P}_{s,t}^{\mathbf{c},\mathbf{d}}(x,E) = E[1_E(\check{\Phi}_{s,t}(x))\check{G}_{s,t}^{\mathbf{c},\mathbf{d}}(x)]$ is the backward transition function with the generator $A_J^{\mathbf{c},\mathbf{d}}(t)$.

Theorem 6.6.4 *Assume that the system of backward jump-diffusions generated by $A_J(t)$ is nondegenerate at the center. Then its backward transition function $\check{P}_{s,t}^{\mathbf{c},\mathbf{d}}(x,E)$ has a rapidly decreasing C^∞-density $\check{p}_{s,t}^{\mathbf{c},\mathbf{d}}(x,y)$ for any $0 \le s < t < \infty$ and $x \in \mathbb{R}^d$. Further, it is a $C^{\infty,1}$-function of $(x,t) \in \mathbb{R}^d \times (s,\infty)$ for any y, s and $\check{p}(x,t;y,s) := \check{p}_{s,t}^{\mathbf{c},\mathbf{d}}(x,y)$ is the fundamental solution of the heat equation associated with the operator $A_J^{\mathbf{c},\mathbf{d}}(t)$.*

Remark We shall consider a jump-diffusion where its Lévy measure may not satisfy the order condition. In this case Theorem 6.6.1 may not hold. We will consider an elliptic jump-diffusion. Suppose that the generator is elliptic and jump-map $\phi_{t,z}$ satisfies Condition (J.2). Then the Malliavin covariance $R^{\Phi_{s,t}(x)}$ of $\Phi_{s,t}(x)$ defined by (6.11) is invertible and the inverse satisfies

$$\sup_{\theta\in S_{d-1},|x|\le M} (t-s)E[R^{\Phi_{s,t}(x)}(\theta)^{-p}]^{\frac{1}{p}} < \infty, \quad \forall p \ge 2 \tag{6.60}$$

for any $0 < M < \infty$. Indeed, discussions of Sect. 6.3 are also valid for jump-diffusions, if we use Lemma 6.5.2 instead of Lemma 6.11. Then the transition function $P_{s,t}^{\mathbf{c},\mathbf{d}}(x,E)$ has a rapidly decreasing C^∞-density. Further, the density function $p_{s,t}^{\mathbf{c},\mathbf{d}}(x,y)$ is a rapidly decreasing C^∞-function of x. See Remark in Sect. 5.11.

If jump-maps $\phi_{t,z}$ do not satisfy Condition (J.2) ($\phi_{t,z}$ are not diffeomorphic), the smoothness of the density is not known even if $A_J(t)$ is elliptic. Indeed, the flow $\Phi_{s,t}$ may not be diffeomorphic and hence its Jacobian matrix $\nabla\Phi_{s,t}$ may not invertible. Then the invertibility of the Malliavin covariance is not clear.

6.7 Short-Time Estimates of Densities

In the remainder of this chapter (Sects. 6.7, 6.8, 6.9, and 6.10) we will study elliptic diffusion and pseudo-elliptic jump-diffusions. We will study *short-time asymptotics* of the smooth densities $p^{\mathbf{c}}_{s,t}(x,y)$ studied in Sect. 6.3 and $p^{\mathbf{c},\mathbf{d}}_{s,t}(x,y)$ studied in Sect. 6.6 as $t \to s$. We will be able to obtain the same short-time estimates for the fundamental solutions for heat equations associated with the elliptic operator $A^{\mathbf{c}}(t)$ and pseudo-elliptic operator $A^{\mathbf{c},\mathbf{d}}_J(t)$. However, we will not repeat these arguments.

We will first state the assertion for a diffusion process.

Theorem 6.7.1 *Assume that the partial differential operator $A^{\mathbf{c}}(t)$ is elliptic. Let $p^{\mathbf{c}}_{s,t}(x,y)$ be the smooth density of the transition function $P^{\mathbf{c}}_{s,t}(x,E)$ generated by $A^{\mathbf{c}}(t)$. Let $\partial^{\mathbf{i}}_x$ and $\partial^{\mathbf{j}}_y$ be any given differential operators with indexes $\mathbf{i}$ and $\mathbf{j}$, respectively.*

1. *For any $T > 0$ and $M > 0$, there exists a positive constant c such that*

$$\left|\partial^{\mathbf{i}}_x \partial^{\mathbf{j}}_y p^{\mathbf{c}}_{s,t}(x,y)\right| \leq \frac{c}{(t-s)^{\frac{|\mathbf{i}|+|\mathbf{j}|+d}{2}}}, \qquad \forall 0 \leq s < t \leq T \tag{6.61}$$

holds for all $|x| \leq M$, $y \in \mathbb{R}^d$.

2. *If $A^{\mathbf{c}}(t)$ is uniformly elliptic, for any $T > 0$, there exists $c > 0$ such that the above inequality holds for all $x, y \in \mathbb{R}^d$.*

Proof We will first consider the case where the diffusion is elliptic. For a given $T > 0$, we consider the Wiener space of time parameter $\mathbb{T} = [0, T]$ and apply the Malliavin calculus on the Wiener space. Let $0 \leq s < t \leq T$ and set $\psi^{\mathbf{c}}_{s,t,x}(v) = E[e^{i(v,\Phi_{s,t}(x))}G_{s,t}(x)]$, where $G_{s,t}(x) = G^{\mathbf{c},\mathbf{d}}_{s,t}(x)$. Then we have from (6.22)

$$\partial^{\mathbf{i}}_x \partial^{\mathbf{j}}_y p^{\mathbf{c}}_{s,t}(x,y) = (-i)^{|\mathbf{j}|}\left(\frac{1}{2\pi}\right)^d \int_{\mathbb{R}^d} e^{-i(v,y)} v^{\mathbf{j}} \partial^{\mathbf{i}}_x \psi^{\mathbf{c}}_{s,t,x}(v)\, dv. \tag{6.62}$$

It holds that

$$\partial^{\mathbf{i}}_x \psi^{\mathbf{c}}_{s,t,x}(v) = E\Big[e^{i(v,\Phi_{s,t}(x))} G^{\mathbf{i}}_{s,t,x,v}\Big],$$

where $G^{\mathbf{i}}_{s,t,x,v}$ is written as a linear sum of terms (6.21). We can rewrite it as

$$G^{\mathbf{i}}_{s,t,x,v} = \sum |v|^{|\mathbf{i}|-|\mathbf{i_0}|} G^{\mathbf{i}_0,\ldots,\mathbf{i_j}}_{s,t,x,\theta},$$

where

$$G^{\mathbf{i}_0,\ldots,\mathbf{i_j}}_{s,t,x,\theta} = i^{|\mathbf{i}|-|\mathbf{i_0}|} \partial^{\mathbf{i}_0} G_{s,t}(x)(\theta, \partial^{\mathbf{i}_1}\Phi_{s,t}(x)) \cdots (\theta, \partial^{\mathbf{i_j}}\Phi_{s,t}(x)),$$

and the summation is taken for multi-indexes $\mathbf{i_l}$ such that $\mathbf{i_l} \le \mathbf{i}$ and $|\mathbf{i_0}| + \cdots + |\mathbf{i_j}| = |\mathbf{i}|$. Setting $\psi^{\mathbf{i_0},\ldots,\mathbf{i_j}}_{s,t,x,\theta}(v) = E[e^{i(v,\Phi_{s,t}(x))} G^{\mathbf{i_0},\ldots,\mathbf{i_j}}_{s,t,x,\theta}]$, we have

$$\begin{aligned} \left| v^{\mathbf{j}} \partial_x^{\mathbf{i}} \psi^{\mathbf{c}}_{s,t,x}(v) \right| &\le \sum |v|^{|\mathbf{i}|+|\mathbf{j}|-|\mathbf{i_0}|} |\psi^{\mathbf{i_0},\ldots,\mathbf{i_j}}_{s,t,x,\theta}(v)| \\ &\le (1 \vee |v|)^{|\mathbf{i}|+|\mathbf{j}|} \sum |\psi^{\mathbf{i_0},\ldots,\mathbf{i_j}}_{s,t,x,\theta}(v)|. \end{aligned} \tag{6.63}$$

Therefore we get

$$\sup_{y \in \mathbb{R}^d} \left| \partial_x^{\mathbf{i}} \partial_y^{\mathbf{j}} p^{\mathbf{c}}_{s,t}(x,y) \right| \le \left(\frac{1}{2\pi} \right)^d \int_{\mathbb{R}^d} (1 \vee |v|)^{|\mathbf{i}|+|\mathbf{j}|} \sum |\psi^{\mathbf{i_0},\ldots,\mathbf{i_j}}_{s,t,x,\theta}(v)| \, dv.$$

Set $w = \sqrt{t-s}\,v$ and change the variable. Since $1 \wedge |v| \le \frac{\sqrt{T}}{\sqrt{t-s}}(1+|w|)$ and $dv = (t-s)^{-d/2}\,dw$, the above is dominated by

$$\frac{c}{(t-s)^{\frac{|\mathbf{i}|+|\mathbf{j}|+d}{2}}} \int_{\mathbb{R}^d} (1 \vee |w|)^{|\mathbf{i}|+|\mathbf{j}|} |\tilde{\psi}(w)| \, dw,$$

where

$$\tilde{\psi}(w) = E[e^{i(w,H_{s,t,x})} G^{\mathbf{i_0},\ldots,\mathbf{i_j}}_{s,t,x,\theta}], \qquad H_{s,t,x} = \frac{\Phi_{s,t}(x)}{\sqrt{t-s}}.$$

Take any $M > 0$ and consider

$$\|DH_{s,t,x}\|_{m,p}, \qquad \sup_{\theta \in S_{d-1}} \|G^{\mathbf{i_0},\ldots,\mathbf{i_j}}_{s,t,x,\theta}\|_{m,p}, \qquad \sup_{\theta \in S_{d-1}} \|R^{H_{s,t,x}}(\theta)^{-1}\|_{m,p}. \tag{6.64}$$

The first and the second term are bounded with respect to $s < t, |x| \le M$ by Theorem 6.1.1. Further, the last term is bounded at the same region by Proposition 6.2.2. Then, for any N there exists a positive constant c_N such that inequalities $|\tilde{\psi}(w)| \le c_N$ (if $|w| \le 1$) and $\le \frac{c_N}{|w|^N}$ (if $|w| \ge 1$) hold for any $s < t$ and $|x| \le M$ in view of Theorem 5.3.1. Consequently, we obtain from (6.62)

$$\begin{aligned} \sup_{y \in \mathbb{R}^d} \left| \partial_x^{\mathbf{i}} \partial_y^{\mathbf{j}} p^{\mathbf{c}}_{s,t}(x,y) \right| &\le \frac{c_N}{(t-s)^{\frac{|\mathbf{i}|+|\mathbf{j}|+d}{2}}} \left(1 + \int_{|w| \ge 1} |w|^{|\mathbf{i}|+|\mathbf{j}|-N} \, dw \right) \\ &\le \frac{c'_N}{(t-s)^{\frac{|\mathbf{i}|+|\mathbf{j}|+d}{2}}} \end{aligned} \tag{6.65}$$

for all $s < t, |x| \le M$, if $N > d + |\mathbf{i}| + |\mathbf{j}|$. Therefore we get the inequality (6.61) for all $|x| \le M$, $y \in \mathbb{R}^d$, proving the first assertion of the theorem.

Finally, if the process is uniformly elliptic, all terms in (6.64) are bounded for all $s < t, x$ in view of Theorem 6.1.1 and Proposition 6.2.1. Then (6.65) holds for all $s < t, x$. This proves the second assertion of the theorem. □

Remark The short-time estimate for fundamental solution for the elliptic operator (4.20) is known in analysis. Assuming its coefficients are bounded and Holder continuous, inequality (6.61) is shown in the case $|\mathbf{i}| \leq 2$ and $\mathbf{j} = 0$ (Il'in–Kalashnikov–Oleinik [42], Dynkin [24]). We got a stronger result using the Malliavin calculus, assuming that coefficients of the operator $A(t)$ is of C_b^∞-class. The short-time estimate of the fundamental solution of a non-elliptic but hypo-elliptic operator seems to be open.

We will next consider jump-diffusions. Suppose first that their generators are elliptic or uniformly elliptic. Then discussions in the proof of Theorem 6.7.1 are valid for these jump-diffusions. Therefore the assertions of Theorem 6.7.1 are valid for elliptic and uniformly elliptic jump-diffusions. We will consider the density for pseudo-elliptic jump-diffusions.

Theorem 6.7.2 *Assume that the operator $A_J(t)$ is pseudo-elliptic. Let $p_{s,t}^{\mathbf{c},\mathbf{d}}(x, y)$ be the smooth density of transition function $P_{s,t}^{\mathbf{c},\mathbf{d}}(x, E)$ generated by $A_J^{\mathbf{c},\mathbf{d}}(t)$. Let α_0 be an arbitrary constant such that $\alpha < \alpha_0 < 2$, where α is the exponent of the Lévy measure ν. Let $\partial_x^{\mathbf{i}}$ and $\partial_y^{\mathbf{j}}$ be any given differential operators with indexes $\mathbf{i}$ and $\mathbf{j}$.*

1. *For any $T > 0$ and $M > 0$, there exists a positive constant c such that*

$$\left|\partial_x^{\mathbf{i}}\partial_y^{\mathbf{j}} p_{s,t}^{\mathbf{c},\mathbf{d}}(x, y)\right| \leq \frac{c}{(t-s)^{\frac{|\mathbf{i}|+|\mathbf{j}|+d}{2-\alpha_0}}}, \qquad \forall 0 \leq s < t \leq T \tag{6.66}$$

   holds for all $|x| \leq M$, $y \in \mathbb{R}^d$.
2. *If the operator $A_J(t)$ is uniformly pseudo-elliptic, for any $T > 0$ there exists $c > 0$ such that the above inequality holds for all $x, y \in \mathbb{R}^d$.*

A similar result is shown in Ishikawa–Kunita–Tsuchiya [46] in the case of jump processes. An interesting point in the inequality (6.66) is that the term $t^{-\beta'}$, where $\beta' = \frac{|\mathbf{i}|+|\mathbf{j}|+d}{2-\alpha_0}$, is common for all jump-diffusions and all Lévy measures with the same exponent $0 < \alpha < 2$ of the order condition, but the constant c should depend on coefficients $V_k(x, t), k = 1, \ldots, d'$, $\tilde{V}_k(x, t), k = 1, \ldots, d'$. It seems that $\alpha_0 = \alpha$ should be the critical value for the estimate (6.66), but we do not know whether the estimate is valid for $\alpha_0 = \alpha$.

It may be an interesting problem to find the short-time asymptotics for non-elliptic or non-pseudo-elliptic, but nondegenerate SDEs such as hypoelliptic SDEs. In such a case the exponent β' in $t^{-\beta'}$ in the inequality (6.66) should be changed. It seems that systematic results are not known.

The proof of Theorem 6.7.2 is not so simple as the proof of Theorem 6.7.1. We need four lemmas. Given $\alpha < \alpha_0 < 2$, let us choose $1 < \delta < \frac{\alpha_0}{\alpha}$. For a given $T > 0$, we set $\mathbb{T} = [0, T]$ and consider the Wiener–Poisson space of time parameter $\mathbb{T}$. With this constant δ, we consider δ-nondegenerate condition with respect to $\bar{Q}^F(\rho)$ defined by (5.146).

Lemma 6.7.1 *For any positive integer $N_0 > 2$, there exist $m, n \in \mathbb{N}$, $p \geq 2$ and a positive constant c such that for $|v| \geq 1$ the inequality*

$$\left|E[e^{i(v,F)}G]\right| \leq \frac{c}{|v|^{(1-\frac{\alpha\delta}{2})\kappa}}\Big(\Big\{\|DF\|^{\sharp}_{m,n,p} + \|\tilde{D}F\|^{\sharp}_{m,n,2(n+1)p}\Big\} \tag{6.67}$$
$$\times\Big\{1 + |v|^{-\frac{\alpha\delta}{2}}\|\tilde{D}F\|^{\sharp}_{m,n,2(n+1)p}\Big\}^n \sup_{(\rho,\theta)}\|\bar{Q}^F_\rho(\theta)^{-1}\|^{\sharp}_{m,n,p}\Big)^{\kappa}\|G\|^{\sharp}_{m,n,p}$$

holds for all $\kappa \in [1, N_0]$ and $G \in \bar{\mathbf{D}}^\infty$.

Proof We modify the proof of Theorem 5.11.1. Let us show that for any positive integer N, there exist m, n, p such that the above inequality holds with $\kappa = N$. Instead of (5.153), we use the inequality

$$\|e^{-i(v,\tilde{D}F)} - 1\|_{m,n,p;A(\rho)}$$
$$\leq c|v|\varphi(\rho)^{\frac{1}{2}}\|\tilde{D}F\|^{\sharp}_{m,n,(n+1)p}\Big\{1 + |v|^{-\frac{\alpha\delta}{2}}\|\tilde{D}F\|^{\sharp}_{m,n,(n+1)p}\Big\}^n$$

which follows from (5.129), replacing the term $|v|\rho\varphi(\rho)^{\frac{1}{2}}$ by the bigger term $|v|^{-\frac{\alpha\delta}{2}}$. Then the discussion in the proof of Theorem 5.11.1 leads to (6.67) with $\kappa = N$.

Since norms $\| \ \|^{\sharp}_{m,n,p}$ etc. are nondecreasing with respect to m, n, p, we can choose m, n, p, c such that the inequality of the lemma holds for any $N = 2, 3, \ldots, N_0$. Then by the interpolation, there exists another constant $c > 0$ such that the inequality of the lemma holds for all $\kappa \in [0, N_0]$. □

Let $0 \leq s < t \leq T$ and $\Phi_{s,t}(x)$ be the solution of a nondegenerate SDE (6.30) starting from x at time s. Let γ be a positive constant. We consider a random variable

$$H_{s,t,x} = \frac{1}{(t-s)^\gamma}\Phi_{s,t}(x).$$

We are interested in the decay property of characteristic functions of $H_{s,t,x}$, as $t \to s$. In the following three lemmas, the operator $A_J(t)$ is assumed to be pseudo-elliptic.

Lemma 6.7.2 *For any $M > 0$ and $\kappa_0 > 2$, there exist m, n, p and a positive constant c such that for any $|x| \leq M$ and $1 < \kappa < \kappa_0$ the inequality*

$$\left|E[e^{i(v,H_{s,t,x})}G]\right| \leq c\Big(\frac{(t-s)^{\gamma-1}}{|v|^{(1-\frac{\alpha\delta}{2})}}\Big)^{\kappa}\|G\|^{\sharp}_{m,n,p} \tag{6.68}$$

holds for all $0 \leq s < t \leq T$ and $|v| \geq (t-s)^{-\frac{2\gamma}{\alpha\delta}}$.

Further, If $A_J(t)$ is uniformly pseudo-elliptic, for any $\kappa_0 > 2$ there exist c, m, n, p such that for any $x \in \mathbb{R}^d$ and $1 < \kappa < \kappa_0$, the inequality (6.68) *holds for all $0 \leq s < t \leq T$ and $|v| \geq (t-s)^{-\frac{2\gamma}{\alpha\delta}}$.*

Proof We will apply Lemma 6.7.1, setting $F = H_{s,t,x}$. Consider the right-hand side of (6.67). Since

$$c_1 := \sup_{s<t, |x| \leq M} \left\{ \|D\Phi_{s,t}(x)\|^\sharp_{m,n,p} + \|\tilde{D}\Phi_{s,t}(x)\|^\sharp_{m,n,2(n+1)p} \right\} \tag{6.69}$$

is finite by Theorem 6.4.1, we have

$$\left\{ \|DH_{s,t,x}\|^\sharp_{m,n,p} + \|\tilde{D}H_{s,t,x}\|^\sharp_{m,n,2(n+1)p} \right\} \left\{ 1 + |v|^{-\frac{\alpha\delta}{2}} \|\tilde{D}H_{s,t,x}\|^\sharp_{m,n,2(n+1)p} \right\}^n \leq \frac{c_1}{(t-s)^\gamma},$$

for any $|v| \geq (t-s)^{-\frac{2\gamma}{\alpha\delta}}$ and $|x| \leq M$. Note that

$$\bar{Q}^{H_{s,t,x}}_\rho(\theta) = \frac{1}{(t-s)^{2\gamma}} \bar{Q}^{\Phi_{s,t}(x)}_{\rho'}(\theta), \quad \text{where } \rho' = (t-s)^{\delta\gamma}\rho. \tag{6.70}$$

Then we have

$$\|\bar{Q}^{H_{s,t,x}}_\rho(\theta)^{-1}\|^\sharp_{m,n,p} = (t-s)^{2\gamma} \|\bar{Q}^{\Phi_{s,t}(x)}_{\rho'}(\theta)^{-1}\|^\sharp_{m,n,p}. \tag{6.71}$$

We know by Proposition 6.5.1

$$\sup_{|x| \leq M, s<t} \sup_{(\rho',\theta)} (t-s) \|\bar{Q}^{\Phi_{s,t}(x)}_{\rho'}(\theta)^{-1}\|^\sharp_{m,n,p} \leq c_2. \tag{6.72}$$

Therefore we have

$$\sup_{|x| \leq M} \sup_{(\rho,\theta)} \|\bar{Q}^{H_{s,t,x}}_\rho(\theta)^{-1}\|^\sharp_{m,n,p} \leq c_2 (t-s)^{2\gamma-1}. \tag{6.73}$$

Then we get (6.68) from Lemma 6.7.1. □

Lemma 6.7.3 *For any $M > 0$ and $N_0 > 2$, there exist m, n, p and a positive constant c such that*

$$|E[e^{i(v,\Phi_{s,t}(x))}G]| \leq \begin{cases} \|G\|^\sharp_{m,n,p} & \text{if } v \in \mathbb{R}^d, \\ \dfrac{c}{(t-s)^{(1-\frac{\alpha\delta\gamma}{2})\kappa} |v|^{(1-\frac{\alpha\delta}{2})\kappa}} \|G\|^\sharp_{m,n,p}, & \text{if } |v| \geq (t-s)^{-c_0\gamma} \end{cases} \tag{6.74}$$

holds for all $1 < \kappa < N_0$, $0 \leq s < t \leq T$ and $|x| \leq M$, where $c_0 = \frac{2}{\alpha\delta} + 1$.

Further, if $A_J(t)$ is uniformly pseudo-elliptic, for any $N_0 > 2$, there exist m, n, p and a positive constant c such that (6.74) *holds for all $1 < \kappa < N_0$, $0 \le s < t \le T$ and $x \in \mathbb{R}^d$.*

Proof Since the equality $E[e^{i(v,\Phi_{s,t}(x))}G] = E[e^{i((t-s)^\gamma v, H_{s,t,x})}G]$ holds, we get from Lemma 6.7.2 the inequality

$$\begin{aligned}|E[e^{i(v,\Phi_{s,t}(x))}G]| &\le c\Big(\frac{(t-s)^{\gamma-1}}{((t-s)^\gamma|v|)^{(1-\frac{\alpha\delta}{2})}}\Big)^\kappa \|G\|^\sharp_{m,n,p}\\ &\le \frac{c}{(t-s)^{(1-\frac{\alpha\delta\gamma}{2})\kappa}|v|^{(1-\frac{\alpha\delta}{2})\kappa}}\|G\|^\sharp_{m,n,p},\end{aligned} \tag{6.75}$$

if $|v| \ge (t-s)^{-(\frac{2}{\alpha\delta}+1)\gamma}$, for any $|x| \le M$, $s < t$ and $1 \le \kappa \le N_0$. Further, if the SDE is uniformly nondegenerate at the center, the above holds for all $x \in \mathbb{R}^d$, $s < t$ and $1 \le \kappa \le N_0$. □

Lemma 6.7.4 *Let $\psi^{\mathbf{c},\mathbf{d}}_{s,t,x}(v)$ be the characteristic function of $P^{\mathbf{c},\mathbf{d}}_{s,t}(x,E)$. For any $M > 0$ and $N_0 > 2$, there exist positive constants c_3, c_4 such that*

$$\Big|\partial^{\mathbf{i}}_x\psi^{\mathbf{c},\mathbf{d}}_{s,t,x}(v)\Big| \le \begin{cases} c_3(1\vee|v|)^{|\mathbf{i}|}, & \text{if } v\in\mathbb{R}^d,\\ \dfrac{c_4}{(t-s)^{(1-\frac{\alpha\delta\gamma}{2})\kappa}}|v|^{|\mathbf{i}|-(1-\frac{\alpha\delta}{2})\kappa}, & \text{if } |v|\ge(t-s)^{-c_0\gamma}\end{cases} \tag{6.76}$$

holds for all $1 \le \kappa \le N_0$, $0 \le s < t \le T$ and $|x| \le M$.

If the SDE is uniformly nondegenerate at the center, there exist $k, p, c > 0$ such that the above holds for all $0 \le s < t \le T$, $x \in \mathbb{R}^d$ and $1 \le \kappa \le N_0$.

Proof The function $\psi^{\mathbf{c},\mathbf{d}}_{s,t,x}(v)$ is infinitely differentiable with respect to x. It holds that

$$\partial^{\mathbf{i}}_x\psi^{\mathbf{c},\mathbf{d}}_{s,t,x}(v) = E\Big[e^{i(v,\Phi_{s,t}(x))}G^{\mathbf{i}}_{s,t,x,v}\Big], \tag{6.77}$$

where $G^{\mathbf{i}}_{s,t,x,v}$ is written as a linear sum of terms written as (6.21). Then the inequality (6.63) holds for $\psi^{\mathbf{c},\mathbf{d}}_{s,t,x}$. Apply (6.75) for each $G = G^{\mathbf{i}_0,\ldots,\mathbf{i}_\mathbf{j}}_{s,t,x,\theta}$ and sum up for $\mathbf{i}_0,\ldots,\mathbf{i}_\mathbf{j}$. Then we get (6.76) for any $|x| \le M$, $s < t$ and $1 \le \kappa \le N_0$. The latter assertion will be obvious. □

Proof of Theorem 6.7.2 It is sufficient to prove (6.66) in the case $0 < t-s < 1$. We want to prove the inequality by applying the following formula:

$$\partial^{\mathbf{i}}_x\partial^{\mathbf{j}}_y p^{\mathbf{c},\mathbf{d}}_{s,t}(x,y) = (-i)^{|\mathbf{j}|}\Big(\frac{1}{2\pi}\Big)^d\int_{\mathbb{R}^d} e^{-i(v,y)}v^{\mathbf{j}}\partial^{\mathbf{i}}_x\psi^{\mathbf{c},\mathbf{d}}_{s,t,x}(v)\,dv.$$

Note that $(t-s)^{-c_0\gamma} > 1$. From Lemma 6.7.4, for any $M > 0$ there exist positive constants c_1, c_2 such that the inequality

$$\left|v^{\mathbf{j}}\partial_x^{\mathbf{i}}\psi_{s,t,x}^{\mathbf{c},\mathbf{d}}(v)\right| \leq \begin{cases} c_0, & \text{if } |v| \leq 1, \\ c_1|v|^{|\mathbf{i}|+|\mathbf{j}|}, & \text{if } 1 < |v| \leq (t-s)^{-c_0\gamma}, \\ \dfrac{c_2}{(t-s)^{(1-\frac{\alpha\delta\gamma}{2})\kappa}}|v|^{|\mathbf{i}|+|\mathbf{j}|-(1-\frac{\alpha\delta}{2})\kappa}, & \text{if } |v| > (t-s)^{-c_0\gamma} \end{cases} \tag{6.78}$$

holds for all $s < t$, $|x| < M$. Note that the right-hand side is an integrable function of v, if and only if κ satisfies $|\mathbf{i}|+|\mathbf{j}|-(1-\frac{\alpha\delta}{2})\kappa < -d$ or equivalently $\kappa > \frac{|\mathbf{i}|+|\mathbf{j}|+d}{1-\frac{\alpha\delta}{2}}$ holds.

We will take $\gamma = \gamma_0$ satisfying $\gamma_0 \geq (2-\alpha\delta) \vee \frac{1}{\alpha\delta}$. Next, set $\kappa_0 = \frac{|\mathbf{i}|+|\mathbf{j}|+d}{1-\frac{\alpha_0}{2}}$. Then the right-hand side of (6.78) is an integrable function of v. In fact, $|\mathbf{i}|+|\mathbf{j}|-(1-\frac{\alpha\delta}{2})\kappa_0 < -d$ holds, because $\alpha\delta < \alpha_0$. Further, if $|x| < M$, we have from (6.78)

$$\begin{aligned} \left|\int_{\mathbb{R}^d} e^{-i(v,y)}v^{\mathbf{j}}\partial_x^{\mathbf{i}}\psi_{s,t,x}^{\mathbf{c},\mathbf{d}}(v)\,dv\right| &\leq c_0\int_{|v|\leq 1} 1\,dv \\ &+ c_1\int_{|v|\leq(t-s)^{-c_0\gamma_0}}|v|^{|\mathbf{i}|+|\mathbf{j}|}\,dv + \frac{c_2}{(t-s)^{(1-\frac{\alpha\delta\gamma_0}{2})\kappa_0}}\int_{|v|\geq 1}|v|^{|\mathbf{i}|+|\mathbf{j}|-(1-\frac{\alpha\delta}{2})\kappa_0}\,dv \\ &\leq c_0' + \frac{c_1'}{(t-s)^{\frac{|\mathbf{i}|+|\mathbf{j}|+d}{c_0\gamma_0}}} + \frac{c_2'}{(t-s)^{(2-\alpha\delta\gamma_0)\frac{|\mathbf{i}|+|\mathbf{j}|+d}{2-\alpha_0}}}, \end{aligned} \tag{6.79}$$

since $\int_{|v|\geq 1}|v|^{|\mathbf{i}|+|\mathbf{j}|-(1-\frac{\alpha\delta}{2})\kappa_0}\,dv$ is finite. We have $\frac{|\mathbf{i}|+|\mathbf{j}|+d}{c_0\gamma_0} \leq \frac{|\mathbf{i}|+|\mathbf{j}|+d}{2-\alpha_0}$ because $\frac{1}{c_0\gamma_0} \leq \frac{1}{2-\alpha_0}$. Furthermore, $2-\alpha\delta\gamma_0 \leq 1$ holds because $\gamma_0 \geq \frac{1}{\alpha\delta}$. Therefore we have

$$\frac{1}{(t-s)^{\frac{|\mathbf{i}|+|\mathbf{j}|+d}{\gamma_0}}} \leq \frac{1}{(t-s)^{\frac{|\mathbf{i}|+|\mathbf{j}|+d}{2-\alpha\delta}}}, \qquad \frac{1}{(t-s)^{(2-\alpha\delta\gamma_0)\frac{|\mathbf{i}|+|\mathbf{j}|+d}{2-\alpha\delta}}} \leq \frac{1}{(t-s)^{\frac{|\mathbf{i}|+|\mathbf{j}|+d}{2-\alpha\delta}}}.$$

Consequently, we get the inequality

$$\begin{aligned} \sup_{y\in\mathbb{R}^d}|\partial_x^{\mathbf{i}}\partial_y^{\mathbf{j}}p_{s,t}^{\mathbf{c},\mathbf{d}}(x,y)| &= \left(\frac{1}{2\pi}\right)^d\int_{\mathbb{R}^d}|v^{\mathbf{j}}\partial_x^{\mathbf{i}}\psi_{s,t,x}^{\mathbf{c},\mathbf{d}}(v)|\,dv \\ &\leq \frac{c_3}{(t-s)^{\frac{|\mathbf{i}|+|\mathbf{j}|+d}{2-\alpha\delta}}}, \end{aligned} \tag{6.80}$$

for all $0 \leq s < t \leq T$, if $|x| < M$.

If $A_J(t)$ is uniformly pseudo-elliptic, we can take constants c_1, c_2 in (6.78) such that the inequality holds for all $0 \leq s < t \leq T$ and $x \in \mathbb{R}^d$. Then the inequality (6.80) holds for all $x, y \in \mathbb{R}^d$ and $0 \leq s < t \leq T$. Then the second assertion of the theorem follows. □

Remark If the generator of a jump-diffusion satisfying Condition (J.2) is elliptic, the same short-time estimate of (6.61) is valid for its transition function. Indeed, the proof of Theorem 6.7.1 can be applied to elliptic jump-diffusions satisfying Condition (J.2).

6.8 Off-Diagonal Short-Time Estimates of Density Functions

We shall next consider the *off-diagonal short-time asymptotics* of density functions for elliptic diffusions and pseudo-elliptic jump-diffusions. These properties are quite different between diffusions and jump-diffusions. We first consider a diffusion process.

Theorem 6.8.1 *Let $p^{\mathbf{c}}_{s,t}(x, y)$ be the smooth density of transition function $P^{\mathbf{c}}_{s,t}(x, E)$, where its generator $A^{\mathbf{c}}(t)$ is elliptic. Let $\partial_y^{\mathbf{j}}$ be any given differential operator of index $\mathbf{j}$. Suppose that U, V are bounded disjoint open subsets of $\mathbb{R}^d$ such that $d(\bar{U}, \bar{V}) > 0$. Then for any $T > 0$ and $N \in \mathbb{N}$, there exists a positive constant $c = c_{T,N,\mathbf{j},U,V}$ such that*

$$\sup_{x\in U, y\in V} \left|\partial_y^{\mathbf{j}} p^{\mathbf{c}}_{s,t}(x, y)\right| \le c(t-s)^N, \quad \forall 0 \le s < t \le T. \tag{6.81}$$

For the proof of the theorem, we will consider a diffusion $X_t^{x,s} = \Phi_{s,t}(x)$ pinned at the point $y \in \mathbb{R}^d$ at time t, or a diffusion conditioned by $\{X_t^{x,s} = y\}$. Later we will consider a similar problem for a jump-diffusion. So we will define the pinned process for a jump-diffusion.

We defined in Sect. 5.12 the composite of a tempered distribution and a non-degenerate smooth Wiener–Poisson functional. Let $\Phi_{s,t}(x)$ be the jump-diffusion determined by a pseudo-elliptic SDE (6.30) and let $\check{\Psi}_{s,t}(y) = \Phi_{s,t}^{-1}(y)$. Then both $\Phi_{s,t}(x)$ and $\check{\Psi}_{s,t}(y)$ are nondegenerate Wiener–Poisson functionals for any $s < t, x, y$. Let δ_x and δ_y be the delta functions of points x and y, respectively. Then the composite $\delta_y(\Phi_{s,t}(x))$ and $\delta_x(\check{\Psi}_{s,t}(y))$ are well defined as generalized Wiener–Poisson functionals. It holds that $p^{\mathbf{c},\mathbf{d}}_{s,t}(x, y) = E[\delta_y(\Phi_{s,t}(x))G_{s,t}(x)]$, where $G_{s,t}(x)$ is given by (6.53) (see Corollary 5.12.1).

Lemma 6.8.1 *We have*

$$E\Big[\prod_{i=1}^n f_i(\Phi_{s,t_i}(x))\delta_y(\Phi_{s,t}(x))\Big] = E\Big[\prod_{i=1}^n f_i(\check{\Psi}_{t_i,t}(y))\delta_x(\check{\Psi}_{s,t}(y))|\det \nabla\check{\Psi}_{s,t}(y)|\Big] \tag{6.82}$$

for any $s < t_1 < \cdots t_n < t$, $x, y \in \mathbb{R}^d$ and C_0^∞-functions $f_i, i = 1, \ldots, n$.

Proof Applying the change-of-variable formula and then taking expectations (Sect. 4.6), we get the dual formula

$$\int_{\mathbb{R}^d} \psi(x) E\Big[\prod_i f_i(\Phi_{s,t_i}(x))\varphi(\Phi_{s,t}(x))\Big] dx$$
$$= \int_{\mathbb{R}^d} E\Big[\prod_i f_i(\check{\Psi}_{t_i,t}(y))\psi(\check{\Psi}_{s,t}(y))|\det \nabla\check{\Psi}_{s,t}(y)|\Big]\varphi(y)\, dy$$

for C_0^∞-functions $f_i, i = 1, \ldots, n$ and φ, ψ. Let φ and ψ tend to δ_y and δ_x, respectively. Then we get the formula of the lemma. □

We need an another lemma. Lemma 6.2.1 is modified as follows, whose proof can be done similarly to the proof of Lemma 6.2.1.

Lemma 6.8.2 *Let U, V be bounded open subsets of $\mathbb{R}^d$ such that $d(\bar{U}, V) > 0$. Let $\tau(x, s)$ be the hitting time of the diffusion process $X_t = X_t^{x,s} := \Phi_{s,t}(x)$ to the set V. Then for any $N \geq 2$, there exists a positive constant c such that*

$$\sup_{x\in U} P(\tau(x, s) < t) \leq c(t - s)^N, \quad \forall s < t. \tag{6.83}$$

Proof of Theorem 6.8.1 We will consider the case $G_{s,t}(x) = 1$ only, since discussions for the case $G_{s,t}(x) > 0$ are similar. We first consider the case $\mathbf{j} = 0$. Let U, V be given sets in the assertion of Theorem 6.8.1. Take a bounded open set W which satisfies $\bar{U} \subset W \subset V^c, d(\bar{U}, W^c) > 0$ and $d(\bar{W}, \bar{V}) > 0$. Let $x \in U$ and $y \in V$. Let $\tau = \tau(x, s)$ be the first leaving time of the diffusion process $X_t = X_t^{x,s} = \Phi_{s,t}(x)$ from the set W. Then it holds that $P(\tau \leq t) \leq c(t - s)^N$ by Lemma 6.8.2. Since $d(\bar{W}, y) > 0$, we have

$$\begin{aligned} p_{s,t}(x, y) &= E\big[\delta_y(\Phi_{s,t}(x)); s < \tau \leq t\big] \\ &= E\big[\delta_y(\Phi_{s,t}(x)); s < \tau \leq (t + s)/2\big] + E\big[\delta_y(\Phi_{s,t}(x)); (t + s)/2 < \tau \leq t\big]. \end{aligned} \tag{6.84}$$

Using the strong Markov property, the first term of the right-hand side is estimated as

$$\begin{aligned} E\big[\delta_y(\Phi_{s,t}(x)); s < \tau \leq (t + s)/2\big] &= E\big[P_{\tau,t}\delta_y(\Phi_{s,\tau}(x)); s < \tau \leq (t + s)/2\big] \\ &= E\big[p_{\tau,t}(\Phi_{s,\tau}(x), y); s < \tau \leq (t + s)/2\big]. \end{aligned}$$

Note that $\Phi_{s,\tau} \in \bar{W}$. Then we have

$$p_{\tau,t}(\Phi_{s,\tau}(x), y) \leq c(t - \tau)^{-d_0} \leq c\Big(\frac{t - s}{2}\Big)^{-d_0}, \tag{6.85}$$

where $d_0 = \frac{d}{2}$, in view of Theorem 6.7.1. Therefore, we get

$$E\big[\delta_y(\Phi_{s,t}(x)); s < \tau \le (t+s)/2\big] \le c_1\Big(\frac{t-s}{2}\Big)^{-d_0} P(s < \tau \le (t+s)/2)$$
$$\le c_2(t-s)^{N-d_0}. \tag{6.86}$$

We shall next consider the second term of (6.84). It holds by Lemma 6.8.1 that

$$E[\delta_y(\Phi_{s,t}(x)); (t+s)/2 < \tau \le t] \tag{6.87}$$
$$= E[|\det \nabla\check{\Psi}_{s,t}(y)|\delta_x(\check{\Psi}_{s,t}(y)); (t+s)/2 < \tau \le t].$$

If $p_{s,t}(x, y) > 0$, the conditional probability measure given $\{\Phi_{s,t}(x) = y\}$ is defined by

$$P(A|\check{\Psi}_{s,t}(y) = x) = \frac{E[1_A\delta_x(\check{\Psi}_{s,t}(y))]}{p_{s,t}(x, y)}.$$

Under this measure, it holds that $\Phi_{s,r}(x) = \check{\Psi}_{r,t}(\Phi_{s,t}(x)) = \check{\Psi}_{r,t}(y)$ a.s. Then τ may be regarded as the last exit time of the backward process $\check{Y}_r := \check{\Psi}_{r,t}(y)$ $(t, y$ are fixed) from $\bar{W}$. Let σ_* be the first exit time of the same process $\check{Y}_r$ from $\bar{W}^c$. It is defined by $\sigma_* = \sup\{r;\ \check{Y}_r \in \bar{W}^c\}$. Then it holds that $t - \tau \ge t - \sigma_*$. We will apply Lemma 6.8.2 to the backward process $\check{Y}_r$. Then, similarly to (6.86), we have

$$E[\delta_y(\Phi_{s,t}(x)); (t+s)/2 < \tau \le t] \tag{6.88}$$
$$\le E[|\det \nabla\check{\Psi}_{s,t}(y)|\delta_x(\check{\Psi}_{s,t}(y)); (t+s)/2 < \sigma_* \le t] \le c_3(t-s)^{N-d_0}.$$

Substituting (6.86) and (6.88) to (6.84), we get the inequality (6.81) of the theorem if we rewrite $N - d_0$ as N.

Finally, if $\mathbf{j} \ne 0$, we consider the composite of the generalized function $(-1)^{\mathbf{j}}\partial_x^{\mathbf{j}}\delta_y$ and $\Phi_{s,t}(x)$. Then we get the assertion (6.81) similarly. □

In the case of a jump-diffusion, the problem is more complicated, since Lemma 6.8.2 does not hold. In order that the inequality (6.83) is valid, the distance of two disjoint sets U and V should be big so that the point $x \in U$ reaches the point $y \in V$ after sufficiently many jumps. It is explained below. Let $\{\phi_{r,z}\}$ be jump-maps of the jump-diffusion. Let U be an open subset of $\mathbb{R}^d$. In Sect. 6.5, we defined the set $\bar{\phi}(U) = \overline{\cup\phi_{r,z}(U)}$ etc.

The following is a consequence of Lemma 6.5.2.

Lemma 6.8.3 *Let N be a positive integer satisfying $d(\bar{\phi}^N(U), \phi^{-N}(V)) > 0$. Let $\tau = \tau(x, s)$ be the hitting time of the process $X_t^{x,s} := \Phi_{s,t}(x), t \in [s, \infty)$ to the set V. Then there exists $c > 0$ such that*

$$\sup_{x\in U} P(\tau(x, s) < t) \le c(t-s)^N, \quad \forall s < t. \tag{6.89}$$

Theorem 6.8.2 *Let $p_{s,t}^{\mathbf{c},\mathbf{d}}(x, y)$ be the smooth density of transition function $P_{s,t}^{\mathbf{c},\mathbf{d}}(x, E)$, where its generator $A_J^{\mathbf{c},\mathbf{d}}(t)$ is pseudo-elliptic. Let α_0 be an arbitrary positive constant such that $\alpha < \alpha_0 < 2$, where α is the exponent of the Lévy measure. Suppose that U, V are bounded open sets of $\mathbb{R}^d$ such that $d(\bar{\phi}^N(U), \bar{\phi}^{-N}(V)) > 0$ holds for some positive integer N. Then for any $T > 0$ there exists a positive constant $c = c_{T,\mathbf{j},U,V}$ such that*

$$\sup_{x\in U, y\in V} \left|\partial_y^{\mathbf{j}} p_{s,t}^{\mathbf{c},\mathbf{d}}(x, y)\right| \leq c(t-s)^{N-\frac{|\mathbf{j}|+d}{2-\alpha_0}}, \quad \forall 0 \leq s < t \leq T. \tag{6.90}$$

Proof We will consider the case $G_{s,t}(x) = 1$, i.e., $\mathbf{c} = 0, \mathbf{d} = 1$, since discussions for the case $G_{s,t}(x) \neq 1$ are similar. We denote $p_{s,t}^{\mathbf{c},\mathbf{d}}(x, y)$ by $p_{s,t}(x, y)$. We first consider the case $\mathbf{j} = 0$. Let $x \in U$ and $y \in V$. Let τ be the first leaving time of the process $X_t = \Phi_{s,t}(x)$ from $\bar{\phi}^N(U)$. Then it holds that $P(\tau \leq r) \leq c(r-s)^N$ by Lemma 6.8.3. Further, we have

$$\begin{aligned} p_{s,t}(x, y) &= E\big[\delta_y(\Phi_{s,t}(x)); s < \tau \leq t\big] \\ &= E\big[p_{\tau,t}(\Phi_{s,\tau}(x), y) : s < \tau \leq (t+s)/2\big] \\ &\quad + E[|\det \nabla\check{\Psi}_{s,t}(y)|\delta_x(\check{\Psi}_{s,t}(y)); (t+s)/2 < \tau < t], \end{aligned} \tag{6.91}$$

as in the proof of the previous theorem. Note that $\Phi_{s,\tau}(x) \in \bar{\phi}^N(U)$ and $\bar{\phi}^N(U)$ is a bounded set, because $g(x, t, z)$ is a bounded function. Then we have

$$p_{\tau,t}(\Phi_{s,\tau}(x), y) \leq c(t-\tau)^{-d_0} \leq c\Big(\frac{t-s}{2}\Big)^{-d_0},$$

where $d_0 = \frac{d}{2-\alpha_0}$, in view of Theorem 6.7.2. Therefore, we get

$$\begin{aligned} E\big[p_{\tau,t}(\Phi_{s,\tau}(x), y) : s < \tau \leq (t+s)/2\big] &\leq c_1\Big(\frac{t-s}{2}\Big)^{-d_0} P(s < \tau \leq (t+s)/2) \\ &\leq c_2(t-s)^{N-d_0}. \end{aligned}$$

We shall next consider the last term of (6.91). Under the conditional measure $P(A|\check{\Psi}_{s,t}(y) = x)$, it holds that $\Phi_{s,r}(x) = \check{\Psi}_{r,t}(\Phi_{s,t}(x)) = \check{\Psi}_{s,t}(y)$ a.s. Then τ may be regarded as the last exit time of the left continuous backward process $\check{Y}_s = \check{\Psi}_{s-,t}(y)$ (t is fixed) from $\bar{\phi}^{-N}(U^c)$. Let σ_* be the first exit time of the same process $\check{Y}_s$ from $\bar{\phi}^{-N}(U^c)$. It is defined by $\sigma_* = \sup\{s;\ \check{Y}_s \in \bar{\phi}^{-N}(V)\}$. Then it holds that $t - \tau \geq t - \sigma_*$, since $d(\bar{\phi}^N(U), \bar{\phi}^{-N}(V)) > 0$. We will apply Lemma 6.8.1 to the backward process $\check{\Psi}_{s,t}(y)$ (t being fixed). Then, similarly to (6.86), we have

$$E[|\det \nabla\check{\Psi}_{s,t}(y)|\delta_x(\check{\Psi}_{s,t}(y)); (t+s)/2 < \sigma_* < t] \leq c_3(t-s)^{N-d_0}.$$

Substituting these two inequalities into (6.91), we get the inequality (6.90) of the theorem in the case $|\mathbf{j}| = 0$.

Finally, if $\mathbf{j} \neq 0$, we consider the composite $(-1)^{\mathbf{j}}\partial_x^{\mathbf{j}}\delta_y$ and $\Phi_{s,t}(x)$. Then we get the assertion (6.90) similarly. □

Note The short-time estimate and the off-diagonal short-time estimate for diffusion processes seems to be known more or less. These estimates for jump processes are studied by Picard [93] and Ishikawa [43], not using the Malliavin calculus. Ishikawa et al. [46] obtained the short-time estimates for jump processes using the Malliavin calculus. Their results are close to our Theorem 6.7.1. The present off-diagonal short-time estimate using the Malliavin calculus might be new.

6.9 Densities for Processes with Big Jumps

Let us return to SDE (6.30). For drift and diffusion coefficients $V_0(x,t)$, $V_1(x,t)$, ..., $V_{d'}(x,t)$, we assumed that they are of $C_b^{\infty,1}$-class; for jump coefficients $\phi_{t,z}$, we assumed that they satisfy Conditions (J.1) and (J.2) stated in Sect. 3.2. In this section we will assume the same condition for drift and diffusion coefficients. But for jump coefficients, we will assume Conditions $(J.1)_K$ and $(J.2)_K$ stated in Sect. 3.9.

For the Lévy measure ν, we assume again that it has a weak drift and that at the center it satisfies the order condition of exponent $0 < \alpha < 2$ with respect to a family of star-shaped neighborhoods. We assume further that its generator is pseudo-elliptic. We should remark that the pseudo-elliptic property is concerned with the diffusion part and small jumps only, since vector fields $\tilde{V}_k(t), k = 1, \ldots, d'$ are determined by jump-maps $\phi_{r,z}$; $|z| < \gamma$ for arbitrary small $\gamma > 0$.

We are again interested in the existence of the smooth density of the law of the solution $\Phi_{s,t}(x)$. However, since equation (6.30) may admit big jumps, the solution may not belong to $L^{\infty-}$. So we cannot apply the Malliavin calculus to the solution directly. To avoid this difficulty, we will first consider an SDE where big jumps are truncated, for which the smooth density of the law exists. Then we will adjoin big jumps as a perturbation and show that such a perturbation should preserve the smoothness of the density.

In this section, we will restrict our attention to the law of the solution $\Phi_{s,t}(x)$ and will avoid some complicated discussions and notations related the weighted laws. Our discussions will be extended to the weighted law of $\Phi_{s,t}(x)$ as in Sects. 6.2, 6.3, 6.4, 6.5, 6.6, 6.7, and 6.8.

Suppose we are given an SDE (6.30), where coefficients of the equation satisfy Conditions $(J.1)_K$ and $(J.2)_K$. We will consider an SDE *truncating jumps which are bigger than* $\delta > 0$. The truncated equation is given by

$$X_t = X_0 + \sum_{k=0}^{d'} \int_{t_0}^{t} V_k(X_r, r) \circ dW_r^k + \int_{t_0}^{t} \int_{\delta \geq |z| > 0+} \{\phi_{r,z}(X_{r-}) - X_{r-}\} N(dr\,dz). \tag{6.92}$$

The solutions of the above equation define a stochastic flow of diffeomorphisms $\{\bar{\Phi}_{s,t}\}$ by Theorem 3.9.1, since the truncated function $g(x,t,z)1_{|z|\le\delta}$ satisfies Conditions (J.1) and (J.2).

For a given $T>0$, we set $\mathbb{T}=[0,T]$. We will fix $0\le s<t\le T$ and a point x. We shall construct the solution of equation (6.30) starting from x at time s, by adding big jumps to $\bar{\Phi}_{s,t}(x)$. Let $\mathbb{U}=\mathbb{T}\times\mathbb{R}_0^{d'}$ and $\mathbf{u}=((t_1,z_1),\ldots,(t_n,z_n))$ be an element of $\mathbb{U}^n$ $(n\ge 1)$ such that $|z_i|>\delta,\ i=1,\ldots,n$ and $s<t_1<\cdots<t_n\le t$. We consider $\bar{\Phi}^{\mathbf{u}}_{s,t}(x):=\bar{\Phi}_{s,t}(x)\circ\varepsilon^+_{\mathbf{u}}$. It is represented by

$$\bar{\Phi}^{\mathbf{u}}_{s,t}(x)=\bar{\Phi}_{t_n,t}\circ\phi_{t_n,z_n}\circ\cdots\circ\bar{\Phi}_{t_1,t_2}\circ\phi_{t_1,z_1}\circ\bar{\Phi}_{s,t_1}(x), \tag{6.93}$$

as in (6.32). We set $\bar{\Phi}^{\phi}_{s,t}(x)=\bar{\Phi}_{s,t}(x)$ if $\mathbf{u}=\emptyset\in\mathbb{U}^0$ (empty set). Now, let $q(\omega)$ be a Poisson point process on $\mathbb{U}'=[s,T]\times\mathbb{R}_0^{d'}$ with intensity measure $dt\,\nu_\delta(dz)$, where $\nu_\delta(dz)=1_{(\delta,\infty)}(|z|)\nu(dz)$. Let $N(dr\,dz)$ be the associated Poisson random measure (Sect. 1.4). Let D_q be the domain of the point process q. Since $\lambda_\delta\equiv\nu(\{|z|>\delta\})$ is finite, the set $D_q\cap(s,t]$ can be written, if it is non-empty, as $\{s=\tau_0<\tau_1<\cdots<\tau_n\}$, where τ_m are stopping times defined inductively by $\tau_0=s$ and by $\tau_m=\inf\{t>\tau_{m-1};\ N_t-N_{\tau_{m-1}}\ge 1\}$ $(=\infty$ if the set $\{\cdots\}$ is empty). Here $N_t:=N((s,t]\times\{|z|>\delta\})$ is a Poisson process with intensity λ_δ. We set $\mathbf{u}(q)=\emptyset$ if $\tau_1>t$ and $\mathbf{u}(q)=((\tau_1,q(\tau_1)),\ldots,(\tau_n,q(\tau_n)))$ if $\tau_n\le t<\tau_{n+1}$. Then we define $\Phi_{s,t}(x):=\bar{\Phi}^{\mathbf{u}(q)}_{s,t}(x)$. It is written as

$$\begin{cases}\bar{\Phi}_{s,t}(x), & \text{if } t<\tau_1,\\ \bar{\Phi}_{\tau_n,t}\circ\phi_{\tau_n,q(\tau_n)}\circ\cdots\circ\bar{\Phi}_{\tau_1,\tau_2}\circ\phi_{\tau_1,q(\tau_1)}\circ\bar{\Phi}_{s,\tau_1}(x), & \text{if } \tau_n\le t<\tau_{n+1}.\end{cases}$$

Therefore $\Phi_{s,t}(x)$ is a solution of the original equation (6.30) starting from x at time s (scc Sect. 3.2 in Chap. 3).

Now, we assume that the generator of the process $\Phi_{s,t}(x)$ is pseudo-elliptic. Then the law of $\bar{\Phi}_{s,t}(x)$; $\bar{P}_{s,t}(x,E)=P(\bar{\Phi}_{s,t}(x)\in E)$ has a rapidly decreasing C^∞-density by Theorem 6.6.1. We denote it by $\bar{p}_{s,t}(x,y)$. Further, the stochastic process $X_t=\bar{\Phi}^{\mathbf{u}}_{s,t}(x)$ is a Markov process with fixed discontinuity, jumping from $\bar{\Phi}^{\mathbf{u}}_{s,t_i-}$ to $\phi_{t_i,z_i}(\bar{\Phi}^{\mathbf{u}}_{s,t_i-})$ at fixed times t_i; $i=1,\ldots,n$. Its transition probability has a C^∞-density given by

$$\bar{p}^{\mathbf{u}}_{s,t}(x,y)=\int\cdots\int_{(\mathbb{R}^d)^n}\bar{p}_{s,t_1}(x,x_1)\bar{p}_{t_1,t_2}(\phi_{t_1,z_1}(x_1),x_2)\times\cdots$$
$$\times\ \bar{p}_{t_n,t}(\phi_{t_n,z_n}(x_n),y)\,dx_1\cdots dx_n. \tag{6.94}$$

Further, setting $\bar{p}^{\phi}_{s,t}(x,y)=\bar{p}_{s,t}(x,y)$, we have

$$P(\Phi_{s,t}(x)\in E)=P(\bar{\Phi}^{\mathbf{u}(q)}_{s,t}\in E)=\int_E E[\bar{p}^{\mathbf{u}(q)}_{s,t}(x,y)]\,dy.$$

Therefore, the function defined by

$$p_{s,t}(x, y) \equiv E[\bar{p}_{s,t}^{\mathbf{u}(q)}(x, y)] \tag{6.95}$$

is the density functions of transition probability $P_{s,t}(x, E)$ of $\Phi_{s,t}(x)$.

We want to prove the smoothness of the above $p_{s,t}(x, y)$ under additional conditions.

Theorem 6.9.1 *Consider a system of uniformly pseudo-elliptic jump-diffusions, which satisfies Conditions $(J.1)_K$ and $(J.2)_K$.*

1. *Assume that $\{\phi_{t,z}\}$ satisfies*

$$\sup_{t,x} \int_{|z|>\delta} \left| \det \nabla \phi_{t,z}^{-1}(x) \right| \nu(dz) < \infty \quad \textit{for some } \delta > 0, \tag{6.96}$$

where $\det \nabla \phi_{t,z}^{-1}(x)$ is the Jacobian determinant of the inverse map $\phi_{t,z}^{-1}$. Then the function $p_{s,t}(x, y)$ given by (6.95) is a continuous function of $x, y \in \mathbb{R}^d$ for any $0 \le s < t \le T$. Further, for any $\alpha < \alpha_0 < 2$, there exists $c_0 > 0$ such that the following holds for all $0 \le s < t \le T$ and $x, y \in \mathbb{R}^d$:

$$p_{s,t}(x, y) \le \frac{c_0}{(t-s)^{\frac{d}{2-\alpha_0}}}. \tag{6.97}$$

2. *Let m_0 be a positive integer. Assume further that*

$$\sup_{t,x} \int_{|z|>\delta} \left| \partial_x^{\mathbf{i}} \phi_{t,z}(x) \right|^p \nu(dz) < \infty \quad \textit{for some } \delta > 0 \tag{6.98}$$

holds for $\partial_x^{\mathbf{i}}$ with $1 \le |\mathbf{i}| \le m_0$ and $p > 1$. Then $p_{s,t}(x, y)$ is a $C_b^{m_0}$-function of x. Further, there exists $c_{m_0} > 0$ such that the following holds for all $0 \le s < t \le T$ and $x, y \in \mathbb{R}^d$ if $|\mathbf{i}| \le m_0$:

$$\left| \partial_x^{\mathbf{i}} p_{s,t}(x, y) \right| \le \frac{c_{m_0}}{(t-s)^{\frac{|\mathbf{i}|+d}{2-\alpha_0}}}. \tag{6.99}$$

Conditions (6.96) and (6.98) indicate that derivatives of jump-maps $\phi_{r,z}^{-1}$ and $\phi_{r,z}$ should not be big but moderate, though $\phi_{r,z}$ can be big. Loosely, the image of a ball by these maps should not be too distorted.

We discuss two lemmas. Let $\mathbf{u} = \{(t_1, z_1), \ldots, (t_n, z_n)\}$, where $s \le t_1 < \cdots < t_n \le t$. We write it as $\mathbf{u} = (\mathbf{t}, \mathbf{z})$, where $\mathbf{t} = (t_1, \ldots, t_n)$ and $\mathbf{z} = (z_1, \ldots, z_n) \in (\mathbb{R}_0^{d'})^n$. We define a probability measure on $(\mathbb{R}_0^{d'})^n$ by $\nu_\delta^{(n)}(d\mathbf{z}) = \nu_\delta(dz_1) \cdots \nu_\delta(dz_n)$, where $\nu_\delta(dz) = \frac{1_{|z|>\delta}}{\lambda_\delta} \nu(dz)$ and $\lambda_\delta = \nu(\{|z| > \delta\})$. We fix $\delta > 0$ stated in the theorem and we set for $t \in \mathbb{T}$ and $x, y \in \mathbb{R}^d$

$$\bar{p}^{\mathbf{t}}_{s,t}(x,y) := \int_{(\mathbb{R}^{d'}_0)^n} \bar{p}^{(\mathbf{t},\mathbf{z})}_{s,t}(x,y)\nu^{(n)}_\delta(d\mathbf{z}),$$

where $\bar{p}^{(\mathbf{t},\mathbf{z})}_{s,t}(x,y)$ is defined by (6.94). Then we get the formula

$$\begin{aligned} p_{s,t}(x,y) &= \sum_{n=0}^{\infty} E\big[\bar{p}^{\mathbf{u}(q)}_{s,t}(x,y)1_{\tau_n \le t < \tau_{n+1}}\big] \qquad (6.100) \\ &\le e^{-\lambda_\delta(t-s)}\bar{p}_{s,t}(x,y) + \sum_{n=1}^{\infty} \frac{(\lambda_\delta(t-s))^n}{n!} e^{-\lambda_\delta(t-s)} \sup_{\mathbf{t}\in(s,t]^n_+} \bar{p}^{\mathbf{t}}_{s,t}(x,y), \end{aligned}$$

where $(s,t]^n_+$ is the set of all $\mathbf{t}=(t_1,\ldots,t_n)$ such that $s < t_1 < \cdots < t_n \le t$. We will show that the above infinite sum is uniformly convergent with respect to x, y under additional conditions for jump coefficients $\phi_{t,z}(x)$.

We set

$$\bar{P}^{\mathbf{t}}_{s,t}f(x) = \int_{\mathbb{R}^d} \bar{p}^{\mathbf{t}}_{s,t}(x,y)f(y)\,dy.$$

We define the supremum norm by $|f|_\infty = \sup_x |f(x)|$.

Lemma 6.9.1 *Assume condition* (6.98). *For any given nonnegative integer* m_0, *there exists a positive constant* C *such that for any* $f \in C^{m_0}_b$, *inequalities*

$$\sup_{s<t}\sup_{\mathbf{t}\in(s,t]^n_+} \left|\partial^{\mathbf{i}}\bar{P}^{\mathbf{t}}_{s,t}f\right|_\infty \le C^{n+1}\sum_{\mathbf{i}'\subset\mathbf{i}} |\partial^{\mathbf{i}'}f|_\infty, \quad n=0,1,2,\ldots \qquad (6.101)$$

hold for any $\mathbf{i}$ *with* $|\mathbf{i}| \le m_0$.

Proof If $\mathbf{i}=0$, inequality (6.101) is obvious. We will show the inequality in the case $|\mathbf{i}|=1$. We first consider the case $n=0$. Since $\bar{P}_{s,t}f(x) = E[f(\bar{\Phi}_{s,t}(x))]$, we have

$$\partial\bar{P}_{s,t}f(x) = E[\partial(f\circ\bar{\Phi}_{s,t}(x))] = E[\partial f\circ\bar{\Phi}_{s,t}(x)\nabla\bar{\Phi}_{s,t}(x)].$$

We know $\sup_{x,s<t} E[|\nabla\bar{\Phi}_{s,t}(x)|^p] < \infty$ for any $p > 1$. Therefore we get the inequality $\sup_{t\in\mathbb{T}} |\partial\bar{P}_{s,t}f|_\infty \le c|\partial f|_\infty$.

We next show the inequality for $n \ge 1$. We will rewrite $\bar{P}^{\mathbf{t}}_{s,t}f$ as $\bar{P}^{\mathbf{t}}_{s,t}f = \int_{\mathbb{R}^{d'}_0} \bar{P}_{s,t_1}(\bar{P}^{\mathbf{t}'}_{t_1,t}f\circ\phi_{t_1,z_1})\nu_\delta(dz_1)$, where $\mathbf{t}'=(t_2,\ldots,t_n)$. Then we get from the above inequality

$$
\begin{aligned}
\sup_x \left|\partial \bar{P}^{\mathbf{t}}_{s,t} f(x)\right| &\le c \sup_x \int_{\mathbb{R}^{d'}_0} |\partial(\bar{P}^{\mathbf{t}'}_{t_1,t} f \circ \phi_{t_1,z_1})(x)| \nu_\delta(dz_1) \\
&\le c \sup_x \int_{\mathbb{R}^{d'}_0} |\partial(\bar{P}^{\mathbf{t}'}_{t_1,t} f) \circ \phi_{t_1,z_1}(x)||\partial \phi_{t_1,z_1}(x)| \nu_\delta(dz_1) \\
&\le c\Big(\sup_x \int_{\mathbb{R}^{d'}_0} |\partial \phi_{t_1,z_1}(x)| \nu_\delta(dz_1)\Big) \sup_x |\partial \bar{P}^{\mathbf{t}'}_{t_1,t} f(x)| \\
&\le c_1 |\partial \bar{P}^{\mathbf{t}'}_{t_1,t} f|_\infty .
\end{aligned}
$$

Therefore we obtain $|\partial \bar{P}^{t_1,\dots,t_n}_{s,t} f|_\infty \le C_1 |\partial \bar{P}^{t_2,\dots,t_n}_{t_1,t} f|_\infty$. Repeating this argument, we get the inequality of the lemma. □

In the following discussion we will use the short-time estimate of Theorem 6.7.2.

Lemma 6.9.2 *If conditions* (6.96) *and* (6.98) *are satisfied, for any given nonnegative integer m_0, there exists a positive constant C such that for any differential operator $\partial^{\mathbf{i}}_x$ with $|\mathbf{i}| \le m_0$, inequalities*

$$
\sup_{x,y\in\mathbb{R}^d,\mathbf{t}\in(s,t]^n_+} \left|\partial^{\mathbf{i}}_x \bar{p}^{\mathbf{t}}_{s,t}(x,y)\right| \le C^{n+1}(n+1)^{\beta_{\mathbf{i}}}(t-s)^{-\beta_{\mathbf{i}}}, \quad n = 1, 2, \dots \tag{6.102}
$$

hold for any $s < t$, where $\beta_{\mathbf{i}} = \frac{|\mathbf{i}|+d}{2-\alpha_0}$.

Proof We will consider the case where $i \ge 1$. The case $i = 0$ will be proved similarly. We set $s = t_0$ and $t = t_{n+1}$, conventionally. There exists $0 \le k \le n$ such that $t_{k+1} - t_k \ge \frac{1}{n+1}(t-s)$. We consider the case $1 \le k \le n$. The case $k = 0$ will be treated similarly. We rewrite $\bar{p}^{\mathbf{t}}_{s,t}$ as

$$
\bar{p}^{\mathbf{t}}_{s,t}(x,y) = \int_{\mathbb{R}^d}\int_{\mathbb{R}^{d'}_0} \bar{p}^{\mathbf{t_1}}_{s,t_k}(x,x_k)\bar{p}^{\mathbf{t_2}}_{t_k,t}(\phi_{t_k,z_k}(x_k),y)\, dx_k \nu_\delta(dz_k),
$$

where $\mathbf{t} = (t_1,\dots,t_n)$, $\mathbf{t_1} = (t_1,\dots,t_{k-1})$ and $\mathbf{t_2} = (t_{k+1},\dots,t_n)$. We set $f(x_k) = \int \bar{p}^{\mathbf{t_2}}_{t_k,t}(\phi_{t_k,z_k}(x_k),y)\nu_\delta(dz_k)$ (y being fixed) and apply the previous lemma. Then we have

$$
\sup_x \left|\partial^{\mathbf{i}}_x \bar{p}^{\mathbf{t}}_{s,t}(x,y)\right| \le C^k \sum_{|\mathbf{i}'|=|\mathbf{i}|} \sup_{x_k} \left|\int_{\mathbb{R}^{d'}_0} \partial^{\mathbf{i}'}_{x_k} \bar{p}^{\mathbf{t_2}}_{t_k,t}(\phi_{t_k,z_k}(x_k),y)\nu_\delta(dz_k)\right| \tag{6.103}
$$

for any $s < t$ and $\mathbf{t} \in (s,t]^n_+$. Set $\mathbf{t_3} = (t_{k+2},\dots,t_n)$. We apply Theorem 6.7.2 for $\bar{p}_{s,t}(x,y)$ in place of $p^{\mathbf{c}}_{s,t}(x,y)$. Setting $k+1 = l$, we have

$$
\begin{aligned}
&\left|\partial_{x_k}^{\mathbf{i}'}\bar{p}_{t_k,t}^{\mathbf{t}_2}(\phi_{t_k,z_k}(x_k),y)\right|\\
&\le \int_{\mathbb{R}^d}\int_{\mathbb{R}_0^{d'}}\left|\partial_{x_k}^{\mathbf{i}'}\bar{p}_{t_k,t_l}(\phi_{t_k,z_k}(x_k),x_l)\right|\bar{p}_{t_l,t}^{\mathbf{t}_3}(\phi_{t_l,z_l}(x_l),y)\,dx_l\nu_\delta(dz_l)\\
&\le C_1(t_l-t_k)^{-\beta_{\mathbf{i}}}\int_{\mathbb{R}^d}\int_{\mathbb{R}_0^{d'}}\bar{p}_{t_l,t}^{\mathbf{t}_3}(\phi_{t_l,z_l}(x_l),y)\,dx_l\nu_\delta(dz_l)\\
&\le C_1(t_l-t_k)^{-\beta_{\mathbf{i}}}\int_{\mathbb{R}^d}\int_{\mathbb{R}_0^{d'}}\bar{p}_{t_l,t}^{\mathbf{t}_3}(x_l',y)|\det\nabla\phi_{t_l,z_l}^{-1}(x_l')|\,dx_l'\nu_\delta(dz_l)\\
&\le C_1(t_l-t_k)^{-\beta_{\mathbf{i}}}\Big(\sup_x\int_{\mathbb{R}_0^{d'}}|\det\nabla\phi_{t_l,z_l}^{-1}(x)|\nu_\delta(dz_l)\Big)\int_{\mathbb{R}^d}\bar{p}_{t_l,t}^{\mathbf{t}_3}(x_l',y)\,dx_l'.
\end{aligned}
$$

In view of condition (6.96), we have $\sup_{t_l,x}\int|\det\nabla\phi_{t_l,z_l}^{-1}(x)|\nu_\delta(dz_l)<\infty$. Since $\sup_{t_l,y}\int\bar{p}_{t_l,t}^{\mathbf{t}_3}(x_l',y)\,dx_l'<\infty$, the above is dominated by $C_2(t_l-t_k)^{-\beta_{\mathbf{i}}}$. Note $(t_l-t_k)\ge\frac{1}{n+1}(t-s)$. Then from (6.103), we can conclude that there exists a positive constant C_3 such that

$$
\sup_{x,y}|\partial_x^{\mathbf{i}}\bar{p}_{s,t}^{\mathbf{t}}(x,y)|\le C_3^{n+1}(n+1)^{\beta_{\mathbf{i}}}(t-s)^{-\beta_{\mathbf{i}}},\quad n=1,2,\ldots
$$

holds for any $s<t$ and $\mathbf{t}\in(s,t]_+^n$. □

Next, we give a simple lemma, which will be used in the proof of Theorem 6.9.1.

Lemma 6.9.3 *Let D be a domain of $\mathbb{R}^d$ and let $\{f_n\}$ be a sequence of functions in $C^{m_0}(D)$. Assume that the sequences $\{f_n\}$ and $\{\partial^{\mathbf{i}}f_n\}$, $|\mathbf{i}|\le m_0$ converge uniformly to f and $f^{\mathbf{i}}$, respectively. Then f belongs to $C^{m_0}(D)$ and the equality $\partial^{\mathbf{i}}f=f^{\mathbf{i}}$ holds for any $|\mathbf{i}|\le m_0$.*

The proof of the lemma is straightforward. It is omitted.

Proof of Theorem 6.9.1 We will prove assertion 2, assuming that the generator is uniformly pseudo-elliptic. Consider

$$
\begin{aligned}
p_{s,t}^{\mathbf{i}}(x,y)&:=E[\partial_x^{\mathbf{i}}\bar{p}_{s,t}^{\mathbf{u}(q)}(x,y)]\\
&=\sum_{n=0}^{\infty}E\big[\partial_x^{\mathbf{i}}\bar{p}_{s,t}^{\mathbf{u}(q)}(x,y)1_{\tau_n\le t<\tau_{n+1}}\big].
\end{aligned}\tag{6.104}
$$

If $|\mathbf{i}|\le m_0$, it holds by Lemma 6.9.2 that

$$
\begin{aligned}
&\big|E\big[\partial_x^{\mathbf{i}}\bar{p}_{s,t}^{\mathbf{u}(q)}(x,y)1_{\tau_n\le t<\tau_{n+1}}\big]\big|\\
&\le\sup_{\mathbf{t}\in\mathbb{T}^n}\left|\partial_x^{\mathbf{i}}\bar{p}_{s,t}^{\mathbf{t}}(x,y)\right|\cdot\frac{(\lambda_\delta(t-s))^n}{n!}e^{-\lambda_\delta(t-s)}
\end{aligned}\tag{6.105}
$$

$$\leq \frac{(C\lambda_\delta(t-s))^n(n+1)^{\beta_{\mathbf{i}}}}{n!}e^{-\lambda_\delta(t-s)}(t-s)^{-\beta_{\mathbf{i}}}.$$

The infinite sum of the last term of (6.105) with respect to $n \in \mathbb{N}$ is convergent. Then the infinite sum (6.104) is uniformly convergent with respect to x, y. Therefore, the limit function $p^{\mathbf{i}}_{s,t}(x, y)$ is bounded and continuous in x, y for any $|\mathbf{i}| \leq m_0$. Consequently, $p_{s,t}(x, y) := p^0_{s,t}(x, y)$ is the continuous density of $P_{s,t}(x, E)$.

Further, $p_{s,t}(x, y)$ is m_0-times continuously differentiable with respect to x and the equality $\partial^{\mathbf{i}}_x p_{s,t}(x, y) = p^{\mathbf{i}}_{s,t}(x, y)$ holds for any $|\mathbf{i}| \leq m_0$ in view of Lemma 6.94. Since (6.105) holds, the infinite sum $p^{\mathbf{i}}_{s,t}(x, y)$ satisfies the inequality $|p^{\mathbf{i}}_{s,t}(x, y)| \leq c_{\mathbf{i}} t^{-\beta_{\mathbf{i}}}$.

Other assertions will be shown similarly. □

Theorem 6.9.2 *In Theorem* 6.9.1, *assume further that for some* $\delta > 0$, *inequality*

$$\sup_{t,y} \int_{|z|>\delta} \left|\partial^{\mathbf{j}}_y \phi^{-1}_{t,z}(y)\right|^p \nu(dz) < \infty \tag{6.106}$$

holds for any $1 \leq |\mathbf{j}| \leq n_0$ *and* $p > 1$. *Then* $p_{s,t}(x, y)$ *is a* $C^{n_0}_b$*-function of* y. *Further, there exists* $c_{n_0} > 0$ *such that if* $|\mathbf{j}| \leq n_0$, *the following inequality holds for all* $0 \leq s < t \leq T$ *and* $x, y \in \mathbb{R}^d$:

$$\left|\partial^{\mathbf{j}}_y p_{s,t}(x, y)\right| \leq \frac{c_{n_0}}{(t-s)^{\frac{|\mathbf{j}|+d}{2-\alpha_0}}}. \tag{6.107}$$

Proof We need the dual property of the jump-diffusion with respect to $L^2(dx)$ on $\mathbb{R}^d$. We will define the dual of $\bar{P}^{\mathbf{t}}_{s,t} f$ by

$$\bar{P}^{\mathbf{t},*}_{s,t} g(y) = \int_{\mathbb{R}^d} \bar{p}^{\mathbf{t}}_{s,t}(x, y) g(x)\, dx. \tag{6.108}$$

Then it holds that $|\bar{P}^{\mathbf{t},*}_{s,t} g|_\infty \leq c^*|g|_\infty$. Further, similarly to Lemma 6.9.1, there exists a positive constant $\check{C}$ such that

$$\sup_{\mathbf{t}\in(s,t]^n_+} \left|\partial^{\mathbf{j}} \bar{P}^{\mathbf{t},*}_{s,t} g\right|_\infty \leq \check{C}^{n+1} \sum_{\mathbf{j}'\subset\mathbf{j}} \left|\partial^{\mathbf{j}'} g\right|_\infty, \quad n = 1, 2, \ldots \tag{6.109}$$

Then we can verify that $p_{s,t}(x, y)$ is a $C^{n_0}_b$-function of y and satisfies the short-time estimate (6.107) by a similar argument. □

In Theorem 6.6.1, we showed that the density function $p_{s,t}(x, y)$ is a rapidly decreasing C^∞-function of x and y, under Conditions (J.1) and (J.2). Hence jumps are assumed to be bounded. However, in Theorem 6.9.2, the density function $p_{s,t}(x, y)$ may not be rapidly decreasing, because of big jumps. For example, any stable process satisfies conditions of the above theorem, and its law has a C^∞-density function, but the density function is not rapidly decreasing if it is not Gaussian.

Assertions of Theorem 6.9.1 can be extended to the smooth density $p_{s,t}^{\mathbf{c},\mathbf{d}}(x, y)$ of a weighted law $P_{s,t}^{\mathbf{c},\mathbf{d}}(x, E) = E[1_E(\Phi_{s,t}(x))G_{s,t}(x)]$, if $G_{s,t}(x)$ is an exponential functional given by (6.53), where $d_{r,z}(x)$ is assumed to be a bounded positive function. For details, see the proof of Theorem 7.5.1 in Sect. 7.5,

Note The existence of the measurable density can be shown without Condition $(J.2)_K$. It means that maps $\{\phi_{t,z}, |z| > \delta\}$ may not be diffeomorphic for some > 0. In fact, $p_{s,t}(x, y)$ of (6.95) is well defined and is the density function of the transition probability, without condition $(J.2)_K$.

However, it seems to be a hard question whether we can remove Condition $(J.2)_K$ for the existence of the smooth density. The continuity and the differentiability are not clear without the diffeomorphic property of maps $\phi_{r,z}$. In Theorem 3.2 in Kunita [61], the existence of the smooth density is asserted without $(J.2)_K$, but his discussion contains gaps.

6.10 Density and Fundamental Solution on Subdomain

In Sect. 4.8, we considered the weighted transition function of a jump-diffusion killed outside of a subdomain D of $\mathbb{R}^d$. The weighted transition function $Q_{s,t}^{\mathbf{c},\mathbf{d}}(x, E)$ is equal to

$$Q_{s,t}^{\mathbf{c},\mathbf{d}}(x, E) = E[1_E(\Phi_{s,t}(x))G_{s,t}^{\mathbf{c},\mathbf{d}}(x)1_{\tau(x,s)>t}],$$

where $\tau(x, s)$ is the first leaving time of the process $X_t = X_t^{x,s} := \Phi_{s,t}(x), t$ in $[s, \infty)$ from the set D. We saw in Sects. 6.3 and 6.6 that, if X_t is nondegenerate, the weighted transition function $P_{s,t}^{\mathbf{c},\mathbf{d}}(x, E) = E[1_E(\Phi_{s,t}(x))G_{s,t}^{\mathbf{c},\mathbf{d}}(x)]$ has a smooth density $p_{s,t}^{\mathbf{c},\mathbf{d}}(x, y)$. Since $Q_{s,t}^{\mathbf{c},\mathbf{d}}(x, E) \leq P_{s,t}^{\mathbf{c},\mathbf{d}}(x, E)$ holds for any s, x, t, the measure $Q_{s,t}^{\mathbf{c},\mathbf{d}}(x, E)$ has always a density $q_{s,t}^{\mathbf{c},\mathbf{d}}(x, y)$ such that $q_{s,t}^{\mathbf{c},\mathbf{d}}(x, y) \leq p_{s,t}^{\mathbf{c},\mathbf{d}}(x, y)$ holds a.e. y. We want to show its smoothness. Suppose that the jump-diffusion is pseudo-elliptic. If there are no jumps entering D^c from D *a.s.*, it should have a C^∞-density. However, if the process has jumps from D^c into D, the result is not simple. The smoothness of the density is related to the off-diagonal short-time estimate studied in Sect. 6.8.

We first consider a diffusion process. Since $\mathbf{d} = \mathbf{1}$, we denote $Q_{s,t}^{\mathbf{c},\mathbf{d}}(x, E)$ and $q_{s,t}^{\mathbf{c},\mathbf{d}}(x, y)$ by $Q_{s,t}^{\mathbf{c}}(x, E)$ and $q_{s,t}^{\mathbf{c}}(x, y)$, respectively.

Lemma 6.10.1 *Suppose that $X_t^{x,s} := \Phi_{s,t}(x)$ is an elliptic diffusion process of initial state (x, s). Then for any $x \in D$ and $s < t$, $Q_{s,t}^{\mathbf{c}}(x, E)$; $E \subset D$ has a C^∞-density $q_{s,t}^{\mathbf{c}}(x, y)$. It is continuously differentiable with respect to t and satisfies*

$$\frac{\partial}{\partial t} q^{\mathbf{c}}_{s,t}(x, y) = A^{\mathbf{c}}(t)^*_y q^{\mathbf{c}}_{s,t}(x, y). \tag{6.110}$$

Further, the density has the following two properties:

1. *Let U be an open subset of $\mathbb{R}^d$ such that $\bar{U} \subset D$. Then for any multi-index $\mathbf{j}$ and $T > 0$, there exists a positive constant c such that the inequality*

$$\left|\partial^{\mathbf{j}}_y q^{\mathbf{c}}_{s,t}(x, y)\right| \leq \frac{c}{(t-s)^{\frac{|\mathbf{j}|+d}{2}}}, \quad \forall 0 \leq s < t \leq T \tag{6.111}$$

   holds for any $x \in D, y \in U$.

2. *Let V be an open subset of D satisfying $\bar{V} \subset U \subset \bar{U} \subset D$. Then for any $\mathbf{j}$ and $T > 0$ there exists a positive constant c' such that the inequality*

$$|\partial^{\mathbf{j}}_y q^{\mathbf{c}}_{s,t}(x, y)| \leq c', \quad \forall 0 \leq s < t \leq T \tag{6.112}$$

   holds for all $x \in U^c \cap D$, $y \in V$.

Proof For a given $T > 0$, we take $0 \leq s < t \leq T$. Let $\tau = \tau(x, s)$ be the first leaving time of the process $X_t = X^{x,s}_t = \Phi_{s,t}(x)$ from the set D. We have by the strong Markov property

$$Q^{\mathbf{c}}_{s,t} f(x) = P^{\mathbf{c}}_{s,t} f(x) - E\Big[P^{\mathbf{c}}_{\tau,t} f(\Phi_{s,\tau}(x)) G^{\mathbf{c}}_{s,\tau}(x) 1_{s<\tau<t}\Big]. \tag{6.113}$$

Set for $x \in \mathbb{R}^d$ and $y \in V$,

$$q^{\mathbf{c}}_{s,t}(x, y) := p^{\mathbf{c}}_{s,t}(x, y) - E\Big[p^{\mathbf{c}}_{\tau,t}(\Phi_{s,\tau}(x), y) G^{\mathbf{c}}_{s,\tau}(x) 1_{s<\tau<t}\Big]. \tag{6.114}$$

Then it is the density function of $Q^{\mathbf{c}}_{s,t}(x, E)$. Further, for a multi-index $\mathbf{j}$ set

$$q^{\mathbf{j}}_{s,t}(x, y) := \partial^{\mathbf{j}}_y p^{\mathbf{c}}_{s,t}(x, y) - E\Big[\partial^{\mathbf{j}}_y p^{\mathbf{c}}_{\tau,t}(\Phi_{s,\tau}(x), y) G^{\mathbf{c}}_{s,\tau}(x) 1_{s<\tau<t}\Big]. \tag{6.115}$$

The last integral is well defined and is continuous with respect to t, y. Indeed, since $\Phi_{s,\tau}(x) \in \partial D$ if $\tau < t$, for any nonnegative integer N, there exists a positive constant c such that

$$|\partial^{\mathbf{j}}_y p^{\mathbf{c}}_{\tau,t}(\Phi_{s,\tau}(x), y)| \leq c(t-\tau)^N \leq c' < \infty,$$

in view of Theorem 6.8.1. Then $q^{\mathbf{j}}_{s,t}(x, y)$ is well defined and is continuous with respect to $y \in U$. Since this is valid for any $\mathbf{j}$, $q^{\mathbf{c}}_{s,t}(x, y)$ is infinitely continuously differentiable with respect to y and the equality $\partial^{\mathbf{j}}_y q^{\mathbf{c}}_{s,t}(x, y) = q^{\mathbf{j}}_{s,t}(x, y)$ holds for any $\mathbf{j}$.

Now, since $Q^{\mathbf{c}}_{s,t}(x, E \cap V)$ has the smooth density $q^{\mathbf{c}}_{s,t}(x, y), y \in V$ for any V such that $\bar{V} \subset D$, $Q^{\mathbf{c}}_{s,t}(x, E \cap D)$ has also a smooth density $q^{\mathbf{c}}_{s,t}(x, y), y \in D$ for any $0 \le s < t \le T$ and $x \in D$. The function $q^{\mathbf{c}}_{s,t}(x, y)$ satisfies

$$\frac{\partial}{\partial t}\int q^{\mathbf{c}}_{s,t}(x, y) f(y)\, dy = \int q^{\mathbf{c}}_{s,t}(x, y) A^{\mathbf{c}}(t) f(y)\, dy = \int A^{\mathbf{c}}(t)^*_y q^{\mathbf{c}}_{s,t}(x, y) f(y)\, dy$$

for any $f \in C_0(D)$ in view of Proposition 4.8.1. Therefore $q^{\mathbf{c}}_{s,t}(x, y)$ is differentiable with respect to t and should satisfy the equation (6.110).

Note that the last term of (6.115) is bounded. Then $\partial^{\mathbf{j}}_y q^{\mathbf{c}}_{s,t}(x, y)$ has the same short-time estimate as $\partial^{\mathbf{j}}_y p^{\mathbf{c}}_{s,t}(x, y)$, which proves (6.111) for $q^{\mathbf{c}}_{s,t}(x, y)$.

We will show the second assertion of the lemma. We know that $|\partial^{\mathbf{j}}_y p^{\mathbf{c}}_{s,t}(x, y)|$ is bounded with respect to $x \in U^c \cap D$, $y \in V$ and $0 \le s < t \le T$, in view of Theorem 6.8.1. Further, the last term of equation (6.115) is also bounded with respect to these x, y, s, t. Consequently, $q^{\mathbf{j}}_{s,t}(x, y)$ is also bounded with respect to these x, y, s, t. □

We are also interested in the regularity of the above $q^{\mathbf{c}}_{s,t}(x, y)$ with respect to x.

Lemma 6.10.2 *Suppose that $X^{x,s}_t = \Phi_{s,t}(x)$ is an elliptic diffusion process of initial state (x, s). Then the density function $q^{\mathbf{c}}_{s,t}(x, y)$ of $Q^{\mathbf{c}}_{s,t}(x, E)$ is a $C^{\infty,1}$-function of $x \in D$ and $s (< t)$ for any y, t, and satisfies*

$$\frac{\partial}{\partial s} q^{\mathbf{c}}_{s,t}(x, y) = -A^{\mathbf{c}}(s)_x q^{\mathbf{c}}_{s,t}(x, y). \tag{6.116}$$

Further, the density has the following two properties:

1. *Let U be an open subset of $\mathbb{R}^d$ such that $\bar{U} \subset D$. Then for any multi-index $\mathbf{i}$ and $T > 0$, there exists a positive constant c such that the inequality*

$$\left|\partial^{\mathbf{i}}_x q^{\mathbf{c}}_{s,t}(x, y)\right| \le \frac{c}{(t-s)^{\frac{|\mathbf{i}|+d}{2}}}, \quad \forall 0 \le s < t \le T \tag{6.117}$$

holds for any $x \in U, y \in D$.

2. *Let V be an open subset of D satisfying $\bar{V} \subset U \subset \bar{U} \subset D$. Then for any $\mathbf{i}$ and $T > 0$ there exists a positive constant c' such that the inequality*

$$|\partial^{\mathbf{i}}_x q^{\mathbf{c}}_{s,t}(x, y)| \le c', \quad \forall 0 \le s < t \le T \tag{6.118}$$

holds for all $x \in U^c \cap D$, $y \in V$.

Proof The smoothness of $q^{\mathbf{c}}_{s,t}(x, y)$ with respect to x is not clear from the proof of Lemma 6.10.1, since in the expression (6.114), the stopping time $\tau = \tau(x, s)$ depends on x. Instead, we will use the dual property. The dual of $P^{\mathbf{c}}_{s,t}(x, E)$ is written as $P^{\mathbf{c},*}_{s,t}(y, E) = E[1_E(\check{\Psi}_{s,t}(y))\check{\mathcal{G}}^{\mathbf{c}}_{s,t}(y)|\det \nabla \check{\Psi}_{s,t}(y)|]$. Since the generator of the backward diffusion $\check{\Psi}_{s,t}(y)$ is $\mathcal{A}(t) = \frac{1}{2}\sum_{k\geq 1} V_k(t)^2 - V_0(t)$, it is elliptic. Then we can apply the same argument to the backward process $\check{\Psi}_{s,t}(y)$. We discussed the dual of the semigroup $Q^{\mathbf{c}}_{s,t}$ in Sect. 4.8. It is given by $Q^{\mathbf{c},*}_{s,t}$ of (4.86). Then its transition function $Q^{\mathbf{c},*}_{s,t}(y, E)$ should have a smooth density $q^{\mathbf{c},*}_{s,t}(y, x)$, $x \in D$.

We want to show that $q^{\mathbf{c},*}_{s,t}(y, x) = q^{\mathbf{c}}_{s,t}(x, y)$ holds for any $s < t$ and $x, y \in D$. We set $Q'_{s,t} f(x) = \int q^{\mathbf{c},*}_{s,t}(y, x) f(y)\, dy$. Then the equality

$$\int_D g(x) Q'_{s,t} f(x)\, dy = \int_D f(y) Q^{\mathbf{c},*}_{s,t} g(y)\, dy = \int_D Q^{\mathbf{c}}_{s,t} f(x) g(x)\, dx$$

holds for any f, g. Therefore we have $Q'_{s,t} f(x) = Q^{\mathbf{c}}_{s,t} f(x)$ almost everywhere. We show that the equality holds for all $x \in D$. For $\delta > 0$, we have

$$Q^{\mathbf{c}}_{s-\delta,t} f(x) = Q^{\mathbf{c}}_{s-\delta,s} Q^{\mathbf{c}}_{s,t} f(x) = Q^{\mathbf{c}}_{s-\delta,s} Q'_{s,t} f(x), \quad \forall x \in D,$$

since kernels $Q^{\mathbf{c}}_{s-\delta,s}(x, \cdot)$ are absolutely continuous. Let δ tend to 0. Since $\tilde{Q}'_{s,t} f(x)$ is a continuous function, we get the equality $Q^{\mathbf{c}}_{s,t} f(x) = Q'_{s,t} f(x)$ for all $x \in D$.

Consequently, for any $x \in D$ $q^{\mathbf{c},*}_{s,t}(y, x) = q^{\mathbf{c}}_{s,t}(x, y)$ holds for almost all $y \in D$. Then, for any x and y we have

$$q^{\mathbf{c},*}_{s,t+\delta}(y, x) = \int Q^{\mathbf{c},*}_{t,t+\delta}(y, dz) q^{*}_{s,t}(z, x)\, dz = \int Q^{\mathbf{c},*}_{t,t+\delta}(y, dz) q^{\mathbf{c}}_{s,t}(x, z)\, dz.$$

Let $\delta \to 0$. Then we obtain $q^{\mathbf{c},*}_{s,t}(y, x) = q^{\mathbf{c}}_{s,t}(x, y)$ for any $s < t$ and $x, y \in D$.

Now, the density function $q^{\mathbf{c},*}_{s,t}(y, x)$ of the dual transition function satisfies assertions of Lemma 6.10.1, interchanging x and y. Then $q^{\mathbf{c}}_{s,t}(x, y)$ should satisfy assertions of the present lemma. □

We will summarize results for the killed diffusion process.

Theorem 6.10.1 *Consider a system of elliptic diffusions on $\mathbb{R}^d$ with the generator $A(t)$ of* (4.4)*. Let $Q^{\mathbf{c}}_{s,t}(x, \cdot)$ be the weighted transition function of the diffusion killed outside of a bounded domain D. For any $s < t$ and $x \in D$, it has a C^∞-density $q^{\mathbf{c}}_{s,t}(x, y)$, $y \in D$. The density function is a $C^{\infty,1}$-function of x, s for any y, t. It satisfies short-time estimates* (6.111), (6.112), (6.117), *and* (6.118)*. Further, it satisfies the backward equation (6.116).*

Corollary 6.10.1 *For any bounded continuous function f and t, $Q^{\mathbf{c}}_{s,t} f(x)$ is a $C^{\infty,1}$-function of x, s. Further, it satisfies the backward heat equation*

$$\frac{\partial}{\partial s} Q^{\mathbf{c}}_{s,t} f(x) = -A^{\mathbf{c}}(s) Q^{\mathbf{c}}_{s,t} f(x).$$

As an application of Corollary 6.10.1, we will consider the *final value problem of the backward heat equation on a bounded domain D of an Euclidean space.*

$$\begin{cases} \dfrac{\partial}{\partial s} v(x,s) = -A^{\mathbf{c}}(s) v(x,s), & x \in D,\ 0 < s < t_1 \\ \lim_{s \to t_1} v(x,s) = f_1(x), & \forall \in D, \end{cases} \tag{6.119}$$

where $A^{\mathbf{c}}(s)$ is the differential operator defined by (4.11) and $f_1(x)$ is a given bounded continuous function. Let $v(x,s)$, $(x,s) \in D \times (0,t_1)$ be a function of $C_b^{2,1}$-class of (x,s). It is called a solution of the final value problem if $v(x,s)$ satisfies (6.119). For the uniqueness of the solution, we need a stochastic Dirichlet boundary condition. A continuous function $v(x,t)$ defined on $D \times (0,t_1)$ is said to satisfy the *stochastic Dirichlet boundary condition*, if for any $x \in D$ and $t_0 \le s$

$$\exists \lim_{t \uparrow \tau(x,s)} v(\Phi_{s,t}(x), t) = 0, \quad \text{if } \tau(x,s) < t, \ a.s.\ P. \tag{6.120}$$

Let $p(x,s;y,t)$, $x,y \in D, 0 < s < t < \infty$ be a continuous function of x,y,s,t. It is called the fundamental solution of equation (6.119) if it is of $C^{2,1}$-class with respect to x,s for any y,t and satisfies $\frac{\partial}{\partial s} p(x,s;y,t) = -A^{\mathbf{c}}(s)_x p(x,s;y,t)$ and further, $v(x,s) := \int p(x,s;y,t_1) f_1(y)\,dy$, $x \in D, 0 < s < t_1$ is a solution of (6.119) satisfying the stochastic Dirichlet boundary condition for any t_1 and a bounded continuous function f_1 on D.

Theorem 6.10.2 *Suppose the operator $A^{\mathbf{c}}(t)$ is elliptic. For any bounded continuous function f_1 on D, the solution of the final value problem of the backward heat equation associated with the operator $A^{\mathbf{c}}(t)$ on the domain D with stochastic Dirichlet boundary condition exists uniquely. It is given by*

$$v(x,s) = E[f_1(\Phi_{s,t_1}(x)) G^{\mathbf{c}}_{s,t_1}(x) 1_{\tau(x,s) > t_1}]. \tag{6.121}$$

Let $q^{\mathbf{c}}_{s,t}(x,y)$ be the density of Theorem 6.10.1. *Then $p(x,s;y,t) := q^{\mathbf{c}}_{s,t}(x,y)$ is its fundamental solution.*

Proof We saw in Corollary 6.10.1 that the function (6.121) is a solution of the final value problem. We will check that the above $v(x,s)$ satisfies the stochastic Dirichlet boundary condition. By the Markov property, we have for $s < t < t_1$

$$v(\Phi_{s,t}(x), t) = E\Big[f_1(\Phi_{s,t_1}(x)) G^{\mathbf{c}}_{s,t_1}(x) 1_{\tau(x,s) \ge t_1} \Big| \mathcal{F}_t \Big].$$

Let t tend to $\tau(x,s)$. Then the above converges to 0 on the set $\{\tau(x,s) < t_1\}$. Therefore, the Dirichlet boundary condition is satisfied.

We want to prove the uniqueness of the solution. Let $v'(x,s)$ be a solution of the final value problem (6.119) satisfying the stochastic Dirichlet boundary condition. Apply Itô's formula to $X_t := v'(\Phi_{s,t}, t) G^{\mathbf{c}}_{s,t}$. Then

$$X_t = v'(x,s) + \int_s^t G^{\mathbf{c}}_{s,r} \Big(A^{\mathbf{c}}(r) + \frac{\partial}{\partial s} \Big) v'(\Phi_{s,r}, r)\,dr + M_t,$$

where M_t is a martingale with mean 0. Since $v'(x,s)$ satisfies equation (6.119), the term $\int_s^t \cdots dr$ in the above expression is 0. Therefore we find that X_t is actually a martingale. Consequently, $E[X_{t_1}]$ is equal to $v'(x,s)$, i.e., we get

$$v'(x,s) = E[f_1(\Phi_{s,t_1}(x))G^{\mathbf{c}}_{s,t_1}(x)1_{\tau(x,s)\geq t_1}].$$

The uniqueness follows.

Finally, $q^{\mathbf{c}}_{s,t}(x,y)$ is the fundamental solution, since it satisfies (6.116). □

Next, the *initial value problem for the heat equation associated with the operator* $A^{\mathbf{c}}(t)$ *on the domain* D is defined in the same way. For this problem, we consider a backward stochastic flow $\{\check{\Phi}_{s,t}\}$ generated by a backward symmetric SDE with coefficients $V_j(x,t)$, $j=0,\ldots,d'$. Let $\check{\tau}(x,t)$ be the exit time of the backward process $\check{X}^{x,t}_s = \check{\Phi}_{s,t}(x)$ from the domain D. Then $\check{Q}^{\mathbf{c}}_{s,t}(x,E)$ defined by $E[1_E(\check{\Phi}_{s,t}(x))\check{G}_{s,t}(x)1_{\check{\tau}(t,x)<s}]$ has a density $\check{q}^{\mathbf{c}}_{s,t}(x,y)$, which is a C^∞-function of x and y. The fundamental solution of the heat equation on D with the stochastic Dirichlet boundary condition is defined similarly.

The following can be verified similarly to the previous theorem.

Theorem 6.10.3 *Suppose that operator* $A^{\mathbf{c}}(t)$ *is elliptic. For any bounded continuous function* f_0*, the solution of the heat equation associated with the operator* $A^{\mathbf{c}}(t)$ *on the domain* D *with stochastic Dirichlet boundary condition exists uniquely. It is given by* $u(x,t) = E[f_0(\check{\Phi}_{t_0,t}(x))\check{G}^{\mathbf{c}}_{t_0,t}(x)1_{\check{\tau}(t,x)<t_0}]$*. Further,* $\check{p}(x,t;y,s) := \check{q}^{\mathbf{c}}_{s,t}(x,y)$ *is its fundamental solution.*

We next consider a pseudo-elliptic jump-diffusion process. We consider again the weighted transition function $Q^{\mathbf{c},\mathbf{d}}_{s,t}(x,E)$. Notations $\bar{\phi}^N(U)$ etc. in the next lemma are introduced in Sect. 6.5. Let α_0 be any positive constant satisfying $\alpha<\alpha_0<2$, where α is the exponent of the Lévy measure.

Lemma 6.10.3 *For a given* $n_0\in\mathbb{N}$*, let* N *be a positive integer such that* $N>\frac{n_0+d}{2-\alpha_0}$*. Suppose that* U *is a bounded open subset of* D *such that*

$$d(\bar{\phi}^N(U),\bar{\phi}^{-N}(D^c))>0. \tag{6.122}$$

Then the measure $Q^{\mathbf{c},\mathbf{d}}_{s,t}(x,\cdot)$ *on the set* U *has a* C^{n_0}*-density* $q^{\mathbf{c},\mathbf{d}}_{s,t}(x,y)$ *for any* $s<t$ *and* $x\in D$*. Further, the density has the following two properties:*

1. *For any* $\mathbf{j}$ *with* $|\mathbf{j}|\leq n_0$ *and* $T>0$*, there exists* $c>0$ *such that the inequality*

$$\left|\partial_y^{\mathbf{j}}q^{\mathbf{c},\mathbf{d}}_{s,t}(x,y)\right|\leq\frac{c}{(t-s)^{\frac{|\mathbf{j}|+d}{2-\alpha_0}}},\qquad\forall 0\leq s<t\leq T \tag{6.123}$$

   holds for all $x\in D$*,* $y\in U$*.*

2. *Let V be an open subset of U such that $d(\bar{\phi}^N(V), \bar{\phi}^{-N}(U^c)) > 0$. Then for any* $\mathbf{j}$ *and $T > 0$ there exists $c' > 0$ such that the inequality*

$$|\partial_y^{\mathbf{j}} q_{s,t}^{\mathbf{c},\mathbf{d}}(x, y)| \le c', \quad \forall 0 \le s < t \le T \tag{6.124}$$

holds for all $x \in U^c \cap D$, $y \in V$.

Proof We will repeat an argument similar to the proof of Lemma 6.10.1. Let U be an open subset of D satisfying properties stated in the lemma. Define the function $q_{s,t}^{\mathbf{c},\mathbf{d}}(x, y)$ by (6.114) replacing $p_{s,t}^{\mathbf{c}}(x, y)$ by $p_{s,t}^{\mathbf{c},\mathbf{d}}(x, y)$. It is the density function of $Q_{s,t}^{\mathbf{c},\mathbf{d}}(x, E)$ for any $x \in \mathbb{R}^d$ and $y \in U$. We can define the function $q_{s,t}^{\mathbf{j}}(x, y)$ by (6.115) replacing $p_{s,t}^{\mathbf{c}}(x, y)$ by $p_{s,t}^{\mathbf{c},\mathbf{d}}(x, y)$. Indeed, since $\Phi_{s,\tau}(x) \in D^c$ if $\tau = \tau(x, s) < t$, we have by Theorem 6.8.2,

$$|\partial_y^{\mathbf{j}} p_{\tau,t}^{\mathbf{c},\mathbf{d}}(\Phi_{s,\tau}(x), y)| \le c(t - \tau)^{N - \frac{|\mathbf{j}|+d}{2-\alpha_0}} \le c < \infty.$$

Then $q_{s,t}^{\mathbf{j}}(x, y)$ is well defined and is continuous with respect to $y \in U$ for any $\mathbf{j}$ with $|\mathbf{j}| \le n_0$. Therefore, $q_{s,t}^{\mathbf{c},\mathbf{d}}(x, y)$ is n_0-times continuously differentiable with respect to $y \in V$ and the equality $\partial_y^{\mathbf{j}} q_{s,t}^{\mathbf{c},\mathbf{d}}(x, y) = q_{s,t}^{\mathbf{j}}(x, y)$ holds for $|\mathbf{j}| \le n_0$.

Since the last term of (6.115) is bounded, $\partial_y^{\mathbf{j}} q_{s,t}^{\mathbf{c},\mathbf{d}}(x, y)$ has the same short-time estimate as $\partial_y^{\mathbf{j}} p_{s,t}^{\mathbf{c},\mathbf{d}}(x, y)$, which proves (6.123) for $q_{s,t}^{\mathbf{c},\mathbf{d}}(x, y)$.

We show the second assertion of the lemma. We know that $|\partial_y^{\mathbf{j}} p_{s,t}^{\mathbf{c},\mathbf{d}}(x, y)|$ is bounded with respect to $x \in U^c \cap D$, $y \in V$ and $s < t$, in view of Theorem 6.8.2. Further, the last term of equation (6.115) is also bounded with respect to these x, y, t. Consequently, $q_{s,t}^{\mathbf{j}}(x, y)$ is also bounded with respect to these x, y, t. □

Now if jump-maps $\phi_{t,z}$ map D onto D, diffeomorphically, both $\phi_{t,z}$ and $\phi_{t,z}^{-1}$ should map D^c into itself. Therefore we have $d(\bar{\phi}^N(U), \bar{\phi}^{-N}(D^c)) > 0$ for any N if $\bar{U}$ is a compact subset of D. Therefore $Q_{s,t}^{\mathbf{c},\mathbf{d}}(x, E)$ has a C^∞-density if $x \in \bar{U}$.

Next, we will apply the same argument to the dual process. Then we find that $q_{s,t}^{\mathbf{c},\mathbf{d}}(x, y)$ is a C^∞-function of x, similarly to the case of diffusion. Therefore, we have the following theorem.

Theorem 6.10.4 *Consider a system of pseudo-elliptic jump-diffusions on $\mathbb{R}^d$ with the generator $A_J(t)$ of (4.52). Assume that jump-coefficients $\phi_{t,z}$ map a bounded domain $D \subset \mathbb{R}^d$ onto itself for all t, z. Then the process killed outside of D is a jump-diffusion. Its weighted transition function $Q_{s,t}^{\mathbf{c},\mathbf{d}}(x, E)$ has a C^∞-density $q_{s,t}^{\mathbf{c},\mathbf{d}}(x, y)$. The density function is a $C^{\infty,1}$-function of $x \in D, s$ $(< t)$ for any y, t. It satisfies the short-time estimates (6.123), (6.124). Further, $p(x, s; y, t) = q_{s,t}^{\mathbf{c},\mathbf{d}}(x, y)$ is the fundamental solution of the backward heat equation with the stochastic Dirichlet boundary condition on D, associated with the operator $A_J^{\mathbf{c},\mathbf{d}}(t)$.*

By the similar argument, we can obtain the fundamental solution for the heat equation on D associated with the integro-differential operator $A^{\mathbf{c},\mathbf{d}}(t)$. Details are omitted.

Note The smoothness of the density function $q_{s,t}^{\mathbf{c},\mathbf{d}}(x, y)$ with respect to y in the case $\mathbf{c} = 0$, $\mathbf{d} = 1$, was shown by Picard–Savona [95] for jump processes.

The existence of fundamental solutions for various type of heat equations are known in analysis and mathematical physics. A well known case is that the operator $A(t)$ is a Laplacian and domain D of the heat equation is an interval, a squares or a ball. Results for arbitrary operators $A^{\mathbf{c}}(t)$, $A_J^{\mathbf{c},\mathbf{d}}(t)$ and arbitrary domains D might be new.

The stochastic Dirichlet boundary condition should be transformed to the usual Dirichlet boundary condition if the boundary ∂D is regular, say if it is of C^1-class. For the proof, further discussions will be needed.

So far, we showed the existence of the smooth density $q_{s,t}^{\mathbf{c}}(x, y)$ for elliptic diffusions (or pseudo-elliptic jump-diffusions) on a subdomain of an Euclidean space. We can extend the result for a nondegenerate diffusion (or a nondegenerate jump-diffusion) if its transition density $p_{s,t}^{\mathbf{c}}(x, y)$ satisfies the short time estimate like (6.81) (or (6.90)). It is expected that a hypo-elliptic diffusion (and pseudo-hypo-elliptic jump diffusions) admits such an estimate, but for the proof further discussions will be needed.

Chapter 7
Stochastic Flows and Their Densities on Manifolds

Abstract In this chapter, we will study stochastic flows and jump-diffusions on manifolds determined by SDEs. If the manifold is not compact, SDEs may not be complete; solutions may explode in finite time. Then solutions could not generate stochastic flow of diffeomorphisms; instead they should define a stochastic flow of local diffeomorphisms. These facts will be discussed in Sect. 7.1. In Sect. 7.2, it will be shown that the stochastic flow defines a jump-diffusion on the manifold. Then, the dual process with respect to a volume element will be discussed. Further, in Sect. 7.3, the Lévy process on a Lie group and its dual with respect to the Haar measure will be discussed.

In Sect. 7.4, we consider an elliptic diffusion process on a connected manifold. We show the existence of the smooth density with respect to a volume element, by piecing together smooth densities on local charts which were obtained in Sect. 6.10. The result of the section can be applied to diffusion processes on Euclidean space with unbounded coefficients, where the explosion may occur. For a pseudo-elliptic jump-diffusion on a connected manifold, we need additional arguments, since sample paths may jump from a local chart to other local charts. It will be discussed in Sect. 7.5.

Finally, in Appendix, we collect some basic facts on manifolds and Lie groups that are used in this chapter.

7.1 SDE and Stochastic Flow on Manifold

Let M be a C^∞-manifold of dimension d. Related notations and terminologies for manifolds are collected in the Appendix at the end of the chapter. If M is not compact, we consider its one point compactification $M \cup \{\infty\}$. If M is compact, ∞ is adjoined to M as an isolated point. Given a function f on M, we set $f(\infty) = 0$, conventionally. A vector field $V(t)$ with parameter $t \in \mathbb{T} = [0, \infty)$ is called a $C^{\infty,1}$-vector field, if $V(t)f(x)$ is a $C^{\infty,1}$-function of (x, t) for any C^∞-function f on M. With a local coordinate $(x_1, \ldots, x_d)$, $V(t)f(x)$ is represented by $\sum_i V^i(x, t)\frac{\partial f}{\partial x_i}(x)$, where $V^1(x, t), \ldots, V^d(x, t)$ are $C^{\infty,1}$-functions.

H. Kunita, *Stochastic Flows and Jump-Diffusions*, Probability Theory and Stochastic Modelling 92, https://doi.org/10.1007/978-981-13-3801-4_7

Let $V_0(t), \ldots, V_{d'}(t)$ be $C^{\infty,1}$-vector fields on M. Let $\{\phi_{t,z}, (t,z) \in \mathbb{T} \times \mathbb{R}^{d'}\}$ be a family of C^∞-maps from M into itself, satisfying:

Condition (J.1)'. $(x,t,z) \to \phi_{t,z}(x)$ *is a* $C^{\infty,1,2}$*-map from* $M \times \mathbb{T} \times \mathbb{R}^{d'}$ *to* M. *Further, it satisfies* $\phi_{t,0}(x) = x$ *for any* x, t.

Condition (J.2)'. $\phi_{t,z}; M \to M$ *are diffeomorphic maps for all* t, z. *Let* $\phi_{t,z}^{-1}$ *be the inverse map. Then* $(x,t,z) \to \phi_{t,z}^{-1}(x)$ *is a* $C^{\infty,1,2}$*-map.*

Let $z = (z^1, \ldots, z^{d'}) \in \mathbb{R}^{d'}$. We set

$$\tilde{V}_k(t) f(x) = \partial_{z^k} f(\phi_{t,z}(x))\Big|_{z=0}, \quad k = 1, \ldots, d'. \tag{7.1}$$

Then $\tilde{V}_k(t), k = 1, \ldots, d'$ are $C^{\infty,1}$-vector fields, called the *tangent vector fields of maps* $\{\phi_{t,z}\}$ at $z = 0$.

We shall define a *symmetric stochastic differential equation on manifold* M associated with the above $V_k(t)$ and $\phi_{t,z}$. Let $\{\mathcal{F}_{s,t}, 0 \le s < t < \infty\}$ be the two-sided filtration generated by a d'-dimensional Wiener process $W_t = (W_t^1, \ldots, W_t^{d'})$, $t \in \mathbb{T}$ and a Poisson random measure $N(dr\,dz)$ on $\mathbb{T} \times \mathbb{R}_0^{d'}$ with the Lévy measure ν, which is independent of W_t. We set $\mathcal{F}_t = \mathcal{F}_{0,t}$. Let $0 \le t_0 < \infty$. Let τ_∞ be a stopping time with values in $[t_0, \infty) \cup \{\infty\}$. Let $X_t, t_0 \le t < \tau_\infty$ be a cadlag M-valued $\{\mathcal{F}_t\}$-adapted process such that $\lim_{t \uparrow \tau_\infty} X_t = \infty$ whenever $\tau_\infty < \infty$. The stopping time τ_∞ is called the *terminal time* or the *explosion time*. Let $C_0^\infty(M)$ be the set of all C^∞-functions of compact supports. X_t is called a solution of a *symmetric SDE with characteristics* $(V_k(t), k = 0, 1, \ldots, d', \phi_{t,z}, \nu)$, starting from X_0 at time t_0, if $f(X_t), t < \tau_\infty$ is a local semi-martingale for any $f \in C_0^\infty(M)$ and satisfies

$$f(X_t) = f(X_0 + \sum_{k=0}^{d'} \int_{t_0}^t V_k(r) f(X_r) \circ dW_r^k \tag{7.2}$$

$$+ \lim_{\epsilon \to 0} \Big\{ \int_{t_0}^t \int_{|z| \ge \epsilon} \Big\{ f(\phi_{r,z}(X_{r-})) - f(X_{r-}) \Big\} N(dr dz) - \sum_{k=1}^{d'} b_\epsilon^k \int_{t_0}^t \tilde{V}_k(r) f(X_{r-}) dr \Big\}$$

for $t_0 < t < \tau_\infty$. Here $\circ dW_r^k, k = 1, \ldots, d'$ mean symmetric integrals by Wiener processes $W_r^k, k = 1, \ldots, d'$ and $\circ dW_r^0$ means the usual integral dr. Further, $b_\epsilon^k = \int_{\epsilon \le |z| \le 1} z^k \nu(dz)$. The terminal time τ_∞ depends on the initial condition of the process X_t. We denote it by $\tau_\infty(X_0, t_0)$.

Let us consider the relation with SDE on Euclidean space $\mathbb{R}^d$ defined in Chap. 3. If X_t is a solution of equation (3.10) on $\mathbb{R}^d$, it satisfies (7.2) for any smooth function f by setting $\phi_{t,z}(z) = g(x,t,z) + x$, as it was shown in (3.69). Conversely, if a process X_t on $\mathbb{R}^d$ satisfies (7.2) for any smooth function f, X_t satisfies (3.10). However, in this chapter, we do not assume the boundedness of coefficients $V_k(x,t), g(x,t,z) := \phi_{t,z}(x) - x$ and their derivatives in equation (7.2). Hence the explosion may occur for the solution in the Euclidean space. Hence our SDE (7.2) may be regarded as a generalization of SDE (3.10).

Theorem 7.1.1 *For the symmetric SDE* (7.2) *on the manifold M, we assume that $V_k(t)$, $k = 0, \dots, d'$ are $C^{\infty,1}$-vector fields and $\{\phi_{t,z}\}$ satisfies Condition (J.1)'. Then, given $x \in M$ and $0 \le s < \infty$, the equation* (7.2) *has a unique solution $X_t^{x,s}(\omega), t < \tau_\infty(x, s, \omega)$, starting from x at time s. Further, for any $s < t$, the solution has a modification $\Phi_{s,t}(x, \omega), t < \tau_\infty(x, s, \omega)$ such that there is a measurable subset $\tilde{\Omega}_{s,t}$ of Ω with $P(\tilde{\Omega}_{s,t}) = 1$ and for any $\omega \in \tilde{\Omega}_{s,t}$ the following properties hold. The terminal time $\tau_\infty(x, s, \omega)$ is lower semi-continuous with respect to x for any s, so that the set $\mathcal{D}_{s,t}(\omega) := \{x; \tau_\infty(x, s, \omega) > t\}$ are open subsets of M. Further, maps $\Phi_{s,t}(\omega); \mathcal{D}_{s,t}(\omega) \to M$ are C^∞.*

The family of pairs $\{\Phi_{s,t}, \mathcal{D}_{s,t}\}$ satisfying properties of the above theorem is called a *stochastic flow of C^∞-maps on the manifold M.*

Proof We will construct the solution of equation (7.2) by piecing together solutions defined on local charts of the manifold M. Let us first choose a family of local charts $U_\alpha \subset D_\alpha, \alpha = 1, 2, \dots$ satisfying the following:

(i) $\bigcup_\alpha U_\alpha = M$.
(ii) The number of α such that $x \in U_\alpha$ is at most finite for any x.
(iii) For any α, $\bar{D}_\alpha$ is compact and satisfies $\bar{U}_\alpha \subset D_\alpha$.

We will consider SDE on D_α. Let ψ_α be a diffeomorphism from D_α to an open set in $\mathbb{R}^d$ (local coordinate). Then the SDE is transformed to an SDE on $\psi_\alpha(D_\alpha)$ ($\subset \mathbb{R}^d$). Set $\acute{V}_k(t) = d\psi_\alpha(V_k(t))$, $\check{V}_k(t) = d\psi_\alpha(\tilde{V}_k(t))$ and $\acute{g}(\acute{x}, t, z) = \psi_\alpha(\phi_{t,z}(x)) - \acute{x}$ if $\phi_{t,z}(x) \in D_\alpha$, where $\acute{x} = \psi_\alpha(x)$. The functions $\{\acute{V}_k(\acute{x}, t), k = 0, \dots, d', \acute{g}(\acute{x}, t, z)\}$ are defined on the set $\psi_\alpha(D_\alpha)$. Since the set is relatively compact in $\mathbb{R}^d$, these functions are extended to $\mathbb{R}^d$ as $C_b^{\infty,1}$-functions and $C_b^{\infty,1,2}$-functions, respectively. Consider an SDE on $\mathbb{R}^d$:

$$\acute{X}_t = \acute{x} + \sum_{k=0}^{d'} \int_s^t \acute{V}_k(\acute{X}_r, r) \circ dW_r^k \tag{7.3}$$

$$+ \lim_{\epsilon \to 0} \left\{ \int_s^t \int_{|z| \ge \epsilon} \acute{g}(\acute{X}_{r-}, r, z) N(dr\, dz) - \sum_{k=1}^{d'} b_\epsilon^k \int_s^t \check{V}_k(\acute{X}_{r-}, r)\, dr \right\}.$$

It generates a stochastic flow $\{\acute{\Phi}_{s,t}\}$ of C^∞-maps on $\mathbb{R}^d$ (Theorem 3.4.4).

Given $x \in M$ and $0 \le s < \infty$, we will define an M-valued process as follows. If $x \in U_\alpha$, we set

$$\Phi_{s,t}^{(\alpha)}(x) = \psi_\alpha^{-1}(\acute{\Phi}_{s,t}(\acute{x})), \quad \text{if } t < \tau_{D_\alpha}(x, s) := \inf\{t > s; \acute{\Phi}_{s,t}(\acute{x}) \notin \psi_\alpha(D_\alpha)\}.$$

Then, by the pathwise uniqueness of the solution, it holds that $\Phi_{s,t}^{(\alpha)}(x) = \Phi_{s,t}^{(\beta)}(x)$ if $x \in U_\alpha \cap U_\beta$ and $t < \tau_{D_\alpha}(x,s) \wedge \tau_{D_\beta}(x,s)$. Now set

$$\tau_1(x,s) := \min_{\{\alpha; x \in U_\alpha\}} \tau_{D_\alpha}(x,s). \tag{7.4}$$

Then $\tau_1(x,s) - s$ is strictly positive for any x, s a.s. We can define the solution of (7.2) up to time $\tau_1(x,s)$ as follows. For $t < \tau_1(x,s)$, take α such that $\tau_{U_\alpha}(x,s) > t$. Then we define $\Phi_{s,t}(x)$ by $\Phi_{s,t}^{(\alpha)}(x)$. Then the process $\Phi_{s,t}(x), t < \tau_1(x,s)$ is indeed a solution of SDE (7.2) on the manifold M.

We will prolong the solution for $t \geq \tau_1 = \tau_1(x,s)$. Define an $\mathbb{R}_0^{d'}$-valued random variable by $S_1 = \int\int_{\{(\tau_1,z); z \in \mathbb{R}_0^{d'}\}} zN(dr\,dz)$. Set $\Phi_{s,\tau_1} = \phi_{\tau_1,S_1} \circ \Phi_{s,\tau_1-}$ and define

$$\Phi_{s,t}(x) = \Phi_{\tau_1,t}(\Phi_{s,\tau_1}(x)), \quad \text{if } \tau_1(x,s) \leq t < \tau_2(x,s) := \tau_1(\Phi_{s,\tau_1}(x), \tau_1).$$

Then it satisfies SDE (7.2) for $s < t < \tau_2(x,s)$. We will continue this procedure. Then we can define $\Phi_{s,t}(x)$ for $s < t < \tau_\infty(x,s) := \lim_{k\to\infty} \tau_k(x,s)$, where $\tau_k(x,s) = \tau_1(\Phi_{s,\tau_{k-1}}(x), \tau_{k-1})$. It satisfies SDE (7.2) for $s < t < \tau_\infty(x,s)$.

The above $\tau_\infty(x,s)$ is a terminal time. Indeed, suppose $\tau_\infty(x,s) < \infty$. Then the sequence of M-valued random variables $\{\Phi_{s,\tau_n-}(x), n = 1, 2, \ldots\}$ should converges to a point $y \in M$ or diverges to ∞. But the former case cannot occur, because of the definition of stopping times $\tau_n(x,s)$. Therefore, the latter case occurs only, proving that $\tau_\infty(x,s)$ is the terminal time.

The functional $\Phi_{s,t}(x), s < t < \tau_\infty(x,s)$ constructed above has the flow property: There exists $\tilde{\Omega}_1 \subset \Omega$ with $P(\tilde{\Omega}_1) = 1$ such that for any fixed $\omega \in \tilde{\Omega}_1$, $\Phi_{s,t}(x,\omega)$ satisfies

$$\Phi_{r,t}(\Phi_{s,r}(x,\omega),\omega) = \Phi_{s,t}(x,\omega), \quad s < \forall r < t < \tau_\infty(x,s,\omega).$$

Set $\mathcal{D}_{s,t}(\omega) = \{x \in M; \tau_\infty(x,s,\omega) > t\}$. Then $\Phi_{s,t}(\omega)$ is a map from $\mathcal{D}_{s,t}(\omega)$ to M. We will prove that it is a C^∞-map for almost all ω. Take any $x_0 \in \mathcal{D}_{s,t}$ and consider the sample path $\Phi_{s,t}(x_0,\omega)$. There is $\tilde{\Omega}_2 \subset \Omega$ with $P(\tilde{\Omega}_2) = 1$ such that for fixed $\omega \in \tilde{\Omega}_2$, we may choose $s = t_0 < t_1 < \cdots < t_n = t$ and local charts $U_{\alpha_i}, i = 1, \ldots, n$ from the family $\{U_\alpha\}$ such that the trajectory $\{\Phi_{s,r}(x_0,\omega); r \in [t_{i-1}, t_i)\}$ is included in U_{α_i} for any $i = 1, \ldots, n$. Then there is an open neighborhood U_0 of x_0 (depending on ω) such that

$$\bigcup_{x \in U_0} \{\Phi_{s,r}(x,\omega); r \in [t_{i-1}, t_i)\} \subset D_{\alpha_i}, \quad i = 1, \ldots, n.$$

If $\omega \in \tilde{\Omega}_1 \cap \tilde{\Omega}_2$, $\Phi_{s,t}(\omega)$ is a C^∞-map, since the flow property

$$\Phi_{s,t}(\omega) = \Phi_{t_{n-1},t}(\omega) \circ \Phi_{t_{n-1},t_{n-2}}(\omega) \circ \cdots \circ \Phi_{s,t_1}(\omega)$$

holds and the right hand side is composites of C^∞-maps. Therefore $\Phi_{s,t}(\omega)$ is a C^∞-map on U_0 if $\omega \in \tilde{\Omega}_1 \cap \tilde{\Omega}_2$. Further, $U_0 \subset \mathcal{D}_{s,t}(\omega)$ and hence $\mathcal{D}_{s,t}(\omega)$ is an open set. We have thus shown the assertion of the theorem. □

Now, for the symmetric SDE (7.2), we will assume further that jump-maps $\phi_{t,z}$ satisfy Condition (J.2)'. We shall study the diffeomorphic property of the flow. We need the backward symmetric SDE on the manifold M. Let $0 < t_1 < \infty$. An $\{\mathcal{F}_{s,t_1}\}$-adapted M-valued backward caglad process $\check{X}_s, t_1 \geq s > \check{\tau}_\infty$ is called a solution of a backward symmetric SDE with characteristics $(-V_k(t), k = 0, \ldots, d', \phi_{t,z}^{-1}, \nu)$ starting from x at time t_1 if $f(\check{X}_s), s \in (\check{\tau}_\infty, t_1]$ is a backward local semi-martingale for any $f \in C_0^\infty(M)$ and satisfies

$$f(\check{X}_s) = f(x) - \sum_{k=0}^{d'} \int_s^{t_1} V_k(r) f(\check{X}_r) \circ d W_r^k \tag{7.5}$$

$$+ \lim_{\epsilon \to 0} \left\{ \int_s^{t_1} \int_{|z| \geq \epsilon} \{ f(\phi_{r,z}^{-1}(\check{X}_{r+})) - f(\check{X}_{r+}) \} N(dr\, dz) + \sum_{k=1}^{d'} b_\epsilon^k \int_s^{t_1} \tilde{V}_k(r) f(\check{X}_r)\, dr \right\},$$

for $t_1 > s > \check{\tau}_\infty$. Here $\circ d W_r^k, k = 1, \ldots, d'$ mean backward symmetric integrals by processes $\check{W}_t^k = W_t^k - W_T^k, t \in \mathbb{T}$, where $W_t = (W_t^1, \ldots, W_t^{d'}), t \in \mathbb{T}$ is a Wiener process and $\circ d W_r^0$ means the usual integral dr. Further, $\tilde{V}_k(t), k = 1, \ldots, d'$ are tangent vector fields of $\{\phi_{t,z}\}$ at $z = 0$. $\check{\tau}_\infty$ is the (backward) terminal time of the backward process $\check{X}_s$; it means that it is a backward stopping time with values in $[0, t_1] \cup \{-\infty\}$ such that $\lim_{t \downarrow \check{\tau}_\infty} \check{X}_t = \infty$ holds if $\check{\tau}_\infty > 0$. The equation has a unique backward solution, which we denote by $\check{X}_s^{x,t_1}(\omega), s > \check{\tau}_\infty$. The backward terminal time is denoted by $\check{\tau}_\infty(x, t_1, \omega)$. Then for any $0 < s < t$, the backward solution has a modification $\check{\Psi}_{s,t}(x, \omega), s > \check{\tau}_\infty(x, t, \omega)$ such that for any $\omega \in \tilde{\Omega}_{s,t} \subset \Omega$ with $P(\tilde{\Omega}_{s,t}) = 1$, the set

$$\check{\mathcal{D}}_{s,t}(\omega) = \{x \in M; \check{\tau}_\infty(x, t, \omega) < s\} \tag{7.6}$$

is open and maps $\check{\Psi}_{s,t}(\omega); \check{\mathcal{D}}_{s,t}(\omega) \to M$ are C^∞-maps. $\{\check{\Psi}_{s,t}(x, \omega), \check{\mathcal{D}}_{s,t}(\omega)\}$ is called the backward stochastic flow of C^∞-maps generated by the backward symmetric SDE (7.5).

We first consider a continuous SDE.

Lemma 7.1.1 *Let $\{\Phi_{s,t}, \mathcal{D}_{s,t}\}$ be the continuous stochastic flow of C^∞-maps on the manifold M defined by the continuous symmetric SDE with coefficients $V_k(x, t), k = 0, \ldots, d'$. Let $\{\check{\Psi}_{s,t}, \check{\mathcal{D}}_{s,t}\}$ be the continuous backward stochastic flow of C^∞-maps defined by a continuous symmetric backward SDE with coefficients $-V_k(x, t), k = 0, \ldots, d'$. Then we have the following:*

1. *For any* $0 < s < t$, $\Phi_{s,t}$ *are* C^∞*-maps from* $\mathcal{D}_{s,t}$ *to* $\check{\mathcal{D}}_{s,t}$ *a.s. Further, these are one to one and onto and inverse maps are* C^∞ *a.s.*
2. *For any* $0 < s < t < T$, *it holds a.s. that* $\check{\Psi}_{s,t}(\Phi_{s,t}(x)) = x$ *if* $x \in \mathcal{D}_{s,t}$ *and* $\Phi_{s,t}(\check{\Psi}_{s,t}(x)) = x$ *if* $x \in \check{\mathcal{D}}_{s,t}$.

Proof Let $\acute{V}_k(\acute{x}, t)$ be vector fields on the Euclidean space $\mathbb{R}^d$ defined in the proof of the previous theorem. Let $\{\acute{\Phi}_{s,t}\}$ be the forward flow on $\mathbb{R}^d$ generated by SDE with coefficients $\acute{V}_k(\acute{x}, t), k = 0, \ldots, d'$ and let $\{\bar{\Psi}_{s,t}\}$ be the backward flow on $\mathbb{R}^d$ generated by the backward SDE with coefficients $-\acute{V}_k(\acute{x}, t), k = 0, \ldots, d'$. Since these coefficients are of $C_b^{\infty,1}$-class, the explosion time of $\acute{\Phi}_{s,t}(\acute{x})$ is ∞ for any $\acute{x}, s$ a.s. and the explosion time of the backward flow $\bar{\Psi}_{s,t}(\acute{x})$ is $-\infty$ for any $t, \acute{x}$ a.s. as we have seen in Chap. 3. Further, we saw in Sect. 3.7 that $\acute{\Phi}_{s,t}(\bar{\Psi}_{s,t}(\acute{x})) = \bar{\Psi}_{s,t}(\acute{\Phi}_{s,t}(\acute{x})) = \acute{x}$ holds for all $0 < s < t$ and $\acute{x} \in \mathbb{R}^d$, a.s.

Let U_α be a local chart of M. We transform $\acute{\Phi}_{s,t}$ and $\bar{\Psi}_{s,t}$ by the inverse map ψ_α^{-1}. Set $\Phi_{s,t} = \psi_\alpha^{-1}(\acute{\Phi}_{s,t})$ and $\check{\Psi}_{s,t} = \psi_\alpha^{-1}(\bar{\Psi}_{s,t})$. Then if $x \in U_\alpha$ and $\tau_{U_\alpha}(x, s) > t$, we have $\check{\tau}_\infty(\Phi_{s,t}(x), t) < s$ and $\check{\Psi}_{s,t}(\Phi_{s,t}(x)) = x$ holds. The same fact holds if $x \in M$ satisfies $\tau_\infty(x, s) > t$. Repeating this argument inductively, we find that

$$\tau_\infty(x, s) > t \Longrightarrow \check{\tau}_\infty(\Phi_{s,t}(x), t) < s \text{ and } \check{\Psi}_{s,t}(\Phi_{s,t}(x)) = x.$$

Consequently, $\Phi_{s,t}$ maps $\mathcal{D}_{s,t}$ into $\check{\mathcal{D}}_{s,t}$, a.s.

Next, apply a similar argument to $\check{\Psi}_{s,t}$ instead of $\Phi_{s,t}$. Then we find that $\check{\Psi}_{s,t}$ maps $\check{\mathcal{D}}_{s,t}$ into $\mathcal{D}_{s,t}$ and the equality $\Phi_{s,t}(\check{\Psi}_{s,t}(x)) = x$ holds for $x \in \check{\mathcal{D}}_{s,t}$. Consequently, the map $\Phi_{s,t} : \mathcal{D}_{s,t} \to \check{\mathcal{D}}_{s,t}$ is one to one and onto, and further, the inverse is a C^∞-map. See the discussion at the beginning of Sect. 3.7. Therefore we get the assertion of the lemma. □

We next consider the case with jumps. Let $\hat{V}_k(t)$ be tangent vector fields of $\{\phi_{t,z}^{-1}\}$ at $z = 0$. Then it holds that $\hat{V}_k(t)f = -\tilde{V}_k(t)f, k = 1, \ldots, d'$, where $\tilde{V}_k(t)f$ are tangent vector fields of $\{\phi_{t,z}\}$ at $z = 0$.

Lemma 7.1.2 *Let* $\{\Phi_{s,t}, \mathcal{D}_{s,t}\}$ *be the stochastic flow of* C^∞*-maps on the manifold* M *defined by symmetric SDE* (7.2). *Let* $\{\check{\Psi}_{s,t}, \check{\mathcal{D}}_{s,t}\}$ *be the backward stochastic flow of* C^∞*-maps on the manifold* M *defined by the backward symmetric SDE* (7.5). *Then assertions 1 and 2 of Lemma* 7.1.1 *are valid.*

Before the proof, we prepare another lemma.

Lemma 7.1.3 *Let* (U_α, ψ_α) *be a local chart of* M. *For* $x \in U_\alpha$, *set* $\acute{x} = \psi_\alpha(x)$, *which is an element of* $\mathbb{R}^d$. *For maps* $\phi_{t,z}(x)$ *satisfying Condition (J.1)', there exist* $\delta_0 > 0$ *and* $\acute{g}(\acute{x}, t, z)$ *such that* $\acute{\phi}_{t,z}(\acute{x}) := \acute{g}(\acute{x}, t, z) + x$ *are diffeomorphisms on* $\mathbb{R}^d$ *for* $|z| < \delta_0, t \in \mathbb{T}$ *and* $\psi_\alpha(\phi_{t,z}(x)) = \acute{\phi}_{t,z}(\acute{x})$ *holds for* $x \in U_\alpha$.

Proof Choose a sufficiently small $\delta_0 > 0$ and a function $\acute{g}(\acute{x}, t, z)$, $\acute{x} \in \mathbb{R}^d$, $|z| < \delta_0$ such that for any z the equality $\acute{g}(\acute{x}, t, z) = \psi_\alpha(\phi_{t,z}(x)) - \acute{x}$ holds if $\acute{x} \in \psi_\alpha(U_\alpha)$ and $|\partial_{\acute{x}}\acute{g}(\acute{x}, t, z)| < 1/2$ holds for any $\acute{x} \in \mathbb{R}^d$. Then the map $\acute{\phi}_{t,z}; \mathbb{R}^d \to \mathbb{R}^d$ is one to one for any t, z, since $|\partial_{\acute{x}}\acute{g}(\acute{x}, t, z)| \leq 1/2$ for any $\acute{x}, t, z$. We will show that maps $\acute{\phi}_{t,z}$ are onto for any z. Let $\bar{\mathbb{R}}^d$ be the one point compactification of $\mathbb{R}^d$. It is homeomorphic to the d-dimensional unit sphere. We set $\acute{\phi}_{t,z}(\infty) = \infty$. Since $\lim_{\acute{x}\to\infty} \acute{\phi}_{t,z}(\acute{x}) = \infty$ holds, the extended map $\acute{\phi}_{t,z}(\acute{x}); \{|z| \leq \delta_0\} \times \bar{\mathbb{R}}^d \to \bar{\mathbb{R}}^d$ is continuous. Further, $\acute{\phi}_{t,0}$ is the identity. Then the map $\acute{\phi}_{t,z}$ is an onto map for any t, z by the homotopy theory. This indicates that the map $\acute{\phi}_{t,z}; \mathbb{R}^d \to \mathbb{R}^d$ is also an onto map for any z. We have thus seen that $\acute{\phi}_{t,z}$ are diffeomorphisms for any t and $|z| < \delta_0$. □

Proof of Lemma 7.1.2 Let δ_0 be the positive number of Lemma 7.1.3 and let $0 < \delta < \delta_0$. We consider a forward SDE with jumps on the manifold M with characteristics $(V_k(x, t), k = 0, \ldots, d', \phi_{t,z}(x), \nu_\delta)$, where the associated Poisson random measure is $N^\delta(dr\,dz) = 1_{(0,\delta]}(z)N(dr\,dz)$ and its Lévy measure ν_δ is given by $\nu_\delta(dz) = 1_{(0,\delta]}(|z|)\nu(dz)$. The forward stochastic flow generated by the SDE is denoted by $\{\Phi^\delta_{s,t}, \mathcal{D}^\delta_{s,t}\}$. Further, we consider a backward SDE with jumps with characteristics $(-V_k(x, t), k = 0, \ldots, d', \phi^{-1}_{t,z}, \nu_\delta)$. The backward stochastic flow generated by the backward SDE is denoted by $\{\check{\Psi}^\delta_{s,t}, \check{\mathcal{D}}^\delta_{s,t}\}$.

The forward equation on M is transformed to a forward equation on $\mathbb{R}^d$ with characteristics $(\acute{V}_k(\acute{x}, t), k = 0, \ldots, d', \acute{g}(\acute{x}, t, z), \nu_\delta)$, where $\acute{g}(x, t, z)$ is the function defined in Lemma 7.1.3. The backward SDE on M is transformed to a backward equation on $\mathbb{R}^d$ with characteristics $(-\acute{V}_k(\acute{x}, t), k = 0, \ldots, d', -\acute{h}(\acute{x}, t, z), \nu_\delta)$, where $\acute{h}(\acute{x}, t, z) = x - \acute{\phi}^{-1}_{t,z}(x)$. Let $\acute{\Phi}^\delta_{s,t}$ be the flow on $\mathbb{R}^d$ generated by the forward SDE and let $\acute{\Psi}^\delta_{s,t}$ be the backward flow on $\mathbb{R}^d$ generated by the backward SDE. Since $\acute{\phi}_{t,z}(\acute{x}) = \acute{g}(\acute{x}, t, z) + \acute{x}$ are diffeomorphic and $\acute{\phi}^{-1}_{t,z}(\acute{x}) = \acute{h}(\acute{x}, t, z) + \acute{x}$, these flows satisfy $\acute{\Phi}^\delta_{s,t}(\acute{\Psi}^\delta_{s,t}(\acute{x})) = \acute{\Psi}^\delta_{s,t}(\acute{\Phi}^\delta_{s,t}(\acute{x})) = \acute{x}$ on the Euclidean space. Further, it holds that $\Phi^\delta_{s,t}(x) = \psi^{-1}_\alpha(\acute{\Phi}^\delta_{s,t}(\acute{x}))$ for $t < \tau^\delta_{D_\alpha}(x, s)$ (the first leaving time of $X_t = X^{x,s,\delta}_t = \Phi^\delta_{s,t}(x)$ from D_α) and $\check{\Psi}^\delta_{s,t}(x) = \psi^{-1}_\alpha(\acute{\Psi}^\delta_{s,t}(\acute{x}))$ for $s < \sigma^\delta_{D_\alpha}(x, t)$ (the backward first leaving time of $\check{X}^{x,t,\delta}_s = \acute{\Psi}^\delta_{s,t}(x)$ from D_α). Therefore we have $\Phi^\delta_{s,t}(\check{\Psi}^\delta_{s,t}(x)) = \check{\Psi}^\delta_{s,t}(\Phi^\delta_{s,t}(x)) = x$ for $x \in D_\alpha$ if $t < \tau^\delta_{D_\alpha}(x, s)$. Then we can prolong the equality $\Phi^\delta_{s,t}(\check{\Psi}^\delta_{s,t}(x)) = \check{\Psi}^\delta_{s,t}(\Phi^\delta_{s,t}(x)) = x$ for $x \in M$ and $t < \tau_1(x, s)$.

Now the stochastic flow $\Phi_{s,t}(x)$ on the manifold M is decomposed as

$$\Phi_{s,t}(x) = \begin{cases} \Phi^\delta_{s,t}(x), & \text{if } t < \tau_1 \\ \Phi^\delta_{\tau_n,t} \circ \phi_{\tau_n,S_n} \cdots \phi_{\tau_1,S_1} \circ \Phi^\delta_{s,\tau_1}(x), & \text{if } \tau_n \leq t < \tau_{n+1}, \end{cases} \tag{7.7}$$

where τ_n are jumping times of the Poisson process $N^\delta_t = N((s, t] \times \{|z| > \delta\})$ and $S_1, \ldots, S_n$ are random variables given by $S_i = \int\int_{\{(\tau_i,z);|z|>\delta\}} zN(dr\,dz)$, $i = 1, \ldots, n$. Further the backward flow $\check{\Psi}_{s,t}(x)$ is represented by

$$
\check{\Psi}_{s,t}(x) = \begin{cases} \check{\Psi}^{\delta}_{s,t}(x), & \text{if } t < \tau_1, \\ \check{\Psi}^{\delta}_{s,\tau_1} \circ \phi^{-1}_{\tau_1,S_1} \cdots \phi^{-1}_{\tau_n,S_n} \circ \check{\Psi}^{\delta}_{\tau_n,t}(x), & \text{if } \tau_n \leq t < \tau_{n+1}. \end{cases} \tag{7.8}
$$

Therefore we get $\Phi_{s,t}(\check{\Psi}_{s,t}(x)) = \check{\Psi}_{s,t}(\Phi_{s,t}(x)) = x$ if $t < \tau_{D_\alpha}(x, s)$. We can prolong the equality $\Phi_{s,t}(\check{\Psi}_{s,t}(x)) = \check{\Psi}_{s,t}(\Phi_{s,t}(x)) = x$ for any x and $t < \tau_1(x, s)$. Repeating this argument, the equality holds for any $x \in M$ and $t < \tau_\infty(x, s)$. Finally, the equality implies that maps $\Phi_{s,t}; \mathcal{D}_{s,t} \to \check{\mathcal{D}}_{s,t}$ are one to one and onto a.s. and further, the inverse maps are C^∞. Hence we established the lemma. □

From the above lemmas, we get the following assertion.

Theorem 7.1.2 *Assume that $V_k(t), k = 0, \ldots, d'$ are $C^{\infty,1}$-vector fields and $\{\phi_{t,z}\}$ satisfy Conditions (J.1)' and (J.2)' for the symmetric SDE* (7.2) *on the manifold M. Then maps $\Phi_{s,t}; \mathcal{D}_{s,t} \to M$ of Theorem* 7.1.1 *are diffeomorphic a.s. for any $s < t$.*

Let $\{\check{\Psi}_{s,t}, \check{\mathcal{D}}_{s,t}\}$ be the backward flow defined by SDE (7.5) *on the manifold M with characteristics $(-V_k(t), k = 0, \ldots, d', \phi^{-1}_{t,z}, \nu)$. Then $\Phi_{s,t}$ are maps from $\mathcal{D}_{s,t}$ onto $\check{\mathcal{D}}_{s,t}$. Further, it holds that $\check{\Psi}_{s,t}(x) = \Phi^{-1}_{s,t}(x)$ if $x \in \check{\mathcal{D}}_{s,t}$.*

The family of pairs $\{\Phi_{s,t}, \mathcal{D}_{s,t}\}$ satisfying properties of the above theorem is called a *stochastic flow of local diffeomorphisms on the manifold M*. SDE (7.2) is called *complete* if $\tau_\infty(x, s) = \infty$ holds a.s. for any x, s. Further, it is called *strongly complete* if $\inf_{y \in M} \tau_\infty(x, s) = \infty$ holds a.s for any s. It is equivalent to that $\mathcal{D}_{s,t} = M$ holds a.s. for any $s < t$.

Corollary 7.1.1 *Assume that SDE* (7.2) *on the manifold M is strongly complete. Then $\mathcal{D}_{s,t} = M$ holds a.s. and $\{\Phi_{s,t}\}$ satisfies the same properties stated in Theorem* 3.9.1, *replacing the state space $\mathbb{R}^d$ by the manifold M.*

Proposition 7.1.1 *If the manifold M is compact and SDE* (7.2) *is time homogeneous, the SDE is strongly complete.*

Proof Let us consider the stopping time $\tau_1(x, s)$ defined by (7.4). We define a stopping time by $\sigma_1(s) = \inf_{x \in M} \tau_1(x, s)$. Then $\sigma_1(s) > s$ a.s. since M is compact. Further, set $\sigma_n(s) = \sigma_1(\sigma_{n-1}(s))$ for $n \geq 2$. Then it holds that $\sigma_n(s) \leq \tau_n(x, s)$ for all x. Set $\sigma_\infty(s) = \lim_{n \to \infty} \sigma_n(s)$. We want to prove $\sigma_\infty(s) = \infty$ for any s a.s. This should imply that the SDE is strongly complete.

Consider a sequence of random variables $\sigma_{n+1}(s) - \sigma_n(s), n = 1, 2, \ldots$. Note that $\sigma_{n+1}(s) - \sigma_n(s) = \sigma_1(\sigma_n(s)) - \sigma_n(s)$. Since $\sigma_1(t) - t$ is $\mathcal{F}_{t,\infty}$-measurable and $1_{\sigma_n(s) \leq t}$ is $\mathcal{F}_{0,t}$-measurable, these two random variables are independent. Therefore, for any bounded continuous function f we have

$$
E[f(\sigma_{n+1}(s) - \sigma_n(s))|\mathcal{F}_{\sigma_n}] = E[f(\sigma_1(t) - t)]|_{t=\sigma_n(s)} = E[f(\sigma_1(0))],
$$

because laws of $\sigma_1(t) - t$ are common for all t. This proves that $\sigma_{n+1}(s) - \sigma_n(s)$ and $\mathcal{F}_{0,\sigma_n}$ are independent.

The above discussion shows that $\sigma_{n+1}(s) - \sigma_n(s), n = 1, 2, \ldots$ are independent and identically distributed. Since

$$\sigma_N(s) - s = \sum_{n=0}^{N-1} \{\sigma_{n+1}(s) - \sigma_n(s)\}, \quad (\sigma_0(s) = s)$$

$(\sigma_N(s) - s)/N$ converges to a positive constant, by law of the large numbers. This proves $\sigma_\infty(s) = \infty$ a.s for all $s > 0$. Finally, $\sigma_N(s)$ is right continuous in s for any N. Then $\sigma_\infty(s) = \infty$ holds for all s a.s. □

Remark Let N be a connected component of the manifold M. Then jump-maps $\phi_{t,z}$ maps N onto itself diffeomorphically. Indeed, since $\phi_{t,0}$ are identity maps and $\phi_{t,z}(x)$ are continuous with respect to t, z, x, we have $\phi_{t,z}(N) \subset N$. Then the stochastic flow $\Phi_{s,t}$ should also map N onto itself diffeomorphically a.s. Hence it is sufficient that we consider SDE and stochastic flow on each connected components.

We will relax Conditions (J.1)' and (J.2)' to the following:

Condition $(J.1)'_K$. *(i) The map $(x, t, z) \to \phi_{t,z}(x)$ is of $C^{\infty,1,2}$-class on the space $M \times \mathbb{T} \times \{|z| \leq c\}$ for some $c > 0$. (ii) It is of C^∞-class on M for any $t \in \mathbb{T}$, $|z| > c$ and is piecewise continuous in $(t, z) \in \mathbb{T} \times \{|z| > c\}$.*

Condition $(J.2)'_K$. *For any t, z, the map $\phi_{r,z}; M \to M$ is a diffeomorphism. Further, the inverse maps $\phi_{t,z}^{-1}$ satisfy Condition $(J.1)'_K$.*

Then the assertion of Theorem 7.1.2 is valid. In this case, the stochastic flow $\Phi_{s,t}$ may jump from a connected component N to other connected components.

Note The continuous SDE on manifolds was studied by Ikeda–Watanabe [41]. Our definition of the SDE on manifold is close to their definition.

We saw in Chap. 3 that the SDE on a Euclidean space is strongly complete, if coefficients $V_k(x, t)$ are of $C_b^{\infty,1}$-class and $g(x, t, z)$ are $C_b^{\infty,2,1}$-class. It is expected to get criteria for vector fields $V_k(t)$ and jump-maps $\phi_{t,z}$ under which an SDE is strongly complete. For continuous SDEs or diffusion processes, we refer to Xue-Mei Li [76].

The SDE with jumps on manifolds was studied by Fujiwara [29] and Kunita [60]. The present definition of the SDE is taken from [60]. For the construction of stochastic flows on a compact manifold, Fujiwara [29] applied the Whitney embedding theorem of a manifold into a higher-dimensional Euclidean space. Assumptions required in the paper are different from ours.

7.2 Diffusion, Jump-Diffusion and Their Duals on Manifold

In Sect. 7.1, we studied a symmetric SDE on a C^∞-manifold M of dimension d. In the sequel, we assume that the Lévy measure has a weak drift. Then the last term (jump part) of (7.2) is split into the difference of two terms:

$$\int_{t_0}^{t}\int_{|z|>0+}\Big\{f(\phi_{r,z}(X_{r-}))-f(X_{r-})\Big\}N(dr\,dz)-\int_{t_0}^{t}\sum_{k=1}^{d'}b_0^k\tilde{V}_k(r)f(X_r)\,dr,$$

where the first term is the improper integral by the Poisson random measure N and $b_0^k=\lim_{\epsilon\to 0}b_\epsilon^k$. Then rewriting $V_0(t)-\sum_k b_0^k\tilde{V}_k(t)$ as $V_0(t)$, SDE (7.2) can be rewritten simply as

$$\begin{aligned}f(X_t)=f(X_0)+&\sum_{k=0}^{d'}\int_{t_0}^{t}V_k(r)f(X_r)\circ dW_r^k\\&+\int_{t_0}^{t}\int_{|z|>0+}\big\{f(\phi_{r,z}(X_{r-}))-f(X_{r-})\big\}N(dr\,dz).\end{aligned}\tag{7.9}$$

We define an integro-differential operator $A_J(t)$ by

$$A_J(t)f=\frac{1}{2}\sum_{k=1}^{d'}V_k(t)^2f+V_0(t)f+\int_{|z|>0+}\{f\circ\phi_{t,z}-f\}\nu(dz).\tag{7.10}$$

Then, using the Itô integral, Eq. (7.9) is rewritten as

$$\begin{aligned}f(X_t)=f(X_0)+&\int_{t_0}^{t}A_J(r)f(X_r)\,dr+\sum_{k=1}^{d'}\int_{t_0}^{t}V_k(r)f(X_r)\,dW_r^k\\&+\int_{t_0}^{t}\int_{|z|>0+}\big\{f(\phi_{r,z}(X_{r-}))-f(X_{r-})\big\}\tilde{N}(dr\,dz).\end{aligned}\tag{7.11}$$

It is called a symmetric SDE with characteristics $(V_k(t), k=0,\ldots,d', \phi_{t,z},\nu)$.

We will define stochastic flow generated by the above SDE slightly different from that defined in Sect. 7.1. Let $M'=M\cup\{\infty\}$, where ∞ is a cemetery adjoined to the manifold M as a one-point compactification if M is noncompact, and as an isolated point if M is compact. For a given function f on M, we extend it to a function on $M\cup\{\infty\}$ by setting $f(\infty)=0$.

Let $X_t^{x,s}, t<\tau_\infty(x,s)$ be the solution starting from x at time s, where $\tau_\infty(x,s)$ is the terminal time. For $t\geq\tau_\infty(x,s)$, we set $X_t^{x,s}=\infty$. Then $X_t^{x,s}, t\in[s,\infty)$ is an $M\cup\{\infty\}$-valued process and satisfies the above equation (7.11) for all $t>s$. Theorems 7.1.1 and 7.1.2 tell us that $X_t^{x,s}, t>s$ has a modification $\{\Phi_{s,t}(x), t>s\}$ satisfying the following properties:

1. The terminal time $\tau_\infty(x,s)$ is lower semi-continuous with respect to x for any s.
2. Set $\mathcal{D}_{s,t}=\{x;\tau_\infty(x,s)>t\}$. These are open subsets of M a.s., for any $s<t$. Further, maps $\Phi_{s,t};\mathcal{D}_{s,t}\to M$ are C^∞ a.s.
3. If jump-maps $\phi_{t,z}$ satisfy Condition (J.2)' or identity maps, then maps $\Phi_{s,t}$: $\mathcal{D}\to M$ are into diffeomorphisms a.s.

We set $\Phi_{s,t}(\infty) = \infty$ for any $t \geq s$. Then $\Phi_{s,t}$ are maps from $M \cup \{\infty\}$ into itself and satisfy $\Phi_{t,u}(\Phi_{s,t}(x)) = \Phi_{s,u}(x)$ for any $x \in M \cup \{\infty\}$ and $s < t < u$. The family of maps $\{\Phi_{s,t}, s < t\}$ is called again the *stochastic flow of local diffeomorphisms* generated by a forward SDE with characteristics $(V_k(t), k = 0, \ldots, d', \phi_{t,z}, \nu)$.

Now, since $\Phi_{s,t}(x)$ and $\Phi_{t,u}(x)$ are independent for any $s < t < u$, the stochastic process $X_t^{x,s} = \Phi_{s,t}(x)$ is a Markov process of initial state (x, s) with transition function

$$P_{s,t}(x, E) = P(\Phi_{s,t}(x) \in E) = P(\Phi_{s,t}(x) \in E, \tau_\infty(x, s) > t)$$

for $x \in M$ and $E \subset M$. The Markov process $X_t^{x,s}$ is called a *jump-diffusion process on the manifold* M with characteristics $(V_k(t), k = 0, \ldots, d', \phi_{t,z}, \nu)$. For $f \in C_b(M)$, we set $P_{s,t} f(x) := \int_M P_{s,t}(x, dy) f(y)$. It is written as $P_{s,t} f(x) = E[f(\Phi_{s,t}(x))] = E[f(\Phi_{s,t}(x)) 1_{\tau_\infty(x,s)>t}]$. Then it is a bounded continuous function of x. Hence $\{P_{s,t}\}$ is a semigroup of linear transformations of $C_b(M)$.

The above Markov process $X_t^{x,s} = \Phi_{s,t}(x), t \in [s, \infty)$ satisfies equation (7.11) with the initial condition $X_0 = x$ and $t_0 = s$. Take expectations for each term of (7.11). Expectations of the third and the fourth terms of the right-hand side are 0, since these are martingales. Consequently, we get

$$E[f(\Phi_{s,t}(x))] = f(x) + E\Big[\int_s^t A_J(r) f(\Phi_{s,r}(x))\, dr\Big].$$

Therefore we have $P_{s,t} f(x) = f(x) + \int_s^t P_{s,r} A_J(r) f(x)\, dr$ for any $f \in C_0^\infty(M)$. This shows that the semigroup $\{P_{s,t}\}$ satisfies Kolmogorov's forward equation with respect to the operator $A_J(t)$.

If $\phi_{t,z}$ are identity maps for any t, z in equation (7.9), the Markov process X_t is called a *diffusion process on the manifold* M with coefficients $(V_k(t), k = 0, \ldots, d')$. In this case, we define a differential operator $A(t)$ by

$$A(t) f(x) = \frac{1}{2} \sum_{k=1}^{d'} V_k(t)^2 f(x) + V_0(t) f(x). \tag{7.12}$$

Then the semigroup satisfies $P_{s,t} f(x) = f(x) + \int_s^t P_{s,r} A(r) f(x)\, dr$ for any $f \in C_0^\infty(M)$. Hence the semigroup $\{P_{s,t}\}$ satisfies Kolmogorov's forward equation with respect to the operator $A(t)$.

It is expected that if f is in $C_0^\infty(M)$, $P_{s,t} f(x)$ should be smooth with respect to x and should satisfy Kolmogorov's backward equation. If the associated stochastic flow is strongly complete, these fact is true, but if it is not strongly complete, these facts are not obvious, since it is not evident whether $P_{s,t} f(x)$ is smooth or not with respect to x. Then it is not clear whether the (backward) heat equation associated the operator $A(t)$ or $A_J(t)$ has a solution. In Sect. 7.4, we will show these facts for elliptic diffusion, by constructing the fundamental solution for the differential operator $A(t)$.

Let us consider the dual processes, assuming that the manifold M is orientable. We will restrict our attention to jump-diffusions. A similar result for diffusions will be obtained with the obvious change. From now, we assume further that jump-maps $\phi_{t,z}$ satisfy Condition (J.2)' in Sect. 7.1. Hence these maps are diffeomorphic. Let dx be a given volume element of the manifold. We will study the dual of the above jump-diffusion with respect to dx. An operator $A_J(t)^*; C_0^\infty(M) \to C^\infty(M)$ is called a dual of the operator $A_J(t)$ with respect to dx if it satisfies

$$\int_M A_J(t)f(x)\cdot g(x)\,dx = \int_M f(x)\cdot A_J(t)^*g(x)\,dx$$

for any $f, g \in C_0^\infty(M)$. It is computed directly, using formulas of the integration by parts and the change of variables. Indeed, similarly to the dual in the Euclidean space (Sect. 4.6), it is written for any $g \in C_0^\infty(M)$ as

$$A_J(t)^*g = \frac{1}{2}\sum_{k=1}^{d'}(V_k(t) + \mathrm{div}\, V_k(t))^2 g - (V_0(t) + \mathrm{div}\, V_0(t))g \qquad (7.13)$$
$$+ \int_{|z|>0+}\Big\{|J_{\phi_{t,z}^{-1}}|g\circ\phi_{t,z}^{-1} - g\Big\}\nu(dz).$$

Here, $\mathrm{div}\, V$ is the divergence of the vector field V with respect to the volume element dx (Appendix). Further, for a diffeomorphic map ϕ; $M \to M$, J_ϕ is defined as a C^∞-function such that the pullback $\phi^*\,dx$ coincides with $J_\phi\,dx$. It coincides with the Jacobian determinant of ϕ if M is a Euclidean space and dx is the Lebesgue measure. We call J_ϕ the Jacobian of ϕ.

Let $\check{\Psi}_{s,t}(x), s > \check{\tau}_\infty(x,t)$ be the backward flow of local diffeomorphisms generated by the backward SDE on the manifold M.

$$f(\check{X}_s) = f(x) - \sum_{k=0}^{d'}\int_s^t V_k(r)f(\check{X}_r)\circ dW_r^k + \int_s^t\int_{|z|>0+}\{f(\phi_{r,z}^{-1}(\check{X}_r)) - f(\check{X}_r)\}N(dr\,dz)$$

for $t > s > \check{\tau}_\infty$ for any $f \in C_0^\infty(M)$. Let $\Phi_{s,t}^{-1}$ be the inverse map of the flow $\Phi_{s,t}$. We saw in Sect. 7.1 that $\Phi_{s,t}^{-1}(x) = \check{\Psi}_{s,t}(x)$ holds for any $t < \tau_\infty(x,s)$. For $s < \check{\tau}_\infty(t,x)$, we set $\check{\Psi}_{s,t}(x) = \infty$ and $\check{\Psi}_{s,t}(\infty) = \infty$ for any $0 < s < t$. Then $\check{X}_s = \check{\Psi}_{s,t}(x)$ is a backward Markov process. Its generator is given by

$$\mathcal{A}_J(t)g = \frac{1}{2}\sum_{k=1}^{d'} V_k(t)^2 g - V_0(t)g + \int_{|z|>0+}\{g\circ\phi_{t,z}^{-1} - g\}\nu(dz). \qquad (7.14)$$

Then the dual operator $A_J(t)^*$ is obtained by the transformation of the operator $\mathcal{A}_J(t)$ as

$$A_J(t)^* g = \mathcal{A}_J(t)^{-\mathbf{div}\,\mathbf{V}, |\mathbf{J}_{\phi^{-1}}|} g. \tag{7.15}$$

For the construction of the dual process, we need an additional condition.

Condition (D) (Dual condition)

(i) div $V_k(t)$, $V_k(t)$(div $V_k(t)$), $k = 0, \ldots, d'$ *are bounded* $C^{\infty,1}$*-functions on the space* $M \times \mathbb{T}$.

(ii) *The improper integral* $\int_{|z|>0+} \{|J_{\phi_{t,z}^{-1}}| - 1\}\nu(dz)$ *is a bounded function of* x, t *and is smooth with respect to* x.

The above condition is always satisfied if the manifold M is compact.

In the case where the manifold M is an Euclidean space and dx is the Lebesgue measure, Condition (D) tells us that coefficients $V_k(x,t)$, $\phi_{t,z}(x) - x = g(x,t,z)$ of the SDE might be unbounded, but div $V_k(x,t)$, $V_k(t)$(div $V_k(t)$)(x) and $\int_{|z|>0}(|\det \nabla\phi_{t,z}(x)| - 1)\nu(dz)$ should be bounded.

The function $c^*(t) = c^*(x,t) = A_J(t)^* 1(x)$ is written as

$$\begin{aligned} c^*(t) &= -\text{div}\, V_0(t) + \frac{1}{2}\sum_{k=1}^{d'}\{V_k(t)(\text{div}\, V_k(t)) + (\text{div}\, V_k(t))^2\} \\ &\quad + \int_{|z|>0+} (|J_{\phi_{t,z}^{-1}}| - 1)\nu(dz). \end{aligned} \tag{7.16}$$

Then the dual condition makes sure that the function $c^*(x,t)$ is a bounded function on $M \times \mathbb{T}$. Then we can construct the dual jump-diffusion with generator (7.13).

We apply the change-of-variable formula (A.4) in the Appendix for open submanifolds $\mathcal{D}_{s,t} \equiv \{x; \tau_\infty(x,s) > t\}$. Replace $f(x)$ by $f(\Phi_{s,t}(x))g(x)$ and ϕ by $\check{\Psi}_{s,t}$. Note that $\check{\Psi}_{s,t}$ are diffeomorphic maps from $\check{\mathcal{D}}_{s,t} \equiv \{x; \check{\tau}_\infty(x,t) < s\}$ to $\mathcal{D}_{s,t}$ by Theorem 7.1.2, and the equality $\Phi_{s,t}(\check{\Psi}_{s,t}(x)) = x$ holds. Then we get the formula of the change of variables:

$$\int_{\mathcal{D}_{s,t}} f(\Phi_{s,t}(x))g(x)\,dx = \int_{\check{\mathcal{D}}_{s,t}} f(x)g(\check{\Psi}_{s,t}(x))|J_{\check{\Psi}_{s,t}}(x)|\,dx, \quad a.s.\ P,$$

for any functions $f, g \in C_0(M)$. On the set $\mathcal{D}_{s,t}^c$, we have $f(\Phi_{s,t}(x)) = f(\infty) = 0$, and on the set $\check{\mathcal{D}}_{s,t}^c$ we have $g(\check{\Psi}_{s,t}(x)) = g(\infty) = 0$. If $\check{\tau}_\infty(x,t) > s$, we set $J_{\check{\Psi}_{s,t}}(x) = 0$. Then we get the equality

$$\int_M f(\Phi_{s,t}(x))g(x)\,dx = \int_M f(x)g(\check{\Psi}_{s,t}(x))|J_{\check{\Psi}_{s,t}}(x)|\,dx, \quad a.s.\ P. \tag{7.17}$$

Lemma 7.2.1 *$|J_{\check{\Psi}_{s,t}}|$, $\check{\tau}_\infty(x,t) < s < t$ is written as a backward exponential functional with coefficients* $-\mathbf{div}\,\mathbf{V} = (-\mathrm{div}\, V_0(x,t), \ldots, -\mathrm{div}\, V_{d'}(x,t))$ *and* $|\mathbf{J}_{\phi^{-1}}| = (|J_{\phi_{t,z}^{-1}}(x)|)$*:*

$$\exp\Big\{\sum_{k=0}^{d'} \int_s^t -\mathrm{div}\, V_k(t)(\check{\Psi}_{r,t}) \circ dW_r^k + \int_s^t \int_{|z|>0+} \log |J_{\phi_{r,z}^{-1}}(\check{\Psi}_{r,t})|\, dN\Big\}. \tag{7.18}$$

Further, under condition (D), $E[|J_{\check{\Psi}_{s,t}}(x)|]$ is bounded with respect to $s < t$ and x.

Proof We showed the above lemma for the stochastic flow on a Euclidean space (Proposition 4.6.2). What we need in the proof are the following three facts:

1. Differential rule for the flow $\{\Phi_{s,t}\}$ with respect to the backward variable s.
2. Change-of-variable formula.
3. Formula of integration by parts.

We know that the first one holds for stochastic flows on manifold. The second and the third are known as we have mentioned above. Therefore, discussions in the proof of Lemma 4.6.1 are valid to the present case and we get the expression (7.18).

Further, (7.18) is written as the product of three backward exponential functionals $G_{s,t}^{(0)}$, $G_{s,t}^{(1)}$ and $G_{s,t}^{(2)}$ as in Sect. 4.6. Since $c^*(x,t)$ is a bounded function, we find that $E[|J_{\check{\Psi}_{s,t}}(x)|1_{\check{\tau}_\infty(x,t)<s}]$ is bounded. □

In view of the above lemma,

$$P_{s,t}^* g(x) := E\big[g(\check{\Psi}_{s,t}(x))|J_{\check{\Psi}_{s,t}}(x)|\big] \tag{7.19}$$

is well defined for any $g \in C_0(M)$. It is a linear transformation from $C_0(M)$ to $C_b(M)$. Taking expectations for both terms of (7.14), we have

$$\int_M P_{s,t} f(x) \cdot g(x)\, dx = \int_M E\big[f(\Phi_{s,t}(x))g(x)\big]\, dx = \int_M f(x) \cdot P_{s,t}^* g(x)\, dx,$$

for any $f, g \in C_0^\infty(M)$. Therefore $P_{s,t}^*$ is the dual of $P_{s,t}$ with respect to dx.

Theorem 7.2.1 *Assume that jump-maps $\phi_{t,z}$ in SDE (7.9) on the orientable manifold M satisfy (J.1)', (J.2)' and Condition (D). Let $\{P_{s,t}\}$ be the semigroup of the stochastic flow $\{\Phi_{s,t}\}$ defined by the SDE (7.9). Then the semigroup $\{P_{s,t}\}$ has the dual semigroup $\{P_{s,t}^*\}$ with respect to a given volume element dx. It coincides with the backward semigroup of the inverse flow $\{\check{\Psi}_{s,t}\}$ transformed by the backward exponential functional with coefficients* $-\mathbf{div}\,\mathbf{V}$ *and* $|\mathbf{J}_{\phi^{-1}}|$.

We will apply the above theorem to the case where M is a Euclidean space. Suppose that coefficients $V_k(t), k = 0, \ldots, d'$ of the operator $A_J(t)$ of (4.52) are C^∞-class. These may be unbounded and may have unbounded derivatives. However, adjoint operator $A_J(t)^*; C_0^\infty(\mathbb{R}^d) \to C_0^\infty(\mathbb{R}^d)$ is well defined. It is given by (4.68). Further, if Condition (D) is satisfied, then the semigroup of the jump-diffusion has a dual semigroup $\{P_{s,t}^*\}$.

7.3 Brownian Motion, Lévy Process and Their Duals on Lie Group

Let $X_t; t \in [0, \infty)$ be a cadlag process with values in a Lie group G, continuous in probability, and $X_0 = e$. It is said to have (left) independent increments if G-valued random variables $X_{t_{m-1}}^{-1} X_{t_m}, m = 1, \ldots, n$ are independent for any $0 \leq t_0 < \cdots < t_n$. It is (left) time homogeneous if the law of $X_s^{-1} X_t$ coincides with the law of X_{t-s} for any $s < t$. If X_t is (left) time homogeneous and has (left) independent increments, it is called a *Lévy process on the Lie group* G. Further, if it is a continuous process, it is called a *Brownian motion on the Lie group* G. Related notations and terminologies for Lie groups are collected in the Appendix at the end of this chapter.

We shall construct a Brownian motion on a Lie group G by solving an SDE on G. Let $V_0, V_1, \ldots, V_{d'}$ be left-invariant vector fields on the Lie group G. Regarding that G is a C^∞-manifold, we consider a continuous SDE on G with coefficients $V_k, k = 0, \ldots, d'$. A solution X_t starting from X_0 at time t_0 should satisfy

$$f(X_t) = f(X_0) + \sum_{k=0}^{d'} \int_{t_0}^{t} V_k f(X_r) \circ dW_r^k, \tag{7.20}$$

for any $f \in C_0^\infty(G)$. We denote the solution by $X_t^{X_0, t_0}$. The solution can be defined up to the explosion time $\tau_\infty = \tau_\infty(X_0, t_0)$.

Let us consider a solution starting from the unit element e at time 0. We denote the solution by $X_t, t < \tau_\infty$, simply. We want to show $\tau_\infty = \infty$ a.s. Let U be an open neighborhood of the origin e of G. We can define a sequence of stopping times σ_n such that $\sigma_n < \tau_\infty$ a.s. for any n and satisfies

$$\sigma_1 = \inf\{t > 0; X_t \notin U\}, \ldots, \sigma_n = \inf\{t > \sigma_{n-1}; X_{\sigma_{n-1}}^{-1} X_t \notin U\}. \tag{7.21}$$

Lemma 7.3.1 *The sequence of random variables $\sigma_1, \sigma_2 - \sigma_1, \ldots, \sigma_n - \sigma_{n-1}$ is independent and identically distributed.*

Proof Set $G' = G \cup \{\infty\}$ and $g \cdot \infty = \infty$ for any $g \in G$. We set $X_t = \infty$ if $t > \tau_\infty$. Then for a function f on G' such that $f(\infty) = 0$ and is C^∞ on G, we have

$$f(X_t) = f(X_{s \wedge \sigma_1}) + \sum_{k=0}^{d'} \int_{s \wedge \sigma_1}^{t} V_k f(X_r) \circ dW_r^k.$$

Since V_k are left-invariant vector fields, we have

$$f(xX_t) = f(xX_{s\wedge\sigma_1}) + \sum_{k=0}^{d'} \int_{s\wedge\sigma_1}^{t} V_k f(xX_r) \circ dW_r^k$$

for any $x \in G$. Setting $x = X_{s\wedge\sigma_1}^{-1}$, we get

$$f(X_{s\wedge\sigma_1}^{-1}X_t) = f(e) + \sum_{k=0}^{d'} \int_{s\wedge\sigma_1}^{t} V_k f(X_{s\wedge\sigma_1}^{-1}X_r) \circ dW_r^k.$$

Consequently, $X_{s\wedge\sigma_1}^{-1}X_t$ coincides with the solution $X_t^{e,s\wedge\sigma_1}$ of equation (7.20), in view of the pathwise uniqueness of the solution. This means that, for any s and σ_1, stochastic processes $\{X_t, t \leq s\wedge\sigma_1\}$ and $\{X_{s\wedge\sigma_1}^{-1}X_{t-s\wedge\sigma_1}, s\wedge\sigma_1 < t < 2s\wedge\sigma_2\}$ are independent and have the same law. Then random variables σ_1 and $\sigma_2 - \sigma_1$ are independent and identically distributed. Repeating this argument, we find that $\sigma_1, \ldots, \sigma_n - \sigma_{n-1}$ are independent and identically distributed. □

Now, the expectation of σ_1 is positive. Then by the law of large numbers, $\lim_{n\to\infty}\sigma_n = \infty$ holds a.s. Therefore we have $\tau_\infty = \infty$ a.s. Further, since the solution X_t is defined for all t, we have the formula

$$f(X_s^{-1}X_t) = f(e) + \sum_{k=0}^{d'} \int_{s}^{t} V_k f(X_s^{-1}X_r) \circ dW_r^k. \tag{7.22}$$

Then $X_s^{-1}X_t$ is independent of $\mathcal{F}_s$ for any $s < t$. It shows that X_t has left independent increments. Therefore we get the following theorem.

Theorem 7.3.1 *Any continuous SDE on a Lie group G with left-invariant vector fields $V_k, k = 0, \ldots, d'$ is strongly complete. The solution is a time homogeneous diffusion on G with the generator $Af = \frac{1}{2}\sum_{k=1}^{d'} V_k^2 f + V_0 f$. Let X_t be the solution starting from the unit element $e \in G$ at time 0. Then it is a Brownian motion on the Lie group G. Further, $\Phi_{s,t}(x) := xX_s^{-1}X_t$ is the stochastic flow of diffeomorphisms.*

We will consider the dual of X_t with respect to a left Haar measure m. Let $\exp sV_k, s \in (-\infty, \infty)$ be the one-parameter subgroup of G generated by the left-invariant vector field V_k. Then the modular function $\Delta(\exp sV_k)$ is a one-parameter subgroup of the multiplicative group $(0, \infty)$. We set $c_k = \frac{d}{ds}\Delta(\exp sV_k)|_{s=0}$ for $k = 0, \ldots, d'$. These constants satisfy

$$\int V_k f(x)g(x)\,dm = -\int f(x)V_k g(x)\,dm - c_k \int f(x)g(x)\,dm$$

for functions $f, g \in C_0^\infty(M)$. Therefore, the constant c_k coincides with $\operatorname{div} V_k$. Then the dual of A is represented by

$$A^* g(x) = \frac{1}{2} \sum_{k=1}^{d'} (V_k + c_k)^2 g(x) - (V_0 + c_0) g(x).$$

In particular, if the Lie group G is unimodular, then $c_k = 0$ holds for all k. Therefore the dual is represented by $A^* f = \frac{1}{2} \sum_{k \geq 1} V_k^2 f - V_0 f$. Hence the dual is a backward Brownian motion.

We shall next consider an SDE with jumps. We assume that the Lévy measure of the Poisson random measure has a weak drift. Let $V_k, k = 0, \ldots, d'$ be left-invariant vector fields on the Lie group G. Let $\phi(z)$ be a smooth map $\mathbb{R}^{d'} \to G$ such that $\phi(0) = e$. We consider an SDE on G with characteristics $(V_k, k = 0, \ldots, d', R_{\phi(z)}, \nu)$, where $R_{\phi(z)}$ is the right translation. Let $X_t, t < \tau_\infty$ be the solution starting from e at time 0, where τ_∞ is the terminal time. It satisfies, for any smooth function f and $0 \leq s < t < \tau_\infty$,

$$f(X_t) = f(X_s) + \sum_{k=0}^{d'} \int_s^t V_k f(X_r) \circ dW_r^k \tag{7.23}$$
$$+ \int_s^t \int_{|z|>0+} \{f(X_{r-}\phi(z)) - f(X_{r-})\} N(dr\, dz).$$

For the process X_t, we define a sequence of stopping times $\sigma_1, \sigma_2, \ldots$ by (7.21). Then the assertion of Lemma 7.3.1 is valid. Therefore $\tau_\infty = \infty$ holds a.s.

We will show that X_t has independent increments. Since V_k are left-invariant vector fields, we have

$$f(xX_t) = f(xX_s) + \sum_{k=0}^{d'} \int_s^t V_k f(xX_r) \circ dW_r^k$$
$$+ \int_s^t \int_{|z|>0+} \{f(xX_{r-}\phi(z)) - f(xX_{r-})\} N(dr\, dz).$$

The above equality holds for $x = X_s^{-1}$. Then we find that $X_s^{-1} X_t$ is a solution of the SDE with characteristics $(V_k, k = 0, \ldots, d', R_{\phi(z)}, \nu)$ starting from e at time s. Then X_s and $X_s^{-1} X_t$ are independent. This proves that X_t is a Lévy process on the Lie group G.

Theorem 7.3.2 *SDE (7.23) on a Lie group G with left-invariant vector fields $V_k, k = 0, \ldots, d'$ and right translation $R_{\phi(z)}$ is strongly complete.*

Let X_t be the solution starting from the unit element $e \in G$ at time 0. Then it is a Lévy process on the Lie group G. Further, $\Phi_{s,t}(x) := xX_s^{-1} X_t$ is the stochastic flow of diffeomorphisms generated by the SDE (7.23). Its generator is given by

$$A_J f(x) = \frac{1}{2}\sum_{k=1}^{d'} V_k^2 f(x) + V_0 f(x) + \int_{|z|>0+} \Big\{ f(x\phi(z)) - f(x) \Big\} \nu(dz). \tag{7.24}$$

We will consider the dual of the operator A_J with respect to the left Haar measure m. Consider the transformation $R_{\phi(z)}$. Regarding m as a positive d-form, we defined the function $J_{R_{\phi(z)}}$ by the relation $R_{\phi(z)}^* \, dm = J_{R_{\phi(z)}} \, dm$ in Sect. 4.8. Since m is a Haar measure, we have $R_{\phi(z)}^* \, dm = \Delta(\phi(z)) \, dm$. Therefore $J_{R_{\phi(z)}}(x)$ is a constant function of x and we have $J_{R_{\phi(z)}}(x) = \Delta(\phi(z))$. If

$$\int_{|z|>\delta} \Delta(\phi(z)) \nu(dz) < \infty \tag{7.25}$$

holds for some $\delta > 0$, the dual of A_J is well defined. It is written as

$$\begin{aligned} A_J^* g(x) &= \frac{1}{2}\sum_{k=1}^{d'} (V_k + c_k)^2 g(x) - (V_0 + c_0) g(x) \\ &\quad + \int_{|z|>0+} \Big\{ \Delta(\phi(z)) g(x\phi(z)^{-1}) - g(x) \Big\} \nu(dz). \end{aligned}$$

For a unimodular Lie group, it holds that $c_k = 0$ and $\Delta(\phi(z)) = 1$. Then the operator A_J^* is written simply as

$$A_J^* g(x) = \frac{1}{2}\sum_{k=1}^{d'} V_k^2 g(x) - V_0 g(x) + \int_{|z|>0+} \Big\{ g(x\phi(z)^{-1}) - g(x) \Big\} \nu(dz).$$

Denote the inverse X_s^{-1} of the Lévy process X_s by $\check{X}_s$. Then $\check{X}_s$ is a backward jump-diffusion with the generator $\mathcal{A}_J$ given by (7.14). It coincides with the above A_J^*. Thus if the Lie group G is unimodular, the inverse process $\check{X}_s$ coincides with the dual process with respect to the left Haar measure m.

Note It was shown by Itô [49] that a stationary Brownian motion on a Lie group is a stationary diffusion with the generator A represented by $Af = \sum_k a^k X_k f + \frac{1}{2}\sum_{i,k} a^{ik} X_i X_k f$, where $\{X_i\}$ is a basis of $\mathcal{L}(G)$ and (a^{ik}) is a nonnegative definite symmetric matrix. By the method of the diagonalization, the above operator is rewritten as $Af = V_0 f + \frac{1}{2}\sum_{k\geq 1} V_k^2 f$, where $V_0, \ldots, V_{d'}$ are elements of $\mathcal{L}(G)$. Therefore all Brownian motions on a Lie group can be obtained by solving SDEs of the form (7.20).

A Lévy process on a Lie group is studied in Hunt [39]. He obtained its generator explicitly, using left-invariant vector fields and a Lévy measure on the Lie group.

7.4 Smooth Density for Diffusion on Manifold

We will consider a diffusion on the orientable manifold M determined by SDE

$$f(X_t) = f(X_0) + \sum_{k=0}^{d'} \int_{t_0}^{t} V_k(r) f(X_r) \circ dW_r^k, \tag{7.26}$$

where f are C^∞-functions on M. Let $\{\Phi_{s,t}(x), t < \tau_\infty(x,s)\}$ be the stochastic flow generated by SDE (7.26), where $\tau_\infty(x,s)$ is the explosion time of the process $X_t^{x,s} := \Phi_{s,t}(x)$. We set $X_t^{x,s} = \Phi_{s,t}(x) = \infty$ if $t > \tau_\infty(x,s)$. Then $X_t^{x,s}$ is a (not necessarily conservative) diffusion process on M. If the tangent vectors $V_k(x,t), k = 1, \ldots, d'$ span the tangent space $T_x(M)$ for any $x \in M$ and t, the SDE is called *elliptic* and the above $X_t^{x,s}$ is called an *elliptic diffusion*.

Let $c_k(x,t), k = 0, \ldots, d'$ be $C^{\infty,1}$-functions on $M \times \mathbb{T}$. We assume that $c_k(x,t)$ and $V_k(t)c_k(x,t)$, $k = 0, \ldots, d'$ are bounded functions. Define the differential operator $A^{\mathbf{c}}(t)$ by

$$A^{\mathbf{c}}(t) f(x) = \frac{1}{2} \sum_{k=1}^{d'} (V_k(t) + c_k(t))^2 f(x) + (V_0(t) + c_0(t)) f(x). \tag{7.27}$$

Its potential part $c(t) = A^{\mathbf{c}}(t)1$ is a bounded function, since it is equal to

$$c(x,t) = c_0(x,t) + \frac{1}{2}\Big\{ \sum_{k=1}^{d'} V_k(t) c_k(x,t) + \sum_{k=1}^{d'} c_k(x,t)^2 \Big\}.$$

We set $c_k(\infty, t) = 0$ for $k = 0, \ldots, d'$ and define the exponential functional $G_{s,t}(x)$ by

$$G_{s,t}(x) = G_{s,t}^{\mathbf{c}}(x) = \exp\Big\{ \sum_{k=0}^{d'} \int_s^t c_k(\Phi_{s,r}(x), r) \circ dW_r^k \Big\}. \tag{7.28}$$

It satisfies $\sup_{s<t,x} E[|G_{s,t}(x)|^p] < \infty$ for any $p > 1$ (see the proof of Lemma 4.2.1).

Let f be a bounded continuous function on M. We set $f(\infty) = 0$. Then the semigroup $P_{s,t}^{\mathbf{c}} f(x) = E[f(\Phi_{s,t}(x)) G_{s,t}^{\mathbf{c}}(x)]$ satisfies Kolmogorov's forward equation associated with the differential operator $A^{\mathbf{c}}(t)$ defined by (7.27).

We will fix a volume element of M and will denote it by dx, dy etc. We will study the smooth density of the weighted transition function of an elliptic diffusion defined by $P_{s,t}^{\mathbf{c}}(x, E) = E[1_E(\Phi_{s,t}(x)) G_{s,t}^{\mathbf{c}}(x)]$, with respect to dy.

For a positive integer i, $d^i \phi$ is the i-th differential of ϕ and for $i = 0$, $d^0 \phi \equiv \phi$. When differentials are operated to the variable x of the $C^\infty(M \times M)$-function

$\phi(x, y)$, these are denoted by $d_x^i\phi$ and their localization at the point (x, y) are denoted by $d_x^i\phi(x, y)$. An objective of this section is to prove the following.

Theorem 7.4.1 *Consider an elliptic diffusion on the orientable manifold M determined by SDE (7.26). Assume that $c_k(x,t)$, $V_k(t)c_k(x,t)$ are bounded $C^{\infty,1}$-functions. Then, for any $0 \leq s < t$ and $x \in M$, the transition function $P_{s,t}^{\mathbf{c}}(x, E)$ weighted by $G_{s,t}^{\mathbf{c}}(x)$ of (7.28) has a C^∞-density $p_{s,t}^{\mathbf{c}}(x, y)$, $y \in M$ with respect to the volume element. Further, for any compact subset K of M, nonnegative integer j and $T > 0$, there exists a positive constant c such that*

$$\sup_{x,y\in K}\left|d_y^j p_{s,t}^{\mathbf{c}}(x,y)\right| \leq \frac{c}{(t-s)^{\frac{j+d}{2}}}, \quad 0 \leq s < t \leq T. \tag{7.29}$$

Assume further Condition (D) (i). Then the density function is a $C^{\infty,1}$-function of x, s $(< t)$ for any y, t. Further, for any compact subset K of M, nonnegative integer i and $T > 0$, there exists a positive constant c' such that

$$\sup_{x,y\in K}\left|d_x^i p_{s,t}^{\mathbf{c}}(x,y)\right| \leq \frac{c'}{(t-s)^{\frac{i+d}{2}}}, \quad 0 \leq s < t \leq T. \tag{7.30}$$

We will first show the existence of the smooth density on each local chart of M. Let V, U, D be a triple of local charts such that $\bar{V} \subset U \subset \bar{U} \subset D$ and $\bar{D}$ is compact. Then D is diffeomorphic to a bounded open set $\acute{D}$ of $\mathbb{R}^d$ by a diffeomorphism $\psi : D \to \acute{D}$. SDE (7.26) on D is transformed to an SDE on $\acute{D}$ by the diffeomorphic map ψ. Indeed, vector fields $V_k(t)$ on the local chart D are transformed to vector fields $d\psi(V_k(t))$ on $\acute{D}$ by the map ψ. Since $\acute{D}$ is relatively compact in $\mathbb{R}^d$, we can extend them to $C_b^{\infty,1}$-vector fields $\acute{V}_k(\acute{x}, t)$ on the whole space $\mathbb{R}^d$ such that these are uniformly elliptic. We consider an equation on $\mathbb{R}^d$:

$$d\acute{X}_t = \sum_{k=0}^{d'} \acute{V}_k(\acute{X}_t, t) \circ dW_t^k. \tag{7.31}$$

We denote the solution starting from $\acute{x}$ at time s by $\acute{\Phi}_{s,t}(\acute{x})$. Let $\acute{c}_k(\acute{x}, t)$ be $C_b^{\infty,1}$-functions on $\mathbb{R}^d$ such that $\psi(c_k(x,t)) = \acute{c}_k(\psi(x), t)$ holds on the local chart D. We define a positive functional $\acute{G}_{s,t}(\acute{x})$ similarly to $G_{s,t}(x)$, making use of functions $\acute{c}_k, k = 0, \ldots, d'$ on $\mathbb{R}^d$ and $\mathbb{R}^d$-valued process $\acute{\Phi}_{s,t}(\acute{x})$.

Let $\acute{\tau}(\acute{x}, s)$ be the first leaving time of the process $\acute{X}_t^{\acute{x},s} = \acute{\Phi}_{s,t}(\acute{x})$ from $\acute{D}$. We shall consider the law of the killed process given by

$$\acute{Q}_{s,t}^{\mathbf{c}}(\acute{x}, E) = E\left[1_E(\acute{\Phi}_{s,t}(\acute{x}))\acute{G}_{s,t}(\acute{x})1_{\acute{\tau}(\acute{x},s)>t}\right]. \tag{7.32}$$

We can apply Lemma 6.10.1 for the killed process. Let $\acute{U} = \psi(U)$. Then the closure of $\acute{U}$ in $\mathbb{R}^d$ is included in the open set $\acute{D}$. Then the measure $\acute{Q}_{s,t}^{\mathbf{c}}(\acute{x}, \cdot)$ restricted to

$\acute{U}$ has a C^∞-density $\acute{q}^{\mathbf{c}}_{s,t}(\acute{x}, \acute{y})$, $\acute{y} \in \acute{U}$ with respect to the Lebesgue measure by Lemma 6.10.1. It satisfies (6.111) and (6.112).

For a given $T > 0$, take $0 \leq s < t \leq T$. Let $\Phi_{s,t}(x)$ be the solution on the manifold M. We shall consider its killed process at D^c. Let $\tau(x, s)$ be the first leaving time of the process $X_t^{x,s} = \Phi_{s,t}(x)$ from D. Then the weighted law

$$Q^{\mathbf{c}}_{s,t}(x, E) = E\big[1_E(\Phi_{s,t}(x))G^{\mathbf{c}}_{s,t}(x)1_{\tau(x,s)>t}\big]$$

has also a C^∞-density $q^{\mathbf{c}}_{s,t}(x, y)$, $y \in U$ with respect to the volume element dy, since $\Phi_{s,t}(x) = \psi^{-1}(\acute{\Phi}_{s,t}(\psi(x)))$ holds for $t < \tau(x, s)$ a.s. P. Further, since $\acute{q}^{\mathbf{c}}_{s,t}$ satisfies (6.111) and (6.112), for any j, i there are positive constants c, c' such that $q^{\mathbf{c}}_{s,t}$ satisfies

$$\sup_{x\in D, y\in U} \left|d_y^j q^{\mathbf{c}}_{s,t}(x, y)\right| \leq \frac{c}{(t-s)^{\frac{j+d}{2}}}, \tag{7.33}$$

$$\sup_{x\in U^c\cap D, y\in V} \left|d_y^i q^{\mathbf{c}}_{s,t}(x, y)\right| \leq c' \tag{7.34}$$

for all $0 \leq s < t \leq T$.

We want to show that for any $x \in M$ the weighted transition function $P^{\mathbf{c}}_{s,t}(x, E)$ has a C^∞-density with respect to the volume element dy. This will be done by piecing together the above density functions $q^{\mathbf{c}}_{s,t}(x, y)$ of the killed process. For this purpose, we will fix a pair (U, D) of the above local charts. For the process $X_t = X_t^{x,s} := \Phi_{s,t}(x)$ we define stopping times $\tau_m = \tau_m(x, s)$, $m = 0, 1, 2, \ldots$ and $\sigma_m = \sigma_m(x, s)$, $m = 1, 2, \ldots$ by induction as $\tau_0 = s$ and

$$\begin{aligned}
\sigma_m &= \inf\{t \geq \tau_{m-1};\ \ X_t \in D^c\}\ \ (= \infty \ \text{ if } \{\cdots\} \text{ is empty}),\\
\tau_m &= \inf\{t \geq \sigma_m;\ \ X_t \in U\}\ \ (= \infty \ \text{ if } \{\cdots\} \text{ is empty}).
\end{aligned} \tag{7.35}$$

Then we have for $E \subset U$,

$$E[1_E(\Phi_{s,t}(x))G^{\mathbf{c}}_{s,t}(x)] = \sum_{m=0}^\infty E\Big[1_E(\Phi_{s,t}(x))G^{\mathbf{c}}_{s,t}(x)1_{\tau_m<t\leq\sigma_{m+1}}\Big].$$

Using the strong Markov property, we can rewrite the above equality as

$$P^{\mathbf{c}}_{s,t}(x, E) = \sum_{m=0}^\infty E\Big[Q^{\mathbf{c}}_{\tau_m,t}(\Phi_{s,\tau_m}(x), E)G^{\mathbf{c}}_{s,\tau_m}(x)1_{\tau_m<t}\Big]. \tag{7.36}$$

Now, $Q^{\mathbf{c}}_{s,t}(x, E \cap U)$ has a C^∞-density function $q^{\mathbf{c}}_{s,t}(x, y)$, $y \in U$ for any $s, t, x \in D$. We set $q^{\mathbf{c}}_{s,t}(x, y) \equiv 0$ if $x \in D^c$. Consider the finite sum for $x \in M$ and $y \in V$:

$$p_{s,t}^{(n)}(x,y) := q_{s,t}^{\mathbf{c}}(x,y) + \sum_{m=1}^{n} E\Big[q_{\tau_m,t}^{\mathbf{c}}(\Phi_{s,\tau_m}(x),y)G_{s,\tau_m}^{\mathbf{c}}(x)1_{\tau_m<t}\Big]. \tag{7.37}$$

We want to show that the above converges to a function $p_{s,t}^{\mathbf{c}}(x,y)$ uniformly with respect to $x \in M$ and $y \in V$. We need three lemmas.

Lemma 7.4.1 *Let $\acute{\Phi}_{s,t}(\acute{x})$ be the solution of equation* (7.31) *on the Euclidean space $\mathbb{R}^d$. Let $B_a(\acute{x}) = \{y; |y-\acute{x}| \le a\}$ and $\sigma(\acute{x},s)$ be the hitting time of the process $\acute{X}_t^{\acute{x},s} = \acute{\Phi}_{s,t}(\acute{x})$ to $B_a(\acute{x})^c$. Then for any $s < T$, there exists $0 < c < 1$ such that*

$$\sup_{\acute{x}\in\mathbb{R}^d} P(\sigma(\acute{x},s) < T) \le c. \tag{7.38}$$

Proof Let f be a C^2-function on $\mathbb{R}^d$ such that $f(x) = |x|^2$ if $|x| \le 1$ and $f(x) = |x|$ if $|x| \ge 2$. By Itô's formula, we have $f(\acute{\Phi}_{s,t}(\acute{x}) - \acute{x}) = M_t + A_t$, where M_t is a continuous martingale with $M_s = 0$ and A_t is continuous process of bounded variation with $A_s = 0$; these are represented by

$$M_t = \sum_{k=1}^{d'} \int_s^t \acute{V}_k(r) f(\acute{\Phi}_{s,r}(\acute{x}))\, dW_r^k,$$

$$A_t = \int_s^t \acute{A}(r) f(\acute{\Phi}_{s,r}(\acute{x}))\, dr,$$

where $\acute{A}(r)$ is the generator. Therefore $\langle M\rangle_t = \int_s^t \sum_k \acute{V}_k(r) f(\acute{\Phi}_{s,r}(\acute{x}))^2\, dr$. Since $\sum_k \acute{V}_k(r) f(\acute{\Phi}_{s,r}(\acute{x}))^2$ is bounded and strictly positive a.s., $\langle M\rangle_t$ is a strictly increasing continuous function of t. Further, the inverse function $\langle M\rangle_t^{-1}$ is also strictly increasing and continuous. Consider the time-changed process $B_t = M_{\langle M\rangle_t^{-1}}$. It is a continuous martingale with respect to the filtration $\mathcal{F}_{\langle M\rangle_t}$, in view of Doob's optional stopping theorem. Its quadratic variation $\langle B\rangle_t$ is equal to t. Therefore B_t is a Wiener process (Proposition 2.2.1).

Let $c_1, c_2 \ge 1$ be constants such that

$$\sum_{k=1}^{d'} \acute{V}_k(r) f(\acute{\Phi}_{s,r}(\acute{x}))^2 \le c_1, \quad |\acute{A}(r) f(\acute{\Phi}_{s,r}(\acute{x}))| \le c_2$$

hold for all $\acute{x}$, $s < r < T$ a.s. Then we have

$$\begin{aligned}\{\sigma(\acute{x},s) < T\} &\subset \{M_t > a - A_t \text{ for some } s < t < T\}\\ &\subset \{B_t > a - c_2 t \text{ for some } s < t < c_1 T\}.\end{aligned}$$

Therefore,

$$\sup_{\acute{x},s} P(\sigma(\acute{x},s) < T) \leq P(B_t > a - c_2 t \quad \text{for some } t < c_1 T).$$

Since B_t is a Wiener process, the last probability denoted by c is less than 1. This proves the assertion of the lemma. □

Lemma 7.4.2 *Let U, D be a pair of local charts satisfying $\bar{U} \subset D$. Let $\tau_m = \tau_m(x,s)$ be stopping times defined by (7.35). Then there exists a positive constant $0 < c < 1$ such that*

$$\sup_{s<t, x\in U} P(\tau_m(x,s) < t) \leq c^m \tag{7.39}$$

holds for any positive integer m.

Proof Let $\sigma(x,s)$ be the hitting time of the process $X_t^{x,s} = \Phi_{s,t}(x)$ to the set D^c, where $d(\psi(U), \psi(D)^c) \geq a > 0$. Then, $\tau_1(x,s) \geq \sigma(x,s)$ a.s. Therefore, we have by Lemma 7.4.1

$$\sup_{x\in U} P(\tau_1(x,s) < t) \leq \sup_{\acute{x}\in \acute{U}} P(\sigma(\acute{x},s) < t) \leq c.$$

Then we have $P(\tau_l - \tau_{l-1} < t) \leq c < 1$. Further, since $\tau_m = \sum_{l=1}^m (\tau_l - \tau_{l-1})$, we have $\{\tau_m < t\} \subset \bigcap_{l=1}^m \{\tau_l - \tau_{l-1} < t\}$. Therefore, using the strong Markov property, we have

$$\begin{aligned}
P(\tau_m < t) &\leq P(\bigcap_{l=1}^{m} \{\tau_l - \tau_{l-1} < t\}) \\
&\leq E\big[P(\tau_m - \tau_{m-1} < t|\mathcal{F}_{\tau_{m-1}}); \bigcap_{l=1}^{m-1} \{\tau_l - \tau_{l-1} < t\}\big] \\
&\leq E\Big[P(\tau_1(y) < t)\big|_{y=\Phi_{\tau_{m-1}}(x)}; \bigcap_{l=1}^{m-1} \{\tau_l - \tau_{l-1} < t\}\Big] \\
&\leq cP(\bigcap_{l=1}^{m-1} \{\tau_l - \tau_{l-1} < t\}) \\
&\leq c^2 P(\bigcap_{l=1}^{m-2} \{\tau_l - \tau_{l-1} < t\}) \\
&\leq c^m,
\end{aligned}$$

proving (7.39). □

Lemma 7.4.3 *The sequence $\{p_{s,t}^{(n)}(x,y)\}$ of functions defined by* (7.37) *converges uniformly with respect to $x \in M, y \in V$. Let $p_{s,t}^{\mathbf{c}}(x,y)$ be its limit. It is a C^∞-function of y and it is the density of transition function $P_{s,t}^{\mathbf{c}}(x, E\cap V)$, so that it does not depend on the choice of local charts U, D. Furthermore, for any $j \in \mathbb{N}$ and $T > 0$, there exists a constant $c_j > 0$ such that*

$$\sup_{x\in D, y\in V} \left| d_y^j p_{s,t}^{\mathbf{c}}(x,y) \right| \le \frac{c_j}{(t-s)^{\frac{j+d}{2}}}, \quad \forall 0 \le s < t \le T. \tag{7.40}$$

Proof Since $\Phi_{s,\tau_m}(x) \in \partial U$, $q_{\tau_m,t}^{\mathbf{c}}(\Phi_{s,\tau_m}(x), y)$ is uniformly bounded with respect to $y \in V$ by (7.34). Further, $E[|G_{s,\tau_m}^{\mathbf{c}}(x)|^p]$ is bounded uniformly with respect to x, s and m for any $p > 1$. Consequently, there is a positive constant K such that

$$\begin{aligned} &E\big[q_{\tau_m,t}^{\mathbf{c}}(\Phi_{s,\tau_m}(x), y) G_{s,\tau_m}^{\mathbf{c}}(x) 1_{\tau_m<t}\big] \\ &\le E\Big[\Big(q_{\tau_m,t}^{\mathbf{c}}(\Phi_{s,\tau_m}(x), y) G_{s,\tau_m}^{\mathbf{c}}(x)\Big)^2\Big]^{\frac{1}{2}} P(\tau_m < t)^{\frac{1}{2}} \le K c^{\frac{m}{2}} \end{aligned}$$

holds for all $x \in M$, $y \in V$ and $m \in \mathbb{N}$, where $0 < c < 1$ is the positive constant of Lemma 7.4.2. Therefore $p_{s,t}^{(n)}(x,y)$ of (7.37) converges uniformly with respect to $x \in M$, $y \in V$. The limit function $p_{s,t}^{\mathbf{c}}(x,y)$ is the density function of $P_{s,t}^{\mathbf{c}}(x, E\cap V)$ in view of (7.36). Hence $p_{s,t}^{\mathbf{c}}(x,y)$ does not depend on the choice of U, D.

Now, differentiating each term of (7.37) with respect to y, we get

$$d_y^j p_{s,t}^{(n)}(x,y) = d_y^j q_{s,t}^{\mathbf{c}}(x,y) + \sum_{m=1}^{n} E\big[d_y^j q_{\tau_m,t}^{\mathbf{c}}(\Phi_{s,\tau_m}(x), y) G_{s,\tau_m}^{\mathbf{c}}(x) 1_{\tau_m<t}\big] \tag{7.41}$$

for any $j \in \mathbb{N}$. Then $d_y^j p_{s,t}^{(n)}(x,y)$ converges uniformly to a function $p_{s,t}^j(x,y)$ with respect to $x \in M$ and $y \in V$, similarly to the convergence of $p_{s,t}^{(n)}(x,y)$. Then $p_{s,t}^{\mathbf{c}}(x,y)$ is a C^∞-function of $y \in V$ and $d_y^j p_{s,t}^{\mathbf{c}}(x,y) = p_{s,t}^j(x,y)$ holds.

$d_y^j p_{s,t}^{\mathbf{c}}(x,y)$ has the short-time estimate (7.40), since $d_y^j q_{s,t}^{\mathbf{c}}(x,y)$ satisfies the inequality (7.33) and the term $\sum_{m=1}^{n}(\cdots)$ in (7.41) is bounded with respect to x, y and n. □

Proof of Theorem 7.4.1 We saw the existence of the density $p_{s,t}^{\mathbf{c}}(x,y)$ and its smoothness with respect to $y \in V$ and its short-time estimate (7.40) in Lemma 7.4.3. The function $p_{s,t}^{\mathbf{c}}(x,y)$ can be extended to $x, y \in M$, since M is covered by a countable number of the above local charts V. It is a C^∞-function of y. Further, since the compact set K is covered by finite numbers of local charts V, the short-time estimate (7.29) holds.

We apply the same argument to the dual transition function, assuming the dual condition. We saw in Sect. 7.2 that the dual transition function of $P_{s,t}^{\mathbf{c}}(x,E)$ is given by $P_{s,t}^{\mathbf{c},*}(y,F) = E[1_F(\check{\Psi}_{s,t}(y))\check{\mathcal{G}}_{s,t}^{\mathbf{c}}(y)|J_{\check{\Psi}_{s,t}}(y)|]$, where $\check{\Psi}_{s,t}(y)$ is the backward flow generated by a continuous backward SDE with coefficients $-V_k(x,t), k =$

$0, \ldots, d'$. It is an elliptic backward diffusion. Further, $\check{\mathcal{G}}^{\mathbf{c}}_{s,t}(y) = G^{\mathbf{c}}_{s,t}(\check{\Psi}_{s,t}(y))$ and $J_{\check{\Psi}_{s,t}}(y)$ is the Jacobian of the map $\check{\Psi}_{s,t}$ on the manifold (see Sect. 7.2). Then the dual transition function $P^{\mathbf{c},*}_{s,t}(y, F)$ has a smooth density $p^{\mathbf{c},*}_{s,t}(y, x)$ for any y. It should coincide with $p^{\mathbf{c}}_{s,t}(x, y)$ for any x, y (see the proof of Lemma 6.10.2). Consequently, the density $p^{\mathbf{c}}_{s,t}(x, y)$ is smooth with respect to x. It satisfies (7.30). □

Corollary 7.4.1 *Assume Condition (D)(i).*

1. *The semigroup* $\{P^{\mathbf{c}}_{s,t}\}$ *maps* $C_0(M)$ *to* $C^\infty(M) \cap C_b(M)$. *Further,* $P^{\mathbf{c}}_{s,t} f$ *satisfies Kolmogorov's backward equation if* $f \in C_0(M)$.
2. *If* $f_1 \in C_0(M)$, *then* $v(x, s) := \int_M p^{\mathbf{c}}_{s,t_1}(x, y) f_1(y)\, dy$ *is a* $C^{\infty,1}$*-function of* x, s *and is a solution of the final value problem of the backward heat equation on the manifold associated with the differential operator* $A^{\mathbf{c}}(t)$.

Proof The dual semigroup $\{P^{\mathbf{c},*}_{s,t}\}$ satisfies $P^{\mathbf{c},*}_{s,t} g = g + \int_s^t P^{\mathbf{c},*}_{r,t} A^{\mathbf{c}}(r)^* g\, dr$. Then we can show that the density function $p^{\mathbf{c}}_{s,t}(x, y)$ satisfies

$$\frac{\partial}{\partial s} p^{\mathbf{c}}_{s,t}(x, y) = -A^{\mathbf{c}}(s)_x p^{\mathbf{c}}_{s,t}(x, y),$$

similarly to the proof of (6.116). Let $f \in C_0(M)$. Then the function $v(x, s) := \int p^{\mathbf{c}}_{s,t}(x, y) f(y)\, dy$ is a bounded $C^{\infty,1}$-function and satisfies

$$A^{\mathbf{c}}(s) v(x, s) = \int A^{\mathbf{c}}(s) p^{\mathbf{c}}_{s,t}(x, y) f(y)\, dy = -\int \frac{\partial}{\partial s} p^{\mathbf{c}}_{s,t}(x, y) f(y)\, dy = -\frac{\partial}{\partial s} v(x, s),$$

since we can change the order of differentials and the integrals for $v(x, s)$. Therefore the semigroup $\{P^{\mathbf{c}}_{s,t}\}$ satisfies Kolmogorov's backward equation. The second assertion will be obvious. □

Let us discuss briefly the backward SDE with coefficients $V_k(x, t), k = 0, \ldots, d'$ on the manifold M. Let $\{\check{\Phi}_{s,t}\}$ be the backward flow defined by the backward SDE. Then $\check{P}^{\mathbf{c}}_{s,t}(x, E) := E[1_E(\check{\Phi}_{s,t}(x)) \check{G}^{\mathbf{c}}_{s,t}(x)]$ has a density $\check{p}^{\mathbf{c}}_{s,t}(x, y)$, which is a C^∞-function of x and y. Further, for any $f_0 \in C_0(M)$, $u(x, t) := \int_M \check{p}^{\mathbf{c}}_{t_0,t}(x, y) f_0(y)\, dy$ is a $C^{\infty,1}$-function and it is a solution of the initial value problem of the heat equation on the manifold M associated with the differential operator $A^{\mathbf{c}}(t)$.

In Sect. 6.3, we showed the existence of the smooth density of transition function of an elliptic diffusion on $\mathbb{R}^d$, if its coefficients $V_k(t), k = 0, \ldots, d'$ are $C_b^{\infty,1}$-functions. The fact can be extended to elliptic diffusions with unbounded coefficients, if we apply Theorem 7.4.1 to the case where the manifold M is an Euclidean space $\mathbb{R}^d$. Indeed, if coefficients $V_k(t)$ are (not necessarily bounded) C^∞-functions, the transition function has a C^∞-density.

It seems that assertions of Theorem 7.4.1 except short-time estimates (7.29) and (7.30) should be valid for hypo-elliptic diffusions on a manifold. However, for the proof, short-time estimates of the density function like Theorem 6.7.1 will be needed for hypo-elliptic diffusions on Euclidean space.

Finally, we will apply Theorem 7.4.1 to a Brownian motion on a Lie group.

Theorem 7.4.2 *Let $X_t, t \in [0, \infty)$ be a Brownian motion on a Lie group G with the generator $Af(x) = \frac{1}{2}\sum_{k=1}^{d'} V_k^2 f(x) + V_0 f(x)$, where $V_k; k = 0, \ldots, d'$ are left-invariant vector fields. Suppose that $V_k; k = 1, \ldots, d'$ spans $\mathcal{L}(G)$. Then the law μ_t of X_t has a C^∞-density $\mu_t(y)$ with respect to the Haar measure m. Further, the transition probability has a C^∞-density and it is given by $p_t(x, y) = \mu_t(x^{-1}y)$.*

For any compact subset K, a positive integer j and $T > 0$, there exists a positive constant c_j such that the following holds for all $0 < t < T$:

$$\sup_{y \in K} \left| d^j \mu_t(y) \right| \le \frac{c_j}{t^{\frac{j+d}{2}}}. \tag{7.42}$$

Note For the existence of the smooth density of the law of a continuous SDE on a manifold, Taniguchi [112] applied the Whitney embedding theorem of a manifold into a higher-dimensional Euclidean space. Assumptions required in the paper are different from ours.

The existence of the smooth density for a diffusion with the explosion seems to be new.

7.5 Density for Jump-Diffusion on Compact Manifold

We will consider a jump-diffusion on the compact and orientable manifold M defined by SDE (7.9). The equation generates a stochastic flow of diffeomorphisms $\{\Phi_{s,t}(x)\}$. See Sect. 7.2. We define vector fields $\tilde{V}_k(t), k = 1, \ldots, d'$ by $\tilde{V}_k(t)f(x) = \frac{\partial}{\partial z^k} f(\phi_{t,z}(x))|_{z=0}$ as before. Let $V_k(x, t)$ and $\tilde{V}_k(x, t)$ be coefficients of vector fields $V_k(t)$ and $\tilde{V}_k(t)$ when these are represented by a local coordinate. Set $V(x, t) = (V_1(x, t), \ldots, V_{d'}(x, t))$ and $\tilde{V}(x, t) = (\tilde{V}_1(x, t), \ldots, \tilde{V}_{d'}(x, t))$. Let Γ_0 be a lower bound of matrices $\{\Gamma_\epsilon\}$ given by (1.21). If, at the tangent space $T_x M$, $V(x, t)V(x, t)^T + \tilde{V}(x, t)\Gamma_0 \tilde{V}(x, t)$ is nondegenerate for any x, t, the SDE and jump-diffusion generated by the SDE are called *pseudo-elliptic*.

Let $G_{s,t}(x) = G_{s,t}^{\mathbf{c},\mathbf{d}}(x)$ be an exponential Wiener–Poisson functional defined by

$$G_{s,t} := \exp\left\{ \sum_{k=0}^{d'} \int_s^t c_k(\Phi_{s,r}, r) \circ dW_r^k + \int_s^t \int_{|z|>0+} \log d_{r,z}(\Phi_{s,r-}) N(dr\, dz) \right\},$$

where $c_k(x, t), k = 0, \ldots, d'$ are $C^{\infty,1}$-functions of x, t on $M \times \mathbb{R}^{d'}$ and $d_{t,z}(x)$ is a $C^{\infty,1,2}$-function of (x, t, z) on $M \times \mathbb{T} \times \mathbb{R}^{d'}$ satisfying $d_{t,0}(x) = 1$ for any x, t.

Suppose we are given a volume element dx on the manifold M. In the following we will fix it. It is also written as dy. We will study the existence of the smooth density of $P_{s,t}^{\mathbf{c},\mathbf{d}}(x, E) = E[1_E(\Phi_{s,t}(x))G_{s,t}^{\mathbf{c},\mathbf{d}}(x)]$ with respect to the volume element

dy. In order to construct the smooth density, we need more careful discussions than those for diffusions, since sample paths may jump to other local charts. Our discussion will be divided into two. First, we show that if the size of jumps is sufficiently small, say smaller than δ, we can construct the density function by a method similar to that of diffusion discussed in Sect. 7.4. Next, we adjoin jumps which are bigger than δ and show the existence of the smooth density by the method of perturbation discussed in Sect. 6.9.

Let δ be a positive number. We will truncate jumps which are bigger than $\delta > 0$ from the above equation: The equation is written as

$$f(X_t) = f(X_0) + \sum_{k=0}^{d'} \int_{t_0}^{t} V_k(r) f(X_r) \circ dW_r^k \qquad (7.43)$$
$$+ \int_{t_0}^{t}\int_{\delta \geq |z| > 0+} \{f(\phi_{r,z}(X_{r-})) - f(X_{r-})\} N(dr\, dz).$$

We denote the stochastic flow of diffeomorphisms generated by the above equation by $\{\Phi_{s,t}^\delta\}$. We set

$$G_{s,t}^\delta := \exp\left\{\sum_{k=0}^{d'} \int_s^t c_k(\Phi_{s,r}^\delta, r) \circ dW_r^k + \int_s^t\int_{\delta \geq |z| > 0+} \log d_{r,z}(\Phi_{s,r-}^\delta) N(dr\, dz)\right\},$$

and define $P_{s,t}^\delta(x, E)(= P_{s,t}^{\mathbf{c},\mathbf{d},\delta}(x, E)) := E[1_E(\Phi_{s,t}^\delta(x)) G_{s,t}^\delta(x)]$. For a given $n_0 \in \mathbb{N}$, we want to show that the measure $P_{s,t}^\delta(x, E)$ has a C^{n_0}-density with respect to the volume element dy, if δ (depending on n_0) is taken sufficiently small. We will discuss how such δ can be taken.

Let D be a local chart of M and let $\tau^\delta(x, s)$ be the first leaving time of the process $X_t^{x,s} = \Phi_{s,t}^\delta(x)$ from D. We consider the transition function of the killed process defined by

$$Q_{s,t}^\delta(x, E) = E[1_E(\Phi_{s,t}^\delta(x)) G_{s,t}^\delta(x) 1_{\tau^\delta(x,s) > t}]. \qquad (7.44)$$

In the first step, we want to show the existence of the smooth density of the above $Q_{s,t}^\delta(x, E)$. Let α_0 be any fixed positive number satisfying $\alpha < \alpha_0 < 2$.

Lemma 7.5.1 *Let n_0 be a positive integer. Let V, U, D be a triple of local charts such that $\bar{V} \subset U \subset \bar{U} \subset D$. Then there exists a positive constant $\delta_{n_0,V,U,D}$ (depending on n_0 and V, U, D) such that for any $0 < \delta < \delta_{n_0,V,U,D}$, the killed transition function $Q_{s,t}^\delta(x, E \cap U)$ has a C^{n_0}-density $q_{s,t}^\delta(x, y)$, $y \in U$ with respect to the volume element dy for any $s < t$ and $x \in D$. Further, for any integer $0 \leq j \leq n_0$ and $T > 0$, there are positive constants $c_{T,\delta,j}$ and $c'_{T,\delta,j}$ such that*

$$\sup_{x\in D, y\in U} \left| d_y^j q_{s,t}^\delta(x,y) \right| \le \frac{c_{T,\delta,j}}{(t-s)^{\frac{j+d}{2-\alpha_0}}}, \qquad \forall 0 \le s < t \le T, \tag{7.45}$$

$$\sup_{x\in U^c\cap D, y\in V} \sup_{0\le s<t\le T} \left| d_y^i q_{s,t}^\delta(x,y) \right| \le c'_{T,\delta,j}. \tag{7.46}$$

Proof Note that D is diffeomorphic to a bounded open set $\acute{D}$ of $\mathbb{R}^d$ by a diffeomorphism $\psi : D \to \acute{D}$. SDE (7.43) on D is transformed to an SDE on $\acute{D}$ by the diffeomorphic map ψ. Indeed, for the vector fields $V_k(t)$ on the local chart D, there are vector fields $\acute{V}_k(\acute{x}, t)$ on $\mathbb{R}^d$ of $C_b^{\infty,1}$-class, which coincide with $d\psi(V_k(t))$ on $\acute{D}$. Further, there exists $\delta_0 > 0$ such that for any $|z| \le \delta_0$ and $t \in \mathbb{T}$, we can construct a family of diffeomorphic maps $\acute{\phi}_{t,z}$ on $\mathbb{R}^d$ such that $\psi(\phi_{t,z}(x)) = \acute{\phi}_{t,z}(\acute{x})$ holds if $x \in D$ (Lemma 7.1.3).

Let N be a positive integer such that $N > \frac{n_0+d}{2-\alpha_0}$. Let $\acute{U} = \psi(U)$ and $\acute{V} = \psi(V)$. There exists a positive constant $\delta_{N,V,U,D}$, less than or equal to δ_0, such that for any $0 < \delta < \delta_{N,V,U,D}$ inequalities

$$d(\acute{\phi}_\delta^N(\acute{U}), \acute{\phi}_\delta^{-N}(\acute{D}^c)) > 0, \quad d(\acute{\phi}_\delta^N(\acute{V}), \acute{\phi}_\delta^{-N}(\acute{U}^c)) > 0 \tag{7.47}$$

hold, where $\acute{\phi}_\delta(\acute{U}) = \overline{\bigcup_{t\in\mathbb{T},|z|<\delta} \acute{\phi}_{t,z}(\acute{U})}$ and $\acute{\phi}_\delta^N(\acute{U})$ are defined by iteration. Indeed, note that $\acute{\phi}_\delta^N(\acute{U}) \downarrow \overline{\acute{U}}$ and $\acute{\phi}_\delta^{-N}(\acute{D}^c) \downarrow \acute{D}^c$ as $\delta \downarrow 0$. Then

$$d(\acute{\phi}_\delta^N(\acute{U}), \acute{\phi}_\delta^{-N}(\acute{D}^c)) \uparrow d(\overline{\acute{U}}, \acute{D}^c) > 0.$$

Therefore the first inequality of (7.47) is valid. The second inequality is verified in the same way. Now, take $0 < \delta < \delta_{N,V,U,D}$ and consider an SDE on $\mathbb{R}^d$;

$$d\acute{X}_t = \sum_{k=0}^{d'} \acute{V}_k(\acute{X}_t, t) \circ dW_t^k + \int_{\delta\ge|z|>0+} \{\acute{\phi}_{t,z}(\acute{X}_{t-}) - \acute{X}_{t-}\} N(dt\,dz). \tag{7.48}$$

It generates a stochastic flow $\{\acute{\Phi}_{s,t}^\delta\}$ of diffeomorphisms on $\mathbb{R}^d$. Let $\acute{\tau}^\delta(\acute{x}, s)$ be the first leaving time of $\acute{X}_t = \acute{\Phi}_{s,r}^\delta(\acute{x})$ from $\acute{D}$. We shall consider the law of the killed process given by

$$\acute{Q}_{s,t}^\delta(\acute{x}, E) = E\Big[1_E(\acute{\Phi}_{s,t}^\delta(\acute{x}))\acute{G}_{s,t}^\delta(\acute{x})1_{\acute{\tau}^\delta(\acute{x},s)>t}\Big]. \tag{7.49}$$

Since the process $\acute{\Phi}_{s,t}^\delta(\acute{x})$ is pseudo-elliptic, the measure $\acute{Q}_{s,t}^\delta(\acute{x}, \cdot)$ restricted to $\acute{U} = \psi(U)$ has a C^{n_0}-density $\acute{q}_{s,t}^\delta(\acute{x}, \acute{y})$, $\acute{y} \in \acute{U}$ with respect to the Lebesgue measure by Lemma 6.10.3. For a given $T > 0$, it satisfies (6.123) and (6.124) in place of $q_{s,t}^{\mathbf{c},\mathbf{d}}$.

Now let $\Phi^\delta_{s,t}(x)$ be the solution of equation (7.43) on the manifold M. Let $\tau^\delta(x,s)$ be the first leaving time of $X_t^{\delta,x,s} = \Phi^\delta_{s,t}(x)$ from D. Then the weighted law $Q^\delta_{s,t}(x,E)$ of the killed process $\Phi^\delta_{s,t}(x), t < \tau^\delta(x,s)$ has also a C^{n_0}-density $q^\delta_{s,t}(x,y),\ y \in U$ with respect to the volume element dy, since $\Phi^\delta_{s,t}(x) = \psi^{-1}(\acute{\Phi}^\delta_{s,t}(\psi(x)))$ holds for $t < \tau^\delta(x,s)$ almost surely. Further, since $\acute{q}^\delta_{s,t}$ satisfies (6.123) and (6.124) (in place of $q^{\mathbf{c},\mathbf{d}}_{s,t}$), $q^\delta_{s,t}$ satisfies (7.45) and (7.46). □

Next, we will show that for a sufficiently small $\delta > 0$, the law $P^\delta_{s,t}(x,E)$ has a C^{n_0}-density with respect to the volume element dy for any $x \in M, s < t$, by piecing together the above density function $q^\delta_{s,t}(x,y)$ of the killed process. Our discussion proceeds in parallel with the discussion for diffusion processes in Sect. 7.4, but we need some rectifications due to jumps of the process.

Given a pair of local charts V, D such that $\bar{V} \subset D$, we can choose open sets U and W such that $\bar{V} \subset U \subset \bar{U} \subset W \subset \bar{W} \subset D$ and a positive constant $\delta_{N,V,U,W,D}$ with $0 < \delta_{N,V,U,W,D} \le \delta_{N,V,U,D}$ such that

$$d(\phi_\delta^N(V), \phi_\delta^{-N}(U^c)) > 0, \quad d(\phi_\delta(U), W^c) > 0, \quad d(\phi_\delta^N(W), \phi_\delta^{-N}(D^c)) > 0$$

hold for any $0 < \delta < \delta_{N,V,U,W,D}$, where $N \ge \frac{n_0+d}{2-\alpha_0}$. We want to prove that $P^\delta_{s,t}(x; E \cap V)$ has a C^{n_0}-density for any $x \in M$ and $0 \le s < t \le T$. For this purpose, we define stopping times $\tau^\delta_m = \tau^\delta_m(x,s), m = 0,1,2,\dots$, and $\sigma^\delta_m = \sigma^\delta_m(x,s), m = 1,2,\dots$ for the process $X^\delta_t = X_t^{\delta,x,s} = \Phi^\delta_{s,t}(x)$ by induction as $\tau^\delta_0 = s$ and

$$\sigma^\delta_m = \inf\{t \ge \tau^\delta_{m-1};\ X^\delta_t \in D^c\} \ (= \infty \text{ if } \{\cdots\} \text{ is empty}),$$
$$\tau^\delta_m = \inf\{t \ge \sigma^\delta_m;\ X^\delta_t \in W\} \ (= \infty \text{ if } \{\cdots\} \text{ is empty}).$$

Then we have for $E \subset U$,

$$E[1_E(\Phi^\delta_{s,t}(x))G^\delta_{s,t}(x)] = \sum_{m=0}^\infty E\Big[1_E(\Phi^\delta_{s,t}(x))G^\delta_{s,t}(x)1_{\tau^\delta_m < t \le \sigma^\delta_{m+1}}\Big].$$

Using the strong Markov property, we can rewrite the above equality as

$$P^\delta_{s,t}(x,E) = \sum_{m=0}^\infty E\Big[Q^\delta_{\tau^\delta_m,t}(\Phi^\delta_{s,\tau^\delta_m}(x), E)G^\delta_{s,\tau^\delta_m}(x)1_{\tau^\delta_m < t}\Big]. \tag{7.50}$$

Now take $0 < \delta < \delta_{N,V,U,W,D}$. Then in view of Lemma 7.5.1, $Q^\delta_{s,t}(x; E \cap V)$ has a C^{n_0}-density function $q^\delta_{s,t}(x,y), y \in V$ for any $0 \le s < t \le T, x \in D$. For $x \in D^c$ and $y \in V$, we set $q^\delta_{s,t}(x,y) = 0$, conventionally.

Lemma 7.5.2 *There is a positive constant $T_{N,U,V,W,D}$ such that if $0 \le s < t \le T$ and $0 < t - s \le T_{N,U,V,W,D}$, the infinite sum*

$$p_{s,t}^{\delta}(x,y) := q_{s,t}^{\delta}(x,y) + \sum_{m=1}^{\infty} E\Big[q_{\tau_m^\delta,t}^{\delta}(\Phi_{s,\tau_m^\delta}^{\delta}(x), y) G_{s,\tau_m^\delta}^{\delta}(x) 1_{\tau_m^\delta < t}\Big] \tag{7.51}$$

is uniformly convergent with respect to $x \in M$ and $y \in V$. Further, it is the density of $P_{s,t}^{\delta}(x, E \cap V)$ with respect to the volume element dy and it does not depend on the choice of sets U, V, W, D.

Proof We need an inequality similar to (7.39). Let us consider Lemma 6.5.2 again. The lemma states a fact for a process on Euclidean space. But the fact is valid for a process on a manifold. Indeed, it tells us that for $0 < c < 1$ there exists $T_{U,V,W,D} > 0$ such that the inequality $\sup_{x \in U} P(\tau_1^\delta(x,s) < t) \le c$ holds for any $0 < t - s < T_{U,V,W,D}$. Then we have the inequality

$$\sup_{x \in U} P(\tau_m^\delta(x,s) < t) \le c^m, \quad \forall 0 < t - s < T_{U,V,W,D},$$

similarly to the proof of Lemma 7.4.2.

Suppose $m \ge 1$. Note that $q_{s,t}^{\delta}$ satisfies (7.45). Since $\Phi_{s,\tau_m^\delta}^{\delta}(x) \in U^c$, $q_{\tau_m^\delta,t}^{\delta}(\Phi_{s,\tau_m^\delta}^{\delta}(x), y)$ is uniformly bounded with respect to $y \in V$ in view of Lemma 6.10.3. Furthermore, $E[\sup_t |G_{s,t}^{\delta}(x)|^p]$ is bounded with respect to x, since $c_k(x,t)$, $V_k(t)c_k(t)(x)$ are bounded on compact set M. Therefore the infinite sum of (7.51) converges uniformly with respect to $x \in M$, $y \in V$, if $0 \le s < t \le T$ and $0 < t - s < T_{U,V,W,D}$. Denote it by $p_{s,t}^{\delta}(x,y)$. It is the density function of $P_{s,t}^{\delta}(x, E \cap V)$ for any $x \in M$ in view of (7.50). Hence $p_{s,t}^{\delta}(x,y)$ does not depend on the choice of U, V, W, D. □

Proposition 7.5.1 *Given a positive integer n_0, there exists $\delta_1 > 0$ such that for any $0 < \delta < \delta_1$, the weighted transition function $P_{s,t}^{\delta}(x, E)$ of the truncated process $\Phi_{s,t}^{\delta}(x)$ has a C^{n_0}-density $p_{s,t}^{\delta}(x,y)$ for any $s < t$ and $x \in M$ with respect to the volume element dy. Further, for any nonnegative integer $j \le n_0$ and $T > 0$, there exists a positive constant $c = c_{\delta,n_0}$ such that the inequality*

$$\left| d_y^j p_{s,t}^{\delta}(x,y) \right| \le \frac{c}{(t-s)^{\frac{j+d}{2-\alpha_0}}}, \quad \forall 0 \le s < t \le T \tag{7.52}$$

holds for all $x, y \in M$.

Proof Let $\{V_\alpha \subset U_\alpha \subset W_\alpha \subset D_\alpha\}$ be a family of quaternions of local charts discussed above, where $\{V_\alpha\}$ covers the manifold M. Since M is compact, it is covered by a finite number of local charts $\{V_{\alpha_i}, i = 1, 2, \ldots, n\}$. We set $\delta_1 =$

$\min_i \delta_{V_{\alpha_i},U_{\alpha_i},W_{\alpha_i},D_{\alpha_i}}$ and $T_1 = \min_i T_{U_{\alpha_i},V_{\alpha_i},W_{\alpha_i},D_{\alpha_i}}$. Suppose $0 \leq s < t \leq T$. Then, if $0 < \delta < \delta_1$, $p^\delta_{s,t}(x, y)$ of (7.51) is uniformly convergent for all $x, y \in M$ and $0 < t - s < T_1$. It is the density function of $P^\delta_{s,t}(x, E)$ with respect to the volume element dy for any $x \in M, 0 < t - s < T_1$.

We will show that $p^\delta_{s,t}(x, y)$ is a C^{n_0}-function of y. Similarly to the proof of Lemma 7.5.2, we can show that for any $j \leq n_0$, the infinite sum

$$d^j_y q^\delta_{s,t}(x, y) + \sum_{m=1}^{\infty} E\Big[d^j_y q^\delta_{\tau^\delta_m,t}(\Phi^\delta_{s,\tau^\delta_m}(x), y) G^\delta_{s,\tau^\delta_m}(x) 1_{\tau^\delta_m < t}\Big] \tag{7.53}$$

is uniformly convergent with respect to $x, y \in M$ if $0 < t - s < T_1$.

Next, if $T_1 < t - s < 2T_1$, using the Chapman–Kolmogorov equality, we can define $p^\delta_{s,t}(x, y)$ by

$$p^\delta_{s,t}(x, y) = \int_M p^\delta_{s,s+T_1'}(x, y') p^\delta_{s+T_1',t}(y', y)\, dy'.$$

It is a C^{n_0}-function of y and is a density function of $P^\delta_{s,t}(x, dy)$. Repeating this argument, we can verify that $P^\delta_{s,t}(x, dy)$ has a C^{n_0}-density for any $0 \leq s < t \leq T$, □

We have seen so far that the transition function of a pseudo-elliptic jump-diffusion on a compact orientable manifold has a C^{n_0}-density, if jumps are sufficiently small. We will show next that the same fact holds for any pseudo-elliptic jump-diffusion on compact orientable manifold, which may allow big jumps.

Theorem 7.5.1 *Suppose that the jump-diffusion on a compact orientable manifold M defined by SDE (7.9) is pseudo-elliptic. Let $P^{\mathbf{c},\mathbf{d}}_{s,t}(x, E)$ be its weighted transition function. Then it has a C^∞-density $p^{\mathbf{c},\mathbf{d}}_{s,t}(x, y)$, $y \in M$ with respect to a given volume element dy. Further, it is a $C^{\infty,1}$-function of x, s for any y and t.*

Let α_0 be any fixed number satisfying $\alpha < \alpha_0 < 2$, where α is the exponent of the Lévy measure. For any $i, j \in \mathbb{N}$ and $T > 0$ there exist positive constants c, c' such that inequalities

$$\sup_{x,y\in M} \left| d^j_y p^{\mathbf{c},\mathbf{d}}_{s,t}(x, y)\right| \leq \frac{c}{(t-s)^{\frac{j+d}{2-\alpha_0}}}, \quad \forall 0 \leq s < t \leq T, \tag{7.54}$$

$$\sup_{x,y\in M} \left| d^i_x p^{\mathbf{c},\mathbf{d}}_{s,t}(x, y)\right| \leq \frac{c'}{(t-s)^{\frac{i+d}{2-\alpha_0}}}, \quad \forall 0 \leq s < t \leq T. \tag{7.55}$$

hold.

Further, $p(x,s;y,t) := p^{\mathbf{c},\mathbf{d}}_{s,t}(x,y)$ *is the fundamental solution of the backward heat equation on the manifold* M*, associated with the operator* $A^{\mathbf{c},\mathbf{d}}_J(t)$ *defined by*

$$A^{\mathbf{c},\mathbf{d}}_J(t)f(x) = \frac{1}{2}\sum_{k=1}^{d'}(V_k(t)+c_k(t))^2 f(x) + (V_0(t)+c_0(t))f(x)$$
$$+\int_{|z|>0+}\{d_{t,z}(x)f(\phi_{t,z}(x)) - f(x)\}\nu(dz). \tag{7.56}$$

Proof We want to apply Proposition 7.5.1 for the proof of the theorem. Given a positive integer n_0, let $\delta_1 > 0$ be a number such that for any $0 < \delta < \delta_1$, the weighted transition function of the truncated process $\Phi^\delta_{s,t}(x)$ has a C^{n_0}-density. We will show the existence of the smooth density of the weighted law of $\Phi_{s,t}(x)$ by the method of perturbation studied at Sect. 6.9. For $\mathbf{u} = \emptyset \in \mathbb{U}^0$ (empty set), we define $\Phi^{\delta,\phi}_{s,t}(x) = \Phi^\delta_{s,t}(x)$. Further, for $\mathbf{u} = ((t_1,z_1),\ldots,(t_n,z_n)) \in \mathbb{U}^n$ such that $s < t_1 < \cdots < t_n \le t$ and $z_i \in \mathbb{R}^{d'}_0$, we define

$$\Phi^{\delta,\mathbf{u}}_{s,t}(x) := \Phi^\delta_{t_n,t}\circ\phi_{t_n,z_n}\circ\cdots\circ\Phi^\delta_{t_1,t_2}\circ\phi_{t_1,z_1}\circ\Phi^\delta_{s,t_1}(x). \tag{7.57}$$

In the following arguments, we will let n run on $n = 0,1,\ldots$. Let q be a Poisson point process with intensity $dt\nu_\delta(dz)$, where $\nu_\delta(dz) = 1_{|z|\ge\delta}\nu(dz)$. Then $\mathbf{D}_q \cap (s,t]$ is an empty set or it is written as $\{\tau_1 < \cdots < \tau_n\}$ (stopping times) as in Sect. 6.9. We set $\mathbf{u}(q) = \emptyset$ or $\mathbf{u}(q) = ((\tau_1,q(\tau_1)),\ldots,(\tau_n,q(\tau_n)))$. Then it holds that $\Phi_{s,t}(x) = \Phi^{\delta,\mathbf{u}(q)}_{s,t}(x)$ a.s. See Sect. 6.9.

Let $p^\delta_{s,t}(x,y), x,y \in M$ be the density function of the weighted transition function $P^\delta_{s,t}(x,E)$ of $\Phi^\delta_{s,t}(x)$. For $\mathbf{u} = \emptyset$, we define $p^{\delta,\phi}_{s,t}(x,y) = p^\delta_{s,t}(x,y)$ and for $\mathbf{u} = ((t_1,z_1),\ldots,(t_n,z_n)) \in \mathbb{U}^n$, we define $p^{\delta,\mathbf{u}}_{s,t}(x,y)$ by

$$p^{\delta,\mathbf{u}}_{s,t}(x,y) = \int_M\cdots\int_M p^\delta_{s,t_1}(x,x_1)\,d_{t_1,z_1}(x_1)p^\delta_{t_1,t_2}(\phi_{t_1,z_1}(x_1),x_2)\times\cdots$$
$$\times\, d_{t_n,z_n}(x_n)p^\delta_{t_n,t}(\phi_{t_n,z_n}(x_n),y)\,dx_1\cdots dx_n. \tag{7.58}$$

Set

$$p^{\mathbf{c},\mathbf{d}}_{s,t}(x,y) := E[p^{\delta,\mathbf{u}(q)}_{s,t}(x,y)]. \tag{7.59}$$

Then it is a density function of the weighted transition function $P^{\mathbf{c},\mathbf{d}}_{s,t}(x,E)$.

We want to show that the function $p^{\mathbf{c},\mathbf{d}}_{s,t}(x,y)$ is of C^{n_0}-class with respect to y. Our discussion is close to those in Theorem 6.9.1. Condition (6.96) should be modified as

$$\sup_{t,x} \int_{|z|>\delta} |J_{\phi_{t,z}^{-1}}| \nu(dz) < \infty.$$

However, the above condition is satisfied in the present case, since M is a compact manifold and the support of the Lévy measure ν is compact. Therefore, $p_{s,t}^{\mathbf{c},\mathbf{d}}(x, y)$ is a continuous function of y. See the proof of Theorem 6.9.1, (1) The n_0-times differentiability with respect to y can be verified similarly to Theorem 6.9.1, (2) Finally, remark that the above fact is valid for any positive integer n_0. Then the density $p_{s,t}^{\mathbf{c},\mathbf{d}}(x, y)$ is in fact a C^∞-function of y. It satisfies (7.54), since $q_{s,t}^{\delta}(x, y)$ satisfies (7.45) and the term $\sum_{i\geq 1} \cdots$ in (7.51) is bounded.

Let $\check{\Psi}_{s,t}$ be the inverse map of $\Phi_{s,t}$ and let $\{P_{s,t}^{\mathbf{c},\mathbf{d},*}\}$ be the dual semigroup of the semigroup $\{P_{s,t}^{\mathbf{c},\mathbf{d}}\}$ with respect to a volume element dx. Apply the above discussion to the backward jump-diffusion $\check{\Psi}_{s,t}$. Then we find that its transition function $P_{s,t}^{\mathbf{c},\mathbf{d},*}(y, E)$ has a C^∞-density $p_{s,t}^{\mathbf{c},\mathbf{d},*}(y, x)$ for any $s < t$ and y. Further, we can show as before that $p_{s,t}^{\mathbf{c},\mathbf{d},*}(y, x) = p_{s,t}^{\mathbf{c},\mathbf{d}}(x, y)$ holds for any $s < t$ and $x, y \in M$. Therefore $p_{s,t}^{\mathbf{c},\mathbf{d}}(x, y)$ is a C^∞-function of x and satisfies (7.55). We can verify that it satisfies $\frac{\partial}{\partial s} p_{s,t}^{\mathbf{c},\mathbf{d}}(x, y) = -A_J^{\mathbf{c},\mathbf{d}}(s)_x p_{s,t}^{\mathbf{c},\mathbf{d}}(x, y)$, making use of the dual semigroup as in Lemma 6.10.2. Then $v(x, s) := \int p_{s,t_1}^{\mathbf{c},\mathbf{d}}(x, y) f_1(y)\, dy$ is a $C^{\infty,1}$-function of x, s and in fact a solution of the final value problem of the backward heat equation associated with the operator $A_J^{\mathbf{c},\mathbf{d}}(t)$, □

The next corollary is an immediate consequence of Theorem 7.5.1.

Corollary 7.5.1 *Let $\{P_{s,t}^{\mathbf{c},\mathbf{d}}\}$ be the semigroup of Theorem 7.5.1. Then it maps $C(M)$ to $C^\infty(M)$ for any $s < t$. Further it satisfies Kolomogorov's backward equation.*

We will study the smooth density of the law of a Lévy process on a compact Lie group. We will consider a Lévy process on a Lie group G with the generator (7.23), where V_k are left-invariant vector fields and $xg(z)$ is the right translation by $g(z)$. We set $\tilde{V}_k = \frac{\partial R_{g(z)}}{\partial z_k}\big|_{z=0}$. Then $\tilde{V}_k, k = 1, \ldots, d'$ are left-invariant vector fields.

Theorem 7.5.2 *Let X_t be a Lévy process on a compact Lie group. We assume that left-invariant vector fields $\{V_k, k = 1, \ldots, d'\} \cup \{\tilde{V}_k; k = 1, \ldots, d'\}$ span the Lie algebra of left-invariant vector fields. Then the law of X_t has a C^∞-density with respect to the Haar measure for any $t > 0$.*

Note It is expected that similar assertions of Theorem 7.5.1 and Theorem 7.5.2 should hold for non-compact manifold and non-compact Lie groups, respectively. It remains open. Picard–Savona [95] studied the smooth density for Markov process of pure jumps on non-compact manifolds or non-compact Lie group with another additional conditions. An analytic approach is taken in Applebaum [2] for the problem on a compact Lie group.

Appendix: Manifolds and Lie Groups

Let M be a Hausdorff topological space with the second countability. It is called a *manifold* of dimension d if each point of M has an open neighborhood that is homeomorphic to an open set in $\mathbb{R}^d$. If U is an open neighborhood of $x \in M$ and ψ is a homeomorphism from M to an open subset of $\mathbb{R}^d$, we call (U, ψ) a *chart* at x. For $y \in U$, we write $\psi(y) = (x_1(y), \ldots, x_d(y))$ and call $x_1, \ldots, x_d$ *local coordinates* at U. An *atlas* is a collection of charts $\{(U_\alpha, \psi_\alpha); \alpha \in I\}$ so that (U_α) covers M. If M has an atlas for which mappings $\psi_\alpha \circ \psi_\beta^{-1}; \mathbb{R}^d \to \mathbb{R}^d$ are C^∞ for all $\alpha, \beta \in I$, then M is said to be a C^∞*-manifold*.

A function $f : M \to \mathbb{R}$ is called smooth or C^∞, if $f \circ \psi_\alpha^{-1}$ are C^∞ from $\psi_\alpha(U_\alpha)$ to $\mathbb{R}$ for any α. Let $C^\infty(M)$ be the collection of smooth functions on M. A *tangent vector* $V(x)$ at $x \in M$ is a linear functional on $C^\infty(M)$ which satisfies $V(x)(fg) = f(x)V(x)(g) + (V(x)f)g(x)$ for any $f, g \in C^\infty(M)$. In local charts, we can write $V(x) = \sum_{i=1}^d V^i(x)\frac{\partial}{\partial x_i}$. The set of all tangent vectors at $x \in M$ forms a d-dimensional vector space called the *tangent space* and it is denoted by $T_x(M)$. The set $T(M) = \bigcup_{x \in M} T_x(M)$ can be regarded as a $2d$-dimensional manifold, called the *tangent bundle* to M.

A C^∞-map V from M to $T(M)$ is called a *vector field*. In each local chart (U, ψ), we can write $V(x) = \sum_{i=1}^d V^i(x)\frac{\partial}{\partial x_i}$, where $V^i(x)$ are C^∞-functions from $\psi(U)$ to $\mathbb{R}^d$. The collection of all vector fields is denoted by $\mathcal{L}(M)$. For $V, W \in \mathcal{L}(M)$, its *Lie bracket* $[V, W]$ is defined by $[V, W]f = V(Wf) - W(Vf)$. It is an element of $\mathcal{L}(M)$. Thus $\mathcal{L}(M)$ is an Lie algebra.

Let M_1 and M_2 be C^∞-manifolds of dimension d_1 and d_2, respectively and let $\{(U_\alpha, \psi_\alpha); \alpha \in I\}$ be an atlas of M_1 and let $\{(U_\beta, \psi_\beta); \beta \in J\}$ be an atlas of M_2. A mapping $f; M_1 \to M_2$ is said to be C^∞ if $\psi_\beta \circ f \circ \psi_\alpha^{-1}$ is C^∞ from $\mathbb{R}^{d_1}$ to $\mathbb{R}^{d_2}$. A C^∞-mapping $f; M_1 \to M_2$ is said to be a *diffeomorphism* if it is a bijection and the inverse f^{-1} is also a C^∞-mapping.

Let $\varphi_t, t \in \mathbb{R}$ be a family of diffeomorphisms of M such that $\varphi_t \circ \varphi_s = \varphi_{t+s}$ and $(t, x) \to \varphi_t(x)$ is a C^∞-map. $\{\varphi_t\}$ is called a *one–parameter group of transformations*. For a given $\{\varphi_t\}$, we set for $f \in C^\infty(M)$

$$Vf(x) = \frac{d}{dt} f \circ \varphi_t(x)\Big|_{t=0}. \tag{A.1}$$

Then V is a vector field. Conversely, for a given vector field V, if there exists a one–parameter group of transformations φ_t satisfying (A.1), the vector field V is called *complete*. The associated $\varphi_t(x)$ is denoted by $\exp(tV)(x)$. If the manifold M is compact, it is known that any vector field is complete, but if M is not compact, non-complete vector fields can exist.

Let M be a connected manifold of dimension d and let $\mathcal{L}(M)$ be the linear space of vector fields on M. For $f \in C^\infty(M)$, we define a linear map $df : \mathcal{L}(M) \to C^\infty(M)$ by $df(V) = Vf$. We call df a *one-form* or the *differential* of f. Localizing df at $x \in M$, df_x is a linear mapping from the tangent space $T_x(M)$ to $\mathbb{R}$, i.e., it is the dual of $T_x(M)$. It is denoted by $T_x(M)^*$ and is called the *cotangent space*. Let $T_x^*(M)^{\otimes^r}$ be the r-times tensor product to $T_x^*(M)$. Its element is called an r-tensor. The alternating r-tensor of $v_1, \ldots, v_r \in T_x^*(M)$ is defined by

$$\mathrm{Alt}(v_1, \ldots, v_r) = \frac{1}{r!} \sum_{\sigma \in \Sigma(r)} \mathrm{sign}(\sigma) v_{\sigma(1)} \otimes \cdots \otimes v_{\sigma(r)}, \tag{A.2}$$

where $v_1, \ldots, v_r \in T_x^*(M)$, $v_{\sigma(1)} \otimes \cdots \otimes v_{\sigma(r)}$ is an *r-tensor*, and $\Sigma(r)$ is the permutation group of r letters.

If an r-tensor v satisfies $v = \mathrm{Alt}(v_1, \ldots, v_r)$, v is called an *alternating tensor*. Let $\Lambda^r(T_x^*(M))$ be the collection of alternating r-tensors in $T_x^*(M)^{\otimes^r}$. It is a vector space of dimension $\binom{d}{r}$ if $r \leq d$ and it is equal to $\{0\}$ if $r > d$. We set $\Lambda^r(M) = \bigcup_{x \in M} \Lambda^r(T_x^*(M))$. A smooth *r-form* ω is a C^∞-mapping from $x \in M$ to $\Lambda^r(M)$.

If $v \in \Lambda^r(M)$ and $w \in \Lambda^s(M)$, the exterior product (wedge product) $v \wedge w$ is defined by

$$v \wedge w = \frac{(r+s)!}{r!s!} \mathrm{Alt}(v_1, \ldots, v_r, w_1, \ldots, w_s).$$

At a local chart U, let $x_1, \ldots, x_d$ be a local coordinate and let $dx_i, i = 1, \ldots, d$ be 1-forms. Then a smooth r-form ω can be written at $x \in U$ as

$$\omega(x) = \sum_{\{i_1, \ldots, i_r\} \subset \{1, \ldots, d\}} h^\omega_{i_1, \ldots, i_r}(x)\, dx_{i_1} \wedge \cdots \wedge dx_{i_r},$$

where $h^\omega_{i_1, \ldots, i_r}$ are smooth maps from U to $\mathbb{R}$. In the case $r = d$, $\omega(x)$ is written as

$$\omega(x) = h^\omega(x)\, dx_1 \wedge \cdots \wedge dx_d. \tag{A.3}$$

If there exists a continuous d-form ω such that $\omega(x)$ is not 0 everywhere, then the manifold is called *orientable*. If $h^\omega(x) > 0$, the d-form ω is called a positive d-form or a *volume element*.

Let $\Omega^d(M)$ be the collection of smooth d-form on M. Let $\phi; M \to M$ be a C^∞-map. The pullback ϕ^* is the linear map from $\Omega^d(M)$ into itself defined by

$$\phi^*(\omega)(V_1, \ldots, V_d)(x) = \omega_{\phi(x)}(d\phi_x(V_1(x)), \ldots, d\phi_x(V_d(x))),$$

for all $V_1, \ldots, V_d \in \mathcal{L}(M)$ and $x \in M$.

We will define the integral by a positive d-form (volume element) ω. Let $C_0^\infty(M)$ be the collection of C^∞-functions on M with compact supports. For $f \in C_0^\infty$, we will define the integral $\int_M f\omega$ by the formula

$$\int_M f\omega = \int_{\mathbb{R}^d} f \circ \phi^{-1}(x_1, \ldots, x_d) h^\omega(x_1, \ldots, x_d)\, dx_1 \cdots dx_d,$$

if the support of f is included in a chart U. If the support of f is not included in a chart, we need a partition of unity. It is a collection of functions $(\psi_i, i \in I)$ in $C^\infty(M)$ for which:

(1) At each $x \in M$, only finite number of ψ_i are nonzero.
(2) For each i, $\mathrm{supp}(\psi_i)$ is compact.
(3) For each $x \in M, i \in I$, $\psi_i(x) \geq 0$ and $\sum_{i \in I} \psi_i(x) = 1$.

We can define the integral of a continuous function f with compact support by

$$\int_M f\omega = \sum_i \int_M f\phi_i \omega.$$

We will fix a positive d-form (volume element) ω and denote it by dx. The integral by dx of a continuous function f on M is denoted by $\int_M f(x)\,dx$ if it exists. It is usually denoted by $\int_M f(y)\,dy$ etc.

Let $\phi : M \to M$ be a locally diffeomorphic C^∞-map. Let $\phi^* dx$ be the pullback of dx by ϕ. Then there exists a positive or negative C^∞-function J_ϕ on M such that $\phi^*\, dx = J_\phi dx$. We have a formula of the change of variables:

$$\int_M f(x)\,dx = \int_M f(\phi(x))|J_\phi(x)|\,dx \tag{A.4}$$

for any continuous functions f of compact supports.

Let V be a complete vector field on M and let $\{\varphi_t\}$ be the one–parameter group of diffeomorphisms generated by V. Then the *Lie derivative* of dx denoted by $L_V\,dx$ is defined by $\lim_{t\to 0} \frac{1}{t}(\varphi_t^*\,dx - dx)$. Since $L_V\,dx$ is a d-form, it is written as $L_V\,dx = h(x)\,dx$ with a smooth scalar function h, which we denote by $\mathrm{div}\,V(x)$. We have for any $f, g \in C_0^\infty(M)$,

$$\begin{aligned}\int f(x)(g(\varphi_t(x)) - g(x))\,dx &= \int (f \circ \varphi_t^{-1}(x) - f(x))g(x)\,dx \\ &\quad + \int f \circ \varphi_t^{-1}(x)g(x)((\varphi_t^{-1})^*\,dx - dx).\end{aligned}$$

Since the inverse map φ_t^{-1} is generated by the vector field $-V$, we have

$$\lim_{t\to 0} \frac{1}{t} \int f(x)(g(\varphi_t(x)) - g(x))\,dx = \int f(x)Vg(x)\,dx,$$

$$\lim_{t\to 0} \frac{1}{t} \int (f \circ \varphi_t^{-1}(x) - f(x))g(x)\,dx = -\int Vf(x)g(x)\,dx,$$

$$\lim_{t\to 0} \frac{1}{t} \int f \circ \varphi_t^{-1}(x)g(x)((\varphi_t^{-1})^*\,dx - dx) = -\int f(x)g(x)\mathrm{div}\,V(x)\,dx.$$

Therefore we have a formula of the integration by parts:

$$\int_M f(x)\cdot Vg(x)\,dx = -\int_M Vf(x)\cdot g(x)\,dx - \int_M \operatorname{div} V(x) f(x) g(x)\,dx \qquad (A.5)$$

for any $f, g \in C_0^\infty(M)$. The above formula holds for any (not necessarily complete) C^∞-vector field on M

We should remark that values of scalar functions J_ϕ and $\operatorname{div} V$ depend on the choice of the volume element dx.

Remark If M is a Euclidean space $\mathbb{R}^d$, we can take dx as the Lebesgue measure. Then the C^∞-function J_ϕ coincides with the Jacobian determinant $\det \nabla\phi$ of the map ϕ. Further, equality (A.4) for $M = \mathbb{R}^d$ is a well known formula for change of variables. Further, $\operatorname{div} V$ coincides with $\sum_i \partial V_i/\partial x_i$. Formula (A.5) for $M = \mathbb{R}^d$ is well known as a formula for the integration by parts.

As a volume element, however, we can take any measure $m(dx) = m(x)\,dx$ on the Euclidean space, where m is a positive C^∞-function. Then J_ϕ is no longer a Jacobian determinant and $\operatorname{div} V$ is non longer equal to $\sum_i \frac{\partial V^i}{\partial x^i}$.

A set G is called a *topological group* if it satisfies the following three properties.

(1) G is a group and, at the same time, it is a topological space.
(2) The map $(x, y) \to xy$ from the product space $G \times G$ into G is continuous.
(3) The map $x \to x^{-1}$ from G into itself is continuous, where x^{-1} is the inverse element of x.

Let G be a topological group and let $g \in G$. Two maps $L_g, R_g; G \to G$ are defined by

$$L_g(x) = gx, \quad R_g(x) = xg, \quad (x \in G).$$

These two maps are homeomorphisms of G. L_g is called the *left translation* of G by g, and R_g is called the *right translation* of G by g.

A topological group G is called a *Lie group* if G is a C^∞-manifold with a countable basis and maps $(x, y) \to xy$ and $x \to x^{-1}$ are differentiable. Let V be a C^∞-vector field on G. It is called *left-invariant* if for any $f \in C_0(G)$, it satisfies

$$V(f \circ L_g)(x) = Vf(L_g x)$$

for any $g \in G$. It is called right invariant if $V(f \circ R_g)(x) = Vf(R_g x)$ holds for any $g \in G$. We denote by $\mathcal{L}(G)$ the set of all left-invariant vector fields on the Lie group G.

Let G be a connected Lie group of dimension d. A nontrivial regular Borel measure m on G is called a *left Haar measure* if $m(A) = m(gA)$ holds for any Borel subset A of G and $g \in G$. A right Haar measure is defined similarly. It is known that the left Haar measure exists and it is unique up to a positive multiplicative constant.

Let m be a left Haar measure on G. For $g \in G$, we define

$$m^g(A) = m(Ag),$$

for each Borel set A. Then m^g is another left Haar measure on G. Then there exists a positive constant $\Delta(g)$ such that $m^g(A) = \Delta(g)m(A)$ holds for all Borel subsets A. We call the mapping $g \to \Delta(g)$ from G to $(0, \infty)$ the *modular function* of G. If the modular function is identically equal to 1, the Lie group is called *unimodular*. Examples of unimodular Lie groups are abelian Lie groups, compact Lie groups, semisimple Lie groups and connected nilpotent Lie groups.

Bibliography

1. Applebaum, D.: Lévy Processes and Stochastic Calculus. Cambridge University Press, Cambridge, MA (2004)
2. Applebaum, D.: Probability on Compact Lie Groups. Springer, Cham (2014)
3. Applebaum, D., Kunita, H.: Lévy flows on manifolds and Lévy processes on Lie groups. J. Math. Kyoto Univ. **33**, 1103–1124 (1993)
4. Arnold, L., Scheutzow, M.: Perfect cocycles through stochastoc differential equations. Prob. Theory Relat. Fields **101**, 65–88 (1955)
5. Arnold, L.: Random Dynamical Systems. Springer, Berlin (1998)
6. Baxendale, P., Harris, T.E.: Isotropic stochastic flows. Ann. Prob. **14**, 1155–79 (1986)
7. Bichteler, K., Gravereau, J.B., Jacod, J.: Malliavin Calculus for Processes with Jumps. Stochastic Monographs, vol. 2. Gordon and Breach, London (1987)
8. Bismut, J.M.: Mécanique Aléatoire. Lecture Notes in Mathematics, vol. 866. Springer, Berlin/Heidelberg/New York (1981)
9. Bismut, J.M.: Martingale, the Malliavin calculus and hypoellipticity under general Hormander's conditions. Z. Wahrsch. verw. Gebiete. **56**, 469–505 (1981)
10. Bismut, J.M.: Calcul des variations stochastique et processus de sauts. Z. Wahrsch. verw. Gebiete **63**, 147–235 (1983)
11. Bismut, J.M.: Jump processes and boundary processes. In: Taniguchi Symposium SA, Kyoto 1982, pp. 53–104. Kinokuniya, Tokyo (1984)
12. Blagoveschenskii, Yu.N., Freidlin, M.I.: Certain properties of diffusions processes depending on parameter. Soviet Math. Dokl. **2**, 633–636 (1961)
13. Bouleau, N., Denis, L.: Dirichlet Forms Methods for Poisson Point Measures and Lévy Processes. Springer, Cham (2015)
14. Brezis, H.: Functional Analysis, Sobolev Spaces and Partial Differential Equations. Springer, New York (2011)
15. Cameron, R.H.: The first variation of an indefinite Wiener integral. Proc. Am. Math. Soc. **2**, 914–924 (1951)
16. Cameron, R.H., Martin, W.T.: Transformations of Wiener integrals under translations. Ann. Math. **45**, 386–396 (1944)
17. Carmona, R.A., Nualart, D.: Nonlinear Stochastic Integrators. Equations and Flows. Stochastic Monographs, vol. 6. Gordon and Breach Science Publishers, New York (1990)
18. Chen, Z.-Q., Fukushima, M.: Symmetric Markov Processes, Time Change, and Boundary Theory. Princeton University Press, Princeton (2011)
19. Chen, Z.-Q., Kim, P., Kumagai, T.: Global heat kernel estimates for symmetric jump processes. Trans. Am. Math. Soc. **363**, 5021–5055 (2011)

H. Kunita, *Stochastic Flows and Jump-Diffusions*, Probability Theory and Stochastic Modelling 92, https://doi.org/10.1007/978-981-13-3801-4

20. Chentzov, N.N.: Limit theorem for some class of functions, ROC. In: All-Union Conference of Theory Probability and Mathematics Statistics, Erevan (1958) (Selected Transl. Math. Stat. Prob. **9**, 11–40 (1970))
21. Dellacherie, C., Meyer, P.A.: Probabilities and Potential B: Theory of Martingales. North Holland, Amsterdam (1982)
22. Doob, J.L.: Stochastic Processes. Wiley, New York (1953)
23. Driver, B.K.: Curved Wiener space analysis. In: Rao, M.M. (ed.) Real and Stochastic Analysis. Birkhauser, Boston (2004)
24. Dynkin, E.B.: Markov Processes, I, II. Springer, Berlin/Heidelberg/New York (1965)
25. Elworthy, K.D.: Stochastic Differential Equations on Manifolds. LMS Lecture Note Series, vol. 70. Cambridge University Press, Cambridge (1982)
26. Elworthy, K.D.: Geometric aspects of diffusions on manifolds. In: Ancona, A., Elworthy, K.D., Émery, M., Kunita, H. (eds.) Stochastic Differential Geometry at Saint-Flour. Probability at Saint-Flour, pp. 113–261. Springer, Berlin (2013)
27. Freidlin, M.L., Wentzell, A.D.: Random Perturbations of Dynamical Systems. Springer, New York (1984)
28. Friedman, A.: Partial Differential Equations of Parabolic Type. Princeton University Press, Princeton (1964)
29. Fujiwara, T.: Stochastic differential equations of jump type on manifolds and Lévy flows. J. Math. Kyoto Univ. **31–1**, 99–119 (1991)
30. Fujiwara, T., Kunita, H.: Stochastic differential equations of jump type and Lévy processes in diffeomorphisms group. J. Math. Kyoto Univ. **25**, 71–106 (1985)
31. Fujiwara, T., Kunita, H.: Canonical SDE's based on semimartingales with spatial parameters. Part II; Inverse flows and backward SDE's. Kyushu J. Math. **53**, 301–331 (1999)
32. Gihman, I.I., Skorohod, A.V.: Stochastic Differential Equations. Springer, Berlin/New York (1972)
33. Girsanov, L.V.: On transforming a class of stochastic processes by absolutely continuous substitution of measures. Theor. Verojatnost. i Primenen. **5**, 314–330 (1960)
34. Gross, L.: Abstract Wiener spaces. In: Proceedings of Fifth Berkeley Symposium Mathematical Statistics and Probability II., Part 1, pp. 31–42. University of California Press, Berkeley (1967)
35. Gross, L.: Potential theory in Hilbert space. J. Funct. Anal. **1**, 123–181 (1967)
36. Harris, T.E.: Coalescing and noncoalescing stochastic flows in R_1. Stoch. Proc. Appl. **17**, 187–210 (1984)
37. Hayashi, M., Ishikawa, Y.: Composition with distributions of Wiener-Poisson variables and its asymptotic expansion. Math. Nachr. **285**, 619–658 (2012)
38. Hergason, S.: Differential Geometry, Lie Groups and Symmetric space. Academic, New York (1978)
39. Hunt, G.A.: Semigroups of measures on Lie groups. Trans. Am. Math. Soc. **81**, 264–293 (1956)
40. Ikeda, N., Watanabe, S.: An introduction to Malliavin's calculus. In: Itô, K. (ed.) Stochastic Analysis, pp. 1–52. Kinokuniya, Tokyo (1982)
41. Ikeda, N., Watanabe, S.: Stochastic Differential Equations and Diffusion Processes, 2nd edn. North Holland, Amsterdam (1989)
42. Il'in, A.M., Karashnikov, A.S., Oleinik, O.A.: Linear equation of the second order of the parabolic type (in Russian). Uspehi Math. Nauk **17**(3), 3–146 (1962)
43. Ishikawa, Y.: Density estimate in small time for jump process with singular Lévy measures. Tohoku Math. J. (2) **53**, 183–202 (2001)
44. Ishikawa, Y.: Stochastic Calculus of Variations for Jump Processes. De Gruyter, Berlin (2013)
45. Ishikawa, Y., Kunita, H.: Malliavin calculus on the Wiener-Poisson space and its application to canonical SDE with jumps. Stoch. Process. Appl. **116**, 1743–1769 (2006)
46. Ishikawa, Y., Kunita, H., Tsuchiya, M.: Smooth density and its short time estimate for jump process determined by SDE. SPA **128**, 3181–3219 (2018)

47. Itô, K.: On stochastic processes (Infinitely divisible laws of probability). Japan. J. Math. **18**, 261–301 (1942)
48. Itô, K.: Stochastic integral. Proc. Imp. Acad. Tokyo **20**, 519–524 (1944)
49. Itô, K.: Brownian motions in a Lie group. Proc. Jpn. Acad. **26**, 4–10 (1950)
50. Itô, K.: On stochastic differential equations. Memoirs Am. Math. Soc. **4**, 1–51 (1951)
51. Itô, K.: On a formula concerning stochastic differentials. Nagoya Math. J. **3**, 55–65 (1951)
52. Itô, K.: Extension of stochastic integrals. In: Proceedings of International Symposium, SDE Kyoto, pp. 95–109 (1976)
53. Itô, K., Mckean, H.P.: Diffusion Processes and Their Sample Paths. Springer, Berlin (1965)
54. Jacob, N.: Pseudo Differential Operators and Markov Processes. Imperial College Press, London (2001)
55. Karatzas, I., Shreve, S.E.: Brownian Motion and Stochastic Calculus. Springer, New York (1991)
56. Karatzas, I., Shreve, S.E.: Methods of Mathematical Finance. Springer, New York (1998)
57. Komatsu, T., Takeuchi, T.: On the smoothness of pdf of solutions to SDE of jump type. Int. J. Differ. Equ. Appl. **2**, 141–197 (2001)
58. Komatsu, T., Takeuchi, T.: Simplified probabilistic approach to the Hörmander theorem. Osaka J. Math. **38**, 681–691 (2001)
59. Kunita, H.: Stochastic Flows and Stochastic Differential Equations. Cambridge University Press, Cambridge (1990)
60. Kunita, H.: Stochastic differential equations based on Lévy processes and stochastic flows of diffeomorphisms. In: Rao, M.M. (ed.) Real and Stochastic Analysis. Birkhäuser, Boston (2004)
61. Kunita, H.: Analysis of nondegenerate Wiener-Poisson functionals and its applications to Itô's SDE with jumps. Sankhya Indian J. Stat. **73-A**(1), 1–45 (2011)
62. Kunita, H.: Itô's stochastic calculus: its surprising power for applications. SPA **120**, 622–652 (2010)
63. Kunita, H.: Chain rules for Lévy flows and Kolmogorov equations for associated jump-diffusions. In: Zhao, H., Truman, A. (eds.) New Trends in Stochastic Analysis and Related Topics. World Scientific, Hackensack (2012)
64. Kunita, H.: Nondegenerate SDE's with jumps and their hypoelliptic properties. J. Math. Soc. Jpn. **65**, 993–1035 (2013)
65. Kunita, H., Oh, J.-P.: Asymptotic properties of Lévy flows. J. Korean Math. Soc. **27**, 255–280 (1990)
66. Kunita, H., Watanabe, S.: On square integrable martingales. Nagoya Math. J. **30**, 209–245 (1967)
67. Kurtz, T.G., Pardoux, E., Protter, P.: Stratonovich stochastic differential equations driven by general semimartingales. Ann. Inst. Henri Poincaré **31**, 351–377 (1995)
68. Kusuoka, S.: The non-linear transformation of Gaussian measure on Banach space and its absolute continuity. J. Fac. Sci. Univ. Tokyo IA **29**(1), 567–697 (1982)
69. Kusuoka, S., Stroock, D.W.: Application of Malliavin calculus, Part I. In: Itô, K. (ed.) Stochastic Analysis. Proceedings Taniguchi International Symposium Katata and Kyoto, 1982, pp. 271–306. Kinokuniya, Tokyo (1984)
70. Kusuoka, S., Stroock, D.W.: Application of Malliavin calculus, Part II. J. Fac. Sci. Tokyo Univ. Tokyo Sect. IA Math. **32**, 1–76 (1985)
71. Kusuoka, S., Stroock, D.W.: Application of Malliavin calculus, Part III. J. Fac. Sci. Tokyo Univ. Tokyo Sect. IA Math. **34**, 391–442 (1987)
72. Lamberton, D., Lapeyre, B.: Introduction to Stochastic Calculus Applied to Finance. Chapman and Hall, London (1996)
73. Léandre, R.: Flot d'une équation differentielle stochastique avec semimartingale directrice discontinue. In: Séminaire Probabilities XIX. Lecture Notes in Mathematics, vol. 1123, Springer, pp. 271–275 (1985)
74. Léandre, R.: Regularities de processus de sauts degeneres (II). Ann. Inst. H. Poincare Probab. Stat. **24**, 209–236 (1988)

75. Le Jan, Y.: Flots de diffusions dans $\mathbb{R}^d$. C.R. Acad. Sci. Paris, Ser. I **294**, 697–699 (1982)
76. Li, X.-M.: Strong p-completeness of stochastic differential equations and the existence of smooth flows on noncompact manifolds. PTRF **100**, 485–511 (1994)
77. Malliavin, P.: Stochastic calculus of variation and hypo-elliptic operator. In: Itô, K. (ed.) Proceedings of International Symposium SDE, Kyoto, 1976. Kinokuniya, Tokyo (1978)
78. Malliavin, P.: Géometrie Différentielle Stochastique. Les Presses de l'Université, Montréal (1978)
79. Malliavin, P.: Stochastic Analysis. Springer, Berlin/Heidelberg (1997)
80. Malliavin, P., Thalmaier, A.: Stochastic Calculus of Variations in Mathematical Finance. Springer Finance, Berlin/Heidelberg/New York (2006)
81. Marcus, S.I.: Modelling and approximations of stochastic differential equations driven by semimartingales. Stochastics **4**, 223–245 (1981)
82. Maruyama, G.: Note on Wiener functionals. Kodao Math. Sem. Rep. **2**, 41–44 (1950)
83. Matsumoto, H., Taniguchi, S.: Stochastic Analysis. Cambridge University Press, Cambridge (2016)
84. Meyer, P.A.: Probability and Potentials. Bleisdell, Waltham (1966)
85. Meyer, P.A.: Un cours sur integrals stochastiques. In: Meyer, P.A. (ed.) Seminaire Probability X. Lecture Notes in Mathematics, vol. 511, pp. 246–400. Springer, Berlin (1976)
86. Meyer, P.A.: Transformations de Riesz pour lois gaussiennes. In: Azéma, J., Yor, M. (eds.) Séminaire de Probabilites XVIII. Lecture Notes in Mathematics, vol. 1059, pp. 179–193. Springer, Berlin (1984)
87. Norris, J.: Simplified Malliavin calculus. In: Azéma, J., Yor, M. (eds.) Seminaire de Probabilités, XX. Lecture Notes in Mathematics, vol. 1204, pp. 101–130. Springer, Berlin (1986)
88. Nualart, D.: The Malliavin Calculus and Related Topics, 2nd edn. Springer, Berlin (2006)
89. Nualart, D., Pardoux, E.: Stochastic calculus with anticipating integrands. Probab. Theory Relat. Fields **89**, 407–422 (1991)
90. Oksendal, B.: Stochastic Differential Equations: An Introduction with Applications, 5th edn. Springer, Berlin (1998)
91. Picard, J.: Formules de dualité sur l'espace de Poisson. Ann. Inst. Henri Poincaré, **32**, 509–548 (1996)
92. Picard, J.: On the existence of smooth densities for jump processes. PTRF **105**, 481–511 (1996)
93. Picard, J.: Density in small time at accessible points for jump processes. SPA **67**, 251–279 (1997)
94. Picard, J., Savona, C.: Smoothness of harmonic functions for processes with jumps. SPA **87**, 69–91 (2000)
95. Picard, J., Savona, C.: Smoothness of the law of manifold-valued Markov process with jumps. Bernoulli **19**(5A), 1880–1919 (2013)
96. Protter, P.: Stochastic Integration and Differential Equations. A New Approach. Applied Mathematics, vol. 21. Springer, Berlin/Heidelberg (1992)
97. Revuz, D., Yor, M.: Continuous Martingales and Brownian Motions, 3rd edn. Springer, Berlin (1999)
98. Rogers, L.G., Williams, D.: Diffusions, Markov Processes, and Martingales, I, II, 2nd edn. Wiley, Chichester (1994)
99. Sato, K.: Lévy Processes and Infinitely Divisible Distributions. Cambridge University Press, Cambridge (1999)
100. Shigekawa, I.: Derivatives of Wiener functionals and absolute continuity of the induced measures. J. Math. Kyoto Univ. **208**, 263–289 (1980)
101. Shigekawa, I.: de Rham-Hodge-Kodaira's decomposition on an abstract Wiener space. J. Math. Kyoto Univ. **20**, 263–289 (1980)
102. Shigekawa, I.: Stochastic Analysis. American Mathematical Society, Providence (2004)
103. Skorohod, A.V.: Studies in the Theory of Random Processes. Dover Publication Inc., Mineola/New York (2017) (Original Russian, 1961)

104. Stratonovich, R.L.: A new form of representing stochastic integrals and equations. Vestnik Moscow Univ. Ser. I. Mat. Mec. **1**, 3–12 (1964)
105. Stroock, D.W.: The Malliavin calcuulus, the functional analytic approach. J. Funct. Anal. **44**, 212–257 (1981)
106. Stroock, D.W.: Malliavin calculus and its applications to second order parabolic differential operators I, II. Math. Syst. Theory **14**, 25–65, 141–171 (1981)
107. Stroock, D.W.: Markov Processes from K. Itô's Perspective. Annals of Mathematics Studies, vol. 155. Princeton University Press, Princeton (2003)
108. Stroock, D., Varadhan, S.R.S.: On the support of diffusion processes, with applications to the strong maximum principle. In: Proceedings of 6th Berkeley Symposium on Probability and Statistics, pp. 333–359. Berkeley University (1970)
109. Stroock, W., Varadhan, S.R.S.: Multidimensional Diffusion Processes. Springer, Berlin/Heidelberg (1979/1998)
110. Sugita, H.: On a characterization of the Sobolev spaces over an abstract Wiener space. J. Math. Kyoto Univ. **25**, 717–725 (1985)
111. Sugita, H.: Positive generalized Wiener functionals and potential theory over abstract Wiener spaces. Osaka J. Math. **25**, 665–696 (1988)
112. Taniguchi, S.: Malliavin's stochastic calculus of variations for manifold-valued Wiener functionals and its applications. Z.W. **65**, 269–290 (1983)
113. Taniguchi, S.: Applications of Malliavin's calculus to time-dependent systems of heat equations. Osaka J. Math. **32**, 307–320 (1985)
114. Totoki, H.: A method of construction of measures on function spaces and its applications to stochastic processes. Memories Fac. Sci. Kyushu Univ. Ser. A. Math. **15**, 178–190 (1962)
115. Tsuchiya, M.: Lévy measure with generalized polar decomposition and the associated SDE with jumps. Stochastics Stochastic Rep. **38**, 95–117 (1992)
116. Watanabe, S.: Stochastic Differential Equations and Malliavin Calculus. Tata Institute of Fundamental Research, Bombay (1984)
117. Watanabe, S.: Analysis of Wiener functionals (Malliavin calculus) and its applications to heat kernels. Ann. Probab. **15**, 1–39 (1987)
118. Watanabe, S.: Fractional order Sobolev spaces on Wiener space. Probab. Theory Rel. Fields **95**, 175–198 (1993)
119. Williams, D.: Probability with Martingales. Cambridge University Press, Cambridge (1991)
120. Wong, E., Zakai, M.: On the relation between ordinary and stochastic differential equations and applications to stochastic problems in control theory. In: Automatic and Remote Control III. vol. 1, pp. 5–13. Institution of Mechanical Engineers, London (1967)

Symbol Index

H. Kunita, *Stochastic Flows and Jump-Diffusions*, Probability Theory and Stochastic Modelling 92, https://doi.org/10.1007/978-981-13-3801-4

Index

H. Kunita, *Stochastic Flows and Jump-Diffusions*, Probability Theory
and Stochastic Modelling 92, https://doi.org/10.1007/978-981-13-3801-4

www.ingramcontent.com/pod-product-compliance
Ingram Content Group UK Ltd.
Pitfield, Milton Keynes, MK11 3LW, UK
UKHW022001270726
14060UKWH00003B/638
* 9 7 8 9 8 1 1 3 3 8 0 0 7 *